Understanding Properties of Atoms, Molecules and Materials

Understanding Properties of Atoms, Molecules and Materials

Pranab Sarkar
Department of Chemistry,
Visva-Bharati, Santiniketan

Sankar Prasad Bhattacharyya
Indian Association for the Cultivation of Science,
Jadavpur, Kolkata

CRC Press
Taylor & Francis Group
Boca Raton London New York

CRC Press is an imprint of the
Taylor & Francis Group, an **informa** business

First edition published 2022
by CRC Press
6000 Broken Sound Parkway NW, Suite 300, Boca Raton, FL 33487-2742

and by CRC Press
4 Park Square, Milton Park, Abingdon, Oxon, OX14 4RN

Library of Congress Cataloging-in-Publication Data
Names: Sarkar, Pranab, author. | Bhattacharyya, Sankar Prasad, author.
Title: Understanding properties of atoms, molecules and materials / Pranab Sarkar [Department of Chemistry, Visva-Bharati, Santiniketan – 731235, India], Sankar Prasad Bhattacharyya [Indian Association for the Cultivation of Science Jadavpur, Kolkata – 700032, India]. Description: First edition. | Boca Raton : CRC Press, 2022. | Includes bibliographical references. | Summary: "The book introduces the readers in material chemistry/engineering to elementary quantum mechanics of atoms, molecules and solids, with methods of statistical mechanics (classical as well as quantum) along with elementary principles of classical MD simulation. The basic concepts are illustrated with easy to grasp examples, thus preparing the readers for an exploration through the world of materials - the exotic and the mundane. The emphasis has been on the phenomena and what shapes them at the fundamental level. A fairly comprehensive description of modern designing principles for materials with examples is a unique feature of the book" – Provided by publisher. Identifiers: LCCN 2021031521 (print) | LCCN 2021031522 (ebook) | ISBN 9780367030346 (hardback) | ISBN 9781032156002 (paperback) | ISBN 9781003244882 (ebook) Subjects: LCSH: Chemistry, Physical and theoretical. | Quantum chemistry. | Molecular dynamics. | Materials science. Classification: LCC QD453.3 .S27 2022 (print) | LCC QD453.3 (ebook) | DDC 530.4–dc23/eng/20211105 LC record available at https://lccn.loc.gov/2021031521 LC ebook record available at https://lccn.loc.gov/2021031522

ISBN: 978-0-367-03034-6 (hbk)
ISBN: 978-1-032-15600-2 (pbk)
ISBN: 978-1-003-24488-2 (ebk)

DOI: 10.1201/9781003244882

Typeset in Times
by Newgen Publishing UK

Contents

Preface

Atoms combine to form molecules. Not only do these molecules vary in size and shape, their properties too vary over a wide spectrum of possibilities. The types of atoms that occur naturally are rather limited. Nevertheless, they span a broad range of properties, starting from the very reactive alkali metal atoms on one hand, to the very unreactive noble gas atoms on the other. The atoms are the fundamental building blocks of many useful bulk three-dimensional (3D) materials like the metallic conductors (copper, silver, aluminium, etc.), the semiconductors (like silicon, germanium, etc.) and the electrical insulators. Alloys provide an enormous variety of technologically useful atom-based materials (e.g. steel). Semiconductors (Si or Ge) when doped with suitable elements (B, As, Sb, etc.) produce the materials for transistors – the work horse of the electronics industry. Metals like iron, cobalt and nickel are the keys to many magnetic materials. Molecule-based materials expand the horizon of technologically important materials almost endlessly. Thus there are 'smart' materials like the halochromic materials (change color reversibly with acidity), electrochromic materials (optical properties change in response to altered light intensity), thermochromic materials (properties change reversibly when thermally stimulated), non-linear optical materials (they are capable of altering the characteristics of electromagnetic radiation passing through them), conducting polymers, etc. are some of the examples of technologically important molecular materials. There are others like piezo-electric materials, shape-memory materials (alloys) and tunneling composites. A new dimension has been added to the ever expanding search for exotic materials with the introduction of 2D matter, graphene, phosphorene and the graphene-likes. Important among these are layered materials with mixed elemental composition, for example, the transition metal dichalcogenides, MXenes and van der Waals heterostructures. The 2D materials demand our attention in view of the range of properties that they exhibit – high carrier mobility, superconductivity, mechanical stability, light absorption in the uv and visible parts of the spectrum, etc. Such properties promise technological spin-off in the fields of electronics, valleytronics, biosensing and catalysis. Nano and nanostructured materials of different dimensionalities have added extra impulse to the growth of materials chemistry and science.

The vast array of materials already developed and those in the making pose a hard challenge to chemists and materials scientists. The quest for new materials is no longer dictated by chance discovery, but propelled by careful reasoning based on understanding why a material behaves the way it does. That understanding calls for interfacing 'chemical intuition' with the principles of quantum and statistical mechanics, hopefully leading to sharpened chemical intuition and refined thumb-rules for designing new materials with targeted properties. The present book is an attempt in that direction, at least partially, by first making the readers acquainted with the basic tools of understanding and analysis – elementary quantum mechanics of atoms, molecules and solids on one hand, and principles of statistical mechanics, on the other. The key concepts of quantum states, symmetry, degeneracy, density of states, conservation laws, distribution functions etc. are introduced and later exploited to rationalize the properties of different classes of materials.

Chapter 1 presents an overview of the topics that are addressed in the book along with a concise introduction to various technologically important materials, and their classifications. Chapter 2 deals with the basic concepts and principles of quantum mechanics exemplified through elementary applications. Chapter 3 focuses on quantum mechanics of atoms while Chapter 4 deals with molecular quantum mechanics with special emphasis on the separation of slow and fast motions (nuclear and electronic) and methods to treat many electron systems. Quantum mechanical description of solids is the focus in Chapter 5 where the key concepts of energy bands, band gap, translational symmetry, etc. are introduced along with the density functional theory. Chapter 6 is devoted to introducing the basic principles of classical statistical mechanics while the quantum generalizations are developed in Chapter 7. The notion of equilibrium, temperature and equilibrium distributions are carefully unraveled in these chapters along with elementary applications of the key concepts to several idealized systems.

Chapter 8 focuses on understanding properties of common materials likes metals, alloys, semiconductors, insulating materials on the basis of quantum and statistical mechanics. The extraordinary phenomena of superconductivity is critically addressed in this chapter. Smart materials are discussed in Chapter 9 with emphasis on understanding how specific properties emerge and evolve. Magnetic materials and materials with giant and colossal magnetic resistance are the focus of chapter 10 while 2D nano and nanostructured materials constitute the subject matter addressed in Chapter 11. Non-linear optical and energy materials are reviewed in Chapter 12 while the problem of designing materials with targeted properties is examined and reviewed in the penultimate chapter (Chapter 13). Chapter 14 addresses some sundry topics in material science and examines the possible directions that future developments could take.

In spite of our best efforts to be exhaustive, we are aware of the limitations of a book that attempts to cover so much. We have wherever needed, referred to the relevant literature to bridge the gaps that exist. The book took shape during the raging pandemic and the grave uncertainties that it brought in its wake. It has affected the work in some ways.

We thank our students for providing the inspiration to write and our colleagues for many fruitful discussions and the members of our families for their silent support.

Pranab Sarkar
(Professor, Department of Chemistry,
Visva-Bharati, Santiniketan, India) and

Sankar Prasad Bhattacharyya
(Professor, School of Chemical Sciences,
Indian Association for the Cultivation of Science,
Kolkata, India).

Authors

Professor Pranab Sarkar received his PhD from Indian Association for the Cultivation of Science, Kolkata, India in 1997. After pursuing his postdoctoral research in the University of Montreal, Canada, Dr. Sarkar started his professional career in the Dept. of Chemistry, Visva-Bharati, Santiniketan, India in the year 1999. He also served as a visiting Scientist at the University of Saarland, Germany during the period of 2001–2003. Presently, he is a Professor in the Dept. of Chemistry, Visva-Bharati, Santiniketan. Professor Sarkar is an expert in the field of Computational Materials Science and his primary research interest is to employ state of the art theoretical methods to understand and predict material properties at nanoscale. He has made significant contributions to the identification of suitable materials for green sustainable energy and also for spintronic applications. He has published more than 165 papers in internationally reputed journals and a number of book chapters and reviews.

Professor Sankar Prasad Bhattacharyya is fellow of the Indian Academy of Sciences and spent 3 years (2012–2015) in the Dept. of Chemistry, IIT Bombay as Raja Ramanna Fellow, Department of Atomic Energy, GOI. He retired as Senior Professor of Physical Chemistry, Indian Association for the Cultivation of Science (IACS), Kolkata. Presently he is a visiting Professor at the School of Chemical Sciences, IACS University, Kolkata, India. His main research interest is in the area of quantum chemistry and his distinguished career in the theoretical chemistry spans a period of over 35 years. His work on the development of new computational techniques, with particular focus on global optimizations, for the treatment of complex problems in molecular chemistry and physics is well documented by an impressive list of highly influential publications in the field, devoted both to traditional mathematical techniques and more recently, to soft computing-oriented approaches. He has supervised more than 20 PhD and 30 undergraduate project students. His publications included more than 200 papers in peer-reviewed journals and a number of book chapters.

1

The Science of Materials

1.1 Introduction: The Age of Materials

We are in the midst of a rather rapidly evolving age dominated by materials and technology. Advanced technology demands advanced materials and advanced materials catalyze the development of appropriately advanced technology that the advanced materials can support. Even an already existing technology may trigger the search for new materials for better or more efficient implementation of the technology. Thus 'materials' and 'technology' have a synergistic relation, one reinforcing the other. This synergy is so important that a particular period of our evolving civilization has often been identified with the dominant material of that age. In the prehistoric ages of the civilization, naturally and abundantly available stone was the dominant material that was used to fashion the tools of everyday life and trade: the corresponding stage of the civilization has been identified as the 'Stone Age'. The prehistoric man, in search of better weapons, lighter and better articles of daily use discovered the art and science of extracting copper from natural sources (minerals) and alloying it with other metals (like Sn, Zn, etc.) ushering the civilization into a new age – the so called 'Bronze Age' roughly about 5000 years ago. The dominant material of the age was copper, a non-ferrous metal and non-ferrous metallurgy was the appropriate technology of that time. Compared to copper smelting of iron ores is a much more demanding process and had to wait for the appropriate furnace technology to evolve. The quest for harder and tougher materials led to the discovery of 'ferrous metallurgy', which involved extraction of iron (Fe) from iron compounds (oxides, sulphides) existing in nature (minerals). The advent of ferrous metallurgy heralded the birth of the 'Iron Age' of the civilization. The dominant material of the age was iron, which was soon replaced by alloys of iron with controlled amounts of carbon and other metals (like Cr, Ni, etc.). Collectively these materials were called steel. The 'Iron Age' thus rolled over in time to the 'Steel Age', which is still alive and flourishing, steel being the most common and versatile structural material of the present age. It is not that iron and copper are the only metals that have left their strong imprints on the civilization.[1-3] Over the last 150 years or so, the late 'Iron Age' had been witnessing another metal (aluminium) rapidly gaining importance and making deep inroads into our everyday life and manufacturing industry. Aluminium is the most abundant metal available in the Earth's crust (as complex silicates or oxide, Al_2O_3). It was, however, scarcely available for large scale industrial or technological consumption primarily due to rather complex process (technology) of extraction of aluminium from bauxite (the chief ore) that demanded a lot of electrical energy as input. In 1886 Charles Hall and Paul Heroult separately discovered a relatively inexpensive process of extraction of pure aluminium from its oxide (Al_2O_3). A year later Kurl Josef Bayer developed a facile process of extracting high grade aluminium from bauxite – the widely available ore of aluminium. These two discoveries together with tremendous growth in cheap hydroelectric power generation specially in America after the second world war catalyzed the production of aluminium and its alloys and technological application of these materials proliferated. Today aluminium and its alloys have virtually invaded all walks of life and industrial activities. As a light weight corrosion resistant structural material for airframes in the aviation industry, for example, and as power carrying cables for electrical power-transmission and utility network, aluminium has virtually replaced copper. Since the 1950s, it has been widely used in consumer product manufacturing units all over the world, be it in producing aluminium frame washers and dryers making light weight robust laptops, ipads and iphones, and so on and so forth. Many feel that we are witnessing the emergence of an 'Aluminium Age' within the 'Steel Age', although the suspected role of aluminium in the onset of some neurodegenerative diseases in human beings has cast a shadow on the

DOI: 10.1201/9781003244882-1

future growth of aluminium industry.[4] Nevertheless, aluminium-based materials continue to flourish and remain technologically very important.

Somewhere down the ages, may be around 2500 BC, the art and science of making silica-based glass was discovered. As a material glass has remained attractive, useful and versatile so much so that one may be tempted to acknowledge the imprint of a 'Glass Age' in the history of evolution of the wonderful material during the last 200 years or so. Indeed, the brittle and transparent 'glass' can now be found as glass sheets that can be rolled into a bundle, or as interactive touch screens of computers, in bullet proof car windows, in space-ships and what not.

Metals, their alloys, metal oxides (ceramic materials) and glass have dominated civilizations over the last 5000 years or so. As a class we may call them inorganic materials; but they are not the only materials. Nature is replete with thousands of organic materials, which are 'polymeric' in nature (polymers are long chain assembly of many covalently bonded subunits or smaller molecules), the most abundant being 'cellulose' – the natural structural material of trees and plants. The proteins that make up our bodies are polymers just as DNA – the bio-materials that carry the genetic code of all living creatures are. They are called bio-polymers. The invention of 'celluloid' by Wesley and Isiak Hyatt (1863) by applying heat and pressure to a mixture of cellulose nitrate (made from natural cellulose) and camphor marked the beginning of an era of man made 'plastics'. The Hyatt brothers were looking for a substitute for ivory out of which billiard balls could be crafted. The first completely synthetic plastic – a man made polymeric material – was made by Leo Backeland (1907) by condensing phenol and formaldehyde in the presence of a small amount of a base. The resulting substance was a thermo-setting plastic material that was amenable to moulding operations and a non-conductor of electricity. Once moulded, it retained the shape even when heated or exposed to the action of different solvents. Backelite found large scale use in the electrical and auto industry.[2] Research and technological development that revolutionized the polymer industry came from the pioneering work by Wallace Carothers in the 1930s. He and his group are credited with the discovery of neoprene (a rubber substitute), nylon (a poly-amide) that became a substitute for silk. Since then thousands of polymers have been made and have found inumerable technological applications. So extensive has the use of synthetic (man made) polymeric materials loosely called 'plastics' been that it will be apt to say that we are living in the 'Plastic Age'. It is difficult now to imagine life without 'plastics' although lack of bio-degradability of these materials is a cause of concern. Plastics like ceramics are generally insulators and are used to provide coatings on conducting materials. Like plastics, 'semiconductors' or 'semiconducting materials' (which fall between conductors and insulators) invaded every aspect of modern lives so much so that it will be wrong not to recognize that the electronics industry, computer science, solid state laser technology, the information science and telecommunication industry have ushered us into a 'Semiconductor Age'. The chief material of this age is understandably silicon (Si): an element belonging to group IV of the periodic table, along with germanium (Ge). Compounds formed by combining group III elements with those in group V or by combining group II elements with elements in group VI result in semiconducting materials, e.g. gallium arsenide (GaAs) or cadmium selenide (CdSe), which have found numerous technological applications. A lot of developments can be anticipated in this field and the semiconductor age could stretch further into the next century.[5]

The materials we have so far referred to are the normal 3D materials. Of late a new class of materials – the new age materials or materials of reduced dimensionality – are making waves in the technological world. It appears that we are on the verge of an unprecedented revolution in technology that could bring in products and functionalities that so far existed only in the realm of fiction. The emergence of these new age materials has been catalyzed by the ever growing industrial demand for substances that would be tougher but lighter, thinner but denser, much more flexible or much more rigid, environmentally benign and could be generated from renewable resources. The most promising and versatile candidate appears to be graphene (Figure 1.1), which was discovered in 2004. Graphene sheets are ultrathin (single atom thick) but almost 200 times as tough as steel, ultralight, flexible, almost transparent, excellent conductor of heat and electricity. No wonder that it is being touted as a game changer in technological innovations that range from purification of water, energy storage, electronics, computer technology, solar energy-based propulsion in aviation, artificial conducting skin (for prosthetic applications), metal-graphene composites (as structural material) and so on. Electronic and structural properties of graphene far transcend those of traditionally used metals and semiconductors. That prompts many to suggest that graphene could become

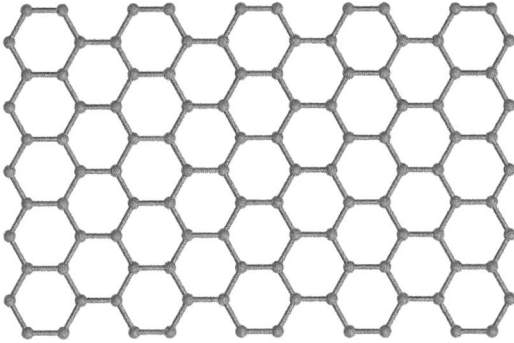

FIGURE 1.1 Schematic representation of a 2D graphene sheet.

the reference material for the post complementary metal oxide semiconductor (CMOS) technology. We are perhaps witnessing the onset of an eclipse of the 'Metal Age' and the breaking of the dawn of a new era of 2D materials dominated by graphene and graphene-likes.

If the particulate constituents of a material have dimensions in the range 1–100 nm, it is called a nano-material. At this length scale, the electrical, chemical, optical and other physical properties of matter can be vastly different from properties of same material at low (atomic) or higher (bulk) length scales. Such materials (nanomaterials) can provide a platform for realizing technology (nanotechnology) and products that never existed before. Many anticipate a second industrial revolution to take shape in the coming decades through the development of nanomaterials and appropriate nanotechnology catapulting the human race into a 'Nano Age' where problems of greenhouse gases, water purification, electronics, energy storage, the scourge of dreaded diseases would all be solved by intelligent applications of nanomaterials and technology. Indeed, nanotechnology may become the basic manufacturing technology of the 21st century wherein it would be possible to control matter and its properties at the length scale of 1–100 nm heralding the emergence of the true 'Nano Age'. At the same time we are also witnessing the onset of an era of 'smart materials'. There are materials that recognize changes in the parameters of the surroundings and respond by altering one or more of their properties. The external parameters can be anything like temperature, pressure, pH, electric field, magnetic field, light intensity, presence of specific chemicals, etc., depending on the class of the 'smart materials' under consideration. We will examine such materials in detail later and review their technological applications already made. We specially mention the case of shape memory alloys (SMA) and shape memory materials (SMM), which appear to remember their original shape even when deformed. There are quite a few self-healing materials that are not just smart, but smart enough to be functionally alive as if with a built in nervous system, which can detect and repair damages caused by 'wear and tear' without external intervention. We are indeed moving fast into an age that will be shaped by smart, advanced smart materials and nanomaterials, be it in medicine, aviation or the auto-mobile industry, in structural engineering or in textile industry and in many other spheres of science and technology. In this 'new age' there will perhaps be increasing emphasis on developing 'composite materials', nano or meso, rather than on creating materials with a unique set of properties, because a single substance with many desirable properties are usually rare. Composite materials or composites are made by combining two or more materials having different sets of properties (the combination does not imply chemical reaction or dissolution). The composite works collectively imparting unique properties to the composite. Such composites exist in the living world. Thus, wood, a hard biomaterial is a composite of cellulose and lignin none of which is hard. Bone in animals is a natural composite made of a hard, brittle material called hydroxy apatite and collagen, a soft and flexible material. They work together to impart the skeleton its stable but flexible structural properties. Mud bricks are the oldest examples of man made composites with high compressive but poor tensile strength. Mud bricks reinforced with straw have good compressive as well as high tensile strength. Concrete is yet another commonly used composite. Of more modern vintage is fiber glass, a composite of a plastic matrix (soft and flexible) reinforced with glass threads, which are hard and brittle. The two components work collectively to impart characteristics that make fiber glass so unique. More advanced versions of fiber glass are now being made with carbon fibers replacing glass threads while much more advanced composites use carbon nanotubes in place of carbon fibers. The materials produced are much lighter but much tougher than steel. It may not be an over statement to say that the age of composites has arrived. We can anticipate exciting developments in the world of materials of the 'new age'. We will take a look at such possibilities in the final chapter (Chapter 14).

1.2 Atoms, Molecules and Solids

Materials can exist in solid, liquid or gaseous states. Most of the technologically useful materials are solids under normal conditions. That makes solids so important. A solid material may have atoms or ions as the fundamental constituents or based on molecules as the fundamental units or particles. In crystalline solids, the constituent particles are arranged regularly in repeating patterns. The simplest unit of the repeating pattern when translated in three-dimensional space generates the 'crystal' or the lattice. It is called the unit cell. The unit cell, whatever may the type be, can be defined as a collection of lattice points, which are occupied by the constituent particles and about which the particles can vibrate. There are only 14 types of the so called Bravais unit cell which can be grouped into seven classes of crystal systems, based on the lengths (a, b, c) of the edges of the unit cells and the three internal angles (α, β, γ) as shown in Figure 1.2 and Table 1.1. Note that opposite faces of the unit cell are parallel and by convention the edges of the unit cell connects equivalent points. Clearly there are a lot of 'voids' in the crystal and the packing efficiency differs from one crystal system to another, or from one kind of lattice to another.

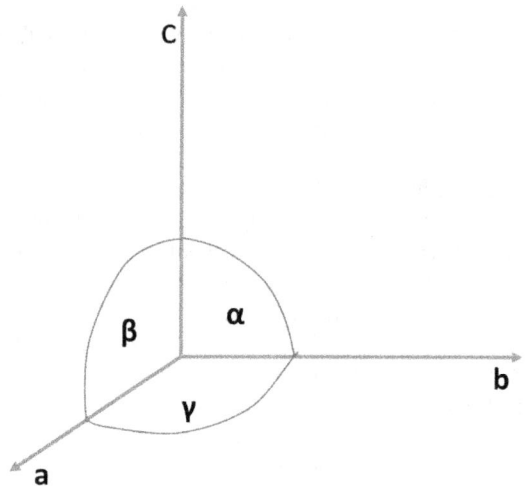

FIGURE 1.2   Crystal axes and their orientations.

TABLE 1.1

The Seven Types of Crystal Systems and Their Unit Cell Parameters

Cubic	$a = b = c$	$\alpha = \beta = \gamma = \frac{\pi}{2}$
Tetragonal	$a = b \neq c$	$\alpha = \beta = \gamma = \frac{\pi}{2}$
Monoclinic	$a = b \neq c$	$\alpha = \gamma = \frac{\pi}{2} \neq \beta$
Orthorhombic	$a \neq b \neq c$	$\alpha = \beta = \gamma = \frac{\pi}{2}$
Rhombohedral	$a = b = c$	$\alpha = \beta = \gamma = \neq \frac{\pi}{2}$
Hexagonal	$a = b \neq c$	$\alpha = \beta = \frac{\pi}{2}, \gamma = \frac{2\pi}{3}$
Triclinic	$a \neq b \neq c$	$\alpha \neq \beta \neq \gamma$

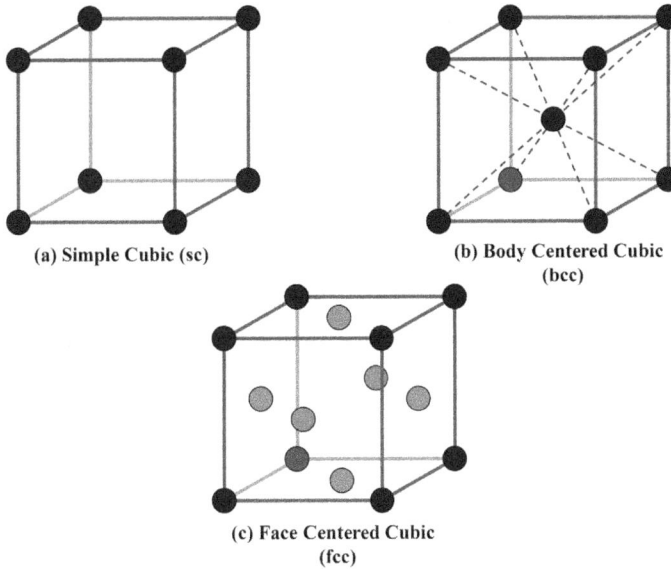

FIGURE 1.3 The three Bravais unit cells under the cubic crystal systems (a) simple cubic (sc), (b) body centered cubic (bcc) and (c) face centered cubic (fcc).

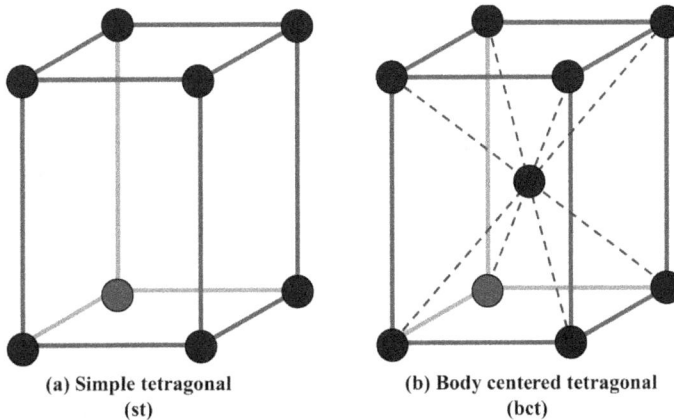

FIGURE 1.4 The unit cells in simple and body centered tetragonal lattices.

It can be anticipated that a given set of particles will arrange themselves into a crystal system that makes the best utilization of the space available to the particles, i.e. they will tend to be packed maximally. We will consider the issue later.

Under the cubic system, we have three Bravais unit cells viz. the simple cubic (sc), the body centered cubic (bcc) and the face centered cubic (fcc) (Figure 1.3). Under the tetragonal system we have the simple tetragonal (ST) and body centered (BC) tetragonal unit cells as shown in Figures 1.4(a–b). The orthorhombic system admits of four Bravais unit cells as shown in Figures 1.5(a–c) called the simple, the body centered, the face centered and the end centered orthorhombic unit cells. Finally, the rhombohedral, hexagonal and triclinic systems each has only one Bravais unit cell as displayed in Figures 1.6(a–c).

Whatever be the type of lattice or packing pattern there are some features common to all the solid crystalline materials – the constituent particles are held fixed at the lattice points occupied by them. The

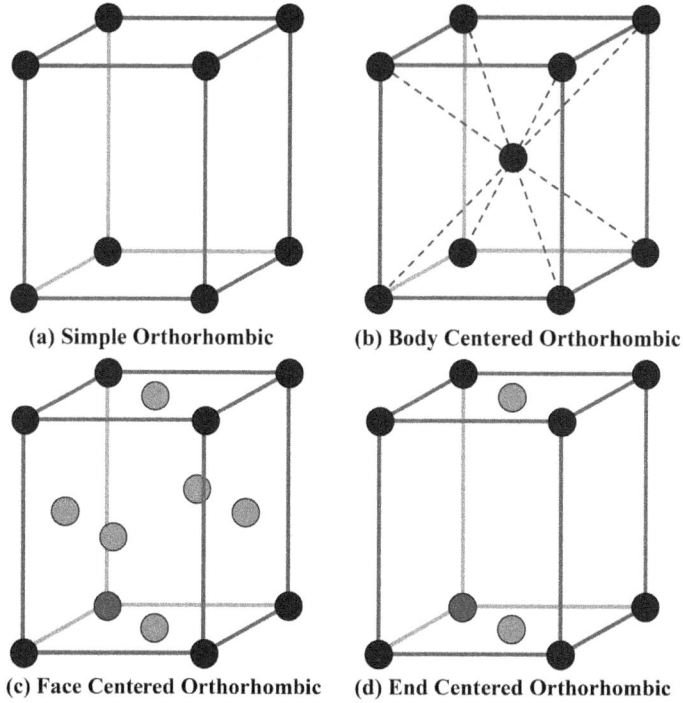

FIGURE 1.5   (a) Simple orthorhombic, (b) body centered orthorhombic, (c) face centered orthorhombic, (d) end centered orthorhombic unit cells.

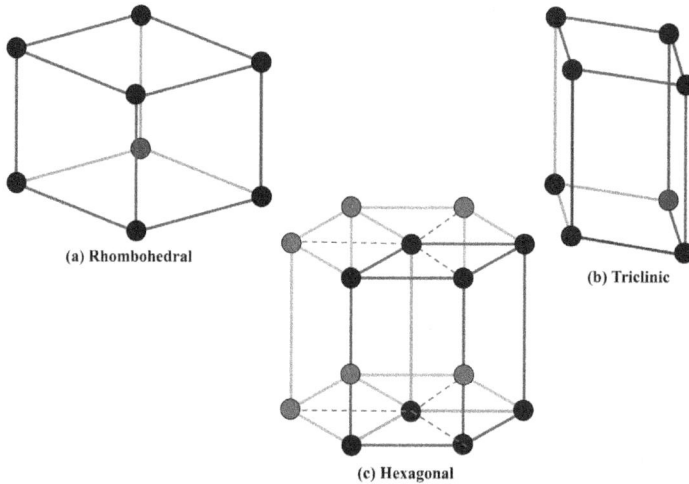

FIGURE 1.6   (a) Rhombohedral, (b) triclinic and (c) hexagonal unit cells.

adjacent particles can not move past each other, but can only oscillate about the fixed lattice positions. This feature is responsible for the incompressibility and fixed shape of the solid. The question that naturally arises is what holds the particles fixed in their lattice positions? The answer obviously is the following: interaction between the adjacent particles (atoms, ions or molecules), which is euphemistically called

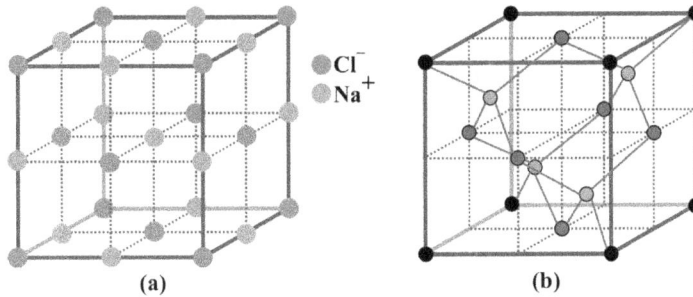

FIGURE 1.7 (a) NaCl (solid with ionic network). (b) Diamond (solid with covalent network).

'bonding'. The solid materials can also be classified into types or categories based on the nature of bonding interaction between adjacent particles in the solid. We may broadly classify the crystalline solids into the following types based on the type of bonding interaction.[5]

(a) Ionic solids in which adjacent atom pairs producing oppositely charged ion pairs with completely occupied electron shells by transfer of electron from one to the other occupy various lattice points and interact electrostatically to provide cohesive energy to the solid, e.g. NaCl (Figure 1.7(a)). NaCl is thus an ionic solid.

(b) Covalent solids in which the adjacent atoms share electron pairs forming strong covalent bonds. It is thus a network of covalent bonds extending through the crystal (solid). Diamond, for example, is a covalently networked solid (Figure 1.7(b)). Understandably, such solids would be hard and difficult to melt. NaCl and diamond are both expected to have high melting points as a large number of ionic (in NaCl) and covalent bonds (in diamond) must be broken simultaneously in order for the phase change to happen. It is difficult to calculate cohesive energy of covalent solids without bringing in the full power of quantum mechanics.

(c) Metallic solids or metals in which the metal atoms can neither transfer electrons to or share electrons between adjacent atoms in the lattice to achieve complete shells of electrons. They do not simply have enough electrons in the valence shell to donate or share. So, normal covalent or ionic bonding do not work to produce metallic bonds. Instead, what happens here is a kind of delocalized covalent bonding. The metal atoms each loses or gives up one or more electrons from their valence shell. The electrons so given up are free to move between the ions thereby producing a delocalized 'sea' of electrons into which positively charged metal ions are embedded. The delocalized electronic charge distribution interacts with the metal ions producing what has become known as metallic bonds. The delocalized nature of binding renders metals malleable and ductile.

(d) Crystalline molecular solids, on the other hand, owe their fixed shapes and volumes to much weaker interactions, known as van der Waals interactions or hydrogen bonding among the molecules occupying the lattice positions. Solid iodine crystals is an example of van der Waals interaction mediated molecular solid. Ice crystals, on the other hand, are a crystalline solid formed by hydrogen bonding interaction between water molecules on adjacent lattice points.

An elemental solid may exist in more than one crystalline form giving rise to allotropism (e.g. dimaond, graphite, fullerenes, etc.) while for a non-elemental solid, it gives rise to what is known as polymorphism. The type of crystal structure and its symmetry are important attributes of a solid as many physical properties like tensile strength, conductivity, refractivity, depend on them, and may display anisotropy (direction-dependence of the property). A solid may be non-crystalline or amorphous. The particles (atoms/ions/molecules) making up the amorphous solid are randomly distributed without any significant long range order (e.g. plastics, wax, etc.). Such solids unlike their crystalline counterparts do not have any well-defined melting point. They are isotropic and do not expose cleavage planes when cut and have edges that could be curved. It is, however, possible to make amorphous solids with different degrees of crystallinity and therefore having different degrees of crystalline-solid like properties like hardness or incompressibility, making them suitable for different types of technological applications.

Between the crystalline (mono-crystalline) and the completely amorphous solids there are poly-crystalline solids. They are aggregates of many tiny crystals or grains. There is complete structural regularity in each grain (giving rise short range order) but the grains themselves are rather randomly distributed so that there is hardly any long range order present. There is neither short range nor long range order in the gaseous state of the same substance, the constituent particles being free to move in any direction anywhere and there is complete isotropy within the mass of the gas. The gases therefore do not have any fixed shape or volume and are highly compressible. In the liquid state, there is only short range but no long range order and the constituent particles can move past one another imparting to the liquid an ability to flow. Such properties are naturally exploited in technological applications.

1.2.1 More on Unit Cells

We will now take a closer look at the three types cubic lattices to learn more about them. Let us first note that that particles (atoms, ions) occupying equivalent lattice positions in Figures 1.3(a–c) are assumed to be tiny spheres, which touch each other along the edges of the body diagonal etc. Even then there is a lot of void in the unit cells. In the simple cubic cells, there is only one atom per unit cell (each corner atom is shared by adjacent cells and contributes 1/8 (there being eight such corners).[5–6] The percent packing efficiency [(volume of atoms in the unit cell/volume of the cell)$\times 100$] is thus poor in simple cubic crystals. Very few metals quite expectedly adopt simple cubic structures. Even if they do, they have as expected, low density. If a metal crystallizes in the form of a body centered cubic (Figure 1.3(b)), the number of atoms in the unit cell becomes 2 (one from the corners and one at the center), the atoms on the lattice touching each other along the body diagonal. The packing efficiency becomes $\approx 68\%$ reducing the void in the unit cell. If the unit cell dimensions ($a = b = c$) remained the same we would expect the density to be twice the density of the metal crystallizing with simple cubic lattice. For the face centered cubic lattice, the number of atoms in the unit cell would be 4 (1/8 from each of the eight corners and 1/2 from each of the six faces). We may thus expect the density to rise to approximately four times the density of the metal achievable with simple cubic lattice. In practice, it does not happen in as much as the dimensions of the unit cell change as one goes from one type of cubic lattice to another, but the trend displayed by density obtained by simple minded calculation tends to agree with what is obtained from experimental data on the density of crystalline solids. In fact many metals, ionic solids and inter metallic compounds crystallize with cubic unit cells, thereby making cubic lattices so important in the study of solid materials. It is a matter of convention that the edges of the unit cell always connect equivalent points or particles. The cubic unit cell therefore must have eight identical particles at the eight corners. The ninth particle can be made to occupy the center of the cell giving rise to the body centered cubic unit cells as in cesium chloride (CsCl), for example (Figure 1.8a). NaCl, on the other hand, crystallizes with the so called face centered cubic unit cell (fcc-structure). The fcc unit cell turns out to be the simplest (repeating) unit of a cubic-closed-packed structure. NaCl, for example, crystallizes in this fashion. There is thus a cubic closest packed array of chloride ions (Figure 1.8b), the large chloride ion occupying each of the 14 lattice points (eight corners and six face centers), with sodium ions occupying the octahedral holes (the tiny depression) between planes of chloride ions. One octahedral hole lies at the center of the fcc unit cell (Figure 1.8c) while three others exist on the three unique edges of the unit cell. The octahedral holes are populated by Na^+ ions, producing the typical 'NaCl' structure (Figure 1.8b). If the cations are packed into the 'tetrahedral holes' that exist at the corners of the fcc unit cell [typically at $(\frac{1}{a}, \frac{1}{a}, \frac{1}{a})$, for example] (Figure 1.9a) it will touch the anions (say S^{2-}) at the centers of the three faces forming the particular corner as well as the anions at the relevant corner giving rise to the well-known zinc blend structure (Figure 1.9b). In fact, the radius ratio (R_+/R_- or $R_</R_>$) determines critically the type of the hole that exists in an ionic solid. A thumb role, based on measurement relates the R_+/R_- ratio to the hole types that can occur in ionic solids[4] (Table 1.2).

Admittedly, these correlation are empirical, but are useful in designing a solid crystalline material. In pure metallic solids the atomic radius of the metal atoms alone can be approximately correlated with the

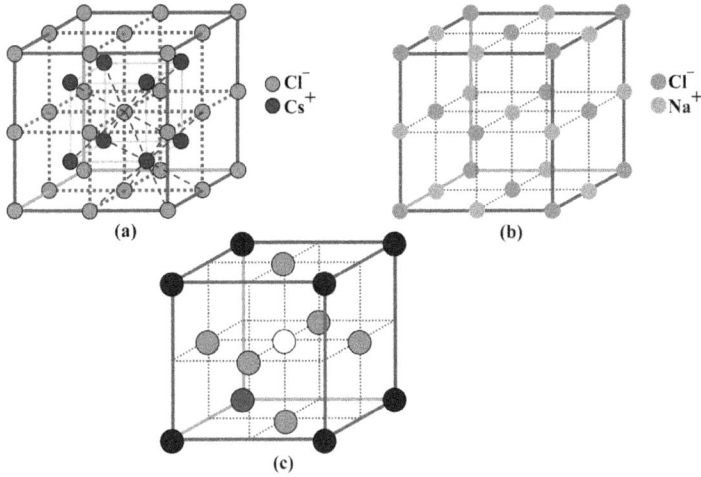

FIGURE 1.8 (a) Body centered cubic unit cell of CsCl. (b) Face centered cubic unit cell of NaCl. (c) An octahedral hole at the center of an fcc unit cell.

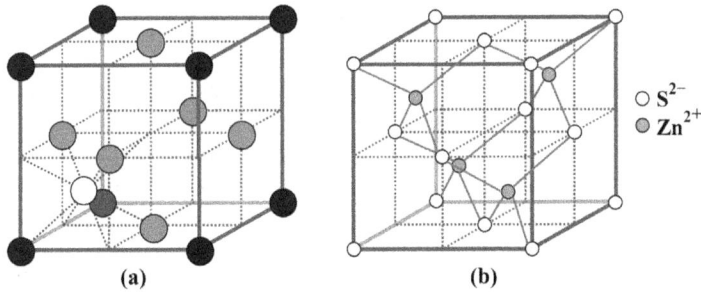

FIGURE 1.9 (a) Tetrahedral holes in an fcc unit cell. (b) Zinc blende structure: note how Zn^{+2} ions occupy the tetrahedral holes.

TABLE 1.2

The Correlation between Hole Type and the Cation to Anion Radius Ratio (R_+/R_-)

Hole type	Radius ratio
Cubic	0.732–1.000
Octahedral	0.414–0.732
Tetrahedral	0.225–0.414

type of unit cell structure that the metal assumes in the lowest energy state. Table 1.3 shows this correlation. The energetic basis of such a correlation needs further study.

1.3 From Atoms and Molecules to Materials

The ultimate constituents of all materials are no doubt atoms. Nevertheless, it is convenient to consider 'materials based on atoms' and 'materials-based on molecules' separately. The 'molecule-based

TABLE 1.3

Correlation between Atomic Radius and Crystal Type in Metals

Atom	Atomic radius (nm)	Crystal structure
Cu	0.128	FCC
Ag	0.145	FCC
Au	0.142	FCC
Pt	0.139	FCC
Ag	0.143	FCC
Co	0.125	HCP
Zn	0.133	HCP
Cd	0.149	HCP
Mg	0.159	HCP
Cr	0.125	HCP

materials' have several (more than one) atoms or groups of atoms interacting via strong binding forces (covalent/ionic/metallic) forming well-defined entities, called molecules, which in turn interact with each other via weak forces of bonding (notably, van der Waals forces and hydrogen bonding) producing the total structural framework of the material. We may assume that the electronic structure, energy levels and other properties of the individual molecules in isolation are known to start with. Even then, it may be difficult to predict the properties of large conglomerate of the molecules – the so called 'molecule – based materials' (MBM), which may have very different properties. The nature, strength and extent of interaction among molecules in the MBM determine the energy and its structural features and properties like melting temperature, compressibility, mechanical properties, machinability, optical response, etc. If the material in question is solid, the crystal symmetry can be a very important attribute in generating or quenching the specific response. The properties of a molecule in isolation is no guarantee that it would be possible to fashion a useful material out of it.

The 'atom-based materials' appear to be a simpler problem. Pure metals like copper or nickel, for example, are made up of atoms of one type only (Cu or Ni, for example) and it is the interaction among these atoms of one type or other that could be expected to determine the behavior of bulk copper or bulk nickel. The energy levels of individual (but interacting) metals/atoms merge into what are called 'energy bands', which ultimately shape up the properties of pure metals. Alloys (which are binary or ternary or quaternary mixtures of a metal with other metals and non-metals) constitute a very important or technologically important class of materials. By varying the types of alloying elements (metals or non-metals) it is possible to 'tune' the properties of the parent metal and create technologically superior materials. Elemental iron and its alloys with carbon and some other metals like Cr, Ni, Mn, etc. can be listed as examples of how technologically superior materials called 'steel or special steels' have been realized by controlled introduction of impurities in pure iron. However, the alloys do not necessarily behave as mixtures. The interaction among different constituents can even lead to the formation of one or more new species of molecules so that no simple interpolation formula is able to predict the outcome of alloying. Detail construction of phase diagrams, identification of coexisting phases and their structural types and information about the energy levels, many body interactions, etc. are needed to understand the properties of alloys of different kinds. Alloying (doping) has a crucial role in the technologically important semiconducting materials. Thus, germanium (Ge) and silicon in their pure states are not very useful. When impurities (dopants) like boron (B) or arsenic (As) are doped into pure Ge or Si in trace, the resulting alloys are found to have band structures and dispositions that turn them into semiconductor materials, which are the basic backbone materials for transistor technology. Alloys of silicon and germanium ($Si_{1-x}Ge_x$) provide materials that are superior to either Si- or Ge-based doped semiconducting substances for transistor technology. Can there still be better materials for the same or superior technology or will silicon (cost effective) continue to dominate the field of semiconductor technology in the near or distant future? Only time will able to answer the question as the path from pure substance to real-life materials or functionalized materials may be quite tortuous and gaps in our understanding must be bridged to enable us to chart out the path without undertaking too many trials or experimentations. It is not anything special for the materials for semiconductor technology. It is equally true for magnetic materials, computer memory materials, atomic or molecular all optical switching materials, shape memory alloys, materials with high non-linear optical response, laser materials, efficient catalytic materials and many others. While the search for better materials of a specific category continues, it is imperative to improve our understanding of the factors responsible for the emergence of specific types of responses, and translate the understanding to design new materials with targetted properties.

1.4 The Need for Theoretical Understanding

A sharpened level of understanding of what shapes specific properties or responses of materials can help the formulation of 'thumb rules', which can be explored to guide the search through the very large dimensional space of features or properties to a useful, probably optimal combination of atoms and bonds (strong as well as weak) that would lead to the emergence of the desired properties. Understanding the interactions among the fundamental units of the materials – their ranges and strengths – is crucial in the entire scheme of designing and realizing targeted materials of one kind or another. Thus, iron and nickel are both magnetic materials. In their pure forms, they have common properties of metals – like thermal expansivity, low specific heat, metallic nature, etc. An alloy of iron and nickel called 'INVAR' (Fe = 65 %, Ni = 35 %), however, does not suffer any volume expansion when heated over a range of temperatures. While an asymmetry in Fe-Fe or Ni-Ni pair potentials can account for the thermal expansivity of either iron or nickel, the new interaction that alloying brings in involves nickel and iron atoms (Fe-Ni pair potential). It is possible that the average interaction potential between atoms in INVAR is somehow more symmetric, resulting in loss of thermal expansivity. The prediction of accurate pair potentials of each kind and their averaging is, however, not a trivial job.[5] Similarly, a metal powder on a specific support may catalyze a reaction, but lose activity on another support.[5] It is the interaction between the metal and the support vis-a-vis the interaction among the metal atoms that must somehow determine the specific activity. A pertinent question that crops up at this point concerns the possibility of realizing a 'single atom' catalyst vis-a-vis catalysts in the form of clusters of a minimum size on a specific support. Designing a successful catalyst is technologically as important as it is scientifically challenging. We are therefore in need of developing good theoretical descriptors that encode the properties of atoms, pair interactions and possibly, many body interactions among different types of atoms. Although correlations among specific properties of materials built up from a certain group of atoms or molecules and related atomic and molecular descriptors can be useful, the complete ability to design materials with targetted properties demand artificial-intelligence-based deep learning networks that can be trained with available atomic and molecular descriptors and properties of corresponding materials. A first step in this scheme would be to understand the atoms, molecules and their energy levels, charge distribution and many other properties using quantum mechanics as the tool and try to model the properties of materials on the basis of statistical mechanics of collections of a large number of atoms or molecules, forming a gas, liquid or solid. The central objects in the analysis of atomic and molecular states are obviously the wavefunctions and the associated energy eigenvalues, angular momenta and other commuting and non-commuting observables of the isolated atomic or molecular systems. For atoms, the spherical symmetry makes the analysis somewhat easy. For molecules, the symmetry, which is no-doubt lower, can still be exploited to build up molecular states by combining atomic states, but the problem is far more difficult than what is encountered in the atomic case. To understand solid materials (crystalline) we have to graft the periodicity (or transalational symmetry) of crystal lattices in the description for constructing energy levels. The transalational symmetry is a key in the analysis along with distribution functions, density of states, band gap, spin angular momentum, etc. The scheme out-lined looks simple but can be expected to face formidable challenges as the systems to be dealt with are many particle systems with many body interactions. Although a complete analysis involving Hamiltonians that take into account every interaction operating in the system may not yet be possible, many types of effective Hamiltonians can be used to understand and model the materials and their responses and exploit the results to design new materials. We have to realize that such effective Hamiltonians are approximations to the actual system and therefore they lead to approximate description and understanding of the systems under investigation.[6] Ab-initio statistical mechanics in which first principles numerical calculations are used to construct effective Hamiltonians (potential energy surfaces) with only the degrees of freedom that are relevant to the process (e.g. phase transition) of interest and also to fix parameters appearing the resulting model. In other words, only the relevant portion of the potential energy surfaces are constructed and the phenomenon of interest is described and explored with reference to the effective potential energy surface. The results may then be analyzed and possibly coded into new rules that could be used to design new materials. We will address some aspects of these issues later.

The number and types of materials already discovered and put to use is large. They can not be described individually. Fortunately such materials can be broadly classified into several classes and each class as a whole can be explored in detail. Even then, not every class can be covered. We have to make our choice and we understand that a subjective bias may thus creep in. With this caution in perspective, we have included the following topics for the discourse.

1.5 Topics Covered

1.5.1 The Mechanics of the Microworld

Quantum mechanics, its basic tenets, the law of quantum evolution, the wave equation of Schrödinger and the matrix formulation of Heisenberg are intriguing. The key concepts of stationary states, commuting and non-commuting measurements and measurability, hermitian operator and unique properties of quantum states are explained. The power of quantum mechanics as a calculational tool is described in a number of model and standard problems including the quantum states of a harmonic oscillator. The concept of delocalization of a quantum state across barriers and the unique phenomenon of quantum tunneling and its importance are analyzed.

1.5.2 Quantum Mechanics of Atoms

Symmetry of a central field potential and quantum states of the hydrogen atom, their symmetries and nodal features are introduced. The 'aufbau principle' is discussed in the context of constructing energy levels of many electron atoms along with angular momentum of atoms, rotational symmetry and level degeneracies. The concept of spin angular momentum of an atom that emerges from Stern-Gerlach experiments is discussed. Classifications of atomic levels based on parity, spin orbital and total angular momentum of the atom is explained.

1.5.3 Quantum Mechanics of Molecules

The quantum states of hydrogen molecule ion and their symmetries, correlation diagrams interpolating between atomic and molecular states are discussed. The central ideas of the Börn-Oppenheimer method of separating electronic and nuclear motion are analyzed in the context of the hydrogen molecule problem. Quantum mechanics of covalent bond formation and formation of ionic bonds are analyzed along with an analysis of weak bonds. The concept of one-electron density as a key quantity or descriptor of many electron systems is examined. The advent of density functional theory, its powers and pitfalls in handling many electron atoms and molecules are analyzed together with some other approaches to many electron atoms and molecules.

1.5.4 Quantum States in Solids

Quantum states in regular crystalline solids are examined. The transalational symmetry in periodic solids and its exploitation in constructing Bloch states or wavefunctions are analyzed. Formation of energy bands from discrete atomic levels, the emergence of band gap, Fermi level and metallic conduction are described in the context of metallic systems. The ideas of band gap, valence and conduction bands and their role in transport of electrons through metallic solids are introduced with applications. Electron states in amorphous solids, the phenomenon of localization and its ramifications are discussed briefly.

1.5.5 Classical Statistical Mechanics

Gibbs formulation of classical statistical mechanics is reviewed with several applications. The concepts of probability distribution functions namely, binomial, Poisson and Gauss distributions are introduced. The partition functions and its connection with different properties of the system are explored with special emphasis on entropy. Classical statistical mechanical modeling of various macroscopic properties of physical systems is reviewed.

1.5.6 Quantum Statistical Mechanics

The discreteness of atomic/molecular energy levels and spin angular momentum of microscopic particles are introduced along with the restrictions on the occupancy level of single particle states depending upon the spin angular momentum. The idea of grand canonical ensemble is developed in the context of defining the equilibrium state and quantum statistics of materials. Statistics of many Fermion and many Boson assemblies are worked out with applications to ideal electron gas, lattice vibrations (phonons) and photons (radiation enclosed in a cavity). Electron states in metals, semiconductor and ceramics are analyzed in this context.

1.5.7 Traditional Materials

Metals, alloys, ceramics, semiconductors are examined with emphasis on understanding the origin of their special properties. The traditional idea of superconductivity and superconducting materials at low temperature is examined along with the rather novel phenomenon of high temperature superconductivity and high temperature superconducting materials in the light of quantum and statistical mechanics.

1.5.8 Smart Materials

The advent of smart materials has boosted advanced technology. Such materials can sense and respond to changes in the environment and are considered under several heading, e.g. electrochromic, ferroelectric, photochromic, piezoelectric and shape-memory materials, thermochromic materials, quantum tunneling composites, etc. The emphasis has been on understanding the roots of these phenomena.

1.5.9 Magnetic Materials

Magnetic matter has attracted the attention of scientists over the ages. These technologically important materials are considered under several categories, e.g. dia and paramagnetic materials, ferro and ferrimagnetic materials and antiferromagnetic materials with emphasis on what causes the emergence of these different types of magnetic order. The relatively new kinds of materials with giant magnetoresistance are introduced separately. The curious case of molecular magnets and their importance is explored. The ongoing search for new magnetic materials is reviewed in several directions.

1.5.10 Low-Dimensional Materials

Traditional materials are all what we call 3D materials. The advent of graphene and graphene-likes (i.e., white graphene, boron nitrides, phosphorenes, Mxenes etc.) has brought in a new class of materials with strong technological potential. The electronic structure and properties of such materials are analyzed. The new age materials – nano or nanostructured materials – and their imprints on the now developing nanotechnology are explored.

1.5.11 NLO and Energy Materials

NLO materials are capable of changing one or more of the characteristics of light passing through them. Different types of such materials are considered with emphasis on molecule-based chromophores for such materials. The roots of different kinds of NLO response are traced. The alternatives to traditional fossil fuels as basic energy materials are in great demand. Hydrogen storage materials for hydrogen-based energy production are in great demand. These materials are considered from thermodynamic and electronic structural view points in the context of demand for fossil fuels and the looming energy crisis that threatens to engulf the civilization.

1.5.12 Materials Design

The problem of designing a material with a preselected set of properties for specific technological applications is addressed in the light of traditional wisdom and computational materials science that has undergone a revolution. We examine if the results of computational materials science have sharpened or reinforced the intution of chemistry in designing new materials. The possibility of cross-talk between electronic structure theory and methods of artificial intelligence leading to newer designer materials is evaluated.

1.6 Classification of Materials

The world of materials is a diverse world. So numerous are the materials already developed and in use, that a proper classification of them into classes becomes both necessary and important as it would help us in understanding each class as a whole and in determining what kind of tuning the class could permit leading to new materials. The classification can be based on chemical composition, type of bonding operating in the material, the dominant common physiochemical features of the entities in a class, the nature of electronic states that the material supports, the technological use that the group of materials can offer, etc. Many subclassifications within a class is also possible. We will, in what follows, consider only a broad classification. According to this scheme, materials (solids, in general) can divided into five main classes: (a) metals, (b) ceramics, (c) polymers, (d) composites, (e) high technology materials.

Metals or metallic materials are made up of relatively dense, ordered arrangement of one or more types of metal atom along with small amounts of non-metallic impurities in some cases. The primary mode of binding in these materials is provided by metallic bonds, which leaves a sea of delocalized mobile electrons not attached to specific atoms. Many types of properties like electrical and thermal conductivities of these materials are shaped by these mobile electrons. They have ductibility and are resistant to fracture. Some of these materials display magnetic properties. We will take a closer look at metallic materials in Chapter 8.

Ceramics, on the other hand, are compounds formed by union of atoms of metals (Fe, Ni, Cr, Al) and non-metals like oxygen, nitrogen, carbon forming oxides, nitrides and carbides. As a class, these materials are hard, but non-ductile and have poor electrical and thermal conductivity (no free electrons). They can bear large thermal stress. Some of these materials (specially the oxides) have a magnetic property that makes them industrially useful. Traditional ceramics like glass, porcelene also belong to this class.

Polymers are made up of large organic molecules of carbon, hydrogen, oxygen and nitrogen. A few inorganic polymers (silicone, for example) are also known. Polymers are generally ductile and easily deformable and can be moulded into complex shapes. These materials as a class have low stiffness and mechanical strength but many of them can withstand a harsh chemical environment. However, they are often not resilient to thermal stress. The atomic constituents of polymeric materials are generally covalently bonded. Thus no free electrons are available in polymeric materials and their thermal and electrical conductivity is very low as a rule although conducting polymers have been realized in the form of doped poly-thiophenes. The conduction mechanism is still not completely understood. Polymeric materials are less dense than either metals or ceramics.

Composite materials are obtained by blending two or more different materials from the three groups, e.g. metal-polymer composites, polymer ceramic composites, etc. The idea is to make blends that have properties not realized in the components and are therefore superior to either of the components. By controlling the concentrations of the components in the blends, it becomes often possible to get a spectrum of materials where almost continuous variation in the targetted property can be realized. Composites therefore occupy a special position in the world of materials today.

High technology materials are also called advanced materials. These are materials for sophisticated technological applications. We may specially mention semiconducting materials (backbone of integrated circuits and electronics), biomaterials (materials for implants in human bodies, e.g. heart), non-linear optical materials (materials for laser generation and optical switching), smart materials (materials that can sense the environment and respond to change in certain environmental parameters) – e.g. halochromic materials, electrochromic and photochronic materials, energy materials (for storage and release of energy), magnetic materials (for computer disks and memory systems) and so on. A new class of materials called 2D materials like graphene, for example, has been receiving serious attention. The nano and nanostructured materials are similarly being projected as futuristic materials for nanotechnology.

1.7 Future Outlook

Designing materials having pre-set properties have long been materials scientists' dream. With the rise of computational material science including the science of simulation as an efficient theoretical tool to predict electronic structure and properties of molecules and materials and the possibility of interfacing these tools with artificial intelligence methods exploring and exploiting structure–property correlations for predicting materials with targetted properties has become a distinct possibility now. The recent developements in this field has been addressed in the penultimate chapter (Chapter 13). In the final chapter (Chapter 14) we have reviewed the present state of development of materials science and pondered over the shape of things to come. The possibility of modeling a material accurately and designing a new material theoretically have been examined. The possible role of artificial intelligencies in the development of designer materials is considered. Two important directions of future development – one concerning energy materials and the other concerning new materials, possibly multiferroics for the electronics industry that is poised to consume almost half of the total energy budget – are reviewed. To an extent the choice of topics has been subjective. We have left out many interesting things just for conserving space. However, the topics covered, we hope, will give the readers a real feel for the subject, develop understanding and help them to explore new windows in the science of atoms, molecules and materials.

REFERENCES

1. R. F. Tylecote, *A History of Metallurgy*, The Materials Society, London (1976).
2. V. C. Pigott, 'Iron versus bronze', *J. Metals*, 42ff, August (1992).
3. J. C. Waldbaum, *The Coming of the Age of Iron*, T. A. Westime and J. D. Muhly, Eds., Yale University Press, New Haven (1980).
4. B. Ehgartner, *The Age of Aluminium: The Dark Side of the Shiny Metal*, Ennsthaler Gesellshaft, Steyr, Austria (2019).
5. W. D. Callister Jr, *Materials Science and Engineering – An Introduction*, Sixth Edition, Wiley, India (2003).
6. S. Glasstoe, *Textbook of Physical Chemistry*, D. Van Nostrand Company, New York (1946).

2

Quantum Mechanics

2.1 Introduction: Mechanics of the Microworld

Mechanics is the science of motion. In the world of our everyday experience, i.e. in the macroscopic world, the motion of an object can be completely described by Newton's equation of motion – a second order differential equation, which for an object of mass 'M' moving along or parallel to the x-axis (coordinate), reads in one spatial dimension

$$M\frac{d^2 X(t)}{dt^2} = F(X) \tag{2.1}$$

where $F(X)$ is the force acting on the particle. In a conservative system, the force is derivable from a potential $V(X)$ and the equation of motion reads

$$M\frac{d^2 X(t)}{dt^2} = -\frac{dV(X)}{dX} \tag{2.2}$$

Being a second order differential equation it requires two initial conditions to fix the two constants of integration, like $X(0) = X_0$ and $X'(0) = v_0$ where $X'(t) = \frac{dX}{dt}$ represents the velocity of the particle at the instant 't'. Thus $v_0 = X'(0)$ is just the velocity at the initial moment ($t = 0$), i.e. the initial velocity. With X_0 and v_0 specified we can integrate Eq. 2.2 for $X(t)$, $X(t)$ being called the trajectory of the particle. Of course we are assuming that the equation of motion can be integrated analytically; if not, a suitable numerical method of integration can be used. Along with $X(t)$ we can also calculate $X'(t) = v(t)$, i.e. the velocity profile along the trajectory and the momentum profile $p(t) = MX'(t)$. Thus, all the features of the motion executed by the macroscopic particle moving in the potential $V(X)$ can be clearly and accurately determined, at least as accurately as the initial conditions are specified and as accurately as the integration can be carried out. In three-dimensional space, the Newtonian equation of motion is more complicated and reads

$$M(\frac{d^2 X(t)}{dt^2} + \frac{d^2 Y(t)}{dt^2} + \frac{d^2 Z(t)}{dt^2}) = -\nabla V(X, Y, Z)$$

and the integration becomes correspondingly difficult. Newtonian Mechanics or any other formulation of classical mechanics (e.g. Lagrangian Mechanics or Hamiltonian Mechanics, etc.), however, turned out to be inadequate to describe the dynamics of a microscopic particle like an electron, a proton and an atom or a molecule, etc. The concept of trajectory itself seemed to break down in the microscopic world and 'determinacy' of classical mechanics had to give way to 'indeterminacy' and 'probabilistic' description.[1] It turned out that the position and the corresponding (conjugate) momentum of the microscopic particle could not be determined or specified simultaneously with absolute precision, however, refined the techniques of measurement of position and the conjugate momentum could be. This inability to determine X and P simultaneously with complete precision or certainty is a fundamental property of the microscopic systems or quantum systems as was first enunciated by Werner Heisenberg in a series of carefully designed thought experiments. The uncertainty has nothing to do with the standard statistical error in a measurement or limitations of the equipment used, which can, in principle, be completely eliminated by carefully designing the setup and carrying out the experiments. It turned out that these uncertainties are due to inherent 'quantum fluctuations', which can not be controlled or eliminated when certain 'observables' called 'non-commuting observables' are being measured.[1–4] On the other hand, there is no such 'uncertainty'

DOI: 10.1201/9781003244882-2

problem in the measurement of 'commuting observables'. The mechanics of the microscopic world there-fore needs to be formulated so as to be consistent with our basic inability to measure the position (X) and the associated (conjugate) momentum (P_X) simultaneously with absolute precision. In fact, Heisen-berg demonstrated that the uncertainty in measuring the position (call it $\triangle X$) and that in the associated momentum (call it $\triangle P_X$) are related by the following inequality:[3]

$$\triangle X \cdot \triangle P_X \geq \frac{\hbar}{2} \tag{2.3}$$

where $\hbar = \frac{h}{2\pi}$, h being the basic quantum of action, called Planck's constant. The constant h was introduced by Max Planck while proposing his quantum theory of black body radiation. The celebrated condition (Eq. 2.3) has become known as Heisenberg's principle of uncertainty. In three-dimensional space, ($\triangle Y$ and $\triangle P_Y$) and ($\triangle Z$ and $\triangle P_Z$) are similarly related by the corresponding uncertainty relations

$$\triangle Y. \triangle P_Y \geq \frac{\hbar}{2}; \triangle Z. \triangle P_Z \geq \frac{\hbar}{2} \tag{2.4}$$

Eqs. 2.3–2.4 do not put any restriction on the simultaneous measurability of X and P_Y or Z and P_X or Y and P_Z, for example. It affects only the simultaneous measurements of conjugate pairs of position and momentum. The uncertainty relations are not the only extraordinary feature of the mechanics of the microworld. It turned out further that the particles of the microscopic world can display a kind of duality in the sense that the same entity can behave like waves as well as particles depending on the situation to which the entity is exposed. In the macroscopic world (the classical world) the wave motion and corpuscular motion are mutually exclusive entities and the same classical system can not display features of both wave and particle motion. They behave either as waves or as particles. The duality mentioned for the microscopic world is not a mere philosophical contention or construct. It is an inherent feature of the microscopic reality (the quantum world) and has been confirmed by experiment on diffraction of an electron beam by thin sheets of metals acting as the diffraction grating by Davison and Germar. De Broglie had earlier proposed on the basis of theoretical arguments that a particle with a linear momentum 'p' can behave as a matter wave of wavelength λ where

$$\lambda = \frac{h}{p} \tag{2.5}$$

Eq. 2.5 has become known as the De Broglie relation and is at the root of the description of microscopic world based on the wave mechanics of matter, which took shape later through the work of E. Schrödinger. The wave-particle duality of matter is not a unique phenomenon. The same fundamental duality has been observed for light.

Light propagates in free space as transverse electromagnetic waves obeying Maxwell's equations and manifests its wave nature in interference and diffraction experiments. It turned out that it interacts with matter and can exchange energy and momentum only in units that are discrete and are called 'quanta'. If we consider monochromatic light of frequency v (angular frequency ω) the basic 'quantum' of energy is hv and energy exchange with matter takes place in integral multiples of hv. If ΔE measures the energy exchanged, $\Delta E = nhv = n\hbar\omega$ ($n = 0, 1, 2, \cdots$). Similarly, momentum exchange can occur in units (quanta) of $\hbar \overrightarrow{k}$, $\overrightarrow{k}$ being the wave vector of light of frequency v.

$$\overrightarrow{k} = \overrightarrow{k_0}\frac{2\pi}{\lambda} = \overrightarrow{k_0}k \tag{2.6}$$

where $\overrightarrow{k_0}$ is a unit vector in the direction of the photon momentum and $k = \frac{2\pi}{\lambda}$ is the wave vector. Evidently, light can therefore be thought of as small 'particles' or quanta of energy (hv) and momentum ($\hbar k$), this corpuscular character being manifested in experiments of light-matter interaction. In some other experi-ments like diffraction, light displays its wave nature as proposed in Maxwell's theory by producing the expected pattern of diffraction. The 'particle of light' are called 'photons'. The concept of light particles or photons was elegantly utilized by A. Einstein in formulating his theory of photoelectric effect[5] many features of which eluded explanation within the framework of wave theory of light. Another experimental confirmation of the photon theory of light come in 1923 from investigations on scattering of X-rays by free electrons. The observation came to be known as Compton scattering or simply the Compton effect and provided an independent verification of the law of conservation energy (already noted by Einstein in his photon theory of photoelectric effect) as well as the law of conservation of momentum in photon-electron scattering. All these experimental evidences paved the way for the acceptance of the idea of 'wave particle duality' being universal and an essential feature of the quantum world. The consequences of the duality can be observed in the microscopic world while they are not noticeable in the macroscopic scale.[4] The mechanics of the microworld that is consistent with the 'duality' (wave-particle) and the requirements of the 'principle of uncertainty' was soon developed and became known as 'Quantum Mechanics'. It was Heisenberg who first proposed a version of this new mechanics, which prescribes a quantum law of evolution for the microworld. Heisenberg's formulation came to be known as 'Matrix Mechanics' as the key dynamical entities like position (X), momentum (P), energy (H), etc. were represented by Hermitian matrices and the law of evolution described how these matrices evolve in time. The second realization of the new mechanics was proposed 3 months later by E. Schrödinger. The key quantities in Schrödinger's formulation were hermitian operators like $\hat{X}$, $\hat{P}$, $\hat{H}$, etc. and the quantum law of evolution proposed took the form of a partial differential equation (Schrödinger equation) involving an entity called the 'wave-function' of the system under consideration. The equation came to be known as the wave equation for matter or material particles and resulting mechanics called wave mechanics or quantum mechanics. The central quantity in this description is a complex function of the coordinates (X, Y, Z) or momenta (P_X, P_Y, P_Z), called the matter waves or wave-function, which itself is not an observable but provides a 'window' to the observable world. ψ represents a wave of probability amplitudes (not an observable) and $|\psi|^2$ the probability density in the coordinate or the momentum space (an observable). We will not adhere to the chronology of development and take up for discussion the version of quantum mechanics propounded by E. Schrödinger, noting that Heisenberg's formulation was proposed earlier. It was Schrödinger who proved the equivalence of the two formulations although in practical problems, Schrödinger's version is much more convenient to apply and has therefore become more popular.

2.2 Law of Quantum Evolution: The Schrödinger Equation

Here we take the first look into the matter-wave equation proposed by Schrödinger. The key quantity in this equation is the wave-function. We will try to guess its form and the equation it obeys through a series of correspondence arguments.

Consider a monochromatic wave of circular frequency $\omega = 2\pi\nu$ propagating in the free space. The wave carries an energy equal to E and in one dimension takes the form

$$f(x,t) = A \cdot \exp(i \cdot (\vec{k} \cdot \vec{x} - \omega t)) \tag{2.7}$$

A microscopic particle (say, an electron) of mass 'm' moving with linear momentum of magnitude p along the x-axis (free space), by De Broglie's hypothesis behaves like a wave (matter wave) of wavelength λ where

$$\lambda = \frac{h}{p} \tag{2.8}$$

p being the magnitude of momentum. The corresponding wave number k is given by

$$k = \frac{2\pi}{\lambda} = \frac{p}{\hbar} \tag{2.9}$$

The momentum carried by the matter wave is therefore

$$p = k\hbar (\vec{k}\,\hbar = \vec{k_0}\,(k\hbar)) \tag{2.10}$$

The energy carried by the wave of frequency ν by the quantum hypothesis is

$$E = h\nu = \hbar\omega \tag{2.11}$$

Let the function of space and time (assuming a one-dimensional system) that represents the matter wave be $\psi(\overrightarrow{x},t)$. The analytical form of the function $\psi(\overrightarrow{x},t)$ or the matter wave can be hopefully obtained from the function $f(x,t)$ of Eq. 2.7 representing a monochromatic wave moving along the x-axis by introducing the quantum relations Eqs. 2.8–2.11, into the right hand side of Eq. 2.7 leading to

$$\psi(\overrightarrow{x},t)=A\cdot\exp\left(i\cdot\left(\frac{\overrightarrow{p_x}\cdot\overrightarrow{x}}{\hbar}-\frac{Et}{\hbar}\right)\right) \tag{2.12}$$

In three-dimensional space, Eq. 2.12 can be easily generalized to obtain

$$\psi(\overrightarrow{r},t)=A\cdot\exp\left(\frac{i\cdot(\overrightarrow{p}\cdot\overrightarrow{r}-Et)}{\hbar}\right) \tag{2.13}$$

'A' being the amplitude of the matter wave at the point $\overrightarrow{x}=0$ ($\overrightarrow{r}=0$) at time $t=0$. Let us defer, for the time being considering what the complex function $\psi(\overrightarrow{x},t)$ or $\psi(\overrightarrow{r},t)$ represents, except mentioning that $\psi(\overrightarrow{x},t)$ does not represent any observable quantity, although it is assumed to carry all the information about the system (the free particle in this case) encoded in it. What is the equation that our matter wave satisfies? To proceed in that direction let us differentiate $\psi(\overrightarrow{x},t)$ with respect to $\overrightarrow{x}$. The differentiation results in $\psi(\overrightarrow{x},t)$ multiplied by $\frac{i}{\hbar}p_x$. It is clear that $\frac{\hbar}{i}\frac{\partial}{\partial x}$ is an 'operator' that acts on $\psi(\overrightarrow{x},t)$ to produce p_x multiplied by $\psi(\overrightarrow{x},t)$. Note that an operator is a mathematical construct that acts on a function to produce another function and in special cases a scalar times the original function. In the latter case, the scalar is called an eigenvalue of the operator and the function is called its eigenfunction. Bearing this in mind, we can identify $\frac{\hbar}{i}\cdot\frac{\partial}{\partial x}$ as the operator $\hat{P}_x$ for generating the x-component of momentum of the particle and $\psi(\overrightarrow{x},t)$ as the corresponding eigenfunction of the momentum operator. Symbolically,

$$\frac{\hbar}{i}\cdot\frac{\partial}{\partial x}A\cdot\exp\left(\frac{i(\overrightarrow{p_x}\cdot\overrightarrow{x}-\omega t)}{\hbar}\right)=p_xA\exp\left(\frac{i(p_x\cdot x-wt)}{\hbar}\right)$$

Generalization to the three-dimensional case is straightforward. The corresponding operator for momentum (p) in 3D space is

$$\hat{P}=\hat{p_x}+\hat{p_y}+\hat{p_z}=\frac{\hbar}{i}(\frac{\partial}{\partial x}+\frac{\partial}{\partial y}+\frac{\partial}{\partial z})=\frac{\hbar}{i}\nabla \tag{2.14}$$

with $p_x=\frac{\hbar}{i}\frac{\partial}{\partial x}$, $p_y=\frac{\hbar}{i}\frac{\partial}{\partial y}$, $p_z=\frac{\hbar}{i}\frac{\partial}{\partial z}$ and $\psi(\overrightarrow{r},t)$ is an eigenfunction of $\hat{P}$ with the eigenvalue p. From the unobservable complex function $\psi(\overrightarrow{r},t)$, $\frac{\hbar}{i}\nabla$ extracts the 'value' of a dynamical quantity – the momentum, a real world observable. It seems plausible therefore to assume that $\frac{\hbar}{i}\nabla$ is an operator that corresponds to the momentum P of the particle.

In a similar way, we may also differentiate $\psi(\overrightarrow{r},t)$ with respect to 't' (time) to obtain the following

$$\frac{\partial}{\partial t}\left\{A\cdot\exp\frac{i(\overrightarrow{p}\cdot\overrightarrow{r}-Et)}{\hbar})\right\}=\frac{-iE}{\hbar}A\cdot\exp\left(\frac{i(\overrightarrow{p}\cdot\overrightarrow{r}-Et)}{\hbar}\right) \tag{2.15}$$

Therefore, $-\frac{\hbar}{i}\cdot\frac{\partial}{\partial t}$ can be taken to correspond to energy operator $\hat{E}$ so that[5]

$$\hat{E}=-\frac{\hbar}{i}\frac{\partial}{\partial t} \tag{2.16}$$

$$\hat{E}\psi(\overrightarrow{r},t)=E\psi(\overrightarrow{r},t)$$

where $\psi(\overrightarrow{r},t)$ by construction is an eigenfunction of our energy operator $\hat{E}=-\frac{\hbar}{i}\frac{\partial}{\partial t}$ with the eigenvalue 'E'. A macroscopic particle of mass 'm' and momentum ($\overrightarrow{p}$) has energy (kinetic), $E=\frac{p^2}{2m}$. For a microscopic (quantum) particle, we can transcribe the same relation among E, P and m in the language of the operators as

$$\hat{E}\psi(\overrightarrow{r},t)=\frac{\hat{p}^2}{2m}\psi(\overrightarrow{r},t) \tag{2.17}$$

Using the explicit forms of operators $\hat{E}$ and $\hat{P}$, we have

$$\frac{-\hbar}{i}\frac{\partial\psi(\overrightarrow{r},t)}{\partial t}=-\frac{\hbar^2}{2m}\nabla^2\psi(\overrightarrow{r},t) \tag{2.18}$$

Eq. 2.18 is the wave equation or the Schrödinger equation for a free particle of mass '*m*' in the time-dependent form (also called the time-dependent Schrödinger equation). If the same particle is moving in a potential $V(x, y, z)$, the corresponding potential energy operator would be $V(\hat{x}, \hat{y}, \hat{z})$ where $\hat{x}$ stands for multiplication by x ($\hat{x} f(x) = x f(x) = g(x)$). The total energy E of the particle is now the sum of its kinetic and potential energy components. In that case we can write classically $E = \frac{p^2}{2m} + V(x, y, z)$ The corresponding equation in the operator language would read

$$\hat{E}\psi(\overrightarrow{r},t) = \frac{\hat{p}^2}{2m}\psi(\overrightarrow{r},t) + V(\hat{x},\hat{y},\hat{z})\psi(\overrightarrow{r},t)$$

$$\text{or, } \frac{-\hbar}{i}\frac{\partial \psi(\overrightarrow{r},t)}{\partial t} = -\frac{\hbar^2}{2m}\nabla^2\psi(\overrightarrow{r},t) + V(\hat{x},\hat{y},\hat{z})\psi(\overrightarrow{r},t) \qquad (2.19)$$

$$= \hat{H}\psi(\overrightarrow{r},t)$$

where $\hat{H}$, the Hamiltonian operator of the system is

$$\hat{H} = -\frac{\hbar^2}{2m}\nabla^2 + V(\hat{x},\hat{y},\hat{z}) \qquad (2.20)$$

Eq. 2.19 is the wave equation or the time-dependent Schrödinger equation of a particle of mass '*m*' moving in a potential $V(x, y, z)$ and is the generalization of the free particle wave equation specified in Eq. 2.18. We must alert the reader to the fact that what we have described is not a derivation of Schrödinger's wave equation or the wave equation for matter or material particles. We have merely explored a kind of correspondence between the classical dynamical quantities and their quantum counterparts – the 'operator' in a roundabout way. Eq. 2.18 or Eq. 2.19 may be regarded as the statement of natural law – the law of quantum evolution or the law of wave mechanics of matter, which has been guessed on the basis of plausible 'correspondence' arguments.

Let us take a closer look at Eq. 2.19 and Eq. 2.20. The operator $\hat{H}$ is called the quantum mechanical Hamiltonian operator of the system and is a sum of kinetic ($\hat{T}$) and potential ($\hat{V}$) energy operators of the system with

$$\hat{T} = -\frac{\hbar^2}{2m} \cdot (\frac{\partial^2}{\partial x^2} + \frac{\partial^2}{\partial y^2} + \frac{\partial^2}{\partial z^2})$$

and

$$\hat{V} = V(\hat{x},\hat{y},\hat{z})$$

and hence $\hat{H}$ is also called the operator for the total energy of the system. It acts on the wavefunction and causes it to evolve. The evolution is trivial for certain states of the system and non-trivial for certain other states. Whatever it is Eq. 2.19 captures the entire dynamics. $\psi(\overrightarrow{r},t)$ or $\psi(x,y,z,t)$ is called the quantum mechanical wavefunction of the system. It is a complex function of space (x, y, z) and time (t) variables that satisfies the Schrödinger equation (Eq. 2.19). In order to be an acceptable solution of the Schrödinger equation, $\psi(x,y,z,t)$ must obey certain boundary condition that quantum mechanics specifies. No other solution is acceptable. So, Eq. 2.19 with the appropriate boundary conditions imposed on $\psi(\overrightarrow{r},t)$ is the time-dependent Schrödinger equation; without the boundary condition. Eq. 2.19 is just a linear partial differential equation. It becomes the Schrödinger equation only when the appropriate boundary conditions are introduced. What does $\psi(\overrightarrow{r},t)$ represent? As mentioned already, we can not identify ψ with a physically observable property of the system. It does, however, provide a 'window' to the real world properties of the system, called observables, through a set of axiomatic rules, to be specified later in this chapter. At this point, let us only mention that $|\psi(\overrightarrow{r},t)|^2$ represents the probability density (ρ) of the system being found at the point $\overrightarrow{r}$ (x, y, z) at time '*t*'. Accordingly $|\psi(\overrightarrow{r},t)|^2 \, dx \, dy \, dz = |\psi(x, y, z, t)|^2 dv$ must represent the probability $P(\overrightarrow{r}, t)$ of the system being found the tiny volume element dv surrounding the point (x, y, z) at time '*t*'. Since the system, a quantum particle for example must be somewhere in the space accessible to it at time '*t*', the probability summed over the entire space must the unity. That means

$$\int |\psi(x,y,z,t)|^2 dv = 1. \qquad (2.21)$$

Such a wavefunction is said to be a square integrable function, which has been normalized to unity. The square integrability only requires the integral on the left hand side of Eq. 2.21 to exist and produce a finite number N. That is

$$\int \psi^*(\overrightarrow{r},t)\psi(\overrightarrow{r},t)dv = N$$

This condition holds if $\psi(\overrightarrow{r},t)$ vanishes at the boundary. $\frac{1}{\sqrt{N}}\psi(\overrightarrow{r},t)$ is then said to have been normalized to unity. While using a square integrable function we will therefore always assume that $\psi(\overrightarrow{r},t)$ is appropriately normalized and Eq. 2.21 holds. The identification of $|\psi(x,y,z,t)|^2$ as the probability density immediately fixes the dimension of $|\psi(x,y,z,t)|^2$ as l^{-3} in three-dimensional space, which in turn fixes the dimension of $\psi(\overrightarrow{r},t)$ as $l^{-3/2}$ in three dimensions and as $l^{-d/2}$ in d-dimensional space. The probability density $|\psi(x,y,z,t)|^2$ may be time-independent in which case $\psi(\overrightarrow{r},t)$ is the wavefunction for a special kind of state called a stationary state. If the probability density is not time-invariant, $\psi(\overrightarrow{r},t)$ describes a non-stationary state (see later). Our major concern in this work will be to look for stationary states of the systems and exploit such solutions to understand the properties of microscopic systems.

2.2.1 Axiomatic Foundation of Quantum Mechanics

We can succinctly summarize the essential features of Schrödinger's new mechanics – the wave mechanics in the form of a set of axioms or postulates. That allows us to develop the new mechanics as a 'calculus' or a 'calculational tool' to understand and compute the properties of microscopic systems, like the electrons, atoms, molecules, solids, etc. leaving aside the deep philosophical questions and implications that the new mechanics confronts us with. We will briefly return to these issues at the end of the chapter. Our main thrust, we reiterate will be to develop the new 'calculus' for microscopic systems.

2.2.2 Postulates of Quantum Mechanics

I. The dynamical state of a quantum system (a particle, for example, like an electron) is uniquely determined by a complex function of space and time variables $\psi(x,y,z,t)$. $|\psi(x,y,z,t)|^2dv$ represents the probability of finding the particle within a small volume element dv centered at the point x,y,z at time t. The description is non-relativistic. $\psi(x,y,z,t)$ is a square integrable function and can be assumed to have been normalized to unity.

II. To every measurable dynamical quantity (observables) of classical mechanics, a, b, c, $\cdots$, etc, there corresponds a suitable operator $\hat{A}$, $\hat{B}$, $\hat{C}$, etc. and

$$\int \psi^*(\overrightarrow{r},t)\hat{A}\psi(\overrightarrow{r},t) = <A>$$

represents the quantum mechanical average value of the dynamical quantity represented by $\hat{A}$, in the state that $\psi(\overrightarrow{r},t)$ represents. For example, the classical position variables x, y, z become operators $\hat{x}$, $\hat{y}$, $\hat{z}$ where $\hat{x}$ implies multiplication by x, etc., while the classical linear momentum components p_x, p_y, p_z become differential operators $\hat{p}_x, \hat{p}_y, \hat{p}_z$ where $\hat{p}_x = \frac{\hbar}{i}\frac{\partial}{\partial x}$, etc. $\hat{x}$ or $\frac{\hbar}{i}\frac{\partial}{\partial x}$ are manifestly linear operators as,

$$\hat{x}(c_1f(x)+c_2g(x)) = c_1xf(x)+c_2xg(x),$$

$$\frac{\hbar}{i}\frac{\partial}{\partial x}\left(c_1f(x)+c_2g(x)\right) = c_1\frac{\hbar}{i}\frac{\partial f}{\partial x}+c_2\frac{\hbar}{i}\frac{\partial g}{\partial x}$$

where c_1 and c_2 are scalars, while $g(x)$ and $f(x)$ are well behaved functions of x. Similarly, $\hat{y}$, $\hat{z}$ or $\frac{\hbar}{i}\frac{\partial}{\partial y}$ and $\frac{\hbar}{i}\frac{\partial}{\partial z}$ are linear operators. These operators or other operators that we will get introduced to in quantum mechanics have an additional property called 'hermiticity'. We will examine this special and very important property in the next section.

III. The time evolution of the wavefunction $\psi(\vec{r}, t)$ is completely described by the matter-wave equation of Schrödinger.

$$\frac{-\hbar}{i} \cdot \frac{\partial \psi(\vec{r}, t)}{\partial t} = \left[\frac{\hat{p}^2}{2m} + V(\hat{x}, \hat{y}, \hat{z}) \right] \psi(\vec{r}, t) \tag{2.22}$$

where $\psi(\vec{r}, t)$ satisfies the appropriate boundary conditions.

2.3 Observables, Operators and Their Eigenfunctions

We have already mentioned in the preceding section while laying out the postulates of quantum mechanics, that every observable quantity in quantum mechanics is represented by an appropriate linear operator, $(\hat{A})$. We have so far introduced quite a few operators: operators for linear momentum: $\hat{P} = \frac{\hbar}{i}\nabla$, with $\hat{p}_x = \frac{\hbar}{i}\frac{\partial}{\partial x}$, $\hat{p}_y = \frac{\hbar}{i}\frac{\partial}{\partial y}$ and $\hat{p}_z = \frac{\hbar}{i}\frac{\partial}{\partial z}$. The kinetic energy operator $\hat{T} = \frac{\hat{p}^2}{2m} = -\frac{\hbar^2}{2m}\nabla^2$, potential energy operator $\hat{V} = V(\hat{x}, \hat{y}, \hat{z})$, the energy operator or the Hamiltonian operator $\hat{H} = \hat{T} + \hat{V}$. Before examining these operators in detail, let us recall that an operator $\hat{A}(x)$ is a mathematical construct, which acts on a function $f(x)$ generally producing a new function, say $g(x)$:

$$\hat{A}f(x) = g(x) \tag{2.23}$$

There may be special functions $\Phi_k(x)$ such that $\hat{A}$ acts on them to produce a scalar (λ_k) times the function on which $\hat{A}$ acts:

$$\hat{A}\Phi_k(x) = \lambda_k \Phi_k(x) \tag{2.24}$$

Such functions are called eigenfunctions of the operator $\hat{A}$ and the scalar λ_k is called the k-th eigenvalue of the operator concerned. Note that k may be discrete or continuous. The nature of the eigenvalues (complex, purely imaginary, real etc.) depends on the boundary conditions imposed on $\Phi_k(x)$. To illustrate, let us consider the operator $\hat{A} = \frac{d}{dx}$. The eigenfunctions of the $\frac{d}{dx}$ satisfy the following linear differential equation:

$$\frac{d\Phi_k(x)}{dx} = \lambda_k \Phi_k(x), \quad (-\infty \leq x \leq \infty) \tag{2.25}$$

We can easily integrate the equation to get

$$\Phi_k(x) = c_k \exp(\lambda_k x) \tag{2.26}$$

where c_k is the constant of integration. If we do not impose any boundary condition or restriction on $\Phi_k(x)$, λ_k may be real (+ve or -ve), complex or even purely imaginary. However, if we demand that $\Phi_k(x)$ must remain 'finite' everywhere in the admissible range of x (-∞ to ∞) λ_k can only be purely imaginary: i.e.,

$$\lambda_k = i\beta_k \quad (\beta_k \text{ real})$$
$$\Phi_k(x) = c_k \exp(i\beta_k x) \tag{2.27}$$

With real β_k values, $\Phi_k(x)$ is a function with sinusoidal (finite) oscillations. Had λ_k been purely real and positive (> 0), $\Phi_k(x) = c_k \exp(\lambda_k x)$ would have become infinitely large as x approaches infinity violating the boundary condition that $\Phi_k(x)$ remains finite everywhere. Had λ_k been purely real but < 0, $\Phi_k(x) \rightarrow$ -∞ as $x \rightarrow$ -∞ thereby violating the restriction imposed on $\Phi_k(x)$. Let λ_k be complex with $\lambda_k = \alpha_k + i\beta_k$; α_k, β_k being both real, then $\Phi_k(x) = \exp((\alpha_k + i\beta_k) \cdot x)$. Such a function can remain finite for $-\infty \leq x \leq \infty$ only if $\alpha_k = 0$, thereby making λ_k a purely imaginary quantity. Hence eigenvalues of the operator $\frac{d}{dx}$ under finiteness restriction on $\Phi_k(x)$ for all space can only be purely imaginary. If the linear operator $\hat{A}$ admits of only real eigenvalues (a_k) the property of the system that $\hat{A}$ represents should be measurable and hence called an 'observable' of the system. If Φ_k is an eigenfunction of $\hat{A}$ with real eigenvalue 'a_k', we say that the system is in the eigenstate (Φ_k) of $\hat{A}$, a_k measuring the value of the property in that state. What guarantees that the eigenvalues of the operator $\hat{A}$ are only real? It turns out that the reality of the eigenvalues of $\hat{A}$ is ensured if $\hat{A}$ is not only linear but also self-adjoint (hermitian). The adjoint of $\hat{A}$ is denoted by the symbol $\hat{A}^\dagger$ and is defined by the following condition:

$$\int (\psi_1^* \hat{A} \psi_2) dv = \int \psi_2 (A^\dagger \Psi_1)^* dv \tag{2.28}$$

where ψ_1 and ψ_2 are two square integrable functions of the coordinates (x, y, z). If A is self-adjoint, $A^\dagger = A$ and the self-adjoint or the hermiticity condition for the operator demands that

$$\int \psi_1^* \hat{A} \psi_2 dv = \int \psi_2 (\hat{A} \Psi_1)^* dv \tag{2.29}$$

for any two square integrable functions (ψ_1, ψ_2). Now the condition presented in Eq. 2.29 must also hold when $\psi_1 = \psi_2$ which means

$$\int \psi_1^* \hat{A} \psi_1 dv = \int \psi_1 (\hat{A} \Psi_1)^* dv \tag{2.30}$$

Suppose now, ψ_1 is an eigenfunction of $\hat{A}$ with the eigenvalue 'a_1', which means

$$\hat{A} \psi_1 = a_1 \Psi_1$$
$$\text{and } (A \psi_1)^* = a_1^* \psi_1^* \tag{2.31}$$

Using these results in Eq. 2.30 we get

$$a_1 \int \psi_1^* \psi_1 dv = a_1^* \int \psi_1 \psi_1^* dv$$
$$\text{or}, (a_1 - a_1^*) \int \psi_1^* \psi_1 dv = 0 \tag{2.32}$$

Since ψ_1 by our assumption is a square integrable function $\int \psi_1^* \psi_1 dv \neq 0$; in fact $\int |\psi_1|^2 dv$ must be a finite number. That means $(a_1 - a_1^*) = 0$ or $a_1 = a_1^*$ proving that the eigenvalue of $\hat{A}$ is real for the eigenfunction ψ_1. However, the same analysis holds true for any eigenfunction of $\hat{A}$, proving that the hermiticity of the operator 'A' ensures that the eigenvalues of $\hat{A}$ are only real.

Hermiticity of the operator $\hat{A}$ also manifests itself in a special property of the eigenfunctions of $\hat{A}$. That property refers to the orthogonality of any two eigenfunctions of the operator $\hat{A}$ if they have different (non-equal) eigenvalues (i.e., they are non-degenerate). We can establish this important feature easily. Let ψ_1, ψ_2 be two eigenfunctions of $\hat{A}$ with the eigenvalues a_1 and a_2, respectively, with $a_1 \neq a_2$. We have the following relation

$$\hat{A} \psi_1 = a_1 \psi_1; \ \hat{A} \psi_2 = a_2 \psi_2;$$
$$a_1 \neq a_2; \ a_1^* = a_1; \ a_2^* = a_2.$$

Using the defination of hermiticity of $\hat{A}$ as laid down in Eq. 2.28 and Eq. 2.29 we have

$$a_2 \int \psi_1^* \psi_2 dv = a_1 \int \psi_1^* \psi_2 dv$$
$$or, (a_2 - a_1) \int \psi_1^* \psi_2 dv = 0 \tag{2.33}$$

Since $a_1 \neq a_2$ by assumption, $\int \psi_1^* \psi_2 dv = 0$ which means ψ_1, ψ_2 are mutually orthogonal. Noticing that ψ_1, ψ_2 are any two eigenfunctions of $\hat{A}$ with unequal eigenvalues we can make the following generalized statement: **Eigenfunctions of a hermitian operator with unequal eigenvalues are mutually orthogonal. The eigenvalues are all real.**[4]

A legitimate question that crops up at this point concerns the degenerate eigenfunction of the hermitian operator $\hat{A}$. Let ψ_1 and ψ_2 be a pair of distinct (linearly independent) eigenfunctions of $\hat{A}$ with the same eigenvalue 'a'. Being linearly independent, ψ_1 and ψ_2 are not multiples of each other. They are doubly degenerate eigenfunctions of the operator $\hat{A}$ with the eigenvalue 'a'. We can make use of the double degeneracy condition in Eq. 2.33 to find that in

$$(a_2 - a_1) \int \psi_1^* \psi_2 dv = 0;$$

$$a_2 - a_1 = 0 \quad (a_1 = a_2 = a)$$

so that $\int \psi_1^* \psi_2 dv$ may or may not be zero, which mean that ψ_1 and ψ_2 are not necessarily orthogonal to each other. If $\int \psi_1^* \psi_2 dv = 0$ even if $(a_2 - a_1) = 0$, the two eigenfunctions of $\hat{A}$ are already mutually orthogonal. If $\int \psi_1^* \psi_2 dv \neq 0$ and $(a_2 - a_1) = 0$, the two eigenfunctions are not mutually orthogonal, but as we will see, they can be made orthogonal to each other. Suppose that $\int \psi_1^* \psi_2 dv = S$, a scalar called the overlap integral of the two eigenfunctions. We can easily construct two new functions ψ_1', ψ_2' where

$$\psi_1' = \psi_1$$
$$\psi_2' = \frac{\psi_1 - S\psi_2}{\sqrt{1 - S^2}} \tag{2.34}$$

The reader can easily verify that the primed functions (ψ_1', ψ_2') satisfy the following conditions

$$A\psi_1' = a\psi_1'$$
$$A\psi_2' = a\psi_2'$$
$$\int \psi_1' \psi_2' dv = 0 \tag{2.35}$$

Thus, the degenerate pair of eigenfunctions of $\hat{A}$ can be easily transformed into a new pair of mutually orthogonal pair of eigenfunctions of $\hat{A}$ with the same eigenvalue in case they are not mutually orthogonal to start with. This result can be generalized for an *n*-fold degenerate set of eigenfunctions $(\psi_1, \psi_2, \cdots, \psi_n)$ of any hermitian operator $\hat{A}$. The scheme of orthogonalization has become known as Schmidt's orthogonalization scheme.

2.3.1 More About Hermiticity and Hermitian Conjugates

Let us take a closer look at the hermiticity condition: an operator $\hat{A}$ is hermitian if it obeys the condition $\hat{A} = \hat{A}^\dagger$, $\hat{A}^\dagger$ being the adjoint of A. How does one find the adjoint $A^\dagger$ of A.

Let us restrict ourselves to working in one dimension and define $\hat{A} = \frac{d}{dx}$. What is $A^\dagger = (\frac{d}{dx})^\dagger$? The answer follows from Eq. 2.28 defining the adjoint of $A(A^\dagger)$. Thus with $A = \frac{d}{dx}$ and $A^\dagger = (\frac{d}{dx})^\dagger$, ψ_1, ψ_2 being two square integrable functions of x, Eq. 2.28 yield the following condition:

$$\int_{-\infty}^{\infty} \psi_1^* \left(\frac{d}{dx}\psi_2\right) dv = \int_{-\infty}^{\infty} \psi_2 \left(\frac{d}{dx}^\dagger \psi_1\right)^* dv \tag{2.36}$$

Integrating the left hand side by parts and noting that $(\psi_1, \psi_2) \to 0$ as $x \to \pm \infty$, we have

$$(\psi_1^* \psi_2)\big|_{-\infty}^{\infty} - \int_{-\infty}^{\infty} \frac{d\psi_1^*}{dx} \psi_2 dx = \int_{-\infty}^{\infty} \psi_2 \left(\frac{d}{dx}^\dagger \psi_1\right)^* dx \tag{2.37}$$

Since ψ_1, ψ_2 vanish at the boundaries the first term on the left hand side of Eq. 2.37 becomes zero (0). The second term on left hand side can be rearranged to give

$$\int_{-\infty}^{\infty} \psi_2 \left(-\frac{d}{dx}\psi_1\right)^* dx = \int_{-\infty}^{\infty} \psi_2 \left(\frac{d}{dx}^\dagger \psi_1\right)^* dx \tag{2.38}$$

That tells us that

$$\frac{d}{dx}^\dagger = -\frac{d}{dx} \tag{2.39}$$

The analysis can be extended to cover higher derivatives $\frac{d^2}{dx^2}$, $\frac{d^3}{dx^3}$, $\cdots$, $\frac{d^n}{dx^n}$ and arrive at the following results

$$\frac{d^2}{dx^2}^\dagger = \frac{d^2}{dx^2}$$

$$\frac{d^3}{dx^3}^\dagger = -\frac{d^3}{dx^3} \qquad (2.40)$$

$$\frac{d^n}{dx^n}^\dagger = (-1)^n \frac{d^n}{dx^n}$$

That means $\frac{d}{dx}$ is non-hermitian ($\frac{d}{dx}^\dagger \neq \frac{d}{dx}$) while $\frac{d^2}{dx^2}$ is $[(\frac{d^2}{dx^2}^\dagger)^\dagger = \frac{d^2}{dx^2}]$. In fact every even order derivative operator is hermitian while the derivative operators of odd order are not. Among the operators of quantum mechanics that we have encountered so far the linear momentum operator $\hat{p}_x = \frac{\hbar}{i}\frac{d}{dx}$ is hermitian since

$$\hat{p}_x^\dagger = \left(\frac{\hbar}{i}\frac{d}{dx}\right)^\dagger = \left(\frac{\hbar}{i}\right)^* \left(-\frac{d}{dx}\right) = -\frac{\hbar}{i}\left(-\frac{d}{dx}\right) = \hat{p}_x \qquad (2.41)$$

Similarly, we can establish that $\hat{p}_y^\dagger = \hat{p}_y$, $\hat{p}_z^\dagger = \hat{p}_z$.

The kinetic energy operator $\hat{T}_x = -\frac{\hbar^2}{2m}\frac{d^2}{dx^2}$ is clearly hermitian as

$$\hat{T}_x^\dagger = \left(-\frac{\hbar^2}{2m}\frac{d^2}{dx^2}\right)^\dagger = -\frac{\hbar^2}{2m}(-1)^2\frac{d^2}{dx^2} = \hat{T}_x \qquad (2.42)$$

Similarly, $\hat{T}_y^\dagger = \hat{T}_y$, $\hat{T}_z^\dagger = \hat{T}_z$. The operator $\hat{x}$ is manifestly hermitian as

$$\int_{-\infty}^{\infty} \psi_1^* \hat{x}\psi_2 \, dx = \int_{-\infty}^{\infty} \psi_2(\hat{x}^\dagger \psi_1)^* \, dx$$

$$= \int_{-\infty}^{\infty} \psi_2 \hat{x}^* \psi_1^* \, dx$$

$$= \int_{-\infty}^{\infty} \psi_2 \hat{x} \psi_1^* \, dx$$

x being real. More generally, $\hat{x} \cdot \hat{x} \cdots \hat{x} = \hat{x}^n$ is also hermitian. Since potential energy $\hat{V}(\vec{r})$ is a function of the coordinates only, $V^\dagger(\hat{x}, \hat{y}, \hat{z}) = V(\hat{x}, \hat{y}, \hat{z})$ so that the operator for potential energy is also hermitian. The total energy operator $\hat{H}$ being a sum of kinetic ($\hat{T}$) and potential energy $\hat{V}(x, y, z)$ operator is also hermitian:

$$\hat{H}^\dagger = \hat{T}^\dagger + \hat{V}^\dagger = \hat{T} + \hat{V} = H$$

The hermiticity of $\hat{H}$ ensures that the energy eigenvalues (eigenvalues of H) are real and the corresponding eigenfunctions are orthogonal as such or can be orthogonalized. In general, the sum of two hermitian operators $\hat{A}$, $\hat{B}$ is also hermitian just as their difference is. The product of two hermitian operators are not generally hermitian as

$$(AB)^\dagger = B^\dagger A^\dagger = \hat{B}\hat{A} \neq AB \quad (A^\dagger = \hat{A}, B^\dagger = \hat{B}) \qquad (2.43)$$

However, if $\hat{A}$ and $\hat{B}$ commute, i.e. $\hat{A}\hat{B} - \hat{B}\hat{A} = [\hat{A}, \hat{B}] = 0$ or $\hat{A}\hat{B} = \hat{B}\hat{A}$, $\hat{A}\hat{B}$ is hermitian if $\hat{A}$ and $\hat{B}$ are themselves hermitian. It is straightforward to find that the symmetrized sum of the operator products of $\hat{A}$

and $\hat{B}$ is hermitian even if $\hat{A}$ and $\hat{B}$ do not commute but are individually hermitian. The symmetrical sum of $\hat{A}$ and $\hat{B}$ is

$$\hat{S}_+ = (\hat{A}\hat{B} + \hat{B}\hat{A})/2$$

clearly, $\quad \hat{S}_+^\dagger = (\hat{B}^\dagger \hat{A}^\dagger + \hat{A}^\dagger \hat{B}^\dagger)/2$

$$= (\hat{B}\hat{A} + \hat{A}\hat{B})/2$$

$$= (\hat{A}\hat{B} + \hat{B}\hat{A})/2$$

$$= \hat{S}_+$$

The notion of the commutator $[\hat{A}, \hat{B}]$ of the operators $\hat{A}$, $\hat{B}$ is very important in quantum mechanics. If the commutator is null-operator $(\hat{0})$, the observables that the operators $\hat{A}$ and $\hat{B}$ represent are called commuting observables – they can be measured simultaneously. If $[\hat{A}, \hat{B}] \neq \hat{0}$, the observables corresponding to $\hat{A}$, $\hat{B}$ are called non-commuting observables – they can not be simultaneously measured with absolute accuracy. These are strong observations and require more extensive discussion. We will therefore turn our attention to commuting and non-commuting hermitian operators and examine the consequences of non-commutativity vis-a-vis commutivity.

2.4 Commuting and Non-Commuting Observables

Let us consider the quantum mechanical operators, $\hat{A} = \hat{p}_x = \frac{\hbar}{i}\frac{d}{dx}$ and $\hat{B} = \hat{x} = x$ and examine if they commute. To do that let us form the commutator of $\hat{A}$, $\hat{B}$, which is $\hat{A}\hat{B} - \hat{B}\hat{A} = [\hat{A}, \hat{B}]$. Let $[\hat{A}, \hat{B}]$ be allowed to act on square integrable function $f(x)$. If we find that $[\hat{A}, \hat{B}]f(x) = 0$, for an arbitrary square integrable function x, the operator pair $[\hat{A}, \hat{B}]$ commute and $[\hat{A}, \hat{B}]$ = null operator. With this definition of commutativity at hand, let us take up the case of $\hat{p}_x = \hat{A} = \frac{\hbar}{i}\frac{d}{dx}$ and $\hat{B} = x$. Thus for the pair $\hat{p}_x$ and $\hat{x}$, we have

$$[\hat{x}, \hat{p}_x]f(x) = (\hat{x} \cdot \hat{p}_x - \hat{p}_x \cdot \hat{x})f(x)$$

$$= \left[x\frac{\hbar}{i}\frac{d}{dx} - \left(\frac{\hbar}{i}\frac{d}{dx}\right)x \right]f(x)$$

$$= -\frac{\hbar}{i}f(x)$$

Thus we arrive at the defination of the basic commutator

$$[\hat{x}, \hat{p}_x] = -\frac{\hbar}{i} \cdot 1 = i\hbar. \qquad (2.44)$$

Similarly, we can establish further that $[\hat{y}, \hat{p}_y] = i\hbar$ and $[\hat{z}, \hat{p}_z] = i\hbar$ revealing that $(\hat{x}, \hat{p}_x)$, $(\hat{y}, \hat{p}_y)$ and $(\hat{z}, \hat{p}_z)$ are non-commuting pairs of hermitian operators. So in quantum mechanics, position and the corresponding (conjugate) momentum components represent non-commuting observables. It is easy to establish that $(\hat{p}_x, \frac{\hat{p}_x^2}{2m})$, $(\hat{p}_y, \frac{\hat{p}_y^2}{2m})$ and $(\hat{p}_z, \frac{\hat{p}_z^2}{2m})$ represent commuting pairs of operators and represent pairs of commuting observables. That is the commutators

$$\left[\hat{p}_x, \frac{\hat{p}_x^2}{2m}\right] = \left[\hat{p}_y, \frac{\hat{p}_y^2}{2m}\right] = \left[\hat{p}_z, \frac{\hat{p}_z^2}{2m}\right] = 0$$

Physically that means the x-component of linear momentum and the kinetic energy in x-direction, are commuting observables. So also are the y-component of linear momentum and kinetic energy component along the y-axis, and z-component of momentum and the kinetic energy in the z-direction. What are the consequences of the commutation of two operators or the lack of it? In general, if two linear hermitian operators, $\hat{A}$, $\hat{B}$ commute, i.e., $[\hat{A}, \hat{B}] = 0$, they have simultaneous or a common set of eigenfunctions. The proof of this assertion is rather simple. Let Φ be a non-degenerate eigenfunction of the operator $\hat{A}$ with the eigenvalue a. So,

$$\hat{A}\Phi = a\Phi$$

Hence

$$\hat{B}\hat{A}\Phi = \hat{B}(a\Phi) = a(\hat{B}\Phi) \tag{2.45}$$

Since $\hat{B}\hat{A} = \hat{A}\hat{B}$, we have

$$\hat{A}(\hat{B}\Phi) = a(\hat{B}\Phi) \tag{2.46}$$

That means $\hat{B}\Phi$ is also an eigenfunction of $\hat{A}$ with the same eigenvalue 'a'. By our assumption Φ is a non-degenerate eigenfunction of $\hat{A}$ with the eigenvalue 'a', hence $\hat{B}\Phi$ and Φ must be linearly dependent forcing us to conclude that $\hat{B}\Phi$ must be a multiple of Φ, i.e. $\hat{B}\Phi$ must be proportional to Φ. Let the constant of proportionality be 'b' so that we have

$$\hat{B}\Phi = b\Phi$$

which reveals that 'Φ' is not only an eigenfunction of $\hat{A}$ with the eigenvalue 'a', it is also an eigenfunction of $\hat{B}$ with the eigenvalue 'b' when $\hat{A}$ and $\hat{B}$ commute. This is true for every non-degenerate eigenfunction of $\hat{A}$, proving the assertion that commuting operators have a common set of eigenfunctions. The assertion also holds for degenerate eigenfunctions, but the proof is more complex and we will not pursue the problem further. Instead, let us enquire if the converse of what we have proved is also true–that is if the linear hermitian operators $\hat{A}$ and $\hat{B}$ have a complete set of common eigenfunctions, they must commute. The proof is simple and proceeds as follows. Let the nth member of the common set of eigenfunctions be Φ_n so that we have

$$\hat{A}\Phi_n = a_n\Phi_n$$
$$\hat{B}\Phi_n = b_n\Phi_n \tag{2.47}$$

Therefore,

$$\begin{aligned}
[\hat{A},\hat{B}]\Phi_n &= \hat{A}\hat{B}\Phi_n - \hat{B}\hat{A}\Phi_n \\
&= (a_n b_n - b_n a_n)\Phi_n \\
&= 0
\end{aligned}$$

Let Ψ be any square integrable function. We can expand Ψ in the complete linearly independent basis Φ_n provided by the common eigenfunction of $\hat{A}$ and $\hat{B}$,

$$\Psi = \sum_n c_n\Phi_n$$

Then

$$[\hat{A},\hat{B}]\Psi = \sum_n c_n(a_n b_n - b_n a_n)\Phi_n$$

$$= 0, \text{ for any such } \Psi$$

Therefore, $[\hat{A}, \hat{B}] = 0$ – that is $\hat{A}$ and $\hat{B}$ commute. The eigenfunctions of hermitian operators occupy a special place in quantum mechanics. Let us investigate why. If Φ_n is an eigenfunction of $\hat{A}$ with the eigenvalue a_n we have

$$\hat{A}\Phi_n = a_n\Phi_n$$
$$\hat{A}^2\Phi_n = \hat{A}(\hat{A}\Phi_n) = a_n^2\Phi_n$$

The average value of the observable $\hat{A}$ in the state Φ_n is

$$<A> = \int \Phi_n^*\hat{A}\Phi_n dv = a_n$$

while the average value of $\hat{A}^2$ is

$$< \hat{A}^2 > = \int \Phi_n^* \hat{A}^2 \Phi_n dv$$

$$= a_n^2$$

Clearly, $< A^2 > - < A >^2 = a_n^2 - a_n^2 = 0$ in the eigenstate Φ_n or any other eigenstate of $\hat{A}$. There is thus no dispersion in the measured quantum mechanical average value of the observable represented by the operator $\hat{A}$ in an eigenstate of $\hat{A}$. The dispersion we are talking about here is purely quantum mechanical in origin. That the dispersion in the average value of $\hat{A}$ in an eigenstate is zero means that the value can be in principle, measured exactly – it will not be vitiated by quantum noise. If the operator $\hat{B}$ commutes with $\hat{A}$, the eigenstate Φ_n of $\hat{A}$ is also an eigenstate of $\hat{B}$ and the average value of $\hat{B}$ in the same state can also be in principle, measured without any dispersion leading to the assertion that properties represented by commuting operators can be measured simultaneously without any uncertainty or dispersion provided that the measurements are made in a state that is a member of the common set of eigenstates of the two commuting operators $\hat{A}$ and $\hat{B}$. In other words, everytime we measure the average value of $\hat{A}$ in its eigenstate Φ_n, we get the value a_n with absolute certainity. In the same state, the measurement of the average value of $\hat{B}$ will produce the eigenvalue 'b_n' with certainty if $\hat{B}$ commutes with $\hat{A}$ (ensuring that $\hat{B} \Phi_n = b_n \Phi_n$). It is but natural to question at this point what happens to the outcome of measured $< A >$ in a state that is not an eigenstate of the commuting set of operators $\hat{A}$, $\hat{B}$. Such a state Ψ can be represented as a linear combination of the complete set of simultaneous eigenfunctions Φ_n of the operators in question. We can therefore write

$$\Psi = \sum_n c_n \Phi_n$$

The measured value of $\hat{A}$ in such a state is called the expectation value of $A (= < A >)$ in the state Ψ and the postulates of quantum mechanics define it as (assuming Φ_n to be a complete orthonormal set)[5]

$$
\begin{aligned}
< A > &= \int \psi^* \hat{A} \psi dv \\
&= \int \left(\sum_n c_n \Phi_n \right)^* \hat{A} \left(\sum_m c_m \Phi_m \right) dv \\
&= \sum_n \sum_m c_n^* c_m \int \Phi_n^* \hat{A} \Phi_n dv \\
&= \sum_n |c_n^2| a_n \delta(a - a_n)
\end{aligned}
\tag{2.48}
$$

where $\delta(x)$ is Dirac's delta function, a discontinuous function with the following set of properties:

$$
\begin{aligned}
\delta(x) &= 0, \text{if } x \neq 0 \\
\delta(x) &= \infty, \text{if } x = 0 \\
\int_{-\infty}^{\infty} &\delta(x) = 1
\end{aligned}
\tag{2.49}
$$

It is clear from the properties of the delta function that the outcome of a single measurement of the property represented by $\hat{A}$ in the state Ψ will always produce one of the set of eigenvalues a_n of $\hat{A}$ with the probability $|c_n|^2$ – there is no possibility of getting any other value of 'a'. An average of many such measurements or the result of measurement on an ensemble of the system each being prepared in the identical state (Ψ) will yield the value of $< A >$ as given by Eq. 2.48. $< A >$ is therefore often called the ensemble average value of the property 'A' of the system in an arbitrary state Ψ. Two curious points about Eq. 2.48 deserve mention:[5]

(i) A single measurement produces one of the eigenvalues of $\hat{A}$ with the probability $|c_n|^2$. That means the act of measurement forces the wavefunction Ψ to collapse onto one of the eigenstates of the (operator) observable in question.

(ii) We need only the probabilities $|c_n|^2$ to calculate $<A>$, but not the probability amplitudes (c_ns) and their phases (the coherences in modern terminology). However, if the arbitrary state function Ψ is expanded in the eigenbasis of an operator $\hat{B}$ that does not commute with $\hat{A}$, the ensemble average or expectation value of $\hat{A}$ is given by the following expression:

$$\Psi = \sum_n c_n \chi_n \qquad (\hat{B}\chi_n = b_n\chi_n; [\hat{A},\hat{B}]=0)$$

$$<A> = \sum_n \sum_m c_n^* c_m \int \chi_n^* \hat{A} \chi_m dv \qquad (2.50)$$

$$= \sum_n \sum_m c_n^* c_m A_{nm}$$

where A_{nm}s are the matrix elements of the operator $\hat{A}$ in the basis χ_n (eigenbasis of $\hat{B}$). Here we need both the probability amplitudes and their phases to complete the calculation of the average value of $\hat{A}$ in an arbitrary state represented by the wavefunction Ψ. It is therefore always convenient to perform the expansion of Ψ in the common eigenbasis of a set of commuting operators. For a pair of non-commuting observables a and b (corresponding operators are $\hat{A}$, $\hat{B}$), the common eigenbasis does not exist. Accordingly the measurements of the averages $<A>$ and $<B>$ in an arbitrary state will be inherently vitiated by the quantum uncertainties (ΔA, ΔB) where

$$(\Delta A)^2 = <A^2> - <A>^2$$

$$(\Delta B)^2 = <B^2> - <B>^2$$

These are essentially measures of mean square values of dispersions due to quantum fluctuations. It can be shown that $(\Delta A)^2$ and $(\Delta B)^2$ are related by an inequality, first noted by W. Heisenberg which reads

$$(\Delta A)^2 \cdot (\Delta B)^2 \geq \frac{\hbar^2}{4} \qquad (2.51)$$

The inequality in Eq. 2.51 embodies the statement of Heisenberg's principle of uncertainty and points to the fundamental impossibility of simultaneously measuring non-commuting observables with absolute precision – if $\Delta A = 0$, then ΔB becomes infinitely large and vice-versa. The minimum uncertainties in $<A>$ and $<B>$ that is attainable simultaneously corresponds to a state for which

$$(\Delta A)_{min}^2 \cdot (\Delta B)_{min}^2 = \frac{\hbar^2}{4}$$

$$\text{or,} \quad \Delta A \cdot \Delta B = \frac{\hbar}{2} \qquad (2.52)$$

the corresponding state being called a state with the minimum uncertainty product.

2.5 Stationary States of Quantum Systems

We are now in a position to write down the Schrödinger wave equation for a microscopic system and elucidate the steps leading to the determination of the quantum states of the system. Let us for the sake of simplicity consider a one-dimensional system representing a microscopic particle of mass 'm' moving in a time-independent potential $V(x)$. The Schrödinger equation for the system in a state $\Psi(x,t)$ is given by

$$ih\frac{\partial \Psi(x,t)}{\partial t} = \left[\frac{\hat{p}_x^2}{2m} + V(\hat{x})\right]\Psi(x,t)$$ (2.53)

$$= \hat{H}(x)\Psi(x,t)$$

Since the potential energy function $V(x)$ does not contain time 'explicitly', we can try to solve Eq. 2.53 for Ψ by separating the variables x and t and writing for the nth quantum state $\Psi_n(x,t)$ as

$$\Psi_n(x,t) = \Phi_n(x)\chi_n(t)$$ (2.54)

Using Eq. 2.54 in Eq. 2.53 and after doing some simple algebra, we get

$$\frac{\hat{H}(x)\Phi_n(x)}{\Phi_n(x)} = \frac{ih}{\chi_n(t)}\frac{\partial \chi_n(t)}{\partial t}$$ (2.55)

Since the left hand side of Eq. 2.55 is a function of 'x' only while the right hand side is a function of 't' only, each side must be equal to a constant, say λ_n, independent of x as well as t. Hence Eq. 2.52 leads to two seperate equations

$$\frac{\hat{H}(x)\Phi_n(x)}{\Phi_n(x)} = \lambda_n$$ (2.56a)

$$ih\frac{1}{\chi_n(t)}\frac{d\chi_n(t)}{dt} = \lambda_n$$ (2.56b)

when λ_n is the seperation constant. The first of the two (Eq. 2.56(a)) is just the eigenvalue equation for the Hamiltonian operator $H(x)$ of the system with $\{\lambda_n\}$ representing the allowed energy eigenvalues (E_n) of the system:

$$\hat{H}\Phi_n = \lambda_n\Phi_n = E_n\Phi_n$$

If $V(x)$ supports only discrete bound state n, the state quantum number can have values $0, 1, 2, \cdots$, etc. Hence $\Phi_n(x)s$ are called the energy eigenfunctions and 'E_n' the corresponding energy eigenvalues admissible for the system. Since $\hat{H}(x)$ is a hermitian operator the eigenvalues $\{\lambda_n\}$ or E_n are real and eigenfunction $\Phi_n(x)$ form a complete linearly independent set of functions, which can be assumed, without any loss of generality, to be a complete orthonormal set. Assuming $\Phi_n(x)s$ have been normalized to unity, these functions satisfy the following conditions:

$$\int_{-\infty}^{\infty} \Phi_n^*(x)\Phi_n(x)dx = 1$$

$$\int_{-\infty}^{\infty} \Phi_n^*(x)\Phi_m(x)dx = 0; \quad \text{if } n \neq m$$

which can be combined into a single condition

$$\int_{-\infty}^{\infty} \Phi_n^*(x)\Phi_m(x)dx = \delta_{nm}$$

where δ_{nm} is the Kroneaker delta function ($\delta_{nm} = 1$ if $n = m$ and $\delta_{nm} = 0$ if $n \neq m$). If $V(x)$ supports non-discrete or continuous energy eigenvalues, we can no longer label the energy eigenstates by discrete numbers 'n'. Instead, we can use the continuous energy eigenvalues as the label for the state and represent Ψ as $\Psi(x,E)$ or $\Psi(x,E')$ or $\Psi(x,E'')$, etc. The orthonormality condition for these energy eigenfunction can be written as

$$\int \Psi^*(x,E')\Psi(x,E'')dx = \delta(E' - E'')$$ (2.57)

where $\delta(E' - E'')$ is the Dirac delta function we have already introduced. It is possible that $V(x)$ supports discrete as well as continuous energy states or energy eigenvalues. An arbitrary wave-function $\Psi(x)$ can be expanded in terms of the eigenfunctions of $\hat{H}$ (or for that matter, in the basis of eigenfunctions of the operator $\hat{A}$ that commutes with $\hat{H}$). Most generally therefore we can write

$$\Psi(x) = \sum_n c_n \Phi_n + \int_{\epsilon_1}^{\epsilon_2} c(E)\Psi(x, E)dE \tag{2.58}$$

where the summation is over the discrete energy eigenstates and the integration is over the energy range for which the eigenenergy is continuous. It is straightforward to find (exploiting orthonormality condition)

$$c_n = \int \Phi_n^*(x)\Psi(x)dx \tag{2.59a}$$

$$c(\epsilon) = \int \Psi^*(x, \epsilon)\Psi(x)dx \tag{2.59b}$$

These coefficients satisfy the normalizing condition

$$\sum_n |c_n|^2 + \int_{\epsilon_1}^{\epsilon_2} |c(\epsilon)|^2 d\epsilon = 1 \tag{2.60}$$

$|c_n|^2$ represents the probability of finding the discrete energy eigenvalues in the state Ψ and $|c(\epsilon)|^2 d\epsilon$ is the probability of finding the energy eigenvalue in the range $\epsilon \leftrightarrow \epsilon + d\epsilon$. The other equation (Eq. 2.56b) yields the solution

$$\chi_n(t) = \chi_n(0)\exp\left(\frac{-iE_n t}{\hbar}\right) \tag{2.61}$$

The stationary states of the system with discrete energy spectrum are therefore representable by wavefunctions $\Psi_n(x, t)$ with

$$\Psi_n(x, t) = \Phi_n(x)\chi_n(t)$$

$$= \Phi_n(x)\chi_n(0)\exp\left(\frac{-iE_n t}{\hbar}\right)$$

$\chi_n(0)$ being the value of $\chi_n(t)$ at $t = 0$ and can be absorbed in $\Phi_n(x)$ before normalization (call it $\overline{\Phi_n}(x)$). Clearly,

$$|\Psi_n(x, t)|^2 = |\overline{\Phi_n}(x)|^2 = P_n(x) \tag{2.62}$$

That means, the probability density of finding the system in the state $\Psi_n(x, t)$ at $x(P_n)$ does not depend on time 't'. This time-independent probability density is a unique feature of stationary states of the system, called the energy eigenstates. An arbitrary state of the system $\Psi_m(x, t)$ can be expanded in terms of the discrete stationary states of the system:

$$\Psi_m(x, t) = \sum_n c_n(\Phi_n(x)\chi_n(t)) \tag{2.63}$$

Such a state will also be a solution of the time-dependent Schrödinger equation (TDSE) as the basis functions employed in the expansion are solutions of the TDSE, which is a linear equation.

2.5.1 The Free Particle

As an example, let us work out the general solution of the TDSE of a free particle of mass 'm' in one dimension (x). The equation reads

$$i\hbar \frac{\partial \Psi(x, t)}{\partial t} = -\frac{\hbar^2}{2m} \frac{\partial^2 \Psi(x, t)}{\partial x^2} \tag{2.64}$$

We write the solution of Eq. 2.64 in a variable separated form: $\Psi(x,t) = \Phi(x)\chi(t)$ and use it in Eq. 2.64 to get

$$i\hbar\frac{1}{\chi}\frac{d\chi}{dt} = -\frac{\hbar^2}{2m}\cdot\frac{1}{\Phi}\frac{d^2\Phi}{dx^2} \tag{2.65}$$

The separation constant ϵ turns out to be the energy of the particle (potential energy $V(x)$ being zero energy). The equation for Φ becomes

$$\frac{d^2\Phi}{dx^2} = -\left(\frac{2m\epsilon}{\hbar^2}\right)\Phi \tag{2.66}$$

Since Φ must remain finite everywhere including $|x|\to\infty$, ϵ must be > 0. There is no other restriction on ϵ. The energy spectrum of the free particle is continuous and > 0. Let us put $\frac{2m\epsilon}{\hbar^2} = k^2$ in Eq. 2.66. which becomes

$$\frac{d^2\Phi}{dx^2} = -k^2\Phi$$

and the solution for the space-part of the wavefunction $\Phi(x) = \exp(ikx)$ (k a real number $-\infty \le k \le \infty$). The time-dependent part of the wavefunction is χ, which is a solution of the following equation

$$i\hbar\frac{1}{\chi}\frac{d\chi}{dt} = \epsilon = \frac{\hbar^2 k^2}{2m}$$

which yields as a solution

$$\chi(t) = \exp\left(\frac{\epsilon t}{i\hbar}\right) = \exp\left(\frac{i\hbar k^2 t}{2m}\right) \tag{2.67}$$

The particular solution of the wave equation is therefore[5]

$$\Psi_k(x,t) = A(k)\exp\left[i\left(kx - \frac{\hbar k^2 t^2}{2m}\right)\right]$$

and the general solution is

$$\Psi(x,t) = \int_{-\infty}^{\infty} A(k)\exp\left[i\left(kx - \frac{\hbar k^2 t^2}{2m}\right)\right]dk \tag{2.68}$$

2.5.2 Stationary States of a Simple Harmonic Oscillator

In simple harmonic motion in one dimension of a particle of mass 'm' moves under a field of force $\vec{F} = -k\vec{x}$, k being the force constant and $\vec{x}$ is the displacement of the particle from the position of equilibrium ($x = 0$). The potential energy of the particle at x is then given by $V(x) = -\frac{1}{2}kx^2$. The classical Hamiltonian of the particle is represented by $H = T + V = \frac{p_x^2}{2m} - \frac{1}{2}kx^2$ and the corresponding quantum mechanical Hamiltonian operator is obtained by replacing p_x by $\hat{p}_x$ and x by $\hat{x}$;

$$\hat{H} = -\frac{\hbar^2}{2m}\frac{\partial^2}{\partial x^2} - \frac{1}{2}k\hat{x}^2 \tag{2.69}$$

The energy eigenstates of $\hat{H}$ satisfy the eigenvalue equation

$$-\frac{\hbar^2}{2m}\frac{d^2\Phi_n(x)}{dx^2} - \frac{1}{2}kx^2\Phi_n(x) = E_n\Phi_n(x)$$

$$\text{or,} \quad \frac{d^2\Phi_n(x)}{dx^2} + \left(\frac{m\omega^2 x^2}{\hbar^2} + \frac{2mE_n}{\hbar^2}\right)\Phi_n(x) = 0 \tag{2.70}$$

where $\Phi_n(x)$s are the energy eigenfunctions and E_ns are the corresponding energy eigenvalues of the oscillator. The stationary states of the oscillator $\Psi_n(x,t)$ are therefore given by

$$\Psi_n(x,t) = \Phi_n(x)\exp\left(-\frac{iE_n t}{\hbar}\right) \tag{2.71}$$

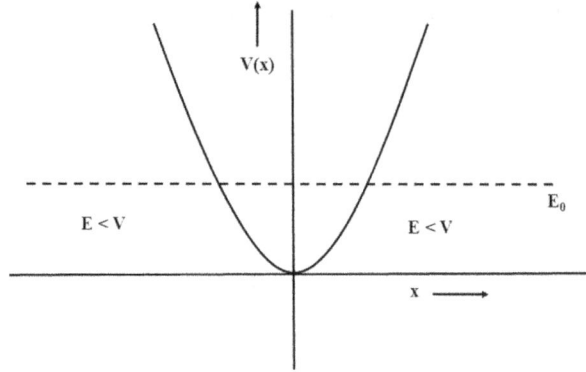

FIGURE 2.1 Potential energy function $V(x) = -\frac{1}{2}kx^2$.

The important task is now to solve the differential Eq. 2.70 for $\Phi_n(x)$ under appropriate boundary conditions and in the process ascertain what the energy eigenvalues 'E_n' are allowed to be. Before attempting the solution, let us examine the potential energy function $V(x)$ displayed in Figure 2.1. Inside the harmonic well where $E > V$ the solutions must be harmonic (oscillating) in nature. Outside the confines of the well, where $E < V$ the solutions must possess two branches – an increasing branch and a diminishing one as $|x| \to \infty$. It is expedient therefore to determine the condition under which the increasing branch of the solution vanishes. As we will see in what follows, this is possible only if the energy 'E' has certain discrete values – that is only if the oscillator energy is quantized and not continuous in these states. Let us start by making some substitution to transform Eq. 2.70 into some simpler form. Thus, replacing E_n by E, Φ_n by Φ, and $a = \frac{2mE}{\hbar}$, $b = \frac{2m\omega}{\hbar} = \frac{1}{x_0^2}$, $\frac{a}{b} = \lambda = \frac{2E}{\hbar\omega}$, and $\xi = x\sqrt{b} = (x/x_0)$, we get[4-5]

$$\frac{d^2\Phi}{dx^2} + (\lambda - \xi^2)\Phi = 0 \tag{2.72}$$

In the limit $\xi \to \pm\infty$, λ in the preceeding equation may be neglected in comparison with 'ξ'. The asymptotic behavior of the wavefunction Φ (call it Φ_∞) can then be found from the equation

$$\Phi_\infty'' - \xi^2\Phi_\infty = 0 \tag{2.73}$$

Φ_∞ in Eq. 2.73 can then be sought in the form $\Phi_\infty = \exp(c\xi^2)$, which allows us to write

$$\Phi_\infty'' = (4c^2\xi^2 + 2c)\exp(c\xi^2)$$
$$\approx 4c^2\xi^2\exp(c\xi^2) \tag{2.74}$$

Using Φ_∞'' as given in Eq. 2.74 in Eq. 2.73 we find that $c = \pm\frac{1}{2}$ and we can write

$$\Phi_\infty = c_1\exp\left(-\frac{1}{2}\xi^2\right) + c_2\exp\left(\frac{1}{2}\xi^2\right) \tag{2.75}$$

c_2 in Eq. 2.75 must set equal to zero to prevent Φ_∞ from growing infinitely as $\xi \to \infty$ so that we have

$$\Phi_\infty \approx c_1\exp\left(\frac{-1}{2}\xi^2\right)$$

We may now seek the solution (Φ) of Eq. 2.72 in the form of a product of the asymptotic solution and a power series $\chi(\xi)$ in ξ:

$$\Phi(\xi) = \exp\left(-\frac{1}{2}\xi^2\right)\chi(\xi) \tag{2.76}$$

It is important now to ensure that the exponentially growing solution is suppressed. That requires us to restrict the power series representing $\chi(\xi)$ to have a finite (n) number of terms – that is $\chi(\xi)$ must be a polynomial of degree 'n'. This restriction as we will see shortly, leads to two important consequences. (i) The energy levels of the oscillator get quantized, (ii) there appears the so called zero-point energy – the energy in the lowest state being not equal to zero. Both are hallmarks of quantum mechanics of an oscillator. To determine $\chi(\xi)$ that we substitute $\Phi(\xi) = \exp(\frac{-1}{2}\xi^2)\chi(\xi)$ in the differential equation Eq. 2.72 to find that $\chi(\xi)$ is a solution of the following differential equation under appropriate constraints or boundary conditions:

$$\chi''(\xi) - 2\xi\chi'(\xi) + (\lambda - 1)\chi(\xi) = 0 \tag{2.77}$$

We seek solution of Eq. 2.77 for $\chi(\xi)$ in the form of a power series

$$\chi(\xi) = \sum_{k=0}^{\infty} c_k\xi^k \tag{2.78}$$

Substituting $\chi(\xi)$ in the form given Eq. 2.78 in Eq. 2.77 leads to the algebraic equation

$$\sum_{k=0}^{\infty} c_k k(k-1)\xi^{k-2} - c_k(2k - \lambda + 1)\xi^k = 0 \tag{2.79}$$

Equating coefficients of like powers of 'ξ' on both sides of Eq. 2.79, we get

$$c_{k+2}(k+2)(k+1) - (2k - \lambda + 1)c_k = 0$$

which leads to the following recursion relation among the expansion coefficients c_{k+2}, c_k:

$$c_{k+2} = \frac{(2k - \lambda + 1)}{(k+1)(k+2)}c_k; \quad k = 0, 1, \cdots \tag{2.80}$$

If we set $c_0 \neq 0$, the recursion formula in Eq. 2.80 generates all the coefficients of even powers of ξ in the power series expansion of $\chi(\xi)$, namely $c_2, c_4, c_6 \cdots$. On the other hand, choosing $k = k + 1$, the same recursive process will generate the coefficients of the odd powers of ξ, viz. $c_{k+1}, c_{k+3}, c_{k+5}, \cdots$, etc. Thus, Eq. 2.80 defines two linearly independent solutions. In order that the solutions thus obtained do not grow infinitely as ξ increases ($\to \pm\infty$), the power series must truncate at a finite value of $k = n$(say), ensuring that while $c_n \neq 0$, $c_{n+2}, c_{n+4}, c_{n+6}, \cdots$ are all zero. That means the infinite power series has been reduced to a polynomial of degree 'n'. The termination demands that the numerator in the recursion relation Eq. 2.80 must be set equal to zero for $k = n$ leading to (we change λ to λ_n to indicate termination for $k = n$)

$$2n - \lambda_n + 1 = 0$$
$$\text{or,} \quad \lambda_n = (2n + 1); \quad n = 0, 1, \cdots \tag{2.81}$$

Since $\lambda = \frac{2E}{\hbar\omega}$, we write $\lambda_n = \frac{2E_n}{\hbar\omega}$ just to take into account that 'n' labels the truncation point where $E = E_n$. That means

$$2E_n = (2n + 1)\hbar\omega$$
$$\text{and,} \quad E_n = (n + \frac{1}{2})\hbar\omega; \quad n = 0, 1, 2..... \tag{2.82}$$

Eq. 2.82 defines the quantized energy level of the harmonic oscillator with 'n' as the quantum number. The ground state (lowest) energy of the oscillator corresponds to $n = 0$ when $E_0 = \frac{1}{2}\hbar\omega \neq 0$. Unlike in the classical case, the quantum harmonic oscillator has a non-zero ground state energy, which is equal to $\frac{1}{2}\hbar\omega$. The truncation condition in Eq. 2.81 however fails to truncate the other solution simultaneously to a finite polynomial of degree 'n'. It can be shown to diverge as $\exp(\xi^2)$ as $|\xi| \to \infty$ and has to be discarded.[6] Let us choose the coefficient of the highest power of ξ, say $k_{max} = n$, in the accepted polynomial solution as $c_n = 2^n$. This is allowed as the wavefunction has not yet been normalized. The recursion relation (Eq. 2.80) then leads us to

$$c_{n-2} = -2^{(n-2)}n(n-1)/1!$$
$$c_{n-4} = -2^{(n-4)}n(n-1)(n-2)(n-3)/2!$$

and so on. $\chi(\xi)$ then turns out to be the following power series:

$$\chi(\xi) = (2\xi)^n - \frac{n(n-1)}{1!}(2\xi)^{n-2} + \frac{n(n-1)(n-2)(n-3)}{2!}(2\xi)^{n-4} + \cdots \qquad (2.83)$$

For the odd series (n odd) the last term in Eq. 2.83 will be $c_1\xi$. It turns out that the right hand side of Eq. 2.83 defines the hermite polynomials $H_n(\xi)$ the first few of them being $H_0(\xi) = 1$, $H_1(\xi) = 2\xi$, $H_2(\xi) = 4\xi^2 - 2$, $H_3(\xi) = 8\xi^3 - 12\xi$, and so on. These polynomials can be represented in the following closed form (Rodrigue's formula)

$$H_n(\xi) = (-1)^n \exp(\xi^2)\frac{d^n}{d\xi^n}\left[\exp(-\xi^2)\right] \qquad (2.84)$$

The reader can easily verify that $H_n(\xi)$ defined in Eq. 2.84 satisfy the differential Eq. 2.77 if we set $\chi(\xi) = H_n(\xi)$ and $\lambda = (2n+1)$ yielding

$$H_n''(\xi) - 2\xi H_n'(\xi) + 2nH_n(\xi) = 0$$

which is the well-known Hermite differential equation. The solutions of the quantum harmonic oscillator equation now can be written down in the form

$$\Phi_n(\xi) = A_n \exp(-\frac{\xi^2}{2})H_n(\xi) \qquad (2.85)$$

when A_ns are the 'yet to be determined' normalization constants. We will normalize $\Phi_n(\xi)$s to unity. The constants A_ns can be determined by considering the general integral.

$$I_{nn'} = A_nA_{n'}x_0 \int_{-\infty}^{\infty} \exp(-\xi^2)H_n(\xi)H_{n'}(\xi)d\xi$$
$$= (-1)^n A_nA_{n'}x_0 \int_{-\infty}^{\infty} H_{n'}(\xi)\frac{d^n}{d\xi^n}[\exp(-\xi^2)]d\xi \qquad (2.86a)$$

After n-fold integration by parts we get

$$I_{nn'} = A_nA_{n'}x_0 \int_{-\infty}^{\infty} \exp(-\xi^2)\frac{d^n}{d\xi^n}H_{n'}(\xi)d\xi \qquad (2.86b)$$

Two cases arise:

(i) If $n > n'$, the n-fold differentiation of $H_{n'}$—a polynomial of degree $n' < n$—will produce zero so that $I_{nn'} = 0$. Similarly, by interchanging n and n' in Eq. 2.86 it can be shown that $I_{nn'} = 0$ when $n < n'$.

(ii) If $n = n'$, we get the normalization integral

$$I_{nn} = A_n^2 x_0 \int_{-\infty}^{\infty} \exp(-\xi^2)\frac{d^n}{d\xi^n}H_n(\xi)d\xi$$
$$= A_n^2 x_0 2^n n!\sqrt{\pi}$$

Normalizing to unity (setting $I_{nn} = 1$) we get

$$A_n = \frac{1}{\sqrt{2^n n!\sqrt{\pi}x_0}}; \quad x_0 = \left(\frac{\hbar}{m\omega}\right)^{\frac{1}{2}} \qquad (2.87)$$

Alternatively, we could make use of the generating function for the Hermite polynomials to evaluate the orthonormality integrals. Once A_ns are determined, we can write down the harmonic oscillator wavefunctions (energy eigenfunctions) $\Phi_0, \Phi_1, \Phi_2, \cdots, \Phi_n$ explicitly along with their energy eigenvalues as shown below for $n = 0, 1, 2, \cdots, n$.

TABLE 2.1

Energy Eigenfunctions and Eigenvalues of Harmonic Oscillator

n	Φ_n	E_n	$\Delta E_n = E_{n+1} - E_n$
0	$\Phi_0 = A_0 \exp(-\frac{\xi^2}{2})$	$\frac{1}{2}\hbar\omega$	$\Delta E_0 = \hbar\omega$
1	$\Phi_1 = A_1(2\xi)\exp(-\frac{\xi^2}{2})$	$\frac{3}{2}\hbar\omega$	$\Delta E_1 = \hbar\omega$
2	$\Phi_2 = A_2(4\xi^2 - 2)\exp(-\frac{\xi^2}{2})$	$\frac{5}{2}\hbar\omega$	$\Delta E_2 = \hbar\omega$
.	.	.	.
.		.	.
.			
n	$\Phi_n = A_n H_n(\xi)\exp(-\frac{\xi^2}{2})$	$(n+\frac{1}{2})\hbar\omega$	$\Delta E_n = \hbar\omega$

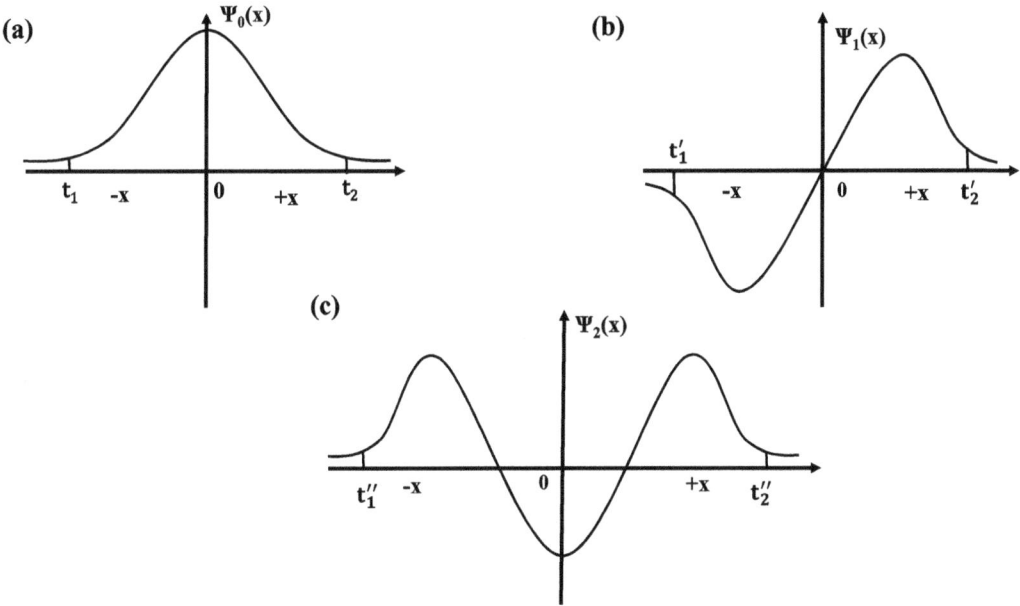

FIGURE 2.2 Plots of harmonic oscillator eigenfunction for (a) the ground state (Ψ_0), (b) the first excited state (Ψ_1) and (c) the second excited state (Ψ_2). Note the increase in number of nodes from Ψ_0 to Ψ_2.

The energy levels are discrete and equally spaced (spacing $= \hbar\omega$). The ground state wavefunction is nodeless and is an even function while the first excited state wavefunction of the harmonic oscillator has one node (at $x = 0$) and is an odd function. The second excited state wavefunction has two nodes and is an even function and so on. A striking feature of the energy eigenfunctions of the harmonic oscillator (displayed in Figure 2.2) concerns their symmetries under spatial inversion. Thus, $\Phi_0(x)$ is even with respect to inversion at the origin ($\Phi_0(-x) = \Phi_0(x)$), $\Phi_1(x)$ is odd under the same operation ($\Phi_1(-x) = -\Phi_1(x)$) while $\Phi_2(x)$ is even under inversion. A look at the general analytical formula for $\Phi_n(\xi)$ and definition of $H_n(\xi)$ (Eq. 2.83 and Eq. 2.84) makes it clear that $H_n(\xi) \equiv H_n(x/x_0)$ can be either odd or even under inversion at the origin. The root of this symmetry of $\Phi_n(x)$ lies in the symmetry of the harmonic oscillator Hamiltonian. $\hat{H}(x)$ which is invariant under the spatial inversion process, i.e. under the transformation $x \rightarrow -x$ obey the equality $\hat{H}(-x) = \hat{H}(x)$. If we represent the inversion operator by $\hat{I}$, we can write the invariance of $\hat{H}$ under $\hat{I}$ as

$$\hat{I}\hat{H}\hat{I}^{-1} = \hat{H} \qquad (2.88)$$

Multiplying both sides by $\hat{I}$ from the right we have

$$\hat{I}\hat{H} = \hat{H}\hat{I}$$

$$\text{or,} \quad [\hat{H},\hat{I}] = 0 \tag{2.89}$$

That means the Hamiltonian operator representing the harmonic oscillator commutes with the inversion operator. (Note: the invariance $\hat{H} = \hat{T} + \hat{V}$ under inversion is obvious as $V(x) = \frac{1}{2}m\omega^2 x^2$ remains unchanged if x is replaced by $-x$. The kinetic energy operator $\frac{\hat{p}^2}{2m} = \frac{\hat{p}\cdot\hat{p}}{2m}$ is a scalar-length of the momentum vector squared divided by the oscillator mass – it does not change under inversion). The commutation of $\hat{H}$ and $\hat{I}$ implies that the oscillator Hamiltonian and space inversion operator must have a common set of eigenfunctions. We have already determined the eigenfunctions (Φ_n) and eigenvalues (E_n) of the harmonic oscillator Hamiltonian. Let us examine if they are eigenfunctions of the inversion operator as well. Suppose that 'f_n' is an eigenfunction of the space inversion operator $\hat{I}$ with the eigenvalue λ. Therefore, we have

$$\hat{I}[f_n(x)] = f_n(-x) = \lambda f_n(x) \tag{2.90}$$

Allowing $\hat{I}$ to operate on both sides of Eq. 2.90, we have

$$I[I(f_n(x))] = \lambda(If_n(x)) = \lambda^2 f_n(x)$$

$$I^2 f_n(x) = \lambda^2 f_n(x) \tag{2.91}$$

$$f_n(x) = \lambda^2 f_n(x) \quad (I^2 = 1)$$

Therefore, the eigenvalue λ must be such that $\lambda^2 = 1$ or $\lambda = \pm 1$. Thus the eigenfunctions of the space inversion operator must belong to one or the other of the two classes. Those with the eigenvalue $\lambda = 1$ are called symmetric, even or gerade (g) eigenfunctions while those with $\lambda = -1$ are called odd, ungerade (u) or antisymmetric eigenfunction. The eigenfunctions of the space inversion operator of the two classes satisfy the following relations

$$f_n(-x) = 1.f_n(x) = f_n(x) - \text{even class;}$$

$$f_n(-x) = (-1).f_n(x) = -f_n(x) - \text{odd class.}$$

The eigenvalues of the space-inversion operator are also called parity eigenvalues with corresponding eigenfunctions being called parity eigenfunctions. From the form of the harmonic oscillator wavefunctions $\Phi_n = A_n H_n(\xi)\exp(-\frac{\xi^2}{2})$ and the analytical form of the Hermite polynomials $H_n(\xi)$, it is clear that $\hat{I}\Phi_n = (-1)^n\Phi_n$ so that the oscillator eigenfunctions are of even parity if 'n' is even and odd if n is odd, n being the quantum number characterizing the oscillator states. That the oscillator states have definite 'parity' quantum number is a reflection or signature of the spatial symmetry of the oscillator Hamiltonians – its invariance under spatial inversion. The oscillator Hamiltonian does not have any other spatial symmetry – had there been any, we would have surely recognized its signature in the eigenfunctions of the oscillator Hamiltonian. The parity is however, a purely quantum property without a classical analog. Apart from parity there is another striking feature that the quantum harmonic oscillator eigenfunctions possess. We must now turn our attention to this feature and its implications. Let us took at Figure 2.2 where some of the lowest energy eigenfunctions (Φ_0, Φ_1, Φ_2, $\cdots$) of the harmonic oscillator are displayed. We may not fail to notice that these wavefunctions do not vanish at the classical turning points (t_1, t_2) – i.e., at the pair of points where the total energy E_n becomes equal to potential energy $V(x)$. A classical particle with total energy $E_n < V(x)$ can not travel to the regions beyond the turning points, i.e. to the regions with $x < t_1$ or $x > t_2$ where $V(x) > E_n$. These regions are classically forbidden and hence inaccessible by the particle. A look at Figure 2.2 reveals that the wavefunctions $\Phi_n(x)$ do not vanish at the respective turning points (t_1, t_2; t_1', t_2'; t_1'', t_2''; etc.) – they continue past the tuning points and decay to zero as $x \to \pm \infty$ as required by the boundary conditions of the problem. Hence $|\Phi_n(x)|^2$ or the probability density of finding the particle in the regions beyond the relevant turning points can be non-zero.[4] The quantum particle, as it tunnels into the classically forbidden regions, gives rise to an entirely non-classical phenomenon called 'quantum tunneling'. It is imperative that we take a closer look at this bizarre phenomenon and try to understand it along with its ramifications or implications in various contexts.

2.6 The Tunnel Effect

Suppose that there are two regions of space (marked I and III) where $E > V$ seperated by a potential energy barrier (region II) where $E < V$ (Figure 2.3). Inside the barrier region, the kinetic energy of the particle is $\frac{p^2}{2m} = (E - V)$, which is < 0 (as $E - V < 0$). The momentum of the particle inside the barrier region $p = \sqrt{2m(E - V)}$ is therefore imaginary. A classical particle with energy E less than the barrier energy, i.e. the potential energy at the barrier top can not pass through the barrier as the particle can not have imaginary momentum. The situation is very different in wave mechanics where the imaginary particle momentum only means that the wavefunction of the particle depends exponentially on the coordinate (x) in the classically forbidden barrier region. Since the wavefunction does not vanish in the barrier region it is quite possible that the quantum particle with energy $E < V$ may even pass through the barrier. This under-barrier passage of the particle through a barrier (not over the barrier) is without any classical parallel and has been known as the quantum mechanical tunnel effect. Before going into the details of the calculation tunnel effect – in specific problems or situation – let us introduce the concepts of charge and current densities in the classical context and try to find their quantum counterparts, which may then be used to calculate the tunnel effect that a charged particle may show up in a particular situation.

In an incompressible charged fluid the charge density ρ and the current density $\vec{j}$ are related by the continuity equation, which demands that

$$\frac{\partial \rho(r,t)}{\partial t} + div\, j(r) = 0 \tag{2.92}$$

Eq. 2.92 is a statement of the law of charge conservation. We have already seen that the wave function Ψ of a quantum system defines probability density of finding the particle at the point 'r' as $\rho(r,t) = |\Psi|^2 = \Psi\Psi^*$. If we multiply $|\Psi|^2$ by the charge 'e' carried by the particle we have a quantum mechanically defined charge density

$$\rho(r,t) = e|\Psi|^2 = e(\Psi\Psi^*).$$

To make the analogy with the classical charge conservation equation complete we also need a probability current density, which together with $\rho(r,t)$ may constitute the quantum counterpart of Eq. 2.92. Let us consider the Schrödinger equation (Eq. 2.93) and its complex conjugate (Eq. 2.94)

$$\frac{-\hbar}{i}\frac{\partial \Psi}{\partial t} = -\frac{\hbar^2}{2m}\nabla^2\Psi + V\Psi \tag{2.93}$$

$$\frac{+\hbar}{i}\frac{\partial \Psi^*}{\partial t} = -\frac{\hbar^2}{2m}\nabla^2\Psi^* + V\Psi^* \tag{2.94}$$

Multiplying Eq. 2.93 by Ψ^* and Eq. 2.94 by Ψ and subtracting one from the other we get

$$\frac{\partial(\Psi^*\Psi)}{\partial t} + \frac{i\hbar}{2m}\nabla(\Psi\nabla\Psi^* - \Psi^*\nabla\Psi) = 0$$

$$\text{i.e.} \quad \frac{\partial(\Psi^*\Psi)}{\partial t} + \frac{i\hbar}{2m}div(\Psi\, grad\Psi^* - \Psi^*\, grad\Psi) = 0 \tag{2.95}$$

Multiplying both sides by the charge carried by the electron (e) we have

$$\frac{\partial(e\Psi^*\Psi)}{\partial t} + \frac{i\hbar e}{2m}div(\Psi\, grad\Psi^* - \Psi^*\, grad\Psi) = 0 \tag{2.96}$$

In Eq. 2.95, $\Psi\Psi^*$ represent the probability density (ρ) at a certain point $\vec{r}$ in space and at a certain time (t). If we define $\frac{i\hbar}{2m}(\Psi\, grad\Psi^* - \Psi^*\, grad\Psi)$ as the probability current density at the same point in space at the same time ($\vec{j}$) we have a continuity like equation for a quantum system in a state described by the wavefunction Ψ:

$$\frac{\partial P}{\partial t} + div\, \vec{j} = 0$$

where the probability density $P = \Psi\Psi^*$ and

$$\vec{j} = \frac{i\hbar}{2m}(\Psi\, grad\Psi^* - \Psi^*\, grad\Psi)$$

gives the probability current density. On the other hand, Eq. 2.96 represents the quantum mechanical continuity equation involving the charge density $\rho = eP$ and, quantum mechanical current density $\vec{j} = \frac{i\hbar e}{2m}(\Psi\, grad\Psi^* - \Psi^*\, grad\Psi) = \vec{j}'e$ and reads

$$\frac{\partial\rho}{\partial t} + div\,\vec{j} = 0 \qquad (2.97)$$

where

$$\rho = e\Psi\Psi^*$$

and

$$\vec{j} = \frac{i\hbar e}{2m}(\Psi\, grad\Psi^* - \Psi^*\, grad\Psi)$$

It is clear that for systems described by real wavefunctions, $\vec{j} = 0$. We will see later how the concept of a probability current density becomes useful in the description and calculation of 'tunnel effect'.

2.6.1 Tunneling Across a Rectangular Potential Barrier

Suppose that there are two regions of space, marked I and III, respectively where the energy E is greater than potential energy (V), separated by potential barrier 'V_0' spread over a region of space marked II lying between regions I and III (Figure 2.3) where the potential energy (V) is greater than the particle energy. Within the barrier region (region II), the kinetic energy of the particle is $\frac{p^2}{2mf} = E - V_0$, which is clearly negative (<0) so that the momentum of the particle in this region is $p = \sqrt{2m(E - V_0)} = i\beta$ (β real), which is purely imaginary. There are thus clearly three regions of space I, II and III. Let us suppose that the potential has a height V_0. Let a particle of mass 'm' be moving in this potential, which is well described by the potential function $V(x)$ where $V(x) = V_0$ for $0 \leq x \leq a$ and $V(x) = 0$, elsewhere. In the regions I and III, $E > V = 0$ while inside the barrier region, $E < V = V_0$. The Schrödinger equation for a particle of mass

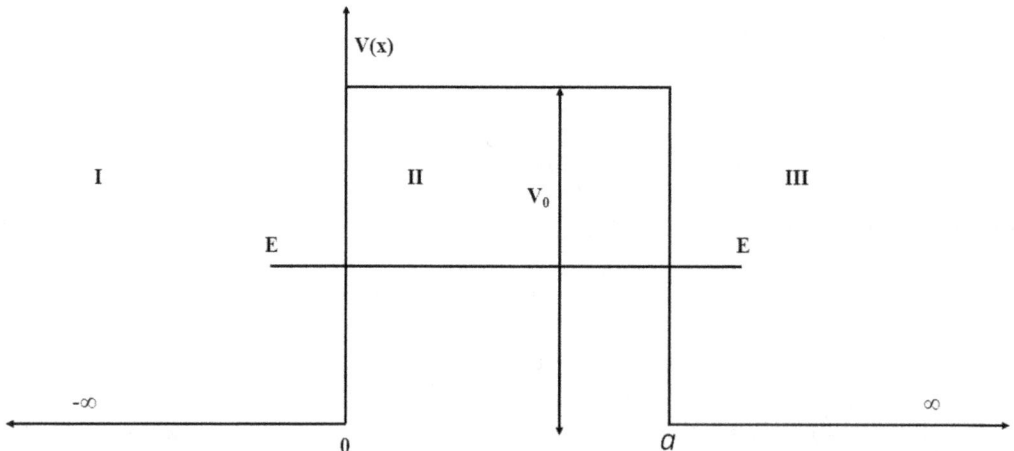

FIGURE 2.3 Potential energy diagram.

'*m*' moving in the just described potential reads

$$-\frac{\hbar^2}{2m}\frac{d^2\Psi(x)}{dx^2} + V(x)\Psi(x) = E\Psi(x)$$

$$i.e., \quad \frac{d^2\Psi(x)}{dx^2} = (-)\frac{2m(E - V(x))}{\hbar^2}\Psi \tag{2.98}$$

In region I and III, $E > V(x) = 0$ so that Eq. 2.98 takes the form

$$\frac{d^2\Psi_{I,III}}{dx^2} = -\frac{2mE}{\hbar^2}\Psi_{I,III} = -k^2\Psi_{I,III} \tag{2.99}$$

where $k = \frac{\sqrt{2mE}}{\hbar}$. $\Psi_I(x)$ can be chosen in the form of a superposition of an incident and a reflected wave:

$$\Psi_I(x) = A_1\exp(ikx) + B_1\exp(-ikx) \tag{2.100}$$

We can set the coefficient of the incident wave $A_1 = 1$ without any loss of generality and write

$$\Psi_I(x) = \exp(ikx) + B_1\exp(-ikx) \tag{2.101}$$

In region III, only the transmitted wave makes up the solution, allowing us to write

$$\Psi_{III}(x) = B_3\exp[ik(x - a)] \tag{2.102}$$

Inside the barrier, i.e. in region II the wave equation takes the form ($E < V(x) = V_0$)

$$\frac{d^2\Psi_{II}}{dx^2} = \frac{2m}{\hbar^2}(V_0 - E)\Psi_{II} = \beta^2\Psi_{III},$$

where $\beta = \frac{\sqrt{2m(V_0-E)}}{\hbar}$ and the solution can be cast in the form

$$\Psi_{III}(x) = A_2\exp(-\beta x) + B_2\exp(\beta x) \tag{2.103}$$

We can now seek the unknown superposition coefficients by imposing the relevant boundary condition that requires the wavefunction (Ψ) and its first derivative to be continuous at every boundary. At the left boundary ($x = 0$) the continuity of Ψ_I and Ψ_{II} demands

$$1 + B_1 = A_2 + B_2 \tag{2.104a}$$

At the right boundary ($x = a$) the continuity of Ψ_{II} and Ψ_{III} leads to

$$A_2\exp(-\beta a) + B_2\exp(\beta a) = A_3 \tag{2.104b}$$

The continuity of the first derivative of Ψ_I and Ψ_{II} at $x = 0$, on the other hand, leads to

$$ik(1 - B_1) = \beta(B_2 - A_2) \tag{2.105a}$$

while the derivative continuity condition on Ψ_{II} and Ψ_{III} at the right boundary ($x = a$) leads to

$$A_2\exp(-\beta a) - B_2\exp(\beta a) = (-)A_3\frac{ik}{\beta} \tag{2.105b}$$

It is straightforward to obtain from Eq. 2.104 and Eq. 2.105, expressions for both A_2 and B_2 in terms of A_3:

$$A_2 = (-)\frac{ik}{\beta}\exp(\beta a)A_3$$

$$B_2 = \frac{1}{2}\left(1 + \frac{ik}{\beta}\right)\exp(-\beta a)A_3 \tag{2.106}$$

These expressions for A_2 and B_2 can be ploughed back in Eq. 2.104a and Eq. 2.105a and the coefficient B_1 eliminated to obtain an expression for the transmitted coefficient

$$A_3 = \frac{2}{2\cosh\beta a + i\left(\frac{\beta}{k} - \frac{k}{\beta}\right)\sinh\beta a} \tag{2.107}$$

The magnitude of the tunnel effect can be estimated in terms a transmission coefficient $T(E)$ through the barrier at energies $E < V_0$. $T(E)$ is conveniently defined as the modulus of the ratio of the density of transmitted flux to the incident flux density. With this defination, we have[6]

$$T(E) = \frac{\vec{j}\text{ transmitted}}{\vec{j}\text{ incident}} \tag{2.108}$$

where $\vec{j} = \frac{ei\hbar}{2m}(\Psi\frac{d\Psi^*}{dx} - \Psi^*\frac{d\Psi}{dx})$ and Ψ is the wavefunction of the tunneling electron (charge = e). In our case, it is easily seen that $T(E) = |A_3|^2$. Making use of Eq. 2.107 for A_3, we get

$$T(E) = \left| \frac{2}{2\cosh\beta a + i(\frac{\beta}{k} - \frac{k}{\beta})\sinh\beta a} \right|^2$$

$$= \frac{4k^2\beta^2}{(k^2 + \beta^2)^2 \sinh^2\beta a + 4k^2\beta^2} \tag{2.109}$$

If the barrier is wide as well as high $\beta a >> 1$ and we can replace the exact expression for the transmission coefficient by an approximate expression[6]

$$T \approx \frac{16k^2\beta^2}{(k^2 + \beta^2)^2}\exp(-2\beta a)$$

$$= T_0\exp(-2\beta a) \tag{2.110}$$

Clearly, as $\beta a \to \infty$, T $\to$ 0 and there no transmission by tunneling. Even to the quantum particle, the barrier appears to be opaque and classical behavior is restored. For low values of βa, we have the situation where the barrier is narrow (small a) as well as low (small $V_0 - E$)$T \neq 0$ and the particle undergoes under barrier transmission. This mode of transmission is entirely quantum mechanical in nature and made possible by the ability of the wavefunction to get delocalized over the entire space if right situations are created.

If the particle has energy $E > V_0$, we expect it to cross over to the other side of the barrier, the barrier appearing to be completely transparent to the incoming particle. That is what we would expect classically. In the quantum mechanical case, the particle no-doubt crosses out to the other side by over barrier transmission but it does not see a completely transparent barrier and suffers a degree of under barrier reflection. This non-classical reflection coefficient (R) can be defined by following the definition of transmission coefficient (T) given earlier in this section (Eq. 2.108). Accordingly, the reflection coefficient becomes

$$R = B_1^2$$

$$= \frac{|j_{reflected}|}{|j_{incident}|}$$

$$= \frac{(k_0^2 - k_1^2)^2 \sin^2 k_1 a}{4k_0^2 k_1^2 + (k_0^2 - k_1^2)\sin^2 k_1 a} \tag{2.111}$$

$$= 1 - T$$

where $k_1 = \frac{\sqrt{2m(E - V_0)}}{\hbar}$ ($E > V_0$), $\beta = ik_1$, $k_0^2 = \frac{2mE}{\hbar^2}$. If the barrier height is zero ($V_0 = 0$), $k_1 = k_0$, and R in Eq. 2.111 becomes zero. The motion is therefore completely free over the entire region of space, there being no reflection at all.

2.7 Heisenberg's Formulation of Quantum Mechanics

Historically, Heisenberg's formulation of quantum mechanics preceded wave mechanical theory of Schrödinger. The two formulation are, however, equivalent as was demonstrated by Schrödinger himself. In what follows, we set out to arrive at the matrix (Heisenberg) formulation of quantum mechanics or the equation of motion of Heisenberg starting from the time-dependent wavefunction of Schrödinger, which are eigenfunctions of the energy operator ($\hat{H}$). Let us consider two such eigenfunctions $\Psi(E,x)$ and $\Psi(E',x)$ in one dimension. Notice that we have labeled the wavefunctions by their energy eigenvalues E and E'. Then, as we have seen already,

$$\Psi(E,x) = \exp\left(\frac{-iEt}{\hbar}\right)\Psi_0(E,x)$$

$$\Psi(E',x) = \exp\left(\frac{-iE't}{\hbar}\right)\Psi_0(E',x)$$

(2.112)

where $\Psi_0(E,x)$ is an energy eigenfunction with the eigenvalue E at time $t = 0$, and $\Psi_0(E',x)$ is the initial ($t = 0$) energy eigenfunction with energy E'. Let $\hat{A}$ be a certain hermitian operator, which is given a matrix representation in the basis of energy eigenfunctions labeled by their energy eigenvalues E, E', E'', $\cdots$ including the time-dependent phase factor associated with each basis function. An element $A_{E,E'}$ of the matrix A is then given in the Heisenberg representation by

$$A_{E',E} = \int \Psi^*(E',x)\hat{A}\Psi(E,x)dx$$

(2.113)

Differentiation with respect to time leads to

$$
\begin{aligned}
\frac{dA_{E',E}}{dt} &= \left[\left(\int \Psi^*(E',x)\frac{\partial\hat{A}}{\partial t}\Psi(E,x)dx\right)\exp\left(\frac{-i}{\hbar}(E-E')t\right)dx - \frac{i}{\hbar}(E-E')\right.\\
&\qquad \left.\int \Psi_0^*(E',x)\hat{A}\Psi_0(E,x)dx \times \exp\left(\frac{-i}{\hbar}(E-E')\right)\right]\\
&= \frac{\partial A_{E',E}}{\partial t} - \frac{i}{\hbar}(E-E')A_{E,E'}\\
&= \frac{\partial A_{E',E}}{\partial t} + \frac{i}{\hbar}E'A_{E,E'} - A_{E,E'}E\\
&= \frac{\partial A_{E',E}}{\partial t} + \frac{i}{\hbar}(E'-E)A_{E',E}
\end{aligned}
$$

(2.114)

The Hamiltonian operator like any other operator is diagonal in its own eigenbasis $\Psi_0(E,x)$, which allows us to write

$$\hat{H}_{E',E} = E'\delta_{E',E}$$

$$\hat{H}_{E'',E} = E''\delta_{E'',E}, \cdots, etc.$$

(2.115)

which when used in Eq. 2.114 yields the following matrix equation:

$$\frac{dA_{E',E}}{dt} = \frac{\partial A_{E',E}}{\partial t} + \frac{i}{\hbar}\left\{(\hat{H}\hat{A})_{E',E} - (\hat{A}\hat{H})_{E',E}\right\}$$

(2.116)

We can write similar equations for every choice of E', E. Therefore, the following matrix equation must hold.[4]

$$
\begin{aligned}
\frac{d\hat{A}}{dt} &= \frac{\partial\hat{A}}{\partial t} + \frac{i}{\hbar}(\hat{H}\hat{A} - \hat{A}\hat{H})\\
&= \frac{\partial\hat{A}}{\partial t} + \frac{i}{\hbar}[\hat{H},\hat{A}]
\end{aligned}
$$

(2.117)

It is the quantum equation of motion for the observable represented by the operator $\hat{A}$ for the quantum system that the Hamiltonian defines. It has been known as Heisenberg's equation of motion. Two observations must now be made:

(i) If $\hat{A}$ does not depend on time explicitly, $\frac{\partial \hat{A}}{\partial t} = 0$;

(ii) If in addition $\hat{H}$ and $\hat{A}$ commute, $[\hat{H}, \hat{A}] = 0$. In that case, $\frac{d\hat{A}}{dt} = 0$ and $\hat{A}$ turns out to be invariant under time displacement and is said to be a conserved quantity for the system. $\hat{A}$ is also called quantum integral of motion. It is also obvious that $\hat{H}$ and $\hat{A}$ will have a common set of eigenfunctions. Let us consider specific hermitian operator for $\hat{A}$: $\hat{A} \equiv \hat{p}$ (linear momentum), $\hat{A} = \hat{H}$ (energy operator). $\hat{p}$ obviously does not contain time explicitly. Also $\hat{H}$ is assumed to have no explicit time dependence. From Eq. 2.117, we can write immediately for $\hat{A} = \hat{H}$

$$\frac{d\hat{H}}{dt} = \frac{\partial \hat{H}}{\partial t} + \frac{i}{\hbar}[\hat{H}, \hat{H}]$$

$$= 0 + 0$$

so that energy is a conserved quantity for any system represented by time-independent Hamiltonian. For $\hat{A} = \hat{p}$, similarly we have

$$\frac{d\hat{p}}{dt} = \frac{\partial \hat{p}}{\partial t} + \frac{i}{\hbar}[\hat{H}, \hat{p}]$$

If the linear momentum operator commutes with the system Hamiltonian $[\hat{H}, \hat{p}] = 0$, $\frac{\partial \hat{p}}{\partial t}$ is also zero so that $\frac{d\hat{p}}{dt} = 0$, making linear momentum a quantum integral of motion. For a free particle $\hat{H} = \frac{\hat{p}^2}{2m}$ and $[\hat{H}, \hat{p}] = [\frac{\hat{p}^2}{2m}, \hat{p}] = 0$. For such a system, linear momentum is a conserved quantity. Generally, however, $\hat{H} = \frac{\hat{p}^2}{2m} + V(x, y, z)$ so that $[\hat{H}, \hat{p}] = [\frac{\hat{p}^2}{2m}, \hat{p}] + [\hat{V}(x, y, z), \hat{p}]$. The commutator $[\hat{V}(x, y, z), \hat{p}]$ will not in general vanish so that linear momentum shall not be a quantum integral of motion for such systems. Thus Heisenberg's equation of motion makes it clear which dynamical properties of a given system could be a conserved quantity. It is generally far harder to solve Heisenberg's equation of motion compared to Schrödinger's wave equation. Fortunately, however, Heisenberg's equation for a one-dimensional linear oscillator can be solved completely algebraically and matrix representations for momentum ($\hat{p}_x$) and position ($\hat{x}$) can be analytically worked out in the energy representation as detailed in what follows.

2.7.1 Matrix Representations for x and p_x

The Hamiltonian operator $\hat{H}$ for the simple harmonic oscillator is given by

$$\hat{H} = \frac{\hat{p}_x^2}{2m} + \frac{1}{2}m\omega^2\hat{x}^2$$

$$= \frac{\hat{p}_x^2}{2m} + \hat{V}(x)$$

The classical equation of motion for $\hat{x}$ and $\hat{p}_x$ are

$$\frac{dx}{dt} = \dot{x} = \frac{p_x}{m} \tag{2.118a}$$

and

$$\dot{p}_x = -\frac{\partial V(x)}{\partial x} = -m\omega^2 x \tag{2.118b}$$

The 'quantum analogs' of these equations of motion (EOM), by 'correspondence argument' should be the corresponding operator equation with $x \rightarrow \hat{x}$, $p_x \rightarrow \hat{p}_x$. Since $\hat{x}$ and $\hat{p}$ do not have explicit time dependence, we have from Eq. 2.117, the following EOM

$$\dot{\hat{x}} = \frac{i}{\hbar}[\hat{H}, \hat{x}] \tag{2.119a}$$

$$\dot{\hat{p}}_x = \frac{i}{\hbar}[\hat{H}, \hat{p}_x] \tag{2.119b}$$

Since $\hat{H} = \frac{\hat{p}_x^2}{2m} + \frac{1}{2}m\omega^2\hat{x}^2$

$$[\hat{H}, \hat{x}] = \left[\frac{\hat{p}_x^2}{2m}, \hat{x}\right] + \left[\frac{1}{2}m\omega^2\hat{x}^2, \hat{x}\right]$$

$$= \frac{1}{2m}\left\{\hat{p}_x^2\hat{x} - \hat{x}\hat{p}_x^2\right\} + \frac{1}{2}m\omega^2\left\{\hat{x}\hat{x} - \hat{x}\hat{x}\right\}$$

$$= \frac{1}{2m}\left\{\hat{p}_x\hat{p}_x\hat{x} - \hat{x}\hat{p}_x\hat{x} + \hat{x}\hat{p}_x\hat{x} - \hat{x}\hat{p}_x\hat{p}_x\right\} + 0$$

$$= \frac{1}{2m}\left\{\hat{p}_x[\hat{p}_x, \hat{x}] + [\hat{p}_x, \hat{x}]\hat{p}_x\right\}$$

Similarly,

$$[\hat{H}, \hat{p}_x] = \frac{1}{2}m\omega^2\left\{\hat{x}[\hat{x}, \hat{p}_x] + [\hat{x}, \hat{p}_x]\hat{x}\right\}$$

Using the obtained expressions for $[\hat{H}, \hat{x}]$ and $[\hat{H}, \hat{p}_x]$ in Eq. 2.119a and Eq. 2.119b respectively, it is straightforward to show that these equations pass over into Eq. 2.118a and Eq. 2.118b respectively with x replaced by $\hat{x}$ and p_x by $\hat{p}_x$, provided we assume that the commutator of $\hat{p}_x$ and $\hat{x}$ has the following form: $[\hat{p}_x, \hat{x}] = \frac{\hbar}{i}\hat{\Lambda}$. $\hat{\Lambda}$ being the identity operator. The assumption $[\hat{p}_x, \hat{x}] = \frac{\hbar}{i}\hat{\Lambda}$ therefore gives us the following EOM. $\dot{\hat{x}} = \frac{\hat{p}_x}{m}$ and $\dot{\hat{p}}_x = -m\omega^2\hat{x}$ as demanded by correspondence argument. Note that we have not explicitly used the coordinate form of the operator $\hat{p}_x = \frac{\hbar}{i}\frac{\partial}{\partial x}$, which would have also led to the same expression for the commutator $[\hat{p}_x, \hat{x}]$.

Let us now try to work out the matrix forms of the equations of motion for $\hat{x}$ and $\hat{p}_x$ in the energy representation. We have already shown that for any operator $\hat{A}$ that does not contain time explicitly, the matrix form of the EOM is (see Eq. 2.114 and Eq. 2.118)

$$\dot{\hat{A}}_{E',E} = \frac{i}{\hbar}(E' - E)A_{E',E} \tag{2.120}$$

with $\hat{A} = \hat{x}$ and $\hat{p}_x$, respectively, we have

$$\frac{i}{\hbar}(E' - E)\dot{x}_{E',E} = \frac{(p_x)_{E',E}}{m} \tag{2.121a}$$

$$\frac{i}{\hbar}(E' - E)(p_{\dot{x}})_{E',E} = -m\omega^2 x_{E',E} \tag{2.121b}$$

Eliminating $p_{E',E}$ from the pair of Eq. 2.121a and b, we arrive at the following equation:

$$[(E' - E)^2 - \hbar^2\omega^2]x_{E',E} = 0 \tag{2.122}$$

This equation immediately tells us that the matrix element $x_{E',E} \neq 0$ only if the labels E' and E such that

$$(E' - E) = \pm\hbar\omega \tag{2.123}$$

which in turn asserts that the neighboring energy eigenvalues or energy levels of the oscillator can differ by $\pm\hbar\omega$ only. The oscillator Hamiltonian $\hat{H} = \frac{\hat{p}_x^2}{2m} + \frac{1}{2}m\omega^2 x^2$ is a sum of the square terms; the average value of energy $<E> = <H>$ of the oscillator must therefore be non-negative, i.e., $<E> = <H> \geq$

0 so also must be the different energy eigenvalues E, E', E'', $\cdots$, etc. Let E_0 (≥ 0) be the lowest energy eigenvalue of the oscillator and E be a higher eigenvalue. Then by subtracting $\hbar\omega$ 'n' times from E we should be able to reach E_0. Similarly from another higher eigenvalue E', we may substract $\hbar\omega$ 'n'' times to reach E_0. That means we have

$$E = n\hbar\omega + E_0$$
$$E' = n'\hbar\omega + E_0$$

(2.124)

where n and n' are integers. We can now use Eq. 2.124 defining E, E' in Eq. 2.122, replace matrix elements labels E, E' by n and n', respectively and obtain[4]

$$[(n' - n)^2 - 1]x_{n',n} = 0$$

(2.125)

Since the quantity within the square brackets on the left hand side of Eq. 2.125 vanishes only if $n' = n \pm 1$, $x_{n',n}$ can be non-zero only and only for $n' = n \pm 1$. That means the only non-vanishing elements of the x matrix are $x_{n \pm 1, n}$ and $x_{n, n \pm 1}$. It is evident that the diagonal elements $x_{n,n}$ of the x-matrix are all zero. The non-vanishing elements of the x-matrix can be conveniently determined by appealing to the basic commutator relation

$$[\hat{p}_x, \hat{x}] = \frac{\hbar}{i}\hat{\Lambda}$$

and equating the diagonal elements on both sides of the commutation relation. By invoking the standard rules of matrix multiplication and noting that the only surviving elements of the x-matrix are $x_{n,n\pm 1}$ and $x_{n\pm 1,n}$ we find that (dropping the subscript in p_x)[4,7]

$$[\hat{p}, \hat{x}]_{n,n} = \frac{\hbar}{i}\hat{\Lambda}_{n,n}$$

(2.126)

$$\text{or, } p_{n,n+1}x_{n+1,n} + p_{n,n-1}x_{n-1,n} - x_{n,n+1}p_{n+1,n} - x_{n,n-1}p_{n-1,n} = \frac{\hbar}{i}$$

We can express the relevant elements of the p-matrix ($p_{n\pm 1,n}$) in terms of the corresponding elements of the x-matrix by making use of the Eq. 2.121 after replacing the labels E', E by integers n' and n, respectively, where we get

$$p_{n+1,n} = im\omega x_{n+1,n}$$
$$p_{n-1,n} = -im\omega x_{n-1,n}$$

(2.127)

Making use of Eq. 2.127 in Eq. 2.126 and exploiting the hermiticity of p- and x-matrices, we arrive at the following equation:

$$2im\omega\left\{|x_{n,n-1}|^2 - |x_{n,n+1}|^2\right\} = \frac{\hbar}{i}$$

(2.128)

Hermitian character of the x-matrix demands that $x_{n,n-1} = x_{n-1,n}^*$, which in turn ensures that $|x_{n,n-1}|^2 = |x_{n-1,n}|^2$ so that Eq. 2.128 can be slightly rearranged to read

$$|x_{n,n+1}|^2 - |x_{n-1,n}|^2 = \frac{\hbar}{2m\omega}$$

(2.129)

Note that n can not be negative. Eq. 2.129 therefore yields the following results:

$$|x_{01}|^2 = \frac{\hbar}{2m\omega} = |x_{10}|^2$$

$$|x_{12}|^2 = 2 \cdot \frac{\hbar}{2m\omega} = |x_{21}|^2$$

$$|x_{23}|^2 = 3 \cdot \frac{\hbar}{2m\omega} = |x_{31}|^2$$

$$|x_{n,n+1}|^2 = (n+1) \cdot \frac{\hbar}{2m\omega} = |x_{n+1,n}|^2$$

(2.130)

Assuming now that the elements of the *x*-matrix are real numbers and ignoring the values of their phases, we can write

$$x_{n,n+1} = \sqrt{(n+1)}\left(\frac{\hbar}{2m\omega}\right)^{\frac{1}{2}}, \quad n = 0, 1, 2, \cdots$$

$$= x_{n+1,n} \text{ and } x_{n,n} = 0$$

With this general expressions for the elements of the *x*-matrix, we can form the *x*-matrix that reads

$$x_{n',n} = \left(\frac{\hbar}{2m\omega}\right)^{\frac{1}{2}} \begin{pmatrix} 0 & \sqrt{1} & 0 & 0 & 0 & \dots & 0 \\ \sqrt{1} & 0 & \sqrt{2} & 0 & 0 & \dots & 0 \\ 0 & \sqrt{2} & 0 & \sqrt{2} & 0 & \dots & \dots \\ 0 & 0 & \sqrt{3} & 0 & \dots & \dots & 0 \\ \cdot \\ \cdot \\ \cdot \end{pmatrix} \quad (2.131)$$

with all the elements of *x*-matrix so determined Eq. 2.127 then assures us that the elements of the *p*-matrix (*x*-component momentum) are purely imaginary quantities. Hermiticity of *p*-matrix ensures on the other hand that

$$p_{n',n} = p_{n,n'}^* = -p_{n,n'}$$

Making use of the Eq. 2.130 we are now in a position of constructing the *p*-matrix, which reads

$$p = i\left(\frac{m\hbar\omega}{2}\right)^{\frac{1}{2}} \begin{pmatrix} 0 & -\sqrt{1} & 0 & 0 & 0 & \dots & 0 \\ \sqrt{1} & 0 & -\sqrt{2} & 0 & 0 & \dots & 0 \\ 0 & \sqrt{2} & 0 & -\sqrt{3} & 0 & \dots & \dots \\ 0 & \sqrt{2} & 0 & -\sqrt{3} & \dots & \dots & 0 \\ 0 & 0 & \sqrt{3} & 0 & \dots & \dots & 0 \\ \cdot \\ \cdot \\ \cdot \end{pmatrix} \quad (2.132)$$

Once the matrix forms of both *x* and *p* have become available without ambiguity we can identify the lowest energy (ground state) E_0 of oscillator as the first element of the corresponding *H*-matrix ($H_{0,0}$) constituted with the *x*- and *p*-matrices just defined. Thus,

$$E_0 = H_{0,0} = \frac{1}{2m}(p^2)_{00} + \frac{1}{2}m\omega^2(x^2)_{00}$$
$$= \frac{1}{2m}(pp)_{00} + \frac{1}{2}m\omega^2(xx)_{00}$$
$$= \frac{1}{2m}p_{01}p_{10} + \frac{1}{2}m\omega^2 x_{01}x_{10}$$
$$= \frac{\hbar\omega}{4} + \frac{\hbar\omega}{4}$$
$$= \frac{\hbar\omega}{2}$$

In a similar vain, it can easily be shown that the first and second excited state energies of the oscillator are,

$$E_1 = H_{11} = \frac{3}{2}\hbar\omega,$$

$$E_2 = H_{22} = \frac{5}{2}\hbar\omega,$$

$$E_n = H_{nn} = \left(n + \frac{1}{2}\right)\hbar\omega.$$

2.7.2 Zero Point Oscillation

The exact energy levels of the harmonic oscillator reveal that (see Figure 2.4) it has an equally spaced discrete energy level with uniform spacing equal to $\hbar\omega$ and no degeneracy. Unlike the classical oscillator, the ground state enegy is not zero, but is $\frac{1}{2}\hbar\omega$. This rather unexpected result is a consequence of the operation of the quantum uncertainty principle according to which the position x and the conjugate momentum p of a quantum system can not be specified simultaneously with absolute precision. It turns out that the uncertainty in position (δx) and the uncertainty in momentum (δp) obey the inequality

$$(\delta x)^2 \cdot (\delta p)^2 \geq \frac{\hbar^2}{4}$$

Making use of the definitions of δx, δp we arrive at

$$(<x^2> - <x>^2) \cdot (<p^2> - <p>^2) \geq \frac{\hbar^2}{4}.$$

We have already seen that the harmonic oscillator states are parity eigenstate. For any state with fixed parity, symmetry dictates that $<x> = 0$, $<p> = 0$ as both the operator $\hat{x}$ and $\hat{p}_x = \frac{\hbar}{i}\frac{\partial}{\partial x}$ are odd under space inversion. The uncertainty relation then needs $<x^2> \cdot <p^2> \geq \frac{\hbar^2}{4}$, which demands $<p^2> \geq \frac{\hbar^2}{4<x^2>}$. The total energy of the oscillator is

$$<E> = <H> = \frac{p^2}{2m} + \frac{1}{2}m\omega^2 <x^2>$$

$$\geq \frac{\hbar^2}{8<x^2>m} + \frac{1}{2}m\omega^2 <x^2>.$$

Minimizing $<E>$ with respect to $<x^2>$ we find that

$$<x^2>_{min} = \frac{\hbar}{2m\omega}$$

$$\text{and } E \geq E_{min} = \frac{\hbar\omega}{4} + \frac{\hbar\omega}{4} = \frac{\hbar\omega}{2}$$

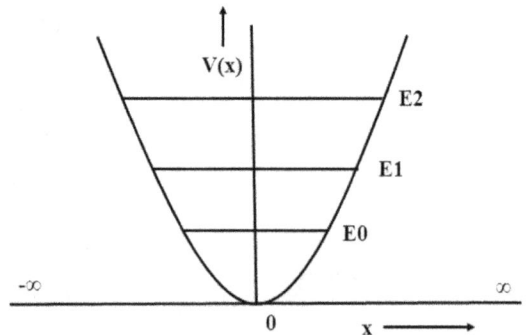

FIGURE 2.4   Energy levels of the harmonic oscillator.

That the minimum energy is not zero directly followed from the position-momentum uncertainty principle and points to the wave nature of the particle moving in a harmonic potential. The minimum non-zero energy is called the zero-point energy of the oscillator. The existence of the zero-point energy was first realized from experiment on the scattering of x-rays by crystals at temperature $T \to 0$ (i.e. at very low temperature). Had E_0 been zero as classically expected, there would not have been lattice vibration at very low temperatures and no scattering of x-ray would have occured. On the other hand, if $E_0 > 0$, the x-ray scattering cross sections of the crystal would tend to attain a limiting value at very low temperatures.[6] Actual experiments confirm that there is scattering even in the limit if $T \to 0$, confirming the existence of zero-point energy.

2.7.3 Harmonic Oscillator in Three Dimensions

We have seen that there is no energy degeneracy for the harmonic oscillator in one-dimension. For that matter, no one dimensional potential supports degenerate energy eigenstates. It is important therefore to investigate what happens in the case of a three-dimensional harmonic oscillator. Let us first consider the case of a completely non-isotropic three-dimensional harmonic oscillator. The Hamiltonian operator for this oscillator reads

$$
\begin{aligned}
\hat{H} &= -\frac{\hbar^2}{2m}\left(\frac{\partial^2}{\partial x^2} + \frac{\partial^2}{\partial y^2} + \frac{\partial^2}{\partial z^2}\right) + \frac{1}{2}k_1 x^2 + \frac{1}{2}k_2 y^2 + \frac{1}{2}k_3 z^2 \\
&= -\frac{\hbar^2}{2m}\left(\frac{\partial^2}{\partial x^2} + \frac{\partial^2}{\partial y^2} + \frac{\partial^2}{\partial z^2}\right) + \frac{1}{2}m\omega_1^2 x^2 + \frac{1}{2}m\omega_2^2 y^2 + \frac{1}{2}m\omega_3^2 z^2
\end{aligned}
$$

where $\omega_i = \frac{\sqrt{k_i}}{m}$, $i = 1, 2, 3$. The energy eigenvalue equation for the oscillator is

$$\hat{H}\Psi(x,y,z) = E\Psi(x,y,z)$$

i.e. $$-\frac{\hbar^2}{2m}\left(\frac{\partial^2\Psi}{\partial x^2} + \frac{\partial^2\Psi}{\partial y^2} + \frac{\partial^2\Psi}{\partial z^2}\right) + \frac{1}{2}m(\omega_1^2 x^2 + \omega_2^2 y^2 + \omega_3^2 z^2)\Psi = E\Psi$$

(2.133)

Eq. 2.133 is separable in cartesian coordinates as can be readily seen by substituting $\Psi = \Psi(x,y,z) = \Phi(x)\chi(y)\xi(z)$ in Eq. 2.133, and dividing both sides by $\Phi(x)\chi(y)\xi(z)$ when we arrive at the following equation:

$$\left(-\frac{1}{\Phi}\frac{\hbar^2}{2m}\frac{d^2\Phi(x)}{dx^2} + \frac{1}{2}m\omega_1^2 x^2\right) + \left(-\frac{1}{\chi}\frac{d^2\chi(y)}{dy^2} + \frac{1}{2}m\omega_2^2 y^2\right) + \left(-\frac{1}{\xi}\frac{d^2\xi(z)}{dz^2} + \frac{1}{2}m\omega_3^2 z^2\right) = E$$

Since x, y, z are independent variables each term in the parenthesis must be equal to a constant, say, ϵ_1, ϵ_2, and ϵ_3, respectively such that $E = \epsilon_1 + \epsilon_2 + \epsilon_3$. That means we have reduced the problem in three dimensions to three one dimensional harmonic oscillator problems:

$$-\frac{\hbar^2}{2m}\frac{d^2\Phi}{dx^2} + \frac{1}{2}m\omega_1^2 x^2 \Phi = \epsilon_1 \Phi$$

$$-\frac{\hbar^2}{2m}\frac{d^2\chi}{dy^2} + \frac{1}{2}m\omega_2^2 y^2 \chi = \epsilon_2 \chi$$

$$-\frac{\hbar^2}{2m}\frac{d^2\xi}{dz^2} + \frac{1}{2}m\omega_3^2 z^2 \xi = \epsilon_3 \xi$$

with the stipulation $E = \epsilon_1 + \epsilon_2 + \epsilon_3$.

Now we can make use of the already known results for the one-dimensional harmonic oscillator and arrive at the energy levels and energy eigenfunctions of the three-dimensional oscillator.

$$E_{n_1,n_2,n_3} = \left(n_1 + \frac{1}{2}\right)\hbar\omega_1 + \left(n_2 + \frac{1}{2}\right)\hbar\omega_2 + \left(n_3 + \frac{1}{2}\right)\hbar\omega_3$$

(2.134)

$$\Psi_{n_1,n_2,n_3}(\theta_1,\theta_2,\theta_3) = C \cdot \exp\left(-\frac{\theta_1^2 + \theta_2^2 + \theta_3^2}{2}\right)H_{n1}(\theta_1)H_{n2}(\theta_2)H_{n3}(\theta_3)$$

where H_{ni} are Hermite polynomial with $n_i = 0, 1, 2, 3$ and $i = 1, 2, 3$. $\theta_1, \theta_2, \theta_3$ are dimensionless variable defined as $\theta_1 = x\frac{\sqrt{m\omega_1}}{\hbar}$, $\theta_2 = x\frac{\sqrt{m\omega_2}}{\hbar}$, $\theta_3 = x\frac{\sqrt{m\omega_3}}{\hbar}$. C is the normalization constant, which can be readily shown to be equal to

$$C = \frac{(m^3\omega_1\omega_2\omega_3)^{\frac{1}{4}}}{(\hbar\pi)^{\frac{3}{4}}(2^{n_1+n_2+n_3}n_1!n_2!n_3!)^{\frac{1}{2}}}$$

If the oscillator is completely isotropic, $k_1 = k_2 = k_3$ and therefore $\omega_1 = \omega_2 = \omega_3 = \omega$ (say) and

$$V(x,y,z) = \frac{1}{2}m\omega^2(x^2 + y^2 + z^2)$$

The energy levels are given (see Eq. 2.134) by

$$E_{n_1,n_2,n_3} = \left(n_1 + n_2 + n_3 + \frac{3}{2}\right)\hbar\omega; \quad n_i = 0, 1, 2, \cdots, \text{ and } i = 1, 2, 3$$

Clearly, the ground state corresponds $n_1 = n_2 = n_3 = 0$ and has energy $\frac{3}{2}\hbar\omega$, which is the zero-point energy of the three-dimensional harmonic oscillator. The next higher energy level is three-fold degenerate and has the energy $\frac{5}{2}\hbar\omega$ corresponding $n_1 = 1, n_2 = n_3 = 0$; $n_1 = n_3 = 0, n_2 = 1$; $n_1 = n_2 = 0, n_3 = 1$. These three-fold degenerate levels correspond to energy eigenfunction $\Psi_{100}, \Psi_{010}, \Psi_{001}$. The degeneracy is a consequence of the full rotational symmetry of the potential $V(x, y, z)$ and therefore of the Hamiltonian, which remains invariant under any rotation about an axis passing through the origin of the cartesian coordinate systems. This rotational symmetry is again reflected in the three-fold degeneracy of the next higher energy level with $E = \frac{7}{2}\hbar\omega$ corresponding to $n_1 = n_2 = 1, n_3 = 0$; $n_1 = n_3 = 1, n_2 = 0$; $n_1 = 0, n_2 = n_3 = 1$, each combination producing the same energy $E = \frac{7}{2}\hbar\omega$. The next higher level (the third excited state) turns out to a non-degenerate level corresponding to the choice $n_1 = n_2 = n_3 = 1$ for which $E = \frac{9}{2}\hbar\omega$. If we take the axis of motion to be z-axis any rotation about this axis leaves the isotropic oscillator Hamiltonian unchanged. We can expect therefore that L_z as also L^2 will be the quantum integrals of motion for the oscillator and hence are conserved quantities of the system. According to Heisenberg's formulation we have

$$\frac{dL_z}{dt} = \frac{i}{\hbar}[\hat{H}, \hat{L}_z]$$

$$\frac{dL^2}{dt} = \frac{i}{\hbar}[\hat{H}, \hat{L}^2]$$

(2.135)

where $\hat{H} = \frac{\hbar^2}{2m}\nabla^2 + \frac{1}{2}m\omega^2(x^2 + y^2 + z^2)$. Note that L_z as well as L^2 have no explicit time-dependence. It is straightforward so show that $[\hat{H}, \hat{L}_z] = 0$ and $[\hat{H}, \hat{L}^2] = 0$ so that z-component of angular momentum as well as the square of angular momentum are conserved quantities for a three-dimensional isotropic oscillator. We would encourage the reader to investigate symmetry properties of the completely general 3D oscillator and try to establish the kind of degeneracy, if any in their energy levels.

2.7.4 Quantum States in Infinitely Deep Potential Wells

Let us consider a particle of mass 'm' confined in a one-dimensional potential well of width 'a'. At the points $x = 0$ and $x = a$ on the base of the well, we can imagine that absolutely impenetrable walls are erected. The particle would therefore get reflected at the walls. We can graphically display the situation with the help of a potential energy curve illustrated in Figure 2.5.

The potential energy $V(x) = \infty$ at the $x < 0$ and $x > a$. Inside the well we have set $v(x) = 0$, for $0 \leq x \leq a$. If the particle has to escape from the region $0 \leq x \leq a$, it must do an infinitely large amount of work, which is an impossibility. Therefore, the probability for the particle to be at $x = 0$ or $x = a$ (i.e. at the walls) is zero. That means, the boundary conditions the wavefunction must obey are

$$\Psi(0) = \Psi(a) = 0$$

(2.136)

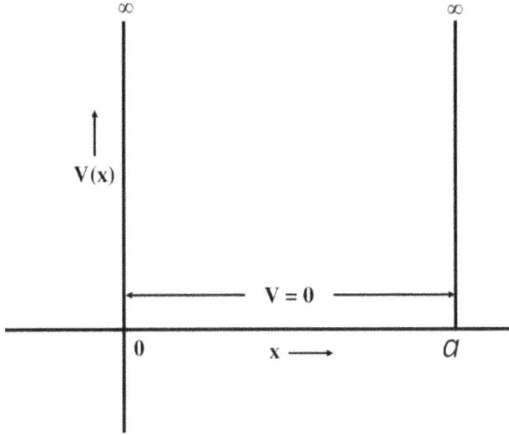

FIGURE 2.5 The potential energy curve of an infinitely deep potential well. Note the abrupt rise of the potential energy at $x = 0$, a (compare potential energy curve of a harmonic well).

The time-independent Schrödinger equation for the problem reads

$$-\frac{\hbar^2}{2m}\frac{d^2\Psi}{dx^2} = E\Psi \tag{2.137}$$

We have used total derivative $\frac{d^2}{dx^2}$ as the problem is in one dimension. We can rearrange Eq. 2.137 to read

$$\frac{d^2\Psi}{dx^2} = -\beta^2\Psi \tag{2.138}$$

where $\beta^2 = \frac{2mE}{\hbar^2}$. The solution of this equation can be immediately written down as

$$\Psi(x) = A_1 \sin\beta x + A_2 \cos\beta x \tag{2.139}$$

In view of the boundary condition that $\Psi(0) = 0$, the cosine component of Ψ must vanish so that we are compelled to set $A_2 = 0$. That means the wavefunction takes the form

$$\Psi(x) = A_1 \sin\beta x \tag{2.140}$$

There is, however, a second boundary condition, which demands $\Psi(a) = A_1 \sin\beta a = 0$. This equation in '$\beta$' has infinitely many solutions in the form

$$\beta a = n\pi \tag{2.141}$$

where n is a non-zero integer with $1 \leq n \leq \infty$. $n = 0$ is discarded as it will force $\Psi(x)$ to be zero everywhere so that the particle can not be found anywhere on the real axis ($-\infty \leq x \leq \infty$). Returning to Eq. 2.141 we find that

$$\beta^2 a^2 = n^2\pi^2$$

$$\text{or, } \frac{2mEa^2}{\hbar^2} = n^2\pi^2$$

$$\text{or, } E_n = \frac{\pi^2\hbar^2 n^2}{2ma^2} \tag{2.142}$$

$$= \frac{h^2 n^2}{8ma^2}; \quad n = 1, 2, 3....$$

Three points worth special mention: (i) The boundary condition imposed on the wavefunction is essential for finding the allowed energy spectrum. (ii) Not all values of energies qualify for the energy spectrum. In this case the spectrum is discrete and is defined by a definite set of numbers (1, 2, 3, ...), which arise from the characteristics of the problem automatically. In other words, the character of the motion itself determines the spectrum. (iii) The allowed energy levels are positive (>0), an outcome that follows from the boundary conditions. Negative energy levels would lead to exponential wavefunctions, which will not vanish at $x = \pm\infty$.

The constant A_1 in Eq. 2.140 is a normalization constant that can be fixed by the equality

$$\int_0^a |\Psi|^2 dx = A_1^2 \int_0^a \sin^2 \beta x\, dx = 1$$

which leads to

$$A_1 = \sqrt{\frac{2}{a}}$$

The normalized stationary state wavefunctions of the problem are therefore completely defined as follows:

$$\Psi_n(x) = \sqrt{\frac{2}{a}} \sin\left(\frac{n\pi x}{a}\right), \quad n = 1, 2, 3.... \tag{2.143}$$

These wavefunctions are real and therefore the probability current in these states $[j' = \frac{i\hbar}{2m}(\Psi\nabla\Psi^* - \Psi^*\nabla\Psi)]$ is zero everywhere. This feature is entirely expected on physical grounds. The wavefunction $\Psi_n(x)$ can be written as a sum of two exponentials

$$\Psi_n(x) = \frac{1}{2i}\left[\exp\left(\frac{in\pi x}{a}\right) - \exp\left(-\frac{in\pi x}{a}\right)\right].$$

When multiplied with the time-dependent phase factor $\exp(\frac{-iE_n t}{\hbar})$, each exponential represents a free particle wavefunction. One of these carries a momentum $p = \hbar\beta$ while the other carries a momentum $p = -\hbar\beta$. Thus $\Psi_n(x)$ is an equal superposition of two states with momentum of equal magnitude but opposite sign. The average quantum mechanical momentum of the particle trapped in an infinite potential well is therefore zero. In the classical case also, the average momentum of the particle can be seen to be zero, because the momentum changes sign at every reflection from the walls of the potential well. There is an important difference between the two cases. In the case of classical motion, the particle has a definite momentum at every instant of time. In the quantum case, at any instant of time, the energy eigenstate is a superposition of two states of equal but opposite momentum, the amplitude of each state in the superposition being equal. Thus no definite value of momentum can be assigned to such states ($\Psi_n(x)$) including zero momentum. A little reflection will assure the reader that this feature is consistent with the 'dictat' of position-momentum uncertainty principle. The coordinate of the particle is restricted only by the condition $0 \leq x \leq a$, so that the uncertainty $\Delta x \simeq a$ is a finite quantity making it impossible for the momentum to have a definite value, i.e. the uncertainty is infinite.

If the particle is a three-dimensional box being confined by edges of length a, b and c respectively along the x, y and z axes. That means, the particle is confined to the space bounded by condition $0 \leq x \leq a$, $0 \leq x \leq b$, $0 \leq x \leq c$. We assume the walls to be absolutely impenetrable so that the wavefunction of the particle must vanish on all the sides of the box. The boundary condition on $\Psi(x, y, z)$ would therefore be

$$i)\Psi(0, y, z) = \Psi(x, 0, z) = \Psi(x, y, 0) = 0;$$

$$ii)\Psi(a, y, z) = \Psi(x, b, z) = \Psi(x, y, c) = 0$$

Inside the box, the potential energy is zero everywhere. The motion being finite, we expect discrete bound states to appear as the eigenvalues (E) of the following wave equation (the so called time-independent Schrödinger equation):

$$-\frac{\hbar^2}{2m}\left(\frac{\partial^2 \Psi}{\partial x^2} + \frac{\partial^2 \Psi}{\partial y^2} + \frac{\partial^2 \Psi}{\partial z^2}\right) = E\Psi(x, y, z)$$

Using the result from the particle in a one-dimensional infinitely deep potential well, we seek the solution in the following form

$$\Psi(x, y, z) = A \cdot \sin\beta_1 x \cdot \sin\beta_2 y \cdot \sin\beta_3 z$$

This choice for Ψ ensures that $\Psi(0,y,z) = \Psi(x,0,z) = \Psi(x,y,0) = 0$ (boundary condition). The satisfaction of the other boundary condition that $\Psi(a,y,z) = \Psi(x,b,z) = \Psi(x,y,c) = 0$ requires $\sin\beta_1 a = 0; \sin\beta_2 b = 0; \sin\beta_3 c = 0$; i.e., $\beta_1 a = n_1\pi; \beta_2 b = n_2\pi; \beta_3 c = n_3\pi$ where n_1, n_2, n_3 are integer none of which are zero. The stationary state wave-function are therefore of the form

$$\Psi = c \cdot \sin\frac{n_1\pi x}{a} \cdot \sin\frac{n_2\pi y}{b} \cdot \sin\frac{n_3\pi z}{c}$$

Partially differentiating both sides twice with respect to x,y,z and multiplying the result with $-\frac{\hbar^2}{2m}$ and adding up we arrive at the following result:

$$-\frac{\hbar^2}{2m}\left(\frac{\partial^2\Psi}{\partial x^2} + \frac{\partial^2\Psi}{\partial y^2} + \frac{\partial^2\Psi}{\partial z^2}\right) = \frac{\hbar^2}{2m}\left(\frac{n_1^2\pi^2}{a^2} + \frac{n_2^2\pi^2}{b^2} + \frac{n_3^2\pi^2}{c^2}\right)\Psi$$

In order for Ψ to be solution of the wave equation we must have

$$\frac{\hbar^2}{2m}\left(\frac{n_1^2\pi^2}{a^2} + \frac{n_2^2\pi^2}{b^2} + \frac{n_3^2\pi^2}{c^2}\right)\Psi = E\Psi$$

$$\text{or, } E(n_1,n_2,n_3) = \frac{h^2}{8m}\left(\frac{n_1^2}{a^2} + \frac{n_2^2}{b^2} + \frac{n_3^2}{c^2}\right);$$

with $n_1 = 1, 2, 3, \cdots$; and $n_2 = 1, 2, 3, \cdots$; and $n_3 = 1, 2, 3, \cdots$.

The ground state corresponds to the choice $n_1 = n_2 = n_3 = 1$ and the energy is $E_{111} = \frac{h^2}{8m}(\frac{1}{a^2} + \frac{1}{b^2} + \frac{1}{c^2})$. Are there degeneracies? That depends on whether a, b, c are equal. The ground state is non-degenerate even if $a = b = c$. If $a = b = c$, the first excited state has three-fold degeneracy corresponding to the choice $n_1 = 2, n_2 = 1 = n_3; n_1 = 1, n_2 = 2, n_3 = 1$ and $n_1 = 1, n_2 = 1, n_3 = 2$, the degeneracy arising because of the cubic symmetry of the box. The energy $E_{211} = E_{121} = E_{112} =$ is given by $E = \frac{h^2}{8ma^2}(2^2 + 1^2 + 1^2) = \frac{6h^2}{8ma^2}$. This degeneracy disappears if the cubic symmetry is destroyed by making $a \neq b \neq c$. A point to note is that $E = 0$ is excluded as that would require $n_1 = n_2 = n_3 = 0$, which would require Ψ to vanish everywhere. That the zero energy solution is impossible also follows from the position momentum uncertainty principle. A particle confined in a well with finite dimension (a, b, c) can not have strictly defined momentum. That means zero-momentum state does not occur in the box which in turn precludes the existence of a zero energy stationary state in the three-dimensional particle in a box problem with completely impenetrable walls (or in an infinitely deep three-dimensional potential well problem).

The allowed quantum states of a particle in an infinitely deep three-dimensional potential well are important quantities as we will see later while calculating the number of possible states in a given energy or momentum interval or the density of states, in the context of statistical mechanical treatment of electron states in metals.

2.8 Representations in Quantum Mechanics

2.8.1 Coordinate Representation

So far, we have used Schrödinger formulation of quantum mechanics in solving particular problems. Let us recall that the wavefunction in the Schrödinger equation is a time-dependent function of coordinates $\Psi(x)$ denoting how probability amplitudes are distributed in space and time. The operators for position and momentum in this formulation are $\hat{x}$ and $\hat{p}_x = \frac{\hbar}{i}\frac{\partial}{\partial x}$, which are both free of time. One of the two, i.e, is $\hat{x}$ is an arbitrary c-number (multiplication by x) that follows the algebra of ordinary numbers ($x\, x' = x' x$) while the other $< \hat{p}_x >$ is a q-number, an operator that follows non-commutative algebra. The basic relation between the two is defined by the commutator

$$(\hat{p}\hat{x} - \hat{x}\hat{p}) = [\hat{p}, \hat{x}] = \frac{\hbar}{i}$$

$$[\hat{p}, \hat{x}]\Psi(x) = \frac{\hbar}{i}\Psi(x)$$

$\Psi(x)$ is thus an eigenfunction of the commutator $[\hat{p}, \hat{x}]$ with the eigenvalue $\frac{\hbar}{i}$. This formulation is referred to as the coordinate representation of quantum mechanics. We have solved the harmonic oscillator problem and a few others in this representation by converting the system Hamiltonian into a hermitian operator the eigenvalue equation for which is the time-independent Schrödinger equation. But this is not the only representation to work with.

2.8.2 Momentum Representation

An equivalent representation for quantum mechanics has been developed in which the characters of x and p have been interchanged. That means, in momentum representation p has been treated as a c-number while x has been treated as a q-number. The position operator in momentum representation takes the form[5]

$$\hat{x} = -\frac{\hbar}{i}\frac{\partial}{\partial p} \quad (q\text{-}number)$$

$$\hat{p} = p \quad (c\text{-}number)$$

with these definitions, it is straightforward to see that the commutator $[\hat{p}, \hat{x}]$ acts on the wavefunction $\Phi(p)$, which is now a function of momentum to produce the following eigenvalue equation:

$$[\hat{p}, \hat{x}]\Phi(p) = \frac{\hbar}{i}\Phi(p)$$

$|\Phi(p)|^2$ in this representation represents the probability density of observing a particle, the momentum of which lies in the range $p \leftrightarrow p + dp$. If we now substitute the new momentum space definition of $\hat{x}$ and $\hat{p}$ in the classical harmonic oscillator Hamiltonian

$$H = \frac{p^2}{2m} + \frac{1}{2}m\omega^2 x^2$$

it is transformed into the following hermitian Hamiltonian operator

$$\hat{H} = \frac{p^2}{2m} - \frac{1}{2}m\omega^2\hbar^2\frac{d^2}{dp^2}$$

The eigenvalue equation for $\hat{H}$ now reads

$$\left(\frac{p^2}{2m} - \frac{1}{2}m\omega^2\hbar^2\frac{d^2}{dp^2}\right)\Phi(p) = E\Phi(p)$$

$$or, \left(E - \frac{p^2}{2m} + \frac{1}{2}m\omega^2\hbar^2\frac{d^2}{dp^2}\right)\Phi(p) = 0$$

$$or, \left(E - ap^2 + b\frac{d^2}{dp^2}\right)\Phi(p) = 0$$

where $a = \frac{1}{2m}, b = \frac{1}{2}m\omega^2\hbar^2$.

Let us introduce now the following quantities,

$$\lambda_0 = \frac{E}{\sqrt{ab}} = \frac{2E}{\hbar\omega}$$

$$\text{and } p_0 = \sqrt{\frac{b}{a}} = \sqrt{m\hbar\omega}$$

$$\eta = \frac{p}{p_0}$$

whereby we arrive at the wave equation in p-representation:

$$\Phi'' + (\lambda_0 - \eta^2)\Phi = 0$$

where $\Phi'' = \frac{d^2\Phi}{d\eta^2}$. We can now proceed to solve the equation as we did in the case of x-representation and find that the eigenfunctions and eigenvalues are

$$E_n = (n + \frac{1}{2})\hbar\omega, \quad n = 0, 1, 2, \cdots$$

$$\text{and,} \quad \Phi_n(p) = \frac{(-i)^n}{\sqrt{2^n n! \sqrt{\pi} p_0}} \exp\left(-\frac{1}{2}\frac{p^2}{p_0^2}\right) H_n\left(\frac{p}{p_0}\right)$$

The eigenfunction $\{\Phi_n\}$ satisfy the orthonormality

$$\int_{-\infty}^{\infty} \Phi_{n'}^*(p)\Phi_n(p)dp = \delta_{n'n}$$

and can be shown to be the Fourier transform of the eigenfunctions $\Psi_n(x)$ of the oscillator in the coordinate representation.

2.8.3 Matrix Representation

We have already presented Heisenberg's matrix formulation of quantum mechanics with the harmonic oscillator as the model system. There we noted that the basic commutation relation between momentum (P) and coordinate (X) can be satisfied if we describe each by appropriate matrices (P) and (X), and consequently transform the classical Hamiltonian of the linear oscillator to a matrix (H). That is

$$(PX) - (XP) = \frac{\hbar}{i}$$

$$H = \frac{P^2}{2m} + \frac{1}{2}m\omega^2 X^2$$

We have already described how such a matrix formulation using the energy eigenfunctions as basis leads to energy levels of the oscillator $E_n = (n + \frac{1}{2})\hbar\omega$ and the X and P matrices. Since energy-basis was used, H turned out to be a diagonal matrix while X and P matrices were non-diagonal. Each matrix turns out to be of infinite dimension. Thus, with the linear oscillator as the example we have been able to show that all the three representations, viz, x, p and the matrix representation lead to the same energy levels for the oscillator. It has also been established that all three formulations are indeed completely equivalent. We are inclined to explore now if the basic laws of quantum mechanics can be formulated independent of any specific representation.

2.8.4 Vector Space Formulation

It begins with the idea of a state vector of a quantum mechanical system, supposedly belonging to an abstract linear space, called the Hilbert space, the state vector being completely characterized by a complete set of quantum numbers $\{n\}$ arising from the eigenvalues of all the operators (observables) that commute with the Hamiltonian operator of the system. The vector representation of the wavefunction Ψ – the state vector, is denoted by $|\Psi\rangle$ (this angular bracket representation is due to Dirac) and called the 'ket vector'.[7] It can also be represented as $|n\rangle$ where n is the collection of the complete set of M quantum numbers, $n_1, n_2, \cdots, n_M$. The conjugate vector, called the bra-vector is denoted as $\langle\Psi|$. Thus, we can make the following assignments

$$\Psi = \begin{pmatrix} a_1 \\ a_2 \\ . \\ . \\ . \end{pmatrix} = |\Psi\rangle \qquad (2.144)$$

$$\Psi^* = \begin{pmatrix} a_1^* & a_2^* & . & . \end{pmatrix} = \langle\Psi|$$

The conjugate vector $\langle\Psi|$ is called the dual vector to $|\Psi\rangle$ and its space is called the 'dual space'. The sum in the dual space results from the following mapping:

$$\Psi_c^* = \alpha^* \Psi_a^* + \beta^* \Psi_b^*$$
$$\langle c| = \alpha^* \langle a| + \beta^* \langle b| \qquad (2.145)$$

The scalar product of two vectors $|\Psi_1\rangle$ and $|\Psi_2\rangle$ or $|\Psi_a\rangle$ and $|\Psi_b\rangle$ in Dirac's notation takes the form: $\langle\Psi_1|\Psi_2\rangle = (\Psi_1, \Psi_2)$; $(\Psi_a, \Psi_b) = \langle a|b\rangle$. Thus defined the scalar product obey the relation $(\langle\Psi_1|\Psi_2\rangle)^* = (\Psi_1, \Psi_2)^* = \langle\Psi_2|\Psi_1\rangle$. Similarly, $(\langle a|b\rangle)^* = \langle b|a\rangle$.

$\Psi_n(x)$ in coordinate representation, as we have already noted, stands for the amplitude of the probability density of the particle being located at x. In the state vector representation

$$\Psi_n(x) = \langle x|n\rangle$$
$$\Psi_n^*(x) = \langle n|x\rangle \qquad (2.146)$$
$$\Psi_n(p) = \langle p|n\rangle$$

The quantities like $\langle x|n\rangle$ can be interpreted mathematically as the components of the vector $|n\rangle$ in the $|x\rangle$ basis. Thus

$$|n\rangle = \int |x\rangle \langle x|n\rangle \, dx$$
$$= \int |x\rangle \Psi_n(x) dx \qquad (2.147)$$

The probability density $|\Psi_n(x)|^2$ in this vector space (bra-ket) notation reads $|\Psi_n(x)|^2 = \langle n|x\rangle \langle x|n\rangle$. If we integrate both sides over 'x' we get the total probability P. Thus

$$P = \int |\Psi_n(x)|^2 dx$$
$$= \int \Psi_n^*(x)\Psi_n(x) dx$$
$$= \int \langle n|x\rangle \langle x|n\rangle \, dx \qquad (2.148)$$
$$= \langle n|n\rangle$$
$$= 1$$

The total probability is representation independent and it can be shown easily that

$$\int |\Psi_n(p)|^2 dp = \langle n|n\rangle = 1.$$

The orthogonality of wavefunction $\Psi'_n(x)$ and $\Psi_n(x)$ when transcribed in the language of the vector-space reads

$$\int \Psi^*_{n'}(x)\Psi_n(x)dx = \int \langle n'|x\rangle \langle x|n\rangle \, dx$$

$$= \langle n'|n\rangle = \delta_{n'n}$$

The all important completeness relation for the system of state vectors $|n\rangle$ is

$$\sum_n |n\rangle \langle n| = 1 \tag{2.149}$$

the sum being over all the possible values of the quantum numbers that 'n' can assume. The completeness ensures that any ket $|\Psi\rangle$ can be constructed as a superposition of the complete set of states:

$$|\Psi\rangle = 1 \cdot |\Psi\rangle = \sum_n |n\rangle \langle n|\Psi\rangle \tag{2.150}$$

We will now show that the harmonic oscillator problem can be solved without adopting any specific representation. The Hamiltonian operator of a harmonic oscillator is

$$H = \frac{p^2}{2m} + \frac{1}{2}m\omega^2 x^2 \tag{2.151}$$

We want to express H in terms of two operators a and $a^\dagger$ where

$$a = \frac{1}{\sqrt{2}}\left(\frac{x}{x_0} + i\frac{ix_0 p}{\hbar}\right)$$

$$a^\dagger = \frac{1}{\sqrt{2}}\left(\frac{x}{x_0} - i\frac{ix_0 p}{\hbar}\right) \tag{2.152}$$

where $x_0 = (\frac{\hbar}{m\omega_0})^{\frac{1}{2}}$. The hermiticity of $\hat{x}$ and $\hat{p}$ operators ensures that $a^\dagger$ and a are hermitian conjugates of each other. Furthermore, the basic commutation relation between $\hat{x}$ and $\hat{p}$ (i.e. $\hat{p}.\hat{x} - \hat{x}.\hat{p} = -i\hbar$) enables us to find that

$$[a,a^\dagger] = aa^\dagger - a^\dagger a = 1 \tag{2.153}$$

The definition of the operators a and $a^\dagger$, together with their commutation relation help us in expressing the oscillator Hamiltonian in terms of a and $a^\dagger$:[8-9]

$$\hat{H} = \hbar\omega\left(a^\dagger a + \frac{1}{2}\right)$$

$$= \hbar\omega\left(\hat{N} + \frac{1}{2}\right) \tag{2.154}$$

The eigenvalues of $\hat{H}$ can therefore be obtained from the eigenvalues of the operator $\hat{N} = a^\dagger a$. It is easy to show that an immediate consequence of the facts that $[a^\dagger,a] = 1$ and $[\hat{u}, \hat{v}\hat{w}] = [\hat{u}, \hat{v}]\hat{w} + \hat{v}[\hat{u}, \hat{w}]$ (u, v, w are operators) is

$$[\hat{N},\hat{a}] = -1$$

$$\text{i.e.} \quad \hat{N}\hat{a} = \hat{a}\hat{N} - 1 \tag{2.155a}$$

$$[\hat{N},\hat{a}^\dagger] = a^\dagger \tag{2.155b}$$

We will now proceed to demonstrate how the eigenvalues and eigenvectors of $\hat{N}$, which are supposed to satisfy the following eigenvalue equation can be found. Thus $\hat{N}$ and $|\lambda\rangle$ satisfy

$$\hat{N}|\lambda\rangle = \lambda|\lambda\rangle \tag{2.156}$$

where $|\lambda\rangle$ is a normalized state vector (ket) satisfying the normalization condition $\langle\lambda|\lambda\rangle = 1$. The demonstration involves the following steps and arguments.[8-9]

1. We prove first that the eigenvalues of $\hat{N}$ are greater than or equal to zero by writing λ as the inner product of $|\lambda\rangle$ and $N|\lambda\rangle$

$$\lambda = \langle \lambda | N | \lambda \rangle = \langle \lambda | a^\dagger a | \lambda \rangle \qquad (2.157)$$

Let $\hat{a}|\lambda\rangle = |\beta\rangle$, whence $\lambda = \langle \beta | | \beta \rangle \geq 0$. Clearly, the eigenvalue $\lambda = 0$ arises if and only if $\hat{a}|\beta\rangle = 0$.

2. In the second step we show that if $|\lambda\rangle$ is an eigenstate of $\hat{N}$, so also is $|\lambda - 1\rangle$ assuming that $\lambda \neq 0$. This follows since

$$\hat{N}\hat{a}|\lambda\rangle = (\hat{a}\hat{N} - 1)|\lambda\rangle = (\lambda - 1)\hat{a}|\lambda\rangle \qquad (2.158)$$

where we have used the result in Eq. 2.155a. Eq. 2.158 asserts that $\hat{a}|\lambda\rangle$ is an eigenvector of $\hat{N}$ associated with the eigenvalue $(\lambda - 1)$ and hence proportional to $|\lambda - 1\rangle$ unless $\lambda = 0$.

3. In this step we show that the series of eigenstates $|\lambda\rangle$, $|\lambda - 1\rangle$, $|\lambda - 2\rangle$, ... must terminate since the eigenvalue must be greater than or equal to zero as established in step 1. The eigenvalue at which the series mentioned terminates must therefore be zero because $a^\dagger a$ acting on the corresponding eigenket produces zero. Thus we have for the terminal eigenket

$$\hat{a}|0\rangle = 0 \qquad (2.159)$$

Note that there is only one linearly independent eigenvector $|0\rangle$ which is a solution of Eq. 2.159 .

4. Now we show that if $|\lambda\rangle$ be an eigenstate of $\hat{N}$, $|\lambda + 1\rangle$ is also an eigenstate of $\hat{N}$. This result follows from the commutation relation $[\hat{N}, a^\dagger] = a^\dagger$ and the eigenvalue equation $\hat{N}|\lambda\rangle = \lambda|\lambda\rangle$. Thus,

$$Na^\dagger \lambda = (a^\dagger N + a^\dagger)|\lambda\rangle = (\lambda + 1)a^\dagger|\lambda\rangle$$

The result above shows that $a^\dagger|\lambda\rangle$ is also an eigenstate of $\hat{N}$ with the eigenvalue $(\lambda + 1)$. $a^\dagger|\lambda\rangle$ must therefore be proportional to the ket $|\lambda + 1\rangle$. Let us note that $a^\dagger|\lambda\rangle$ can not be equal to zero because

$$aa^\dagger|\lambda\rangle = (1 + a^\dagger a)|\lambda\rangle = (\lambda + 1)|\lambda\rangle \neq 0$$

Summing up, we have been able to show that the eigenvalues of $\hat{N}$ are non-degenerate integers. We can assert that there are no eigenstates of $\hat{N}$ other than those associated with non-degenerate integers; had there been any such state, we could have easily generated eigenstates of $\hat{N}$ with negative eigenvalues by allowing $\hat{a}$ to act on such a state a sufficient number of times, contradicting the result obtained in step 1. The eigenvalues of $\hat{N}$ are therefore confirmed to be non-negative integer $(0, 1, 2, ..)$. From Eq. 2.154, it turns out therefore that the eigenvalues of $\hat{H}$ are given by

$$E_n = \hbar\omega\left(n + \frac{1}{2}\right), \quad n = 0, 1, 2, \cdots$$

as $\hat{H} = \hbar\omega(\hat{N} + \frac{1}{2})$ echoing what we found with coordinate, momentum or energy representations. The ground state energy is $E_0 = \frac{1}{2}\hbar\omega$ corresponding to the choice $n = 0$, reflecting the zero-point energy arising out of the operation of Heisenberg's uncertainty principle. The representation independent formulation is very appealing though in many occasions, a particular representation may be more convenient for handling a specific problem.

REFERENCES

1. L. Pauling and E. B. Wilson, *Introduction to Quantum Mechanics: With Application to Chemistry*, McGraw-Hill, New York and London (1935).
2. H. Eyring, J. Walter and G. E. Kimball, *Quantum Chemistry*, John Wiley, New York and London (1944).
3. H. A. Kramers, *Quantum Mechanics*, (Translated by D. Ter Haar), Dover Publications, Mineola, New York (1957).
4. A. S. Kompaneyets, *A Course of Theoretical Physics*, Vol-I (Translated from Russian by V. Talmy), MIR Publishers, Moscow (1978).
5. F. Schwebl, *Quantum Mechanics*, (Translated by Ronald Kates), Springer International Student Edition: First Narosa Publishing House Reprint, New Delhi (1992).

6. A. A. Sokolov, I. M. Ternov and V. Ch. Zhukovskii, *Quantum Mechanics*, (Translated from the Russian by Ram S. Wadhwa), MIR Publishers, Moscow (1984).

7. P. Dirac, *Principles of Quantum Mechanics*, Oxford University Press, London (1930).

8. H. Smith, *Introduction to Quantum Mechanics*, World Scientific, Singapore (1991).

9. A. Mesiah, *Quantum Mechanics*, Dover Publications, Mineola, New York (1999).

3

Quantum Mechanics of Atoms

3.1 Introduction

The ultimate constituents of all materials – elemental or those based on chemical compounds are atoms or ions. The diverse properties of materials may therefore be rooted in the properties of the constituent atoms (ions) and interactions (bonding) among them. Different elements have no-doubt different properties and distinct chemical and physical signatures. Nevertheless, in such groups of elements it is found that within the group the elements are much more similar than across the groups. Thus, the group called alkali metals have elements (Li, Na, K, Rb, Cs, $\cdots$), which are very similar to each other while the group called halogens contains elements (F, Cl, Br, I, At) that are quite similar. The alkali metals are however very different from the elements in the halogen group and interact across the group forming ionic compounds, called metallic halides (NaCl, KCl, for example), which are important materials. There are the boron group of elements (B, Al, Ga, In, Tl), carbon group (C, Si, Ge, Sn, Pb), oxygen group (O, S, Se, Te, $\cdots$), the transition metals (Sc, Ti, V, Cr, Mn, Fe, Co, Ni, Cu, $\cdots$), the group of 14 very similar elements called rare earths, the alkali earth group of elements (Be, Mg, Ca, Sr, Ba) and so on. Why are certain elements so similar? How do properties of elements vary within a group and why? The search for ways to systematize the empirically observed similarities leads to the developement of the most important classification system in chemistry. This classification system has been known as the periodic table of elements.

3.2 The Periodic Table of Elements

Mendeleev successfully arranged the 63 elements known at that time in increasing order of their atomic weights (A) and discovered that in the table the same chemical properties reappear after a certain number of elements.[1] He asserted that physical and chemical properties of elements are periodic functions of their atomic weights and his table of elements came to be known as the periodic table. Mendeleev also assigned an atomic number (Z) to each element, which purportedly defined the position of the element in his table. Mendeleev used his table with telling effect and succeeded in predicting the existence of ten new elements. For three of them, namely Scandium ($Z = 21$), Gallium ($Z = 31$) and Germanium ($Z = 32$) he even predicted their physical and chemical properties. Subsequent investigations of X-ray spectra of elements and scattering of alpha particles by atoms proved conclusively that the atomic number 'Z' defines the nuclear charge carried by the atoms of the elements and fixes the number of extra nuclear electrons carried by the atom. Although the increase in Z and A occurs simultaneously, Mendeleev's table did have some awkward inversions Ar ($Z = 18$, $A = 40$), K ($Z = 19$, $A = 39$), Te ($Z = 52$, $A = 128$), I ($Z = 53$, $A = 127$). Discovery of isotopes and their abundances in naturally occuring elements however, soon resolved the issue. It became clear that properties of elements are a periodic function of their atomic number (Z) and not atomic mass (A). Since chemical properties of elements are shaped by the number of extranuclear electrons, specially the so called valence electrons, and how they are distributed within the atoms, the observed periodicity of elemental properties appear to be connected with recurrence of similar types of electron distribution at regular intervals of Z. It is imperative therefore to try to understand variations in the electronic structure of atoms across the periodic table. Such an understanding can only come from application of quantum mechanics to elucidate the stationary electronic states of different atoms, specially

DOI: 10.1201/9781003244882-3

the ground state and how the Z number of electrons are distributed among such states. Much of this understanding follows from our knowledge of the electronic structure of hydrogen – the simplest of all atoms the Schrödinger equation for which has been exactly solved.

3.3 The Quantum States of the Hydrogenic Atoms: Symmetry

The single electron in a hydrogen atom moves in the Coulomb potential $V(r)$ produced by the interaction of negatively charged electron $(-e)$ with the positively charged nucleus $(+Ze)$ where

$$V(r) = -\frac{Ze^2}{r} \tag{3.1}$$

We assume the massive positively charged nucleus to be at the origin of the coordinate system used and held at rest, r being the distance separating the electron from the nucleus (Figure 3.1). Clearly, $r = \sqrt{x^2 + y^2 + z^2}$.

The quantum mechanical Hamiltonian operator describing the system in this system of coordinate reads

$$\hat{H}(\hat{x}, \hat{y}, \hat{z}) = -\frac{\hbar^2}{2m_e}\left(\frac{\partial^2}{\partial x^2} + \frac{\partial^2}{\partial y^2} + \frac{\partial^2}{\partial z^2}\right) - \frac{Ze^2}{\sqrt{\hat{x}^2 + \hat{y}^2 + \hat{z}^2}} \tag{3.2}$$

The potential $V(r)$ is manifestly spherically symmetric and called a central field potential as the magnitude of it is entirely determined by the distance of the electron from the center of a sphere. It does not matter where on the surface of a sphere of radius r and center at O $(0, 0, 0)$, the electron is – the value of the potential is the same for all the values of the spherical angles θ and ϕ for any given value of r. The potential $V(r)$ is displayed in Figure 3.2. Note that $V(r) \to 0$ as $r \to \infty$ but $V(r)$ has a singularity at $r = 0$ where $V(r) \to -\infty$. The singularity does not however destroy bound states as can be justified from a simple argument based on the uncertainty principle; close to the center of the potential the uncertainty in position of the electron is small, say 'a' (a being the radius of a very small sphere with the center at $r = 0$). The corresponding uncertainty Δp in momentum is $\frac{\hbar}{2a}$. The symmetry of the system dictates that $<p> = 0$ so that $\Delta p^2 = \frac{\hbar^2}{4a^2} = <p^2>$. The average kinetic energy is $\frac{<p^2>}{2m} = \frac{\hbar^2}{8ma^2}$ while the average potential energy is $-\frac{Ze^2}{a}$. The average total energy $<E>$ is therefore given by $<E> = \frac{\hbar^2}{8ma^2} - \frac{Ze^2}{a}$. $<E>$ clearly does not become $-\infty$ as $a \to 0$ so that all bound states are not destroyed and the system does not collapse. Minimization $<E>$ with respect to 'a', leads to

$$E_{min} = -\frac{mZ^2e^4}{\hbar^2} < 0 \tag{3.3}$$

corresponding to the value of a

$$a_{min} = \frac{\hbar^2}{2me^2Z} \tag{3.4}$$

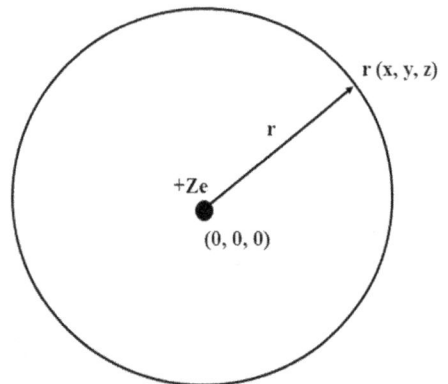

FIGURE 3.1 An electron (e^-) in a spherically symmetric potential produced by its interaction with a nucleus carrying charge +Ze located at the origin $(0, 0, 0)$ of the cartesian coordinate system.

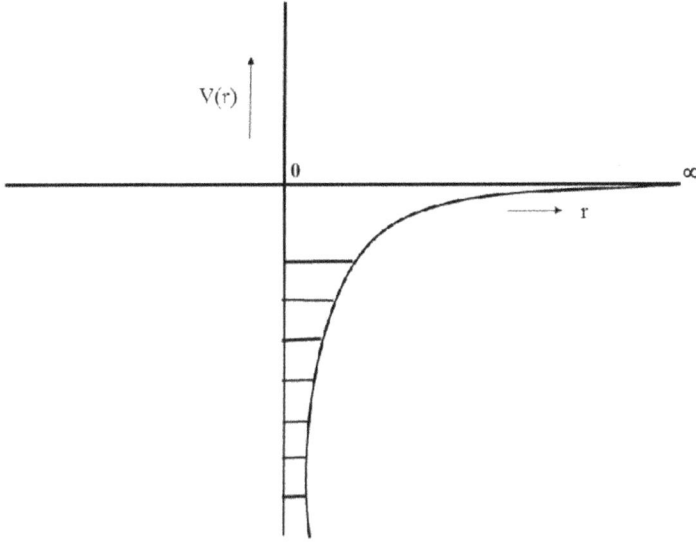

FIGURE 3.2 Coulomb potential $V(r)$ felt by an electron in an atom and states supported by it.

We can anticipate therefore that many bound states of hydrogen or hydrogenic atoms would exist between 0 and E_{min}. Before we turn our attention to the problem of solving the Schrödinger equation of the hydrogen atom and find the stationary states of the system, let us consider a special symmetry of the hydrogen atom, called the inversion symmetry. The signature of the symmetry can be recognized in the Hamiltonian $H(x, y, z)$. Let us consider the spatial Inversion operator I which takes (x, y, z) into $(-x, -y, -z)$. With this definition of $\hat{I}$, and $\hat{H}(x,y,z)$ as defined in equation Eq. 3.2, we have the result that[2]

$$\hat{I}H(x,y,z) = H(-x,-y,-z) \tag{3.5}$$

The Coulomb potential $V(x, y, z)$ is invariant under inversion as can be seen easily:

$$\hat{I}V(x,y,z) = V(-x,-y,-z)$$

$$= -\frac{Ze^2}{\sqrt{\hat{x}^2 + \hat{y}^2 + \hat{z}^2}} \tag{3.6}$$

$$= V(x,y,z)$$

The same holds true for the kinetic energy operator: $\hat{I}\hat{T}(x,y,z) = -\frac{\hbar^2}{2m}(\frac{\partial^2}{\partial x^2} + \frac{\partial^2}{\partial y^2} + \frac{\partial^2}{\partial z^2}) = \hat{T}(x,y,z)$. Therefore, $\hat{H}(x,y,z)$, which is a sum of $\hat{T}$ and $\hat{V}$, is also invariant under inversion:

$$\hat{I}\hat{H}(x,y,z) = \hat{H}(-x,-y,-z) = \hat{H}(x,y,z) \tag{3.7}$$

As seen before, this invariance of $\hat{H}$ under $\hat{I}$ implies that $\hat{H}$ and $\hat{I}$ commute and have common or simultaneous eigenfunctions. That means the stationary states of an electron in an atom like hydrogen are characterized by a 'property' that does not have a classical analogue. The 'property' is intrinsic to the wavefunction $\Psi(x,y,z)$ of the electron itself. When the signs of x, y, z are all changed simultaneously in a process called spatial inversion, the process transforms a right-handed coordinate system into one that is left-handed and vice-versa. The two coordinate systems can not be brought into coincidence by any imaginable spatial rotations. Therefore, the symmetry being probed is not a rotational symmetry, but something new, called inversion symmetry. Let us consider the energy eigenvalue of equation

$$H\Psi(x,y,z) = E\Psi(x,y,z) \tag{3.8}$$

The equation is manifestly linear, and does not change form under inversion. The solutions of Eq. 3.8 determined by the imposition of appropriate boundary condition upto a constant factor can only acquire an additional factor under inversion. That is

$$\hat{I}\Psi(x,y,z) = a\Psi(x',y',z') = a\Psi(-x,-y,-z) \tag{3.9}$$

The primed system (left-handed) is in every way similar to the unprimed right-handed system. Therefore, the reverse transformation must also involve the same transformation[3] factor 'a'. That means

$$\Psi(x',y',z') = \Psi(-x,-y,-z) = a\Psi(x,y,z) \tag{3.10}$$

Using Eq. 3.10 in Eq. 3.9, we get

$$\Psi(x,y,z) = a^2\Psi(x,y,z) \tag{3.11}$$

whence we conclude that $a^2 = 1$, $a = \pm 1$.

Thus, the inversion operator I has two kinds of eigenfunctions; one with $a = +1$ whence $\Psi(-x,-y,-z) = 1 \cdot \Psi(x,y,z)$ (Eq. 3.10) and the other with $a = -1$, which means $\Psi(-x,-y,-z) = -1 \cdot \Psi(x,y,z) = -\Psi(x,y,z)$.

The values of 'a' define the parity quantum label or number with $a = +1$ being called even parity and $a = -1$ being termed odd parity eigenfunction of the inversion operator. Since $\hat{H}$ and $\hat{I}$ commute, the eigen functions of $\hat{H}$ must also carry these parity labels or quantum numbers. That means, the eigenfunctions of the hydrogen or hydrogenic Hamiltonian will fall into two classes, even and odd parity, a conclusion we have reached by examining the inversion symmetry of the Coulomb Hamiltonian without solving Eq. 3.8. The parity labels are good quantum numbers as long as the inversion symmetry remains intact. Does the hydrogen atom have any other symmetry? Spherical symmetry of the potential suggests that the hydrogen atom and its states have rotational symmetries that a sphere has. The investigation of rotational symmetries can be undertaken conveniently by changing over from Cartesian to the spherical coordinate system. In fact, Eq. 3.8 is not separable in Cartesian coordinate system, but is separable in spherical coordinates. The process of separation brings into open the rotational symmetries of the eigenstates of the hydrogen and hydrogenic atoms.

3.4 Rotational Symmetry, Angular Momentum, Eigenstates and Parity

In Cartesian coordinates, the energy eigenvalue equation of the Hamiltonian ($H(x, y, z)$) of the hydrogen and hydrogenic atoms reads

$$\left\{ -\frac{\hbar^2}{2m}\left(\frac{\partial^2}{\partial x^2} + \frac{\partial^2}{\partial y^2} + \frac{\partial^2}{\partial z^2}\right) - \frac{Ze^2}{\sqrt{x^2+y^2+z^2}} \right\}\Psi(x,y,z) = E\Psi(x,y,z)$$

$$-\frac{\hbar^2}{2m}\nabla^2\Psi(x,y,z) - \frac{Ze^2}{\sqrt{x^2+y^2+z^2}}\Psi(x,y,z) = E\Psi(x,y,z) \tag{3.12}$$

The potential energy function is spherically symmetric and is invariant under any rotation about any axis passing through the center of the potential. We can anticipate that the transformation to spherical polar coordinate will reveal the rotational symmetry. The transformation to the spherical polar system of coordinates (r, θ and ϕ) (see Figure 3.3) can be performed using the following relations

$$x = r.\cos\theta\sin\phi$$

$$y = r.\sin\theta\sin\phi$$

$$z = r.\cos\theta$$

$$r = \sqrt{x^2+y^2+z^2} \tag{3.13}$$

$$dxdydz = r^2\sin\theta drd\theta d\phi$$

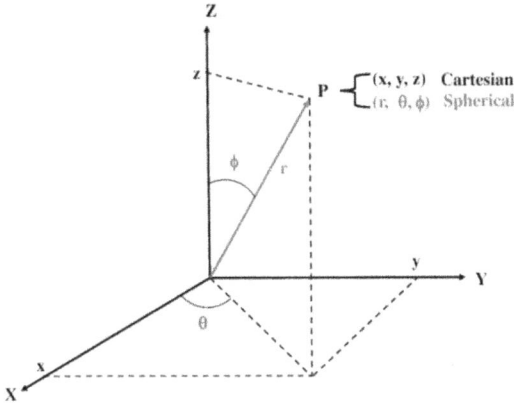

FIGURE 3.3 Spherical polar system of coordinates.

Eq. 3.12 thus gets converted into an equivalent equation in spherical coordinates. The steps are straight-forward:

Transforming the Laplacian $\nabla^2(x,y,z)$ to spherical polar form, we get

$$\nabla^2 = \frac{1}{r^2}\frac{\partial}{\partial r}\left(r^2\frac{\partial}{\partial r}\right) + \frac{1}{r^2}\left[\frac{1}{\sin\theta}\frac{\partial}{\partial\theta}\left(\sin\theta\frac{\partial}{\partial\theta}\right) + \frac{1}{\sin^2\theta}\frac{\partial^2}{\partial\phi^2}\right]$$

$$\nabla^2 = \nabla_r^2 + \frac{1}{r^2}\nabla_{\theta,\phi}^2 \qquad (3.14)$$

where $\nabla_r^2 = \frac{1}{r^2}\frac{\partial}{\partial r}(r^2\frac{\partial}{\partial r})$, $\nabla_{\theta,\phi}^2 = \frac{1}{\sin\theta}\frac{\partial}{\partial\theta}(\sin\theta\frac{\partial}{\partial\theta}) + \frac{1}{\sin^2\theta}\frac{\partial^2}{\partial\phi^2}$.

The Schrödinger equation for hydrogenic atoms now reads

$$-\frac{\hbar^2}{2m}\left[\nabla_r^2 + \frac{1}{r^2}\nabla_{\theta,\phi}^2\right]\Psi(r,\theta,\phi) + V(r)\Psi(r,\theta,\phi) = E\Psi(r,\theta,\phi) \qquad (3.15a)$$

with $V(r) = \frac{-Ze^2}{r}$

$$\left(-\frac{\hbar^2}{2m}\nabla_r^2 - \frac{\hbar^2}{2mr^2}\nabla_{\theta,\phi}^2 + V(r)\right)\Psi(r,\theta,\phi) = E\Psi(r,\theta,\phi) \qquad (3.15b)$$

In a more compact notation Eq. 3.15 takes the form

$$\nabla^2\Psi(r,\theta,\phi) + k^2(r)\Psi(r,\theta,\phi) = 0 \qquad (3.16)$$

where

$$k^2(r) = \frac{2m}{\hbar^2}(E - V(r)) \qquad (3.17)$$

Eq. 3.16 can be separated in r, θ, ϕ coordinates by writing

$$\Psi(r,\theta,\phi) = R(r)Y(\theta,\phi) \qquad (3.18)$$

where $R(r)$ is the radial part of the wave function and $Y(\theta,\phi)$ is the part of the wavefunction that depends on the spherical angles θ, ϕ. To proceed further with the separation of radial and angular motions, we multiply Eq. 3.16 by $\frac{r^2}{RY}$ and use the fact that $\nabla^2 = \nabla_r^2 + \frac{1}{r^2}\nabla_{\theta,\phi}^2$ whence we arrive at

$$r^2\frac{\nabla_r^2 R(r)}{R(r)} + r^2 k^2 = -\frac{\nabla_{\theta,\phi}^2 Y}{Y} \qquad (3.19)$$

Note that the left hand side of Eq. 3.19 depends only on the radial coordinate (r) while the right hand side depends only on the angular variables θ, ϕ. The equality of two sides in Eq. 3.19 therefore can hold only if each side is equal to a constant 'β' independent of r, θ, ϕ called the separation constant. This separation process thus leads to the following two equations[4]

$$\nabla_r^2 R(r) + \left(k^2 - \frac{\beta}{r^2} \right) R(r) = 0 \tag{3.20a}$$

$$(\nabla_{\theta,\phi}^2 + \beta) Y(\theta, \phi) = 0 \tag{3.20b}$$

Although we have defined $V(r) = -\frac{Ze^2}{r}$, Eq. 3.15 and Eq. 3.16 and therefore Eq. 3.20 do not depend on the exact form of the potential function $V(r)$, $V(r)$ is only required to be a central field potential like the Coulomb potential. A further separation of the spherical angular variables can be carried by substituting $Y(\theta, \phi) = \chi(\theta)\Phi(\phi)$ in Eq. 3.20b leading to Eq. 3.21a,b.

$$\left\{ \frac{1}{\sin\theta} \frac{d}{d\theta} \left(\sin\theta \frac{d}{d\theta} \right) + \left(\beta - \frac{m^2}{\sin^2\theta} \right) \right\} \chi(\theta) = 0 \tag{3.21a}$$

$$\frac{d^2\Phi(\phi)}{d\phi^2} + m^2 \Phi(\phi) = 0 \tag{3.21b}$$

In a more compact notation the two equations read

$$\nabla_\theta^2 \chi + \left(\beta - \frac{m^2}{\sin^2\theta} \right) \chi = 0 \tag{3.22a}$$

$$(\nabla_\phi^2 + m^2)\phi = 0 \tag{3.22b}$$

where $\nabla_\phi^2 = \frac{d^2}{d\phi^2}$ and $\nabla_\theta^2 = \frac{1}{\sin\theta} \frac{d}{d\theta}(\sin\theta \frac{d}{d\theta})$.

Note that we have used total derivative $\frac{d}{d\theta}$ and $\frac{d}{d\phi}$ in place of partial derivatives in defining ∇_ϕ^2 and ∇_θ^2 as the former operates on functions of θ and the latter on functions on ϕ only. m^2 is the separation constant. Let us first take up Eq. 3.21b as it contains only one separation constant (m^2). It is easy to find that $\Phi(\phi) = A \cdot \exp(im\phi)$ is a particular solution of the equation concerned. Since Φ should be unique (single valued), $\Phi(\phi)$ must return to the same value as $\phi \to \phi + 2\pi$. That requires $\Phi(\phi + 2\pi) = \Phi(\phi)$, i.e. $A \cdot \exp(im(\phi + 2\pi)) = A \cdot \exp(im\phi)$, which is possible if $\exp(im2\pi) = 1$. That holds only if m = 0, ± 1, ± 2, ... The normalization constant 'A' can be determined by the requirement

$$A^2 \int_0^{2\pi} \Phi^*(\phi)\Phi(\phi)d\phi = 1$$

whence $A = \frac{1}{\sqrt{2\pi}}$.

With this result we have completely specified the ϕ-dependent part of the hydrogenic wavefunction:

$$\Phi(\phi) = \frac{1}{\sqrt{2\pi}} \exp(im\phi), \quad m = 0, \pm 1, \pm 2, \ldots$$

m is clearly the eigenvalue of the angular moment operator $\frac{d^2}{d\phi^2}$ and is called the magnetic quantum number (see later). $\Phi(\phi) = \frac{1}{\sqrt{2\pi}} \exp(im\phi)$ represents a wave traveling along the circumference of a circle.

Next we turn to determine the θ-dependent part of the wavefunction $\chi(\theta)$, which satisfies Eq. 3.22a. Note that both the separation constants β and m appear in this equation, but 'm' has already been determined. We proceed by letting $y = \cos\theta$ in Eq. 3.21a whence a new[4] differential equation is obtained:

$$\frac{d}{dy}\left[(1-y^2)\frac{d\chi}{dy} \right] + \left(\beta - \frac{m^2}{1-y^2} \right)\chi = 0 \tag{3.23}$$

Eq. 3.23 has singularities at $y = \pm 1$. The divergence at $y = \pm 1$ can be eliminated by seeking the solution in the following forms

$$\chi = (1 - y^2)^{\frac{t}{2}} u \tag{3.24}$$

Using Eq. 3.24 in Eq. 3.23 and dividing both sides by $(1 - y^2)^{\frac{t}{2}}$, we arrive at the following differential equation for the unknown function u:

$$(1 - y^2)\frac{d^2u}{dy^2} - 2t(t+1)\frac{du}{dy} + \left[\beta - t^2 - t + \frac{t^2 - m^2}{1 - y^2}\right]u = 0 \tag{3.25}$$

Eq. 3.25 has singularities at $y = \pm 1$. The singularities at $y = \pm 1$ can be dealt with by taking 't' = $\pm|m|$. We must not fail to note that solution for these two values of 't' will satisfy the same differential equation as the parent Eq. 3.23 depends on m^2. The two solutions can therefore differ from each other utmost by a constant factor (say B) so that we can write

$$\chi(|m|) = B \cdot \chi(-|m|) \tag{3.26}$$

It is sufficient to seek the solution of Eq. 3.25 for χ for 't' $= m \geq 0$ only. For $t = m < 0$, the solutions can be obtained from Eq. 3.26. For $t = m \geq 0$, Eq. 3.25 simplifies to the following equation free of singularities:

$$(1 - y^2)\frac{d^2u}{dy^2} - 2y(m+1)\frac{du}{dy} + \left[\beta - m(m+1)\right]u = 0 \tag{3.27}$$

A series solution of Eq. 3.27 can be sought in the form

$$u = \sum_{k=0}^{\infty} a_k y^k \tag{3.28}$$

which when used in Eq. 3.27 leads to

$$\sum_{k=0}^{\infty}\left[k(k-1)a_k y^{k-2} + \left\{\beta - (k+m)(k+m+1)\right\}a_k y^k\right] = 0 \tag{3.29}$$

Equating coefficients of y^k on both sides of Eq. 3.29 generates the following recurrence relation

$$(k+2)(k+1)a_{k+2} + [\beta - (k+m)(k+m+1)]a_k = 0$$
$$a_{k+2} = -\frac{[\beta - (k+m)(k+m+1)]}{(k+2)(k+1)}a_k \tag{3.30}$$

which effectively provides connections among all terms of series. Note, however that coefficients a_k are connected only with a_{k+2}, the function modeled by the series will be either odd or even depending on the higher order term being odd or even. If we demand that 'u' must be a polynomial of degree k = 'q' the following truncation condition must be introduced:

$$a_{q+2} = 0, a_q \neq 0$$

That immediately leads us (Eq. 3.30) to

$$\beta = (q+m)(q+m+1) \tag{3.31}$$

where q can assume integer values 0, 1, 2, ... upto the highest power of the terms retained in the series expansion for 'u'. Note that we have two sets of numbers q and m, both integers ≥ 0. We define a composite number[4]

$$l = (q+m) \tag{3.32}$$

which too can assume integer values 0, 1, 2, ... with the stipulation that $l \geq m$ (l being equal to $q+m$). Eq. 3.31 therefore leads to

$$\beta = l(l+1)$$

Turning back to Eq. 3.20b, we can clearly see that β is the eigenvalue of the operator $-\nabla^2_{\theta,\phi}$, therefore the eigenvalues of $-\nabla^2_{\theta,\phi} = l(l+1)$, $l = 0, 1, 2$, etc. Eq. 3.25 becomes

$$(1-y^2)u'' - 2y(m+1)u' + [l(l+1) - m(m+1)]u = 0 \tag{3.33}$$

where $u'' = \frac{d^2u}{dy^2}, u' = \frac{du}{dy}$ and

$$u = \begin{cases} a_{l-m}y^{l+m} + a_{l-m-2}y^{l-m-2} + \cdots + a_0 \\ a_{l-m}y^{l+m} + a_{l-m-2}y^{l-m-2} + \cdots + a_1 y \end{cases} \tag{3.34}$$

We can represent the solution in a more compact form by introducing the function $v = (y^2 - 1)^l$, which on differentiation with respect to y yields the following equation satisfied by v:[2-5]

$$(1-y^2)v' + 2ylv = 0 \tag{3.35}$$

Let us differentiate both sides of Eq. 3.35 $(l+m+1)$ times exploiting Leibnitz's rule to arrive at a new differential equation

$$(1-y^2)w'' - 2y(m+1)w' + [l(l+1) - m(m+1)]w = 0 \tag{3.36}$$

where we have used

$$v^{(l+m)} \equiv \frac{d^{l+m}}{dy^{l+m}}(y^2 - 1)^l = w$$

and

$$w'' \equiv \frac{d^2w}{dy^2}, w' = \frac{dw}{dy}$$

The reader must not fail to note that the differential Eq. 3.36 for w is identical with the differential Eq. 3.33 for u. Hence 'u' and 'w' must be multiplies of each other and we can therefore write

$$u = c \cdot w$$

We note here that we have so far not determined the normalization constant for the function $\chi(\theta)$, we choose $c = \frac{1}{2^l l!}$ so that for $m = 0$, we get

$$u = c.w = \frac{1}{2^l l!} \frac{d^l}{d\chi^l}(y^2 - 1)^l \equiv P_l(y), say$$

The values of function $\chi(\theta)$ then can be identified easily by introducing a new function $P_l^m(y)$ and defining

$$\chi_l^m(y) = \beta_l^m P_l^m(y)$$

β_l^m being the normalization constant and

$$P_l^m(y) = (1-y^2)^{\frac{m}{2}} \frac{d^{l+m}}{d\chi^{l+m}} \frac{[y^2 - 1]^l}{2^l l!}; (m \geq 0) \tag{3.37}$$

The above expression for P_l^m also holds automatically in view of the fact that

$$P_l^m(y) = (-1)^m \frac{(l+m)!}{(l-m)!} P_l^{-m}(y) \tag{3.38}$$

Eq. 3.37 and Eq. 3.38 together fix the range of values that the quantum number 'm' can have:

$$m = 0, \pm 1, \pm 2, \ldots, \pm l$$

as $P_l^m(y)$ becomes zero for $|m| > l$. The normalizing factor β_l^m can be found from the condition

$$\int_0^{\pi} \chi_l^m(\theta)\chi_l^m(\theta)\sin\theta d\theta = \int_{-1}^{+1} \chi_l^m(y)\chi_l^m(y)dx = 1$$

which yields after the definite integrals are evaluated

$$\beta_l^m = \sqrt{\frac{(2l+1)(l+m)!}{2(l+m)!}}$$

and hence

$$\chi_l^m(\theta) = \sqrt{\frac{(2l+1)(l-m)!}{2(l+m)!}} P_l^m(\theta)$$

Thus we have obtained the full expression for spherical function $Y(\theta, \phi)$, which are eigenfunction of the $\nabla_{\theta,\phi}^2$ operator for angular motion on the surface of a sphere of radius r. Each eigenfunction contains two quantum numbers l and m ($m = 0, \pm 1, \pm 2, \cdots, \pm l$) and we can identify these eigenfunctions correctly as the spherical harmonics $Y_l^m(\theta, \phi)$ with

$$Y_l^m(\theta, \phi) = \chi_l^m \Phi_m = \sqrt{\frac{(2l+1)(l-m)!}{4\pi(l+m)!}} \cdot P_l^m(\cos\theta) \cdot \exp(im\phi) \qquad (3.39)$$

The orthonormalization condition of the spherical harmonics ($Y_l^m(\theta, \phi)$) reads

$$\int\limits_{\theta,\phi} \{Y_l^m(\theta, \phi)\}^* Y_{l'}^{m'}(\theta, \phi) \sin\theta \, d\theta \, d\phi = \delta_{ll'} \delta_{mm'} \qquad (3.40)$$

The expression for spherical harmonics as presented in Eq. 3.39 can be presented in an alternative form, by making use of Eq. 3.38:

$$Y_l^m(\theta, \phi) = c_m \sqrt{\frac{(2l+1)(l-|m|)!}{4\pi(l+|m|)!}} P_l^{|m|}(\cos\theta) \cdot \exp(im\phi)$$

$$\text{where,} \quad c_m = \begin{cases} 1, \text{for } m \geq 0 \\ (-1)^m, \text{for } m < 0 \end{cases}$$

$$(3.41)$$

This choice for c_m is not unique and many choose the coefficient $c_m = 1$ for all values of 'm'. If we confine our task to simply determining the spherical harmonics (functions) subject to the orthonormalization condition stipulated in Eq. 3.41, the two choices are equivalent as c_m^2 is equal to unity in each case. The choice of phase assumes importance if we are required to make use of certain recurrence relation among spherical harmonics with different values of 'm' for example in the calculation of transition moment integrals or while developing a relativistic description of an electron in a central field potential. So far, we have concentrated on the angular part of the wavefunction $\Psi(r, \theta, \phi)$, which was written as a product of the radial wavefunction $R(r)$ and the angular part $Y(\theta, \phi)$. The angular part has been identified as spherical harmonics $Y_l^m(\theta, \phi)$ – as they appear in all central field potentials and in vibrations of a sphere. The determination of the angular part of the wavefunction has led to the identification of two quantum numbers l and m, l being called the orbital and m the magnetic quantum number. Let us recall that any central field potential is invariant under spatial inversion, which does not affect the scalar r, but alters θ, ϕ. It is necessary therefore to examine how $Y_l^m(\theta, \phi)$ transforms under inversion in order to determine parity of the atomic states described by the wavefunctions $R(r)Y_l^m(\theta, \phi)$. We have already noted that spatial inversion (reflection through the origin) reverses the directions of all the three cartesian coordinates ($x \to -x$, $y \to -y$, $z \to -z$). In spherical coordinates, inversion changes the spherical angles in the following way: $\theta \to \pi - \theta$; $\phi \to \pi + \phi$; $\cos\theta \to -\cos\theta$.

$P_l^m(y)$ (see Eq. 3.37) therefore ($y = \cos\theta$) becomes $P_l^m(-y) = (-1)^{l+m} P_l^m(y)$. The other component of the angular function $e^{im\phi}$ transforms under inversion as follows: $e^{im\phi} \to e^{im(\phi+\pi)} = e^{im\phi} \cdot e^{im\pi} = (-1)^m e^{im\phi}$. The spherical function $Y_l^m(\theta, \phi)$ is therefore transformed upon inversion in the following manner:

$$Y_l^m(\theta, \phi) = \sqrt{\frac{1}{2\pi}} \cdot \beta_l^m P_l^m(y) e^{im\phi} \to \frac{1}{\sqrt{2\pi}} (-1)^l \beta_l^m P_l^m \cdot e^{im\phi} = (-1)^l Y_l^m(\theta, \phi)$$

Thus the parity of the spherical harmonics is given by the number $(-1)^l$, l being the orbital quantum number, which ranges over all positive integers including zero. The parity is odd if l is odd $(1, 3, 5, 7, \cdots)$ and even if l is even or zero. In any central field potential, the quantum states will carry parity labels 1 or -1, i.e. even or odd. The 'parity' serves as a good quantum number and has a purely quantum mechanical origin. It has no classical analog. For many electron problems, the parity of a state is determined by the product of parity labels of all the occupied one electron states. Having seen that the quantum number l (not m) alone determines the parity it is necessary now to interpret the physical meanings of the two quantum numbers 'l' and 'm' characterizing the spherical harmonics. It turns out that they are physically connected to the important dynamical quantity called angular momentum of the electrons in an atom.

3.5 Orbital Angular Momentum of Electron

We have seen that the quantum mechanical operator for the Hamiltonian function of hydrogenic atom can be written as in spherical coordinates (Eq. 3.42)

$$\hat{H} = -\frac{\hbar^2}{2m}\nabla_r^2 - \frac{\hbar^2}{2mr^2}\nabla_{\theta,\phi}^2 + V(r) \tag{3.42}$$

where $V(r) = -\frac{Ze^2}{r}$.

We have also seen that the operator $\nabla_{\theta,\phi}^2$ have eigenvalues $l(l+1)$ with $l = 0, 1, 2, \cdots$. For physical interpretation, let us also consider the classical Hamiltonian function H_c for the same system:

$$\begin{aligned} H_c &= \frac{mv^2}{2} + V(r) \\ &= \frac{1}{2m}(m\dot{r})^2 + \frac{(mr^2\dot{\phi})^2}{2mr^2} + V(r) \\ &= \frac{p_r^2}{2m} + \frac{L^2}{2mr^2} + V(r) \end{aligned} \tag{3.43}$$

p_r is the classical radial momentum and L the angular momentum. On comparing Eq. 3.42 and Eq. 3.43 we find the following correspondences:

$$p_r^2 = -\hbar^2\nabla_r^2 = \hat{P}_r^2 \tag{3.44a}$$

$$L^2 = -\hbar^2\nabla_{\theta,\phi}^2 = \hat{L}^2 \tag{3.44b}$$

The square of the angular momentum operator $(\hat{L}^2)$ can therefore be assumed to be

$$L^2 = -\hbar^2\nabla_{\theta,\phi}^2 \tag{3.44c}$$

We have already established that $\nabla_{\theta,\phi}^2$ has eigenfunctions $Y_l^m(\theta,\phi)$ with eigenvalue $\beta = l(l+1)$. That immediately leads to the result that

$$L^2 Y_l^m(\theta,\phi) = \hbar^2 l(l+1) Y_l^m(\theta,\phi); \quad l = 0, 1, \cdots \tag{3.44d}$$

The spherical harmonics are thus the eigenfunctions of the quantum mechanical operator representing the square of the angular momentum $(\hat{L}^2)$ with eigenvalues $l(l+1)\hbar^2$, l being the orbital quantum number. The corresponding value of the angular momentum itself is $\sqrt{l(l+1)}\hbar$.

Let us now try to obtain more information about quantum mechanical angular momentum operator $\hat{L}$ and its components from the classical notion of angular momentum $(\vec{L})$. In classical mechanics,

$$\vec{L} = \vec{r} \times \vec{p} = iL_x + jL_y + zL_z$$

$$\vec{L} = \begin{pmatrix} i & j & k \\ x & y & z \\ p_x & p_y & p_z \end{pmatrix}$$

$$L^2 = \vec{L} \cdot \vec{L} = L_x^2 + L_y^2 + L_z^2$$

The classical quantum correspondence suggests that we could obtain $\hat{L}$ by replacing x, y, z and p_x, p_y, p_z by the corresponding quantum mechanical operators $(\hat{x}, \hat{y}, \hat{z}, \hat{p}_x, \hat{p}_y, \hat{p}_z)$. Thus

$$\hat{L} = \begin{pmatrix} i & j & k \\ \hat{x} & \hat{y} & \hat{z} \\ \hat{p}_x & \hat{p}_y & \hat{p}_z \end{pmatrix}$$

It is straightforward to find that

$$\hat{L}_x = (\hat{y}\hat{p}_z - \hat{p}_z\hat{y})$$
$$\hat{L}_y = (\hat{z}\hat{p}_x - \hat{p}_x\hat{z}) \tag{3.45}$$
$$\hat{L}_z = (\hat{x}\hat{p}_y - \hat{p}_y\hat{x})$$

By making use of the commutation relations among the operators for coordinates and momenta, the reader may verify that angular momentum operators $\hat{L}_x$, $\hat{L}_y$, $\hat{L}_z$ do not commute with each other. It turns out that

$$[\hat{L}_x, \hat{L}_y] = i\hbar\hat{L}_z$$
$$[\hat{L}_y, \hat{L}_z] = i\hbar\hat{L}_x \tag{3.46}$$
$$[\hat{L}_z, \hat{L}_x] = i\hbar\hat{L}_y$$

For many purposes, it is useful to obtain the angular momentum operators in spherical polar coordinates. Using Eq. 3.13 defining transformation from (x, y, z) to (r, θ, ϕ), the definitions of the operator $(\hat{x}, \hat{y}, \hat{z})$, $(\hat{p}_x, \hat{p}_y, \hat{p}_z)$ in Eq. 3.45 defining the angular momentum operator, and exploiting rules of partial differentiation, one can arrive at the following expressions for $\hat{L}_x$, $\hat{L}_y$, $\hat{L}_z$ in spherical polar coordinates:

$$\hat{L}_x = -\frac{\hbar}{i}\left(\sin\phi\frac{\partial}{\partial\theta} + \cos\phi\cot\theta\frac{\partial}{\partial\phi}\right)$$
$$\hat{L}_y = \frac{\hbar}{i}\left(\cos\phi\frac{\partial}{\partial\theta} - \sin\phi\cdot\cot\theta\frac{\partial}{\partial\phi}\right) \tag{3.47}$$
$$\hat{L}_z = \frac{\hbar}{i}\frac{\partial}{\partial\phi}$$

Now allowing $\hat{L}_z$ to operate on $Y_l^m(\theta,\phi)$ of Eq. 3.41, we can arrive at the result $\hat{L}_z Y_l^m(\theta,\phi) = m\hbar Y_l^m(\theta,\phi), m = 0, \pm1$, etc. revealing thereby that $Y_l^m(\theta,\phi)$s are eigenfunctions of $\hat{L}_z$ with eigenvalues $m\hbar$ $(m = 0, \pm1,..)$. We have already seen that $Y_l^m(\theta,\phi)$s are also eigenfunctions of L^2 with eigenvalues $l(l+1)\hbar^2$ $(l = 0, 1, \cdots)$. The result is indeed entirely expected as $\hat{L}^2$ and $\hat{L}_z$ commute with each other, i.e. $[\hat{L}^2, \hat{L}_z] = 0$ (the proof is elementary). The allowed range of 'm' values can be easily established for a given value of l: $m = 0, \pm1, \pm2, \cdots, \pm l$ as we can see from Eq. 3.37 and Eq. 3.38 that P_l^m vanishes for $|m| > l$. L^2 also commute with L_x^2 and L_y^2, i.e. $[L^2, L_x] = [L^2, L_y] = 0$. Does it mean L^2, L_z and also L_x and L_y can have the same common set of eigenfunctions? The answer is 'no', because L_x, L_y, L_z do not commute amongst themselves. Thus, we will always work with the common set of eigenfunctions of $\hat{L}^2$ and $\hat{L}_z$ operators ($Y_l^m(\theta,\phi)$) when dealing with central field (spherically symmetric) potentials. The choice of $\hat{L}_z$ as the commuting counterpart of $\hat{L}^2$ operator is nothing unique. It is the simplest, that we could choose. We could have chosen linear combination of the spherical harmonics that are, for example, eigenfunctions of L^2, L_x. These linear combinations will not be eigenfunctions of L_y or L_z. It is important now to analyze these results a little further, from the point of view of the uncertainty principle as non-commuting observables are involved. The angular momentum eigenfunctions of our choice are simultaneous eigenfunctions of $\hat{L}^2$ and $\hat{L}_z$ operators.

$$\hat{L}^2 Y_l^m(\theta, \phi) = l(l+1)\hbar^2, \quad l = 0, 1, 2...$$
$$\hat{L}_z Y_l^m(\theta, \phi) = m\hbar, \quad -l \leq m \leq l$$

(3.48)

Since the maximum value of m is l and $l^2 < l(l+1)$, the maximum projection of the angular momentum (except for $l = 0$) is always smaller than its momentum. The angular momentum vector can therefore be fully aligned with the z-axis.

If the system, the atom for example is in the state $l = 0$ (the lowest angular momentum state), $m = 0$, the values of L^2 as well as L_z vanish. That means the mechanical angular momentum (L) and its projection on z-axis (L_z) are both zero. Classically, such a state can not exist as it would imply that either the linear momentum $\vec{p}$ is zero or the motion takes place through the center of the potential ($\vec{r} = 0$), $\vec{L}$ being equal to $\vec{r} \times \vec{p}$. Classically, we can write $L^2 = L^2_{z,max}$ where $L_{z,max}$ is the maximum value of the z-projection of L in the state being considered. $L_{z,max}$ has the value $l\hbar$, thereby demanding

$$L^2 = l^2 \hbar^2$$

(3.49)

assuming complete classical-quantum correspondence. We have already seen that the quantum mechanical angular momentum squared L^2 has eigenvalues $l(l+1)\hbar^2$, i.e.

$$L^2 = l^2 \hbar^2 + l\hbar^2 = l^2_{z,max} + l\hbar^2$$

The additional term $l\hbar^2$ appears in the full quantum estimate of L^2 due to the non-commutivity of L_x, L_y, L_z, which makes it impossible to define or measure exact values of L_x, L_y and L_z simultaneously (unlike in classical mechanics). If $L_z = L_{z,max} = l\hbar$ (precisely defined) the quantum mean value of L_x and L_y, i.e. $< L_x >$ and $< L_y >$ are equal to zero. Their mean square deviation $< (\Delta L_x)^2 >$ and $< (\Delta L_y)^2 >$ are accordingly equal to $< L_x^2 >$ and $< L_y^2 >$, respectively, which do not vanish, but takes up some minimum values compatible with uncertainty principle. It is possible then to write

$$< L^2 > \; = \; < L_x^2 > + < L_y^2 > + < L_z^2 >$$
$$= L^2_{z,max} + < (\Delta L_x)^2 >_{min} + < (\Delta L_y)^2 >_{min}$$
$$= \hbar^2 l^2 + < (\Delta L_x)^2 >_{min} + < (\Delta L_y)^2 >_{min}$$

The minimum values of $< (\Delta L_x)^2 >$ and $< (\Delta L_y)^2 >$ can be estimated by appealing to the uncertainty relation, which allows us to write[4]

$$< (\Delta L_x)^2 >_{min} < (\Delta L_y)^2 >_{min} = \frac{1}{4}[L_x, L_y]^2 = \frac{1}{4}\hbar^2 L^2_{z,max} = \frac{1}{4}\hbar^4 l^2$$

In the spherically symmetric problems that we are dealing with here x and y are equivalent. That enables us to write $< (\Delta L_y^2) >_{min} = < (\Delta L_x^2) >_{min}$ so that

$$< (\Delta L_x)^2 >_{min} < (\Delta L_y)^2 >_{min} = < (\Delta L_x^2) >^2_{min} = < (\Delta L_y^2) >^2_{min} = \frac{1}{4}\hbar^4 l^2$$

Hence $< (\Delta L_x^2) >_{min} = < (\Delta L_y^2) >_{min} = \hbar^2 \frac{l}{2}$.

The sum of $< (\Delta L_x^2) >_{min}$ and $< (\Delta L_y^2) >_{min}$ is thus equal to $\hbar^2 l$—the extra angular momentum appearing in the quantum mechanical definition of eigenvalues L^2. This additional angular momentum term behaves as the zero-point angular momentum.[4]

In dealing with angular momentum of atoms apart from L^2 and L_z^2, two other operators $\hat{L}_+$ and $\hat{L}_-$, respectively called the raising and lowering operators, are found to be useful where

$$\hat{L}_+ = \hat{L}_x + i\hat{L}_y$$

$$\hat{L}_- = \hat{L}_x - i\hat{L}_y$$

Now, making use of the expression for $\hat{L}_x$ and $\hat{L}_y$ in Eq. 3.47 in spherical coordinates and introducing a variable 'μ' = $\cos\theta$, one finds that

$$\hat{L}_x \pm i\hat{L}_y = \hbar e^{\pm i\phi}\left\{ i\frac{\mu}{\sqrt{1-\mu^2}}\frac{\partial}{\partial\phi} \mp \sqrt{1-\mu^2}\frac{\partial}{\partial\phi}\right\}$$

Before determining what happens when $\hat{L}_x \pm i\hat{L}_y$ are allowed to act on the spherical harmonics[5] let us note that $Y_l^m(\theta,\phi)$ can be represented in the equivalent forms:

$$Y_l^m(\theta,\phi) = a_m\sqrt{\frac{(2l+1)(l-|m|)!}{4\pi\,(l+|m|)!}}\,P_l^{|m|}(\cos\theta)\exp(im\phi)$$

$$\text{with, } a_m = \begin{cases} 1, \text{for } m \geq 0 \\ (-1)^m, \text{for } m < 0 \end{cases} \tag{3.50}$$

and

$$Y_l^m(\theta,\phi) = (-1)^m\sqrt{\frac{(2l+1)(l+m)!}{4\pi\,(l-m)!}}\,P_l^{-m}(\cos\theta)\exp(im\phi) \tag{3.51}$$

We can now determine the action of $\hat{L}_x \pm i\hat{L}_y$ on spherical harmonics easily by choosing $\hat{L}_x + i\hat{L}_y$ to act on $Y_l^m(\theta,\phi)$ given in Eq. 3.41 while $\hat{L}_x - i\hat{L}_y$ is chosen to act on equivalent form of $Y_l^m(\theta,\phi)$ given in Eq. 3.51. After some simple but tedious algebra, we arrive at the following relation

$$(L_x \pm iL_y)Y_l^m = -\hbar\sqrt{(l+1\pm m)(l\mp m)}\,Y_l^{m\pm1} \tag{3.52}$$

Now L^2 can be expressed in a convenient form by making use of L_+ and L_- operators as shown below.

$$L^2 = \left[\frac{1}{2}(L_x+iL_y)(L_x-iL_y) + \frac{1}{2}(L_x-iL_y)(L_x+iL_y) + L_z^2\right] \tag{3.53}$$

Using Eq. 3.52 and Eq. 3.53 we then find that

$$L^2 Y_l^m(\theta,\phi) = \hbar^2 l(l+1)Y_l^m(\theta,\phi),$$

a result we have already arrived at in a different manner (see Eq. 3.44a–d).

3.5.1 Spherical Harmonics and Eigenstates of Rigid Rotator

Let us consider a particle of mass 'm' moving on the surface of a sphere of fixed radius 'r' (r_0). The potential energy is constant everywhere on the surface of the sphere and we can set $V(r_0) = C$. The corresponding dynamical system constitutes what is called the rigid rotator. It is clear that the rigid rotator is just a special case of particle motion in a central field potential where $V(r_0) = 0$. The angular part of the wavefunction will thus be described correctly by the spherical harmonics $Y_l^m(\theta,\phi)$, which have been shown to be eigenfunctions of $\hat{L}^2$ and $\hat{L}_z$ operators. The radial part of the wavefunction, $R(r)$, must satisfy the following equation (see Eq. 3.20a–b and note that here we have to put $V(r_0) = 0$, $\beta = l(l+1)$)

$$\nabla_r^2(R(r)) + \left[\frac{2mE}{\hbar^2} - \frac{l(l+1)}{r^2}\right]R(r) = 0 \tag{3.54}$$

For the rigid rotator r is by defination a constant equal to r_0 so that $R(r) = R(r_0)$ is also a constant. Hence we can set $\nabla_r^2(R(r)) = 0$. Eq. 3.54 then leads to the following expression for the energy of the rotator

$$E = E_l = \frac{l(l+1)\hbar^2}{2mr_0^2} = \frac{l(l+1)\hbar^2}{2I}, \quad l = 0,1,2,\cdots \tag{3.55}$$

mr_0^2 is the moment of the inertia (I) of the rotator. Eq. 3.55 confirms that the energy of the rigid rotator depends only on the orbital quantum number 'l' and is independent of the quantum number 'm', which can assume integer values from $-l$ to $+l$, including zero. The energy of the rotator $E_l = \frac{l(l+1)\hbar^2}{2I}$ is therefore $(2l+1)$ - fold degenerate. The $(2l+1)$ number of spherical harmonics $Y_l^{-l}, Y_l^{-l+1}, \cdots, Y_l^{-l}, Y_l^0, Y_l^1, Y_l^2, \cdots,$ Y_l^l form a linearly independent set of eigenfunctions of L^2 operator. The energy depends only on the square of the angular momentum [$l(l+1)\hbar^2$] but not on the allowed projections ($m\hbar, -l \le m \le l$) of the angular momentum on the z-axis. This $2l+1$-fold energy degeneracy of the rotator eigenfunction ($Y_l^m(\theta,\phi)$) is a signature of the rotational symmetry of the potential and the rigid rotator Hamiltonian, which remain unchanged for any rotation about any axis passing through the center (origin of the coordinates in this case). In free space there is then no preferred direction or axis. All lines passing through the origin of the coordinate system, i.e. the center of the sphere is equivalent and the angular momentum L does not have any preferred direction or orientation in space. This full rotational symmetry breaks down when an external magnetic field is applied in a particular direction (call it the z-direction). Now, all the directions are no longer equivalent so that we can anticipate reduction or even complete lifting of the $(2l+1)$-fold degeneracy of spherical harmonics.

Let us examine different states of the rotator in some detail as these states arise in any central field problem of the electrons including the Coulomb field. The lowest state Y_0^0 has $l = 0$, $m = 0$ and is devoid of any energy degeneracy. E_0 is equal to zero and since $Y_0^0 = \frac{1}{\sqrt{4\pi}}$, the probability density is given by $|Y_0^0|^2 = \frac{1}{4\pi}$. This state is called a s-state. In the first excited state of rotator, $l = 1$ and m assumes values -1, 0 and 1. The energy is $E_1 = \frac{\hbar^2}{I}$. The state, called a p-state, has three-fold energy degeneracy, Y_1^0, Y_1^{-1} and Y_1^1 all having the energy E_1. The three-degenerate spherical harmonics are the following:

$$Y_1^0 = \sqrt{\frac{3}{4\pi}}\cos\theta$$

$$Y_1^{-1} = -\sqrt{\frac{3}{8\pi}}e^{-i\phi}\sin\theta \qquad (3.56a)$$

$$Y_1^1 = \sqrt{\frac{3}{8\pi}}e^{i\phi}\sin\theta$$

while the probability densities for these states turn out to be

$$|Y_1^{-1}|^2 = |Y_1^{+1}|^2 = \frac{3}{8\pi}\sin^2\theta$$
$$\qquad (3.56b)$$
$$|Y_1^0|^2 = \frac{3}{4\pi}\cos^2\theta$$

The probability densities are seen to be independent of the angle ϕ. It depends only on the angle θ. This feature is generally true for $|Y_l^m|^2$. The probability densities for the $l = 1$ state are graphically displayed in Figure 3.4 only on the yz-plane in view of the fact that they are independent of ϕ. It is clear from Figure 3.4a as well as formula $|Y_0^0|^2 = \frac{1}{4\pi}$ that in the s ($l = 0$) state direction of the angular momentum $\vec{L}$ relative to the z-axis is independent of the angle 'θ' as it should be in view of the fact that the value of the angular momentum is zero. The particle may therefore be found anywhere on the surface of the sphere with the same probability. Figure 3.4b as well as formula for $|Y_1^{\pm 1}|^2$ in Eq. 3.56 indicates that in the p-states ($l = 1$, $m = \pm 1$) the most probable trajectory of the particle would be one lying on the xy-plane. The particle in the state with $m = 1$ has right-handed rotation, the angular momentum being parallel to Z-axis. The motion in the state $l = 1$, $m = -1$ is just the opposite.[4] Figure 3.4c and the formula for $|Y_1^0|^2$ in Eq. 3.56 suggests that the most probable trajectory of the rotator in $l = 1$, $m = 0$ state must lie in the plane passing through the z-axis the direction of the angular momentum vector being perpendicular to the z-axis. It is possible to analyze the higher states of the rotator with $l = 2, 3, \cdots$, etc. in a similar manner.

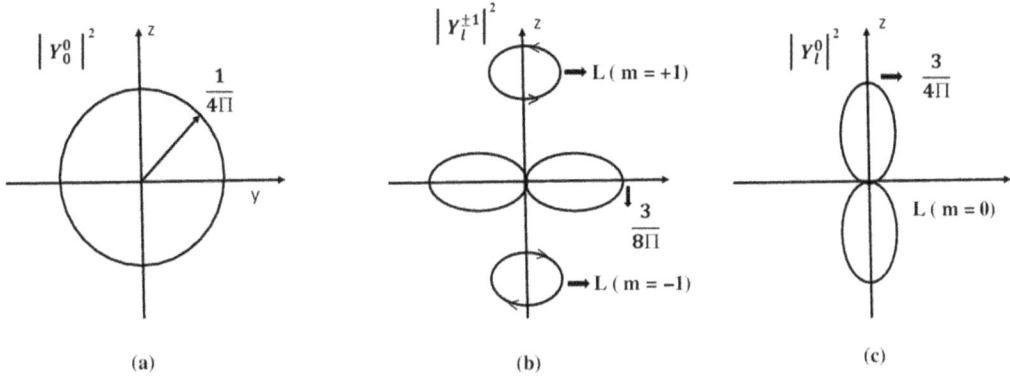

FIGURE 3.4 (a)–(c) Probability density distribution for *s* and *p* states of a rotator.

3.5.2 Radial Motion of the Electron in H-Atom

Let us recall that the full wavefunction in any central field potential $V(r)$ including the Coulomb potential $[V(r) = -\frac{Ze^2}{r}]$ can be written as a product of the radial and angular wavefunction $\Psi(r,\theta,\phi) = R(r)Y_l^m(\theta,\phi)$. We have seen that the angular part $Y_l^m(\theta,\phi)$ is characterized by two quantum numbers l and m. It is now necessary to focus on the determination of the radial part of $R(r)$ of the wavefunctions and ascertain what other quantum numbers, if any, are required to characterize the full wavefunctions and energy eigenvalues completely. To determine $R(r)$ we start by substituting the wavefunction $\Psi(r,\theta,\phi)$ in the product form $\Psi = R(r)Y_l^m(\theta,\phi)$ in the Schrödinger equation $\hat{H}\Psi = E\Psi$ where $\hat{H}$ is given by

$$\hat{H} = -\frac{\hbar^2}{2m}\nabla_r^2 - \frac{\hbar^2}{2mr^2}\nabla_{\theta,\phi}^2 + V(r)$$

which leads to the following equation

$$-\frac{\hbar^2}{2m}\left[\frac{1}{r^2}\frac{\partial}{\partial r}\left(r^2\frac{\partial}{\partial r}\right)\right]\Psi - \frac{\hbar^2}{2mr^2}\nabla_{\theta,\phi}^2\Psi + V(r)\Psi = E\Psi$$

$$\text{or,} \ -\frac{\hbar^2}{2m}\left[\frac{1}{r^2}\frac{\partial}{\partial r}\left(r^2\frac{\partial}{\partial r}\right)\right]R(r)Y_l^m + \frac{L^2}{2mr^2}R(r)Y_l^m + V(r)R(r)Y_l^m = ER(r)Y_l^m \quad (3.57)$$

Let us note the following points:

(1) The operator $\hat{L}^2$ acts on Y_l^m to produce a constant term $l(l+1)\hbar^2$ (i.e. the eigenvalue of $\hat{L}^2$, Y_l^m being being the eigenfunctions) times Y_l^m; (2) The partial derivative with respect to 'r' acts only on $R(r)$ and can be replaced by the total derivative.

Thus the equation determining $R(r)$ becomes

$$-\frac{\hbar^2}{2m}\frac{1}{r^2}\frac{d}{dr}\left(r^2\frac{dR}{dr}\right) + \frac{\hbar^2}{2mr^2}l(l+1)R(r) + V(r)R(r) = ER(r) \quad (3.58)$$

Eq. 3.58 can be reduced to a more convenient one-dimensional form with the substitution $R(r) = \frac{\chi(\rho)}{r}$ leading to the final equation for determining the energy eigenvalue E.

$$-\frac{\hbar^2}{2m}\frac{d^2\chi}{dr^2} + \frac{\hbar^2 l(l+1)\chi}{2mr^2} + V(r)\chi = E\chi \quad (3.59)$$

3.5.3 Asymptotic Forms of $R(r)$ and Continuous Energy Spectrum

Even without specifying the form of the central field potential $V(r)$, we can try to find $\chi(r)$ in two limiting cases:

(i) r is very small (the electron is very close to the nucleus);

(ii) r is very large (far away from the nucleus).

(i) In close proximity to the nucleus, the electron experiences only the pull of the nuclear charges so that $V(r) = -\frac{Ze^2}{r}$, which is much smaller in magnitude compared to the term $\frac{l(l+1)\hbar^2}{2mr^2}$ so that $\chi(r)$ is determined by a very simple differential equation

$$\frac{d^2\chi}{dr^2} = \frac{l(l+1)\chi}{r^2} \tag{3.60}$$

It can be solved by the substitution $\chi = r^k$, which leads to the result

$$k(k-1) = l(l+1) \tag{3.61}$$

Eq. 3.61 for k has two roots

$$k = -l \quad \text{and} \quad k = l+1$$

The root $k = -l$ must be abandoned as it leads to $R(r) = r^{-(l+1)}$, which blows up to infinity at $r = 0$ for all values of 'l' including $l = 0$. The other root leads to

$$\chi(r) = A \cdot r^{l+1}$$
$$R(r) = A \cdot r^l \tag{3.62}$$

For $l = 0$, however, the Coulomb term has to be retained in the equation for $\chi(r)$ and the equation solved by substituting

$$\chi(r) = b_1 r + b_2 r^2 + b_3 r^3 + \cdots \tag{3.63}$$

It ensures that the form of dependence of the wavefunction on r close to the nucleus takes the same form given in Eq. 3.62. It may be noted that the higher the angular momentum, the higher is the order of the zero of the wavefunction at the origin of the coordinates ($r = 0$). It is only for the $l = 0$ state (s-state) that the wave function remains non-zero and finite for small r (i.e. close to the nucleus). Thus with $r \to 0$, $R = \frac{\chi(r)}{r} = b_1$ (Eq. 3.63). The behavior of the wavefunction close to the origin can be physically explained as follows. A repulsive centrifugal force acts on the electron, which corresponds to an effective potential $\frac{l(l+1)\hbar^2}{2mr^2}$. This restricts the classically allowed region of motion for small values of r. In quantum mechanics, the particle can penetrate into this region where the classical velocity would have to acquire imaginary values. The wavefunction in this region decays rapidly as the particle approaches the origin of the potential – that is the origin of the coordinate system. The decrease takes place in this case in accordance with a power law: $\Psi(r) \approx r^l$ as $r \to 0$, meaning thereby that the higher the centrifugal barrier, the higher the damping. For $l = 0$, the barrier does not exist (zero) and therefore the electron can move into a region infinitely close to the nucleus.

(ii) We will now consider the other situation. As the electron moves into regions with large values of r, we can drop the terms in the Hamiltonian that decreases with increase in r - that is the terms $\frac{l(l+1)\hbar^2}{2mr^2}$ (centrifugal) and $-\frac{Ze^2}{r}$ (Coulomb term). The equation determining $\chi(r)$ then takes a particularly simple form

$$\frac{d^2\chi}{dr^2} = (-)\frac{2mE}{\hbar^2}\chi \tag{3.64}$$

The general solution of the Eq. 3.64 takes the form:

$$\chi = c_1 e^{-\frac{(-2mE)^{\frac{1}{2}}}{\hbar}r} + c_2 e^{\frac{(-2mE)^{\frac{1}{2}}}{\hbar}r} \tag{3.65}$$

Two cases arise: (i) $E > 0$; (ii) $E < 0$.

(i) For $E > 0$, χ appears in the form

$$\chi = c_1 e^{-i\frac{(-2mE)^{\frac{1}{2}}}{\hbar}r} + c_2 e^{i\frac{(-2mE)^{\frac{1}{2}}}{\hbar}r} \tag{3.66}$$

Note that both the terms remain finite for every value of r for $E > 0$ so that both the constants c_1, c_2 must be retained in the general solution. This solution for x applies only when r is very large. What happens at $r \to 0$ region? As we have already seen, $\chi(r)$ for small r takes the form $A \cdot r^{l+1}$. Imagine that this $\chi(r)$ is continued in the region where r is large and where χ can not be represented by the simple r^{l+1} form. Nevertheless, $\chi(r)$ in this region satisfies the original Eq. 3.59 and an integral curve can be obtained for the equation. Any such integral curve can be described by properly choosing the constants c_1, c_2 appearing in the general solutions, which assumes the asymptotic form (Eq. 3.66) as $r \to \infty$ where $E > 0$. However, χ remains finite in limit $r \to \infty$ irrespective of the constants c_1, c_2 chosen. We may therefore conclude that the relevant wave equation has a finite solution for all values of r for any value of energy $E > 0$. These positive energy solutions thus constitute a continuous energy spectrum.

(ii) We now turn our attention to the situation when $E < 0$. The solution (Eq. 3.65) must now be presented in the following form, noting that $E = -|E|$:

$$\chi(r) = c_1 e^{\frac{(2m|E|)^{\frac{1}{2}}}{\hbar}r} + c_2 e^{-\frac{(2m|E|)^{\frac{1}{2}}}{\hbar}r} \tag{3.67}$$

We must set $c_1 = 0$ in Eq. 3.67 since the coefficient of c_1 blows up as $r \to \infty$, with the result that the solution now carries only one arbitrary constant, c_2 and reads

$$\chi(r) = c_2 e^{-\frac{(2m|E|)^{\frac{1}{2}}}{\hbar}r} \tag{3.68}$$

Now let us consider the integral curve from $r = 0$ to $r \to \infty$. For small values of r (near $r = 0$) the curve has a form $\sim r^{l+1}$, which will not automatically change over into the form of $\chi(r)$ given in Eq. 3.68 for large r. For all energy values $E < 0$, except for certain special ones the integral curve will be represented at infinity in the form specified in Eq. 3.66 and fail to satisfy the boundary conditions that we have imposed on the wavefunction. Only for the special energy values $E < 0$ for which the integral curve conforms to the condition $c_1(E) = 0$, the wave equation will have a solution consistent with the required boundary conditions. Thus for negative values of E, the energy spectrum is discrete, and the wavefunction $\chi(r)$ vanishes at $r \to \infty$, as required.

3.5.4 Discrete Spectrum of Energy

We are now in a position to try to determine the discrete energy spectrum of an electron in a purely Coulomb potential $V(r) = \frac{-Ze^2}{r}$, which is encountered in the hydrogen atom ($Z = 1$) or hydrogenic atoms ($Z > 1$). The radial wavefunction now reads (for $E < 0$)

$$-\frac{\hbar^2}{2m}\frac{d^2\chi}{dr^2} + \frac{\hbar^2 l(l+1)\chi}{2mr^2} - \frac{Ze^2\chi}{r} = -|E|\chi \tag{3.69}$$

In the above equation E is negative and we have seen that the spectrum is discrete. Before proceeding further, it will be very covenient to change the units of length and energy, expressed in terms of the electronic charge (e) and mass (m) and the basic quantum of action h. In terms of the quantities mentioned the unit of length is defined as

$$\frac{\hbar^2}{me^2} = 0.529172 \times 10^{-8} \text{ cm}$$

and the unit of energy as

$$\frac{me^4}{\hbar^4} = 27.20976 \text{ eV}$$

If we now put $\hbar = 1 = e = m$ in Eq. 3.69, the length and energy are measured in this unity. To that effect let us denote the length ξ as

$$\xi = \frac{me^2 r}{\hbar^2}$$

and energy ϵ as

$$\epsilon = \frac{\hbar^2 |E|}{me^4}$$

and rewrite the equation in terms of ξ and, ϵ, which straightway leads to the equation

$$-\frac{d^2\chi(\xi)}{d\xi^2} + \frac{l(l+1)\chi(\xi)}{\xi^2} - \frac{2Z\chi(\xi)}{\xi} = -2\epsilon\chi \qquad (3.70)$$

We have already seen how χ behaves in small ξ and large ξ regions. Using this knowledge we can write χ in a form that smoothly interpolates between the limiting forms and obeys the boudary condition. To this end, we choose

$$\chi(\xi) = \xi^{l+1} e^{-\xi(2\epsilon)^{\frac{1}{2}}} (a_0 + a_1\xi + a_2\xi^2 + a_3\xi^3 + \cdots) \qquad (3.71)$$

The first factor fixes the form that χ takes as $\xi \to 0$ while the second determines the form of χ as $\xi \to \infty$. The polynomial $\sum_{n=0}^{\infty} a_n\xi''$ present in χ of Eq. 3.71 allows smooth interpolation between the two asymmetric regions. $\chi(\xi)$ of Eq. 3.71 can be expressed in a more compact form

$$\chi(\xi) = e^{-\xi(2\epsilon)^{\frac{1}{2}}} \sum_{k=0}^{\infty} a_k\xi^{k+l+1} \qquad (3.72)$$

Differentiating χ of Eq. 3.72 twice with respect to ξ and substituting the expression for double derivative $\frac{d^2\chi}{d\xi^2}$ in Eq. 3.70, we carry out obvious cancellations and regrouping of terms leading to the following inequality:

$$\sum_{k=0}^{\infty} \left[l(l+1) - (k+l+1) \right](k+l)a_k\xi^{k+l+1} = \sum_{k=0}^{\infty} \left[2Z - 2(2\epsilon)^{\frac{1}{2}}(k+l+1) \right]a_k\xi^{k+l} \qquad (3.73)$$

Such an equality implies that the coefficient of the same powers of ξ on the two sides of Eq. 3.73 must be equal. By equating coefficients of ξ^{k+l} on both sides we get the recursion relation

$$a_{k+1} = a_k \frac{2[Z - (k+l+1)(2\epsilon)^{\frac{1}{2}}]}{l(l+1) - (k+l+1)(k+l+2)} \qquad (3.74)$$

From the recursion relation (Eq. 3.74) let us note that all the coefficients a_0 ($\neq 0$), a_1, $\cdots$, a_k can be determined one by one. As $k \to \infty$, Z and l can be neglected compared to 'k' leading to, in the limit of large k

$$a_{k+1} = a_k \frac{2(2\epsilon)^{\frac{1}{2}}}{1} \qquad (3.75)$$

Using the result, it is possible to show that the series $\sum_{k=0}^{\infty} a_k\xi^k$ reduces to an exponential form:

$$\sum_{k=0}^{\infty} a_k\xi^k \simeq e^{2\xi(2\epsilon)^{\frac{1}{2}}}$$

This infinite series clearly diverges as $\xi \to \infty$ so that $\chi(\xi)$ in Eq. 3.72 will diverge at ∞ violating the imposed boundary condition on the wavefunction. In order that this kind of divergence does not take place, the recursion relation in Eq. 3.74 must be truncated – that is all the coefficients of the power series must become zero from a certain a_{k+1} onwards. When this happens, the series in Eq. 3.72 becomes a polynomial and $\chi(\xi) \to 0$ as $\xi \to \infty$ satisfying the required boundary condition. Now looking at Eq. 3.74, we can see that such a truncation happens when

$$Z - (k+l+1)(2\epsilon)^{\frac{1}{2}} = 0 \qquad (3.76a)$$

or,

$$\epsilon = \frac{Z^2}{2(k+l+1)^2} \tag{3.76b}$$

By changing over to the conventional units of energy and taking note of its sign (negative energy) we have

$$E = -\frac{Z^2 m e^4}{2\hbar^2 (k+l+1)^2} \tag{3.77}$$

What does the number 'k' in Eq. 3.77 represent? Apparently it is the highest power of ξ in the polynomial $\sum_{t=0}^{k} a_t \xi^t$. It can be shown that the k-degree polynomial has k real and unequal roots and one of these roots is either zero or infinity. Thus, if we plot the radial wavefunction from $r = 0$ to $r \to \infty$ the function will have exactly k-zeros or nodes when the zero at $r = \infty$ (the node at $r = \infty$ signifies the boundedness of the motion) or node at ξ or $r = 0$ (this zero occurs for any $l \neq 0$) are not counted. Let the number of zeros (nodes) in the radial wavefunction be 'n_r' (k in Eq. 3.77 can be replaced by n_r, the radial quantum number). These nodes occur at finite distances from the origin and serve to characterize the state of an electron in a hydrogen atom and atoms other than hydrogen as well at least from a qualitative point of view. In the latter case, we can imagine that the effect of all other electrons on the electron of our choice has been absorbed in the effective potential $V_{eff}(r)$ and n_r can be thought of as quantifying the state of radial motion of the electron in the atom. In the same vain, we can describe the state of an individual electron with the help of the number 'l', which describes the square of the angular momentum that this electron carries in the atom under consideration – and therefore quantifies the state of angular motion of the electron. Apart from 'n_r' and 'l' there is another integral number 'n' that serves to quantify the energy of the electron. It is defined as

$$n = n_r + l + 1 \quad (n_r = 0, 1, 2, \cdots ; l = 0, 1, 2, \cdots)$$

n is obviously an integer and takes on values 1, 2, 3, $\cdots$, etc. and is called the principal quantum number. The minimum value of n_r is 0 and therefore for a given 'n' the maximum value of $l = n - 1$. Looking at the expression for energy of the hydrogenic atoms given in Eq. 3.77, we see that the energy is uniquely determined by the principal quantum number 'n', which is just the sum of the radial and angular quantum numbers characterizing the states of radial and angular motion of the electron in an atom. Thus

$$E_n = -Z^2 \frac{m e^4}{2\hbar^2 (n_r + l + 1)^2} = -Z^2 \frac{m e^4}{2\hbar^2 n^2} \tag{3.78}$$

For hydrogen atom, $Z = 1$ and

$$E_n^H = -\frac{m e^4}{2\hbar^2} \cdot \frac{1}{n^2} = \frac{-13.6}{n^2} eV \tag{3.79}$$

3.5.5 Energy Degeneracy: Discrete Spectrum

For many electron atoms, energy can not be written in the form given in Eq. 3.78 as the electrons interact to produce an n-electron state in which the energy depends on n and l of individual electrons in a very complicated way. However, we can still qualitatively characterize the individual electrons in many electron atoms in terms of their hydrogenic principal quantum number n (= 1, 2, $\cdots$) and hydrogenic orbital angular momentum quantum numbers $l (l = 0, 1, 2, \cdots)$. We have already seen that the projection of angular momentum of an atom on some axis is $m\hbar$ with $-l \leq m \leq l$. The integer 'm' is called the magnetic quantum number (magnetic because the reference is always to the axis (call it z-axis) along which an external magnetic field is applied). The quantum number 'm' can also be used to characterize the state of an individual electron with a given n and l and the corresponding wavefunction, called an orbital, will carry these quantum numbers as identifiers (Ψ_{nlm}). We must note here that the energy of one-electron atoms (hydrogen or hydrogenic) depends only on the principal quantum number n, which is just the sum of the radial and angular momentum quantum number n_r and l plus one. The quantum number 'm' determining the z-projection of the angular momentum does not enter into the expression of energy. Therefore the $(2l+1)$ number of orbitals of a given n_r and l quantum numbers constitute a $(2l+1)$-fold degenerate set of energy eigenfunctions – they have the same energy eigenvalues but different (mutually orthogonal or linearly independent) wavefunctions. For a given fixed principal quantum number 'n', l can assume values 0, 1, 2, $\cdots$, $n-1$ and for each 'l' here is $(2l+1)$-fold degeneracy because of rotational symmetry of the potential. The total degeneracy can be calculated easily as the sum $\sum_{l=0}^{n-1}(2l+1) = n^2$. In field free space, this degeneracy carries a signature of the rotational symmetry of all the central field potentials including the Coulomb potential, which guarantees that all rotations around the origin of the coordinate system leaves the Hamiltonian invariant or unchanged.

For the Coulomb potential we encounter a rather unexpected additional energy degeneracy, over and above the m-degeneracy. We have noted that the energy of the hydrogenic atom does not depend on the two quantum numbers n_r and l, separately but depends only on the sum n_r+l of the two. That means, states with different values of n_r and l such that their sum n_r+l is still the same, will have the same energy eigenvalue. Since different 'l' values correspond to very different spherical harmonics, they represent physically distinct states, they are indeed degenerate states. Such an extra degeneracy in Coulomb potential has been called accidental degeneracy. This degeneracy is special and does not exist in any other central field potential. We may mention that a profound connection exists between the accidental degeneracy of states in Coulomb potential in classical and quantum mechanics. In classical mechanics it has been shown that the Keplarian motion in a central field produces classical states with energy depending only on the sum of what are known as the adiabatic invariants, J_r and J_ϕ, but not on J_ϕ and J_r individually. It can be shown that a rather simple relationship exists between the classical adiabatic invariants and quantum numbers in a central field problem. One may therefore claim that the energy formula given in Eq. 3.77 is the quantum analogue of the classical formula for energy in terms of the two adiabatic invariants.

Let us consider the specific case of the $n=2$ level. We have already established that the degree of degeneracy of energy for a given n is n^2. For $n=2$, the degeneracy is therefore four-fold, meaning thereby that the energy levels corresponding to $n=2$ and $l=0$ and $n=2$ and $l=1$ are degenerate. The $n=2$, $l=1$ level is three-fold degenerate because of rotational symmetry but the degeneracy of states with $l=0$ and $l=1$ for $n=2$ is not rooted in any rotational symmetry. It can be shown that for the Coulomb potential, the Lenz vector

$$\vec{A} = \frac{1}{2m}(\vec{P}\times\vec{L} - \vec{L}\times\vec{P}) - \frac{l^2}{r}\vec{r} \tag{3.80}$$

is an additional conserved quantity (over and above $\vec{L}$) and the extra degeneracy is linked to this additional conserved quantity.[2,6]

3.5.6 Complete Wavefunction of a Hydrogen Atom

The full wavefunction of the hydrogen atom in the state characterized by the three quantum number n, l and m now reads

$$\Psi_{n,l,m}(r,\theta,\phi) = R_{nl}(r)Y_{lm}(\theta,\phi)$$

$$= \frac{\chi_{nl}(\zeta)}{r}Y_{lm}(\theta,\phi)$$

$$= N_{nlm}\frac{\zeta^{l+1}}{r}e^{-\zeta(2\epsilon)^{\frac{1}{2}}}\sum_{k=0}^{n-1}(a_k\zeta^k)Y_{lm}(\theta,\phi)$$

where $\zeta = \frac{me^2 r}{\hbar^2}$ and $\epsilon = \frac{\hbar^2}{me^2}|E_n|$ and N_{nlm} is the normalization constant.

The polynomial coefficients (a_ks) can be easily calculated by involving the recursion relation Eq. 3.74. However, a more elegant and practicable way to construct the radial wavefunctions of hydrogenic atom is to convert the radial wave equation (Eq. 3.58) into a well-known differential equation called the associated Laguerre equation, through a few substitution $R(r) = \frac{\chi(r)}{r}$ and use of the truncation condition leading to

$n = n_r + l + 1$. When these steps are carefully worked out, we arrive at the normalized radial wavefunction[7-8]

$$R_{nl}(r) = \frac{\chi(r)}{r}$$

$$= \left[\frac{(n-l-1)!(2k)^3}{2n[(n+l)!]^3}\right]^{\frac{1}{2}} (2kr)^l e^{-kr} L_{n+l}^{2l+1}(2kr) \tag{3.81}$$

where

$$k = \frac{\sqrt{2m|E_n|}}{\hbar} = \frac{mze^2}{n\hbar^2} = \frac{Z}{na_0}.$$

Here $a_0 = \frac{\hbar^2}{me^2} = 0.529$ Å, is called the first Bohr radius, E_n being given by the already derived formula (Eq. 3.78)

$$E_n = -\frac{mZ^2 e^4}{2\hbar^2 n^2} \tag{3.82}$$

$L_{n+l}^{2l+1}(2kr)$ is the associated Laguerre polynomial. The associated Laguerre polynomial $L_r^s(x)$ are solutions of the differential equation (associated Laguerre differential equation)

$$x(L_r^s)'' + (s+1-x)(L_r^s)' + (r-s)L_r^s = 0 \tag{3.83}$$

$L_r^s(x)$ can be shown like representable in the form

$$L_r^s(x) = \frac{d^s}{dx^s} e^x \frac{d^r}{dx^r}(e^{-x}x^r) \tag{3.84}$$

Before giving the analytical expressions for a few radial wavefunctions of the hydrogenic atoms, let us note that

(a) $R_{nl}(r)$ by virtue of the properties of the associated Laguerre polynomials has exactly $n_r = n - l - 1$ positive zeros or nodes excluding zeros at $r = 0, \infty$.

(b) $R_{nl}(r)$ does not depend on the quantum number 'm' – it depends only on n and l.

(c) $|R_{nl}(r)|^2 r^2 dr$ gives the radial position probability at a distance r from the origin.

(d) $|\Psi_{n,l,m}(r,\theta,\phi)|^2 r^2 dr \sin\theta d\theta d\phi$, on the other hand gives the probability of finding the electron in the volume element $drd\Omega(d\Omega = \sin\theta d\theta d\phi)$ at a distance r from the origin. The radial probability can be obtained by performing the integration over angular variables (θ, ϕ).

We list below a few of the lowest normalized radial wavefunctions specified by n and l quantum numbers (shell and subshell quantum numbers):

1. $n = 1$, $l = 0$ corresponds to K-shell and s-subshell (orbital): $R_{nl} \equiv R_{10}(r) = 2\left(\frac{Z}{a_0}\right)^{\frac{3}{2}} e^{-\frac{Zr}{a_0}}$.

2. $n = 2$, $l = 0$ corresponds to L-shell and s-subshell (orbital): $R_{20}(r) = 2\left(\frac{Z}{2a_0}\right)^{\frac{3}{2}}\left(1 - \frac{Zr}{2a_0}\right)e^{-\frac{Zr}{2a_0}}$.

3. $n = 2$, $l = 1$ corresponds to L-shell and p-subshell (orbitals): $R_{21}(r) = \frac{1}{\sqrt{3}}\left(\frac{Z}{2a_0}\right)^{\frac{3}{2}}\frac{Zr}{a_0}e^{-\frac{Zr}{2a_0}}$.

4. $n = 3$, $l = 0$ corresponds to M-shell and s-subshell (orbital): $R_{30}(r) = 2\left(\frac{Z}{3a_0}\right)^{\frac{3}{2}}\left\{1 - \frac{2Zr}{3a_0} + \frac{2(Zr)^2}{27a_0^2}\right\}e^{-\frac{Zr}{3a_0}}$.

5. $n = 3$, $l = 1$ corresponds to M-shell p-subshell (orbitals): $R_{31}(r) = \frac{4\sqrt{2}}{3}\left(\frac{Z}{3a_0}\right)^{\frac{3}{2}}\frac{Zr}{a_0}\left(1 - \frac{Zr}{6a_0}\right)e^{-\frac{Zr}{3a_0}}$.

6. $n = 3$, $l = 2$ corresponds to M-shell d-subshell (orbitals): $R_{32}(r) = \frac{2\sqrt{2}}{27\sqrt{5}}\left(\frac{Z}{3a_0}\right)^{\frac{3}{2}}\left(\frac{Zr}{a_0}\right)^2 e^{-\frac{Zr}{3a_0}}$.

In each case the complete wavefunction $\Psi_{n,l,m}(r,\theta,\phi)$ can be constructed by multiplying $R_{nl}(r)$ by the corresponding normalized spherical harmonics $N_{lm}Y_{lm}(\theta,\phi)$ already introduced. The transition energies in hydrogen atom ($Z = 1$) involving the initial state with principal quantum number (PQN) $= m$ and the final state with PQN $= n$ is given by $\Delta E_{mn} = \hbar\omega_{mn} = (\frac{1}{n^2} - \frac{1}{m^2})$ Rydberg $= 13.6(\frac{1}{n^2} - \frac{1}{m^2})eV$.

TABLE 3.1

Spectral Transition in Hydrogen Atom

Series name	PQN of final state	Orbital involved	Energy range
Lymer	$n = 1$	$p \rightarrow s$	ultra violet
Balmer	$n = 2$	$s \rightarrow p$	uv-visible
Paschen	$n = 3$	$d \rightarrow p$	IR
Bracket	$n = 4$	$f \rightarrow l$	IR

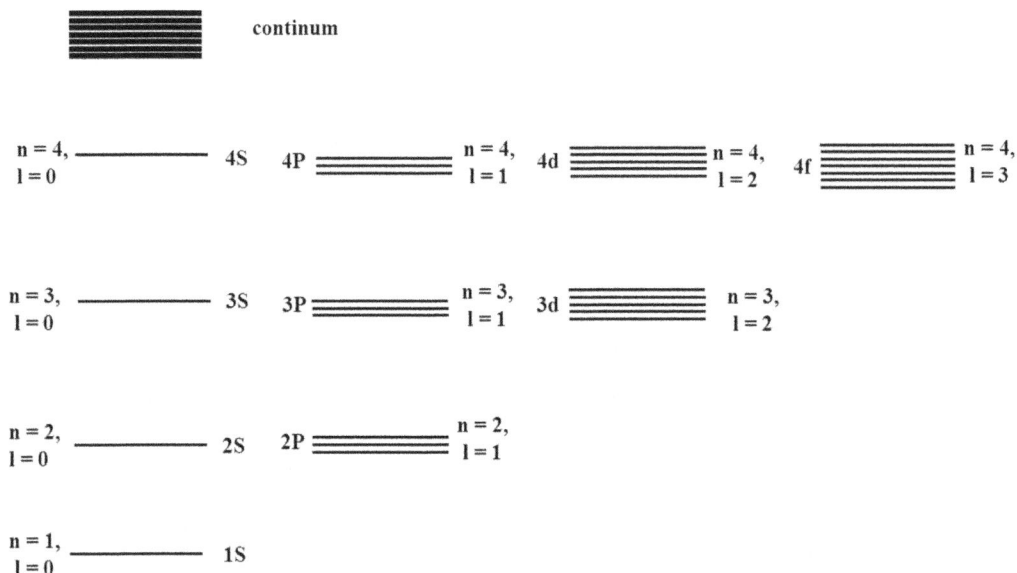

FIGURE 3.5   Energy levels of hydrogen atoms and their m and l degeneracies.

Under the electric dipole mechanism, only transitions that involve a change Δl in l-quantum number equal to ± 1 is allowed (it follows from parity selection rule). Using these facts, we can apply the transition energy formula to identify the spectral series of the hydrogen atom quite easily (Table 3.1).

The one-electron states in hydrogen atoms (orbitals) are designated by the principal quantum number (n) and angular momentum quantum number (l). These orbitals are ($2l + 1$)-fold degenerate (m-degeneracy). In addition, their is an additional l-degeneracy as orbitals with the same n-value but different l-values have the same energy. These energy levels are depicted in Figure 3.5.

3.6 Spin Angular Momentum

The electron in the ground state of a hydrogen atom has an orbital angular momentum $l = 0$. Does it have a magnetic moment? Now classical electromagnetic theory tells us that an electric current 'i' in a loop subtending an area 'A' has a magnetic moment 'μ_i',

$$\mu_i = iA.$$

An electron carrying charge 'e' moving with velocity 'v' in a circle of radius 'r' can be likened to an electric current $i = \frac{ev}{2\pi r}$, which in turn produces a magnetic moment

$$\vec{\mu_i} = \frac{e\,\vec{v}}{2\pi\,r} \cdot \pi\,r^2 = \frac{e(m\,\vec{v}\,r)}{2m} = \frac{e\,\vec{L}}{2m}$$

Changing over into operator notation we have

$$\vec{\mu_l} = \frac{e}{|e|} \cdot \frac{g_l \mu_B}{\hbar} \hat{L}$$

Where $g_l = 1$ is called the orbital gyromagnetic ratio, $\mu_B = \frac{|e|\hbar}{2m}$ is known as the Bohr magneton. A particle with magnetic moment $\vec{\mu}$ exposed to a magnetic field $\vec{B}$ experiences a torque (τ): $\tau = \vec{\mu} \times \vec{B}$ and its potential energy is given by

$$U(\theta) = -\vec{\mu} \cdot \vec{B}$$

Therefore, due to the presence of a magnetic field, the energy of the particle would be expected to change by an amount proportional to the angular momentum (L) projected on the magnetic field direction (Z-axis, L_Z). Classically, this change should be a continuous function of the orientation of the angular momentum vector. In quantum mechanics, as we have already seen, the projection of the angular momentum on the field axis are discrete and takes on values $m_l\hbar$ ($-l \le m_l \le l$, zero inclusive, and in steps of 1). A particle with angular momentum $l\hbar$ in an external magnetic field would therefore produce ($2l + 1$) non-degenerate energy eigenstates with energy

$$\Delta E_{m_l} = g_l \mu_B m_l B$$

This splitting of energy levels goes by the name of Zeeman effect. If the electron is in a state with $l = 0$, there should be no Zeeman splitting as $m_l = 0$. The uniform magnetic field exerts no net force on the particle with a magnetic dipole, and no displacement takes place. However, if the applied field is non-uniform (inhomogeneous) the situation changes. Assuming that the field has a gradient along the Z-axis, the net force on the particle in the Z-direction is

$$F_Z = -\frac{\partial}{\partial Z}(-\vec{\mu} \cdot \vec{B}) = \mu_Z \frac{\partial B}{\partial Z}$$

In the Stern-Gerlac experiment a collimated beam of the Ag-atoms was passed through a non-uniform magnetic field. The schematic sketch of the set up is displayed in Figure 3.6. In the ground state the silver atoms carry a net electronic angular momentum $L = 0$ (neglecting contributions from the nuclei). Therefore, the Z-projection of L is zero ($m_l\hbar = 0$ as $m_l = 0$). The beam was therefore expected to remain unsplit. If L were not equal to zero, the beam would have split into $2l + 1$ beams corresponding to $2l + 1$ values of m_l. If the atoms behaved classically, the beam would have continuous spread. In practice, they noted the beam was split into two sub-beams (Figure 3.6). The experiments indicate that the silver atoms in the ground state has a net angular momentum $j = \frac{1}{2}\hbar$ and the splitting into two-beams correspond to

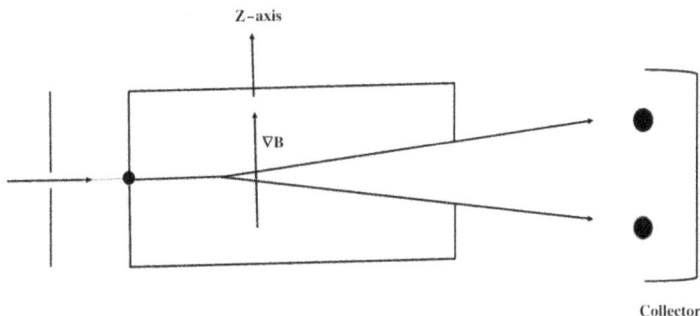

FIGURE 3.6 Sketch of a Stern-Garlach set up.

its Z-projections $m_j\hbar$ with $m_j = -\frac{1}{2}$ and $\frac{1}{2}$. Since Ag atoms have only one s-electron in the outermost shell, while all the inner electrons together produce a spherically symmetric electronic charge distribution with net electronic angular momentum $L = 0$, the observed half-integer angular momentum ($\frac{1}{2}\hbar$) of the outermost electron in the Ag atom must have nothing to do with orbital motion. It must be an additional intrinsic angular momentum that every single electron carries, which has no classical analog or root. It is an extra, essentially a quantum degree of freedom, called spin with spin angular momentum quantum number $s = \frac{1}{2}$. An atomic electron in an orbital with angular momentum $l\hbar$ will couple to spin degree of freedom, with angular momentum $\frac{1}{2}\hbar$, producing a total angular momentum $j\hbar = (l + \frac{1}{2})\hbar$ and its discrete $(2j + 1)$ number of Z-projections (or projections on any suitable defined directions) would be $m_j\hbar$ with $-j \leq m_j \leq l$. Just as the orbital angular momentum is represented by the operator $\hat{L}$ acting on the coordinates the spin angular momentum can be represented by an operator $\hat{S}$ that acts only on the spin degrees of freedom or spin coordinate(s) leaving the physical coordinates unaffected. The total angular momentum is then described by an operator $\hat{J}$ where

$$\hat{J} = \hat{L} + \hat{S}$$

As already mentioned spin is a purely quantum obervables having classical counterpart we must adopt matrix representation for the spin part. For half-integer spin angular momentum, $s = \frac{1}{2}\hbar$, we can use a basis set represented by two-component vectors:

$$\chi_+^z \equiv \chi_{\frac{1}{2},\frac{1}{2}} \equiv \begin{pmatrix} 1 \\ 0 \end{pmatrix} \equiv \alpha \tag{3.85a}$$

$$\chi_-^z \equiv \chi_{\frac{1}{2},-\frac{1}{2}} \equiv \begin{pmatrix} 0 \\ 1 \end{pmatrix} \equiv \beta \tag{3.85b}$$

The following representation for the spin operators can then be constructed in the χ-basis:

$$\hat{s}_x = \frac{\hbar}{2}\sigma_x; \quad \hat{s}_y = \frac{\hbar}{2}\sigma_y; \quad \hat{s}_z = \frac{\hbar}{2}\sigma_z$$

and

$$s_i^2 = \frac{\hbar^2}{4}I, \quad i = x, y, z$$

σ_x, σ_y, and σ_z are known as Pauli matrics where

$$\sigma_x = \begin{pmatrix} 0 & 1 \\ 1 & 0 \end{pmatrix}, \sigma_y = \begin{pmatrix} 0 & -i \\ i & 0 \end{pmatrix}, \sigma_z = \begin{pmatrix} 1 & 0 \\ 0 & -1 \end{pmatrix}, I = \begin{pmatrix} 1 & 0 \\ 0 & 1 \end{pmatrix} \tag{3.86}$$

Note that $\sigma_i^2 = I$ for $i = x, y, z$. Thus defined the eigenvalue equation for $\hat{s}^2$ and $\hat{s}_z$ operators of an electron become

$$\hat{s}^2 \chi_{s,m_s} = \hbar^2 s(s + 1)\chi_{s,m_s}$$

$$\hat{s}_z \chi_{s,m_s} = \hbar m_s \chi_{s,m_s}$$

where the possible values of the quantum numbers s and m_s are $s = \frac{1}{2}$, $-s \leq m_s \leq s$. For $s = \frac{1}{2}$, m_s has only two possible values, $m_s = \pm\frac{1}{2}$ which serve as important quantum numbers for labeling the single particle states in an atom. The spin operators introduced here satisfy the following commutation relations

$$\hat{s}_x\hat{s}_y - \hat{s}_y\hat{s}_x = \hbar\hat{s}_z$$
$$\hat{s}_y\hat{s}_z - \hat{s}_z\hat{s}_y = \hbar\hat{s}_x \tag{3.87}$$
$$\hat{s}_z\hat{s}_x - \hat{s}_x\hat{s}_z = \hbar\hat{s}_y$$

echoing the commutation relations among the components $\hat{L}_x$, $\hat{L}_y$ and $\hat{L}_z$ of the orbital angular momentum. An additional important feature of the spin operator concerns their anticommutation relations:

$$s_x s_y + s_y s_x = [s_x, s_y]_+ = 0$$

$$s_y s_z + s_z s_y = [s_y, s_z]_+ = 0 \tag{3.88}$$

$$s_z s_x + s_x s_z = [s_z, s_x]_+ = 0$$

The $[s_x, s_y]_+$, etc. are called the anticommutator. The anticommutation relations are characteristic of particles carrying half integer angular momentum $s = \frac{1}{2}\hbar$. They are called Fermions. The square of the spin angular momentum operator (matrix) of one electron is defined as

$$s^2 = s_x^2 + s_y^2 + s_z^2 = s(s+1)\hbar^2 = \frac{3}{4}\hbar^2$$

The spin (s) has its own associated magnetic moment (μ_s). Expressed in operator notation, we have for electrons

$$\hat{\mu}_s = \frac{g_s \mu_B \hat{s}}{\hbar} \tag{3.89}$$

The gyromagnetic ratio for electronic spin is $g_s = 2.000$.

The total magnetic moment operator is obtained by combining $\hat{\mu}_s$ and $\hat{\mu}_l$:

$$\begin{aligned} \hat{\mu} &= \hat{\mu}_s + \hat{\mu}_l \\ &= \frac{\mu_B}{\hbar}(g_s \hat{s} + g_l \hat{l}) \end{aligned} \tag{3.90}$$

Thus the quantum mechanic magnetic moment of an atom is not proportional to the angular momentum.

3.7 Total Angular Momentum (J): General Addition of Angular Momentum

Let us consider two independent angular momentum operators $\hat{J}_1$ and $\hat{J}_2$. Being independent they satisfy the following commutation condition:

$$[\hat{J}_1, \hat{J}_2] = 0 \tag{3.91}$$

Let χ_{j_1, m_1} and χ_{j_2, m_2} be the eigenfunctions of J_1^2 and J_{1z}, and J_2^2 and J_{2z}, respectively. The product states like

$$\chi_{j_1, m_1 j_2, m_2} = \chi_{j_1, m_1} \cdot \chi_{j_2, m_2} \tag{3.92}$$

are easily seen to be the simultaneous eigenstates of operators $\hat{J}_1^2$, $\hat{J}_{1z}$, $\hat{J}_2^2$ and $\hat{J}_{2z}$. The number of such product eigenstates are clearly $(2j_1+1) \cdot (2j_2+1)$ and together they constitute a complete basis for representing states labeled by the quantum numbers j_1, m_1, j_2 and m_2. For certain purposes, it may be more convenient to work in a different basis that carries as labels quantum numbers generated from the total angular momentum (J) where

$$\hat{J} = \hat{J}_1 + \hat{J}_2 \tag{3.93}$$

The components of $\hat{J}$ ($= \hat{J}_x, \hat{J}_y, \hat{J}_z$) satisfy the following conditions:

$$[\hat{J}_x, \hat{J}_y] = i\hbar \hat{J}_z$$

$$[\hat{J}_y, \hat{J}_z] = i\hbar \hat{J}_x$$

$$[\hat{J}_z, \hat{J}_x] = i\hbar \hat{J}_y$$

In addition, the following commutation condition also holds:

$$[\hat{J}^2, \hat{J}_1^2] = [\hat{J}^2, \hat{J}_2^2] = [\hat{J}^2, \hat{J}_z] = [\hat{J}_1^2, \hat{J}_z] = [\hat{J}_2^2, \hat{J}_z] = 0$$

There must therefore exist another basis set constructed from the eigenstates of $\hat{J}^2$ and $\hat{J}_z$, which are labeled by the quantum number j_1, j_2, j and m as follows:

$$\chi_{j_1,j_2,j,m} = \left[\chi_{j_1}\chi_{j_2}\right]_m^j \tag{3.94}$$

The two basis sets are equivalent and equally valid, a unitary transformation connecting one with the other. Thus,[9-10]

$$\chi_{j_1,j_2,j,m} = \sum_{m_1 m_2} C(j_1,m_1;j_2,m_2;j,m)\chi_{j_1,m_1,j_2,m_2} \tag{3.95}$$

The amplitudes or coefficients $C(j_1,m_1;j_2,m_2;j,m)$ are real numbers, called Wigner or Clebsch-Gordan coefficients. The sum over m_1, m_2 is subject to the restriction $m = m_1 + m_2$; j, j_1 and j_2 are also bounded by the following inequalities:

$$|j_1 - j_2| \geq j \leq j_1 + j_2 \tag{3.96}$$

If j_1 and j_2 are both integers or half-integers, j is an integer; j is however a half integer only if one of the two components j_1 and j_2 is a half-integer. The inverse unitary transformation[9-10] exists and is defined as follows:

$$\chi_{j_1,m_1}\chi_{j_2,m_2} = \sum_{j=|j_1-j_2|}^{(j_1+j_2)} C(j_1,m_1;j_2,m_2;j,m) \cdot \left[\chi_{j_1}\chi_{j_2}\right]_m^j \tag{3.97}$$

The Clebsch-Gordan coefficients have interesting and useful symmetry properties, which we list below:

$$C(j_1 m_1; j_2 m_2; jm) = (-1)^{j_1+j_2-j} \cdot C(j_1(-m_1);j_2(-m_2);j(-m))$$

$$= (-1)^{j_1+j_2-j} \cdot C(j_2 m_2; j_1 m_1; jm)$$

$$= (-1)^{j_1-m_1} \sqrt{\frac{2j+1}{2j_2+1}} \cdot C(j_1 m_1; j(-m); j_2(-m_2))$$

Let us note that the expansions in Eq. 3.96 or Eq. 3.97 follow from purely geometrical arguments. They therefore remain valid even if one or both the vectors are replaced by operators labeled by appropriate angular momentum quantum numbers. This results in selection rules for operator matrix elements of the type $\langle j_f m_f | A_{\lambda\mu} | j_i m_i \rangle$ where $\lambda\mu$ are angular momentum labels. One such rule for a spherically symmetric operator $\hat{A}$ tells us that the matrix element vanishes unless the angular momentum quantum numbers for both the initial and final states are the same. Since parity remain a good quantum number in a central field problem, a second selection rule (parity selection rule) demands that the product of parities $p_f\, p_{\hat{O}}\, p_i$ must be even. Many such selection rules can be worked out even when $A_{\lambda\mu}$ is not spherically symmetric. In Figure 3.7 we have displayed the coupling of two angular momentum vectors $j_1 = \frac{3}{2}$ and $j_2 = \frac{1}{2}$ producing the vector $j = 2$.

3.8 Many Electron Atoms: Aufbau Principle

The energy levels of hydrogen atoms, as we have seen already are entirely and completely determined by the principal quantum number 'n', which is also called shell-quantum number, $n = 1$ corresponds to K-shell, $n = 2$ corresponds to L-shell and so on. For a given value of n, we have n-sub shells corresponding to the 'n' possible value of orbital angular momentum quantum number 'l' for which the values allowed are $l = 0, 1, 2, \cdots, n-1$. $l = 0$ corresponds to s-subshell, $l = 1$ to p-subshell, $l = 2$ to d-subshell and $l = 3$ to the f-subshell. For a given n, and l, the angular momentum L may have non-zero projections on a chosen axis usually identified as the z-axis, leading to the appearance of the so called magnetic quantum number m_l ($-l \leq m_l \leq l$). The one-electron states (wavefunctions) in hydrogen atoms are characterized by the set of quantum numbers n, l, m_l. Now an electron in (n, l, m_l) state has spin angular momentum $s = \frac{1}{2}$, which can have only two well defined z-projections ($m_s\, \hbar$) corresponds to the quantum number $m_s = \frac{1}{2}$ or $-\frac{1}{2}$. The-one electron states are fully characterized by the set of quantum numbers n, l, m_l, m_s.

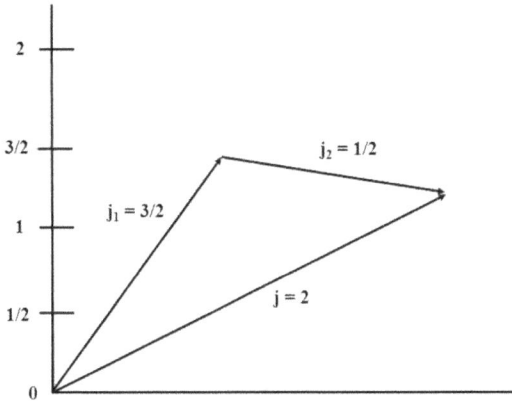

FIGURE 3.7   Coupling of two angular momentum vectors.

The Schrödinger equation for many electron atoms can not be solved analytically for the energy levels and the corresponding wavefunctions. Nevertheless, we can work out an approximate description of the ground state of many electron atoms (atomic number Z) by making use of the hydrogenic one-electron states (orbitals), which are supposed to be gradually filled up by adding 'Z' number of electrons. They tend to occupy the lowest available one-electron states (orbitals). This gradual filling of orbitals takes place so as to minimize the ground state energy of the Z-electron atoms and leads to the ground state electronic configuration of the atoms. It turns out that such configurations can be generated by following three principles.

(a) **Pauli exclusion principle**[11]: This principle states that no two electrons in an atom can occupy the same one-electron state labeled by the four quantum numbers: n, l, m_l and m_s. If the n, l, m_l quantum numbers of the two electrons are the same, the Z-projection of spin angular momentum of one must be $\frac{1}{2}\hbar$ ($m_s = \frac{1}{2}$) while for the other it would be $-\frac{1}{2}\hbar(m_s = -\frac{1}{2})$. The exclusion principle is a strict and fundamental principle and has roots in the requirement that many fermionic wavefunctions must be antisymmetric under the exchange of space and spin coordinates of any two electrons. It is easy to show that the exclusion principle immediately restricts the capacity of an atomic subshell with azimuthal quantum number 'l' to accommodate a maximum of $2(2l+1)$ number of electrons and a shell of principal quantum number 'n' to accommodate only $2n^2$ electrons. Thus the capacity of s, p, d, f subshells are fixed at 2, 6, 10 and 14 electrons, respectively while the capacity for K-shell ($n=1$) is 2, that of L-shell ($n=2$) is 8, M-shell ($n=3$) is 18, N-shell ($n=4$) is 32, so on.

(b) **Madelung's Rule**: The electrons tend to occupy the lowest available one-electron states (energy levels). That means the shell with $n=1$ is filled first, to be followed by the shells with $n=2$, 3 and so on. Such simple order of filling assumes that the Zth electrons being added to the atom of nuclear charge $+Ze$ feels the potential of the nucleus fully screened by $Z-1$ remaining electrons all of which are assumed to be located at the nucleus. Only in this rather ideal situation we obtain the hydrogenic one-electron states or energy levels with l degeneracy. That is, in a shell of a given principal quantum number 'n', all the subshells are degenerate in energy. However, if mutual interactions of the electrons are taken into account the l-degeneracy is lifted and the subshells in a particular shell are arranged in increasing order of l, i.e. within the same shell, s subshell is filled first, then the p, d and f subshells receive the electrons being added in that order. Electron interactions make the $4s$ subshell lower in energy than the $3d$ subshell or the $5s$ subshell. Similarly it turns out that the $6s$ subshell is lower in energy than the $5d$ as well as $4f$ subshells while $7s$ subshell lies lower in energy than the $5f$ subshell. An outcome of this ordering is that in many electron atoms the outer-most shell mostly contain s and p subshells. The d and f subshells are filled only when they belong to the first and the second inner shells, respectively. The sequence in which electronic subshells are filled is aptly coded by Madelung's law, which states that the filling of the one-electron states or energy levels proceeds in the increasing order of '$n+l$', n being the shell and l the subshell quantum

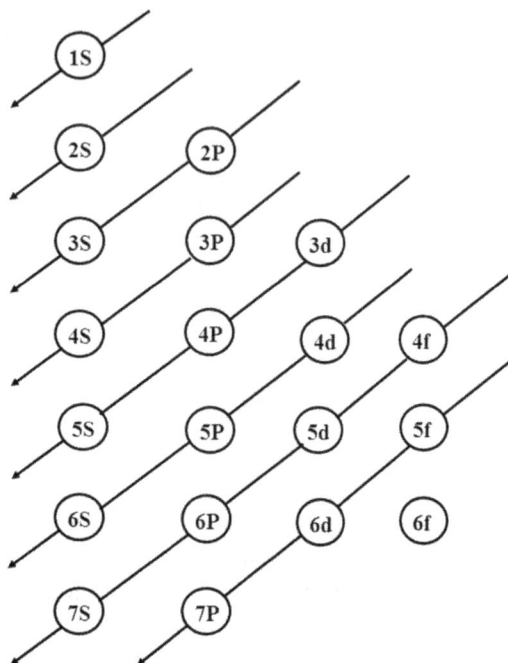

FIGURE 3.8   Filling order of orbitals under Madelung's rule.

numbers. In the case of levels having the same value of '$n + l$,' the filling follows increasing order of 'n' The diagram (Figure 3.8) explains how Madelung's rule operates.

(c) **Hund's Rule**[12]: It states that electrons tend to occupy degenerate orbitals (one-electron states) in an atom so as to keep the maximum number of spins parallel in the ground state. In other words, the ground state is chosen to maximize the spin multiplicity. Thus two electrons in $3d$ or $2p$ orbitals will lead to a spin triplet ($s = 1$, $2s + 1 = 3$) ground state rather than a singlet ($s = 0$, $2s + 1 = 1$) assuming all other lower energy orbitals are fully occupied by electrons of opposite spins.

3.8.1 Periods and Shells

We are now in a position to examine how different periods of Mendeleev's table are connected with the filling pattern of different shells and subshells. Thus, the first period begins and ends with the filling of the K-shell with $n = 1$ where the only subshell, the s-subshell is filled. Only two elements H ($1s$) and He ($1s^2$) constitute this period. The second period consists of eight elements (Li - Ne) that correspond to the filling of L-shell with principal quantum number $n = 2$ where of the two subshells the s-subshell is filled first followed by the p-subshell. The third period corresponds to the filling of the M-shell with $n = 3$ giving rise to eight elements (Na - Ar; $Z = 11$-18) in which $3s$ and $3p$ subshells are filled, just like what is seen in the second period. Although a part of M-shell, the $3d$ subshell remains vacant throughout the third period till the filling of the N-shell (n = 4) begins. Only after the $4s$ subshell of N shell is filled at $Z = 20$ (Ca), the $3d$, the first inner subshell (shell M) starts receiving electrons at $Z = 21$ (Sc). The filling of $3d$ subshell continues till Ni ($Z = 28$) and only then the normal filling order of the subshells begins at Cu ($Z = 29$), continuing upto $Z = 30$ (Zn). The fourth period therefore contains 18 elements with electrons distributed among the outer $4s$, $4p$ and inner $3d$ subshells. Note that $4f$' subshell remains unoccupied all along the fourth period in conformity with Madelung's rule. The next period ($n = 5$) is a rehearse of the previous (the fourth) period in which we see the gradual filling of the $5s$, $4d$ and $5p$ subshells giving rise to 18 elements (Rb - Xe). The sixth period ($n = 6$) is longer one with 32 elements [Cs ($z = 55$) – Rn ($Z = 86$)]. Here we see the filling of the $6s$ and $6p$ subshells (eight electrons) plus the first inner $5d$ subshell (ten electrons) and the second inner $4f$ subshell (14 electrons). This period is marked by the presence of 14 rather closely similar elements of the Lanthanide group of rare earths in which the inner $4f$ subshell is gradually filled. The seventh period witnesses the filling of the $7s$, $5f$, $6d$ and $7p$ subshells (Madelung's order) and is expected to be a repetition of the sixth period. But out of the expected 32 elements only 18 have been fully characterized so far. The actinide elements that are a part of the seventh period arise from the filling of the second inner $5f$ subshells [Th ($Z = 99$) – Lr ($Z = 103$)]. We may expect them to behave like the Lanthanide group of elements.[13] This period remains incomplete as of now.

3.8.2 Groups and Outer Shells

After establishing a clear connection between the electronic structure (filling up shells and subshells) and the occurrence of different periods in Mendeleev's table, we turn our attention to the periodicity of properties that Mendeleev discovered. The quantum mechanics of atoms quite naturally explains the root of the observed periodicity, which arises because of the periodicity noticed in the filling of the outer shells (ns, np), which can accommodate a maximum of eight electrons (two in ns and six in np) due to the operation of Pauli's principle. The outer shell electronic configuration largely determines the chemical as well as spectroscopic properties of atoms. No wonder then that periodicity in properties are seen to occur in Mendeleev's table at regular intervals where similar outer shell electronic configuration recur. In fact the elements in the table can be classified into eight groups, the group number reflecting the number of outer shell s and p electrons (Mendeleev's table, Figure 3.9).

Gr I of this table contains hydrogen (H) and the alkali metals (Li, Na, K, Rb, Cs, Fr), all having only one electron in the outermost shell (ns^1). The spectroscopic terms have doublet ($l = 0$, $s = \frac{1}{2}$) structure and these elements are all monovalent. Gr II, known as alkaline earths group consists of Be, Mg, Ca,

Periodic Table of Mendeleev (old form)

Periods	Group numbers of Elements									
	I	II	III	IV	V	VI	VII	VIII	IX	X
1	H 1						(H) 1	He 2		
2	Li 3	Be 4	B 5	C 6	N 7	O 8	F 9	Ne 10		
3	Na 11	Mg 12	Al 13	Si 14	P 15	S 16	Cl 17	Ar 18		
4	K 19 / Cu 29	Ca 20 / Zn 30	Sc 21 / Ga 31	Ti 22 / Ge 32	V 23 / As 33	Cr 24 / Se 34	Mn 25 / Br 35	Fe 26 / Kr 36	Co 27	Ni 28
5	Rb 37 / Ag 47	Sr 38 / Cd 48	Y 39 / In 49	Zr 40 / Sn 50	Nb 41 / Sb 51	Mo 42 / Te 52	Tc 43 / I 53	Ru 44 / Xe 54	Rh 45	Pd 46
6	Cs 55 / Au 79	Ba 56 / Hg 80	La 57 * / Tl 81	Hf 72 / Pb 82	Ta 73 / Bi 83	W 74 / Po 84	Re 75 / At 85	Os 76 / Rn 86	Ir 77	Pt 78
7	Fr 87	Ra 88	Ac 89 **							

* Lanthanide series	Ce 58	Pr 59	Nd 60	Pm 61	Sm 62	Eu 63	Gd 64	Tb 65	Dy 66	Ho 67	Er 68	Tm 69	Yb 70	Lu 71
** Actinide series	Th 90	Pa 91	U 92	Np 93	Pu 94	Am 95	Cm 96	Bk 97	Cf 98	Es 99	Fm 100	Md 101	No 102	Lr 103

Rare Earths

FIGURE 3.9   Mendeleev's periodic table (old form).

Ba, Sr and Ra all having two electrons in the valence (outermost shell). Chemically, they are divalent and the electronic configuration ns^2 suggests that the spectral terms would be singlets ($s = 0$) and triplets ($s = 1$). Gr III consists of the elements B, Al, Ga, In and Tl, all having three electrons in the outermost shell (configuration $ns^2 np^1$), which are maximally trivalent and give rise to the spectral terms with the maximal multiplicity of four (quartets). Similarly, group IV accommodates elements like carbon, silicon, germanium, stannum (tin/Sn) and lead (Pb), all having four electrons in the outer shell (configuration $ns^2 np^2$). Gr IV elements are maximally quadrivalent. Passing on to group V, we encounter elements like nitrogen, phosphorus, arsenic, antimony and bismuth, all having five electrons in the outermost shell (configuration $ns^2 np^3$) and have the maximum valency of 5. Gr VI contains elements like oxygen, sulphur, selenium, tellurium and polonium, all having six electrons in the outermost shell (configuration $ns^2 np^4$) and is followed by the group of halogens in Gr VII (following chlorine, bromine, iodine and astatine) all having seven electrons in the outermost shell, they have a valency of one (negative valency) exhibited in the ionic compounds. The maximum positive valency is, as expected 7. Halogens are followed by the group of chemically very inert elements – they are the monotonic inert gases, Helium, neon, argon, krypton, xenon and radon in each of which the outer shell has been completely occupied by electrons while the new shell is still without any electrons added to it. These elements are thus naturally assigned group VIII. A reference to Mendeleev's table makes it clear that the filling of the inner $3d$ subshell begins at $Z = 21$ (scandium), continuing upto nickel ($Z = 28$). Unlike in the previous periods, we may now define the group number by the number $3d$ electrons plus the number of electrons in the outer $4s$-subshell, which demands adding two new groups, group IX and X, in a very formal manner. Then Fe with $3d^6 4s^2$ configuration is assigned to group VIII, cobalt with $3d^7 4s^2$ to group IX and nickel with $3d^8 4s^2$ configuration to group X. This is a purely formal arrangement that makes Mendeleev's table more readable. In fact, Fe, Co and Ni are very similar in physical and chemical properties and should be in Gr VIII. The trio constitutes the first set of elements displaying ferromagnetic order (see Chapter 10), which has root in the unpaired spins of $3d$ electrons. Irrespective of their group numbers, the maximum valency of these elements is determined by the number of uncompensated spins.

After the ferromagnetic elements (Fe, Co, Ni) the $4s$ and then the $4p$ subshells start filling up from Cu ($Z = 29$) to krypton ($Z = 36$), the inert gas krypton marking the completion of the M-shell. The fifth period begins as expected, with an alkali metal rubidium (Rb, $Z = 37$) and repeating the pattern seen in the fourth period, terminates at another inert gas called xenon (Xe, $Z = 54$). The period accommodates ten transition metals, beginning with yttrium (Y, $Z = 39$) corresponding to the filling of the inner $4d$ subshell. As can be seen from Figure 3.9, group VIII accommodates three metals, ruthenium, rhodium and palladium, but unlike the Fe-Co-Ni group, the Ru-Rh-Pd triplet is not ferromagnetic. The $4d$ subshell is complete at Pd ($Z = 46$). From Ag ($Z = 47$) to Xe ($Z = 54$), we see the filling up of the outer $4s$ and $4p$ subshells, which is complete at Xe ($5s^2 5p^6$). The next period begins with the filling of the $6s$ subshell at Cs ($Z = 55$), which is complete at Ba ($Z = 56$). There are now two possibilities: either the inner $5d$ subshell starts filling up or second inner $4f$ subshell lying completely vacant starts receiving electrons. It turns out that after receiving one electron in the $5d$ subshell at Lanthanum ($Z = 57$) the filling of the second inner subshell $4f$ of the N-shell begins, giving rise to a group of very similar elements, called rare earths. The Pauli principle limits the capacity of f subshells to 14 and the rare earth group has exactly 14 elements beginning with cerium ($Z = 58$) and terminating at lutetium ($Z = 71$). These 14 elements have the same outermost subshell occupancy and are therefore chemically very similar. Just like the rare earth group, the seventh (the last) period is seen to accommodate 14 actinides elements, which appear just after actinium ($Z = 89$) and begin with thorium ($Z = 90$), ending at lawrencium (Lr, $Z = 103$). These 14 elements see the gradual filling of the low lying $5f$ orbitals of the O-shell with the $6s$, $6p$ and $7s$ subshells already fully occupied. We have to wait to see what happens to the accommodation of elements beyond Lr.

3.8.3 A Case Study of Two-Electrons Atoms: He

We have seen how the ground state electronic configurations of many electron atoms can be qualitatively determined. The next question concerns determination of their wavefunctions and energies. As a case study, we will take up the case of helium atom in the ground state. The Hamiltonian operator of this

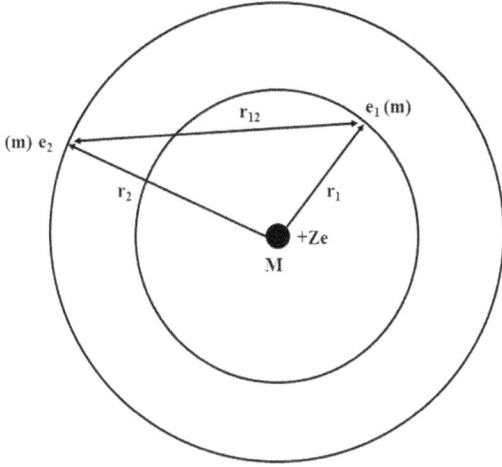

FIGURE 3.10 The Coulomb interactions in two-electron atoms of nuclear charge +Ze.

two-electron atom with atomic number $Z = 2$ can be written down easily following the rules of quantum mechanics (non-relativistic) and referring to Figure 3.10.

We assume the massive nuclei ($M = 4$) to be at rest at the origin of the coordinate system and $\vec{r_1}$, $\vec{r_2}$ are position vectors of electrons numbered 1 and 2 (they are indistinguishable).

$$\hat{H} = -\frac{\hbar^2}{2m}\nabla_1^2 - \frac{\hbar^2}{2m}\nabla_2^2 - \frac{Ze^2}{|\vec{r_1}|} - \frac{Ze^2}{|\vec{r_2}|} + \frac{e^2}{|\vec{r_1} - \vec{r_2}|}$$

$$= H_1 + H_2 + V_{12}$$

$$= T_1 + V_1 + T_2 + V_2 + V_{12}$$

$$\text{where} \quad \hat{H}_1 = -\frac{\hbar^2}{2m}\nabla_1^2 - \frac{Ze^2}{|\vec{r_1}|} = T_1 + V_1 \qquad (3.98)$$

$$\hat{H}_2 = -\frac{\hbar^2}{2m}\nabla_2^2 - \frac{Ze^2}{|\vec{r_2}|} = T_2 + V_2$$

$$V_{12} = \frac{e^2}{r_{12}}$$

$\hat{H}_1$ is the Hamiltonian describing the motion of one electron (e_1) in the field provided by the helium nucleus ($Z = 2$); while $\hat{H}_2$ represents the same for the other electron. $\hat{T}_1$ and $\hat{T}_2$ are the kinetic energy operators for electron 1 and 2, respectively while V_1 and V_2 are the potential energy operators for the electrons due to the interaction with the nucleus. V_{12} represents the potential energy operator, which accounts the mutual repulsive interaction between the two electrons. The task here is to find out the stationary state wavefunctions and energy eigenvalues of the Hamiltonian $\hat{H}$. The problem can not be solved exactly. It is possible however to obtain very accurate solutions by adopting systematically improvable approximation schemes, specially for the ground and low lying excited states. Before reviewing such methods, let us first consider the angular momentum of two-electron atoms and possible classification of two-electron states based on angular momenta. Since the Hamiltonian $\hat{H}$ does not contain spin explicitly, the eigenstates of $\hat{H}$ must also be the eigenstates of S^2 and S_Z, S being the total spin angular momentum and S_Z its projection on the Z-axis. Because of the purely central field character of $\hat{H}$, L and L_Z (L = total orbital angular momentum, L_Z being its projection on the Z-axis), may provide the quantum numbers for labeling the states. The calculation of the total angular momentum of two-electron atoms however, poses a very important question concerning the order in which the four angular momenta should be added. These four angular momenta are the two orbital angular momentum vectors $\vec{l_1}$ and $\vec{l_2}$ and two spin angular momentum vector $\vec{s_1}$ and $\vec{s_2}$.

In classical theory, the order would have been inconsequential. It is, however, not so in quantum mechanics. The vector addition here must take place only at the angles that ensure integral or half-integral values of their geometric sum, depending on whether the algebraic sum is integral or half-integral. Accordingly, the addition can be carried out in two ways:

(i) We add the orbital (l_1, l_2) and spin angular (s_1, s_2) momenta separately, producing the total orbital angular momentum (L) and total spin angular momentum (S), each producing integer values. Thus

$$\vec{L} = \vec{l_1} + \vec{l_2}$$

and

$$\vec{S} = \vec{s_1} + \vec{s_2}.$$

This coupling scheme, called the *LS* or Russell-Saunders coupling is commonly encountered in light elements and corresponds to the existence of two independent conservation laws for the orbital and spin angular momenta, which are conserved quantities and lead to good quantum numbers.

(ii) The second way of adding the angular momenta is to add up the orbital and spin angular momentum $(\vec{l_i}, \vec{s_i})$ for each electron (producing half-integral angular momentum value): $\vec{J_1} = \vec{l_1} + \vec{s_1}$; $\vec{J_2} = \vec{l_2} + \vec{s_2}$; and then add up $\vec{J_1}$ and $\vec{J_2}$ to get the total angular momentum $\vec{J}$ for the two electrons: $\vec{J} = \vec{J_1} + \vec{J_2}$, $\vec{J}$ having integral value. Such a coupling scheme is encountered in heavy elements and is known as $j - j$ coupling. Whether it is the *LS* coupling or $j - j$ coupling that prevails is decided by the relative strength of spin-orbit (magnetic) and Coulomb (electrostatic) interaction energies of the two electrons. It turns out that the spin-orbit (SO) interaction energy in atoms grows as Z^4 while the Coulomb energy remains proportional to Z (Z being the nuclear charge). Thus for heavy atoms the SO interaction energy dominates over Coulomb interaction energy, leading to the preference for $j - j$ coupling. A more detailed discussion is out of the scope of this work.

3.8.4 Designating Electronic States of a He Atom

The designation scheme for two-electron states of He is similar to what is used for designating states of one electron, except that the idea of principal quantum number now loses significance and we use uppercase Latin letters to represent the total orbital angular momentum L as S, P, D, F, etc., depending on what L is equal to ($L = 0 \equiv S$, $L = 1 \equiv P$, etc.). The value $2S + 1$ where S is the total spin angular momentum of the electrons is written as a superior prefix to the 'state symbol' and the total parity of the electrons $[(-1)^{l_1+l_2}$ and not $(-1)^L]$ is added as a superscript to the state symbol. Taking the case of two electrons with $n = 1$, $l = 0$, the configuration $(1s)^2$ produces the resultant $L = 0$, $S = 0$ ($2S + 1 = 1$) and $J = 0$. The resultant state of the shell is written as $(1s)^2 : {}^1S_0^g$. This case is particularly simple as only one resultant state can be realized. Similarly, the configuration $1s\,2s$ leads to $L = 0$, $S = 0$, 1 and $J = 0$, 1 (since the two electrons differ in their principal quantum number, we can disregard Pauli's principle in building up the two-electron state). Thus the resultant state designations are $L = 0$, $S = 0$, $J = 0$; $L = 0$, $S = 1$, $J = 1$ and we write them as $(1s\,2s): {}^1S_0^g$ and ${}^3S_1^g$. Similarly, $(1s\,2p)$ configuration produces the two-electron states with $L = 1$, $S = 0$, 1, $J = 2$, 1, 0 ($L+S$, $L+S-1$, $\cdots$, L-S). Thus four states are possible $(1s\,2p): {}^1P_1^u$, ${}^3P_2^u$, ${}^3P_1^u$, ${}^3P_2^u$, all being consistent with Pauli's principle as the principal quantum numbers of the two electrons are different ($n_1 = 1$, $n_2 = 2$). The case of $(np)^2$ configuration needs careful consideration as the principal quantum numbers are identical ($n_1 = n_2$) along with the orbital angular momentum quantum number ($l_1 = l_2 = 1$). Therefore, their magnetic or spin quantum numbers or both should differ, in order to be consistent with Pauli's principle. Let us note that each p-electron can be in one of the following six states $\Phi_i(m_l, m_s)$ designated by their m_l and m_s quantum numbers:

$$\Phi_1 : \left(1, \frac{1}{2}\right); \ \Phi_2 : \left(0, \frac{1}{2}\right); \ \Phi_3 : \left(-1, \frac{1}{2}\right); \ \Phi_4 : \left(1, -\frac{1}{2}\right); \ \Phi_5 : \left(0, -\frac{1}{2}\right); \ \Phi_6 : \left(-1, -\frac{1}{2}\right).$$

The two np electrons, can occupy any two of the six electron states $\Phi_1, \cdots, \Phi_6$ in ${}^6c_2 = 15$ different ways. There 15 states clearly differ in total orbital angular momentum L and total spin S as well as in

```
─────────────  1S 2P  $^1P_1^u$
```

```
                                        ─────────────  1S 2P  $^3P_0^u$
                                        ─────────────  1S 2P  $^3P_1^u$
                                        ─────────────  1S 2P  $^3P_2^u$
```

```
─────────────  1S 2S  $^1S_0^g$
```

```
                                        ─────────────  1S 2S  $^3S_0^g$
```

```
─────────────  1S²    $^1S_0^g$
```

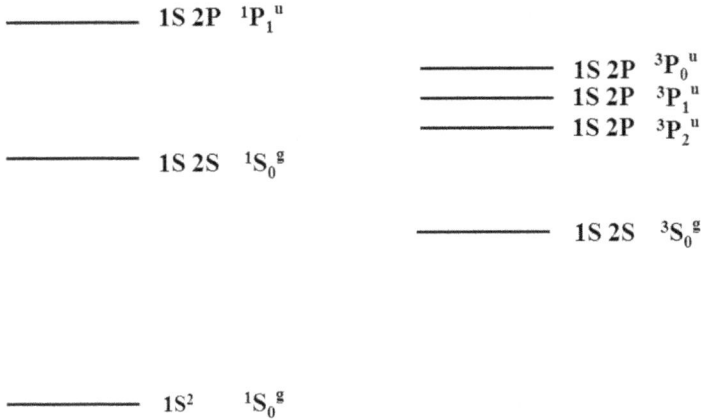

FIGURE 3.11 Energy levels of He atom (not drawn to scale).

their Z-projections. We know that a state with a given angular momentum is characterized by its maximum projection, that is, by the maximum possible value of L consistent with Pauli's principle, and the corresponding largest projection of S. Therefore, out of the 15 states, we must select only those with total projections, L_Z or $S_Z \geq 0$ as negative projection can not certainly be the largest. On that basis, we can choose the following eight (8) out of the fifteen possible two-electron states; each having non-negative Z-projections

$$\Phi_1\Phi_2 : (1,1); \Phi_1\Phi_3 : (0,1); \Phi_1\Phi_4 : (2,0); \Phi_1\Phi_5 : (1,0);$$

$$\Phi_1\Phi_6 : (0,0); \Phi_2\Phi_4 : (1,0); \Phi_2\Phi_5 : (0,0); \Phi_3\Phi_4 : (0,0).$$

We will now have to consider states having the largest angular momentum projections from among the states listed above. Let us consider the state that $\Phi_1\Phi_2$ represents. This state has an orbital angular momentum projection as well as a spin angular momentum projection equal to unity. Each of them can take on the value equal to zero (-ve values have been discarded by choice). That means we can discard states represented by $\Phi_1\Phi_3$, $\Phi_1\Phi_5$, $\Phi_1\Phi_6$ without any further argument. Let us next consider the state $\Phi_1\Phi_4$ with the maximum orbital angular momentum projection of 2. It has positive projections of 2, 1 and 0, which suggests that the states $\Phi_2\Phi_4$ and $\Phi_2\Phi_5$ can be left out considering them as possible projections of $\Phi_1\Phi_4$. That leaves out the state $\Phi_3\Phi_4$, which has no other valid projection. It is immaterial whether we accept $\Phi_1\Phi_5$ or $\Phi_2\Phi_4$ as projection of $\Phi_1\Phi_2$ (it will not affect counting the number of states). Thus the spectroscopic terms arising out of $(2p)^2$ configuration of He atom, for example are

$$\Phi_1\Phi_2 : (2p)^2 \ [^3P_2^g, {}^3P_1^g, {}^3P_0^g]$$

$$\Phi_1\Phi_4 : (2p)^2 \ [^1D_2^g]$$

$$\Phi_3\Phi_4 : (2p)^2 \ [^1S_0^g]$$

The same terms would be generated by any $(np)^2$ configuration. They will, however, be still higher in energy. An energy level diagram that includes the ground as well as few excited states of the He atom is sketched in Figure 3.11. The spectroscopic terms and orbital configuration leading to the terms are also indicated.

3.8.5 Constructing Wavefunctions for the Two-Electron States of the He Atom

We have seen how angular momentum coupling leads to designation of different spectroscopic states of helium atom. The next question is how do we construct the wavefunctions (approximate) representing the state? Since we are dealing with a two-electron atom we have to be careful that the wave-functions we construct conform to the requirement of Pauli's principle that such wavefunctions must be antisymmetric under the exchange of space and spin coordinates of the two electrons. The Hamiltonian does not contain spin in the non-relativistic description we are following, so spin has to be grafted into the wavefunction. For two-electron systems the task is simple:

(i) Choose a two-electron function of the spatial coordinates of the two-electron atom – call it Φ ($\vec{r_1}$, $\vec{r_2}$). It represents the space-part of the total wavefunction.

(ii) Construct eigenfunctions of the $\hat{S}^2$ operator ($\hat{S}$ is the total spin angular momentum operator of the two electrons) – call it η_{spin} (s_1, s_2), s_1 and s_2 being the spin coordinates of the two electrons. The spin coordinates, let us emphasize, are discrete and can assume only two values ($\pm\frac{1}{2}\hbar$). Since $\hat{H}(r_1,r_2)$ does not contain spin, H and S^2 commute. Therefore, eigenfunctions of $H(r_1,r_2)$ must also be eigenfunctions of $\hat{S}_1$.

(iii) Now we can construct the total wavefunction $\Psi(r_1,r_2,s_1,s_2)$ as a product of $\Phi_{space}(r_1,r_2)$ and $\eta_{spin}(s_1,s_2)$ such that Ψ is antisymmetric under the exchange of space and spin coordinates of the two electrons. That means $\Psi(r_2,r_1;s_2,s_1)=-\Psi(r_1,r_2;s_1,s_2)$. We note, however, that $\Phi_{space}(r_1,r_2)$ is an approximate eigenfunction of $H(r_1,r_2)$ while $\eta_{spin}(s_1,s_2)$ is an eigenfunction of S^2.

Let us suppose that $\alpha(1)$ and $\beta(1)$ are the eigenfunctions of the Z component of the spin operator S_z (1) of electron 1 with eigenvalues $\frac{\hbar}{2}$ and $-\frac{\hbar}{2}$, respectively, while $\alpha(2)$ and $\beta(2)$ are the corresponding eigenfunctions of S_z (2). Then it is easy to establish the following facts. $\eta_{spin}=\frac{1}{\sqrt{2}}[\alpha(1)\beta(2)-\alpha(2)\beta(1)]$ is an eigenfunction of S^2 operator ($S^2=S_1^2+S_2^2+2S_1S_2$) with the eigenvalue S = 0. Its multiplicity is (2S+1) = 1 and it represents a spin-singlet. It is also an eigenfunction of $\hat{S}_z(=\hat{S}_{z1}+\hat{S}_{z2})$ with eigenvalue $M_S\hbar$ with $M_S = 0$. On the other hand, the following three spin eigenfunctions are the components of a triplet:

$$\eta^1_{spin}=\frac{1}{\sqrt{2}}[\alpha(1)\beta(2)+\alpha(2)\beta(1)]$$
$$\eta^2_{spin}=\alpha(1)\alpha(2)$$
$$\eta^3_{spin}=\beta(1)\beta(2)$$

(3.99)

Each is an eigenfunction of S^2 operator with eigenvalue $s(s+1)\hbar^2$ with $s = 1$ (in units of $\hbar$) while η^1_{spin} is an eigenfunction of S_z with eigenvalue $M_S\hbar$ with $M_S = 0$. η^2_{spin} and η^3_{spin} are eigenfunctions of S_z with with eigenvalues $M_S\hbar$ where $M_S = 1$ and -1, respectively.

Let us now consider the space-part of the ground state wavefunction of He atom ($^1S_0^g$). The orbital configuration for the spectroscopic state is $1s^2 \equiv 1s(1)\ 1s(2) \equiv \Phi_{1s}(r_1)\ \Phi_{1s}(r_2)$ where $\Phi_{1s}(r_1)$ is a hydrogenic 1s orbital for the helium atom accommodating electron 1 described by the radial coordinate r_1 while $\Phi_{1s}(r_2)$ is the same function for electron 2 in the ground state of the helium atom. So, $\Phi_{space}(r_1,r_2) \equiv \Phi_{1s}(\vec{r_1})\Phi_{1s}(\vec{r_2}) \equiv \Phi_{space}(\vec{r_1},\vec{r_2})$. Note that $\Phi_{space}(\vec{r_1},\vec{r_2})$ thus defined is symmetric under interchange of the spatial coordinates of electrons 1 and 2 [$\Phi_{space}(\vec{r_1},\vec{r_2}) = \Phi_{1s}(\vec{r_2})\Phi_{1s}(\vec{r_1}) \equiv \Phi_{1s}(\vec{r_1})\Phi_{1s}(\vec{r_2})$] as the order of appearances of $\Phi_{1s}(\vec{r_1})$ and $\Phi_{1s}(\vec{r_2})$ is immaterial, it does not change the value of the function at any point in space. This symmetric space function must have to be multiplied with a η_{spin} (s_1, s_2) function that represent a spin-singlet ($M_s = 0$, $s = 0$) and is antisymmetric under exchange of the discrete spin coordinates s_1, s_2 of the two electrons. There is only one two-electron antisymmetric spin eigenfunction of S^2 and S_z with eigenvalues $S = 0\hbar$ and $S_z = 0\hbar$. That function is

$$\eta_{spin}(s_1,s_2;s=0,\mu_s=0)=\frac{1}{\sqrt{2}}[\alpha(1)\beta(2)-\alpha(2)\beta(1)] \equiv \eta_{spin}(s=0,M_s=0)$$

Note that $\eta_{spin}(s=0,M_s=0)$ is antisymmetric under the interchange of spin coordinates s_1, s_2. The total two-electron wave-function (approximate) of the helium atom in the ground state can therefore be represented as a product of $\Phi_{space}(r_1,r_2)$ and $\eta_{spin}(s=0,M_s=0)$:

$$\Psi^g(\vec{r_1},\vec{r_2},s=0,M_s=0)=\Phi^g_{spin}(r_1,r_2)\times\eta^u_{spin}(s=0,M_s=0)$$
$$=\Phi_{1s}(r_1)\Phi_{1s}(r_2)\times\frac{1}{\sqrt{2}}[\alpha(1)\beta(2)-\alpha(2)\beta(1)]$$

(3.100)

which is manifestly antisymmetric under the interchange of space-spin coordinates of the electrons 1 and 2. Thus $\Psi(\vec{r_1}, \vec{r_2}, s_1, s_2; s = 0, M_s = 0)$ can be accepted as an approximate ground state wavefunction of the helium atom with the correct exchange symmetry as required by Pauli's exclusion principle.

Let us now consider an orbitally excited configuration of the He atom, for example, (1s 2s). From angular momentum coupling rules, we have already seen that the orbital configuration (1s 2s) leads a $^1S_0^g$ and a $^3S_0^g$ states. For the singlet, the two-electron antisymmetric spin eigenfunction $\eta_{spin}(s_1, s_2; s = 0, M_s = 0)$ has already been determined. The space-part of the total wavefunction must obviously be a symmetric two-electron wavefunction $\Phi(\vec{r_1}, \vec{r_2})$ built up from approximate 1s and 2s hydrogenic wavefunctions or orbitals of He atom, which can be assumed to be mutually orthogonal and individually normalized to unity. Calling them $\Phi_{1s}(\vec{r})$ and $\Phi_{2s}(\vec{r})$, we can easily verify that the following function of $\vec{r_1}$ and $\vec{r_2}$ is symmetric under $1 \leftrightarrow 2$ interchange:

$$\Phi^g(\vec{r_1}, \vec{r_2}) = \{\Phi_{1s}(\vec{r_1})\Phi_{2s}(\vec{r_2}) + \Phi_{1s}(\vec{r_2})\Phi_{2s}(\vec{r_1})\}.$$

The total (approximate) wavefunction $\Psi(\vec{r_1}, \vec{r_2}, s_1, s_2; s = 0, M_s = 0)$ for the $^1S_0^g$ (1S 2S) state is

$$\Psi(\vec{r_1}, \vec{r_2}, s_1, s_2; s = 0, M_s = 0) = \Phi^g_{space}(\vec{r_1}, \vec{r_2}) \times \eta^u_{spin}(s_1, s_2, s = 0, M_s = 0)$$

$$= \{\Phi_{1s}(\vec{r_1})\Phi_{2s}(\vec{r_2}) + \Phi_{1s}(\vec{r_2})\Phi_{2s}(\vec{r_1})\} \times \qquad (3.101)$$

$$\frac{1}{\sqrt{2}}[\alpha(1)\beta(2) - \beta(1)\alpha(2)]$$

which is clearly antisymmetric and $1 \leftrightarrow 2$ exchange of both space and spin coordinates of the two electrons. Approximate wavefunctions for any other singlet spectroscopic state of He atom can be similarly constructed.

For the triplet state arising from the 1S 2S orbital configuration, the space-part of the wavefunction $\Phi(\vec{r_1}, \vec{r_2})$ must be antisymmetric as the three triplet spin-eigenfunctions corresponding to M_S equal to 0, 1, -1 are each symmetric under $1 \leftrightarrow 2$ transformation. Accordingly, the total wavefunctions for the triplet state (1S 2S: $^3S_0$) can be chosen as

$$\{\Phi_{1s}(\vec{r_1})\Phi_{2s}(\vec{r_2}) - \Phi_{1s}(\vec{r_2})\Phi_{2s}(\vec{r_1})\}\frac{1}{\sqrt{2}}[\alpha(1)\beta(2) + \alpha(2)\beta(1)](M_s = 0)$$

$$\frac{1}{\sqrt{2}}\{\Phi_{1s}(\vec{r_1})\Phi_{2s}(\vec{r_2}) - \Phi_{1s}(\vec{r_2})\Phi_{2s}(\vec{r_1})\}\alpha(1)\alpha(2)(M_s = 1) \qquad (3.102)$$

$$\frac{1}{\sqrt{2}}\{\Phi_{1s}(\vec{r_1})\Phi_{2s}(\vec{r_2}) - \Phi_{1s}(\vec{r_2})\Phi_{2s}(\vec{r_1})\}\beta(1)\beta(2)(M_s = -1)$$

The space-part, as noted earlier, is approximate as the orbitals Φ_{1s} and Φ_{2s} are only approximate one-electron wavefunctions for the He atom. We will now have to worry about the choice of these wavefunctions, the kind of energy values of these functions can generate and how to construct better solutions.

3.8.6 Calculating Energy of He Atom in the Ground State: Variational Approximation

We have already constructed an approximate wavefunction $\Psi(\vec{r_1}, \vec{r_2})$ (see Eq. 3.100) for the ground state. Let us note that the spatial functions $\Phi_{1s}(\vec{r_1})$ or $\Phi_{1s}(\vec{r_2})$ have not been completely specified. If we assume that the second electron has been removed from the He atom, $\Phi_{1s}(\vec{r_1})$ simply becomes the 1S orbital of He$^+$ ion, which is a hydrogenic ion for which exact wavefunctions and energies can be calculated (see Chapter 2). In that case, we can write

$$E_{1s}^{He^+} = -\frac{Z^2 e^2}{2a_0} \qquad (3.103a)$$

$$\Phi_{1s}^{He^+}(\vec{r}) = \frac{1}{\sqrt{\pi}}\left(\frac{Z}{a_0}\right)^{\frac{3}{2}} e^{-\frac{Z\vec{r}}{a_0}} \qquad (3.103b)$$

where Z is the nuclear charge carried by the He nucleus. If we disregard the presence of the second electron the obvious choice for Z is 2 as the single electron in the He atom feels the pull of the entire positive charge located in the He nucleus; for He$^+$, the choice of $Z = 2$ is correct and exact. But we are considering He atom and not He$^+$. When the second electron is there, an electron in the He atom does not feel the pull of the entire positive charge $(+2e)$ located in the nucleus as the other electron will partially screen it, leading to each electron seeing a somewhat lower value of positive charge – say $(2-\sigma)$ where σ is the screening constant. Thus a better choice for the one-electron function $\Phi_{1s}(\vec{r_1})$ for describing the He atom in the ground state would be

$$\Phi'_{1s}(\vec{r}) = \frac{1}{\sqrt{\pi}}\left(\frac{Z'}{a_0}\right)^{\frac{3}{2}} e^{\frac{-Z'r}{a_0}} \tag{3.104}$$

where Z' is the screened value of the nuclear charge that each electron sees in the He atom in the ground state. The question now is how do we choose the optimum value of Z'? One possibility is to choose the variational route as developed by Ritz, Hylleraas, and others.[14,15] The basic idea hinges on the fact the quantum mechanical average value of energy $< E >$ of a system described by a Hamiltonian (H), which has an energy spectrum that is bounded from below (i.e. it supports a ground state with energy E_0 such that $-\infty < E_0$), satisfies the inequality

$$< E >= \frac{\int \Psi^* H \Psi dv}{\int \Psi^* \Psi dv} \geq E_0 \tag{3.105}$$

where Ψ commonly called a trial function, is a normalizable function of coordinates appearing in the Hamiltonian. If the trial Ψ is already normalized to unity the inequality takes a simpler form

$$< E >= \int \Psi^* H \Psi dv \geq E_0 \tag{3.106}$$

That means the mean value of energy calculated with normalizable wavefunction Ψ is an upper bound to the exact ground state energy 'E_0' of the system. The result is entirely expected because the average value of energy $< E >$ calculated with approximate wavefunctions (normalized or normalizable) can not be lower than the lowest eigenvalue of the Hamiltonian 'H'. The equality sign holds only when the trial wavefunction happens to coincide with the exact ground state wavefunction (Ψ_0) of the system. A more comprehensive discussion on the variational principle and theorem will be presented later in the next chapter. For our purpose here, the inequality (Eq. 3.105 and Eq. 3.106) holds the key. Thus if Ψ- the so called trial wavefunction contains variable parameters $a_1, a_2, a_3, \cdots, a_n$, i.e. $\Psi \equiv \Psi(\vec{r}; a_1, a_2, \cdots, a_n)$. The average value of energy calculated with such a trial wavefunction will be a function of all such parameters $(< E >= E(a_1, a_2, \cdots, a_n))$. We can therefore find the optimum values of such parameters by setting up and solving the following equation:

$$\frac{\partial E}{\partial a_i} = 0 \quad (i = 1, 2, \cdots, n) \tag{3.107}$$

Let the solution of the equation above the designated as $a_1^*, a_2^*, \cdots, a_n^*$. Then $\Psi(\vec{r}; a_1^*, a_2^*, ..., a_n^*)$ is the best approximate wavefunction (of the chosen form) and the corresponding energy is the best approximation (upper bound) to the exact ground state energy (E_0). This is what is known as the Ritz variational method. For the He ground state, a normalized trial wavefunction including spin can be (Eq. 3.100) chosen as

$$\Psi(\vec{r_1}, \vec{r_2}) = \Phi_{1s}(\vec{r_1})\Phi_{1s}(\vec{r_2}) \times \frac{1}{\sqrt{2}}[\alpha(1)\beta(2) - \alpha(2)\beta(1)]$$

where (cf. Eq. 3.104)

$$\Phi_{1s}(\vec{r}) = \frac{1}{\sqrt{\pi}}\left(\frac{Z'}{a_0}\right)^{\frac{3}{2}} e^{-\frac{Z'r}{a_0}} \tag{3.108a}$$

$$\Psi(\vec{r_1}, \vec{r_2}) = \frac{1}{\pi}\left(\frac{Z'}{a_0}\right)^{\frac{3}{2}} e^{-\frac{Z'(\vec{r_1}+\vec{r_2})}{a_0}} \tag{3.108b}$$

The trial wavefunction in our case has only one optimizable parameter Z', the screened nuclear charge. The task now is to calculate the energy E as a function of Z', which can be done quite easily. The He Hamiltonian can be written as

$$H = \hat{T}_1 + \hat{V}_1 + \hat{T}_2 + \hat{V}_2 + \hat{V}_{12}$$

Considering the fact that H is free of spin coordinates and $\eta_{spin}(s_1, s_2)$ is normalized to unity, we can immediately write

$$< H > = < \Psi(\vec{r_1}, \vec{r_2})|H|\Psi(\vec{r_1}, \vec{r_2}) > \cdot < \eta_{spin}|\eta_{spin} > \ = \hat{T}_1 + \hat{V}_1 + \hat{T}_2 + \hat{V}_2 + \hat{V}_{12}$$

Since $\hat{H}$ does not contain spin, spin variables factor out and are integrated separately, the 'memory' of spin survives only in the symmetry of the spatial wavefunction under $1 \leftrightarrow 2$ exchange. Both the electron are in the same quantum state ($1s$), so we have the following equalities:

$$< T_1 > = < T_2 > ; < V_1 > = < V_2 >$$

so that

$$< H > = 2 < T_1 > + 2 < V_1 > + < V_{12} >$$

where

$$< T_1 > = \int \Phi_{1s}(\vec{r_1}, Z') \left(-\frac{\hbar^2}{2m} \nabla_1^2 \right) \Phi_{1s}(\vec{r_1}, Z') dv \tag{3.109a}$$

$$< V_1 > = - \int \{\Phi_{1s}(\vec{r_1}, Z')\}^2 \frac{Ze^2}{r_1} dv \tag{3.109b}$$

$$< V_{12} > = \int \Phi_{1s}^2(\vec{r_1}, Z') \frac{e^2}{|\vec{r_1} - \vec{r_2}|} \Phi_{1s}^2(\vec{r_2}, Z') dv \tag{3.109c}$$

$< T_1 >$ represents the average kinetic energy of an electron in the $1s$ state of a hydrogenic atom with nuclear charge $Z'e$. From quantum mechanical viral theorem

$$< T_1 > = -E_1 = \frac{Z'^2 e^2}{2a_0}$$

where E_1 is the ground state energy of the hydrogenic atom carrying nuclear charge of $Z'e$. Similarly, virial theorem leads to the relation $< V_1 > = 2E_1$. Replacing Z by Z' in the expression for $< V_1 >$ in Eq. 3.109b, and dividing by Z' to compensate we have

$$< V_1 > = \frac{Z}{Z'} E_1(Z') = \frac{-ZZ'e^2}{a_0}$$

The last hurdle is to calculate $< V_{12} >$, which can be done in a number of ways. The two-electron Coulomb integral in Eq. 3.109c when evaluated turns out to be

$$< V_{12} > = \frac{5}{8} \frac{Z'e^2}{a_0}$$

The total energy E as a function of Z' becomes

$$E(Z') = 2 < T_1 > + 2 < V_1 > + < V_{12} > \ = \frac{e^2}{a_0} (Z'^2 - 2ZZ' + \frac{5}{8} Z')$$

Differentiating with respect to Z' and setting the resulting expression to zero we have $Z' = (Z - \frac{5}{16}) = (Z - \sigma)$. For He, $Z = 2$ and thus Z' equals $2 - \frac{5}{16} = \frac{27}{16}$. σ the screening constant being equal to $\frac{5}{16}$. The calculated ground state energy is

$$E_0(\text{variational}) = -0.85 \frac{e^2}{a_0}$$

which is an upper bound to the exact ground state energy ($E_0 = -0.9\frac{e^2}{a_0}$). Considering the rather simple choice made for the trial wavefunction, the predicted energy is quite good. With a more flexible choice Ψ (trial) with additional parameters the quality of calculated energy becomes much better and approaches the exact ground state energy of He, always remaining above the exact energy as expected on variational ground. Hylleraas used a trial function, which was a product of $\Psi(\vec{r_1}, \vec{r_2})$ of Eq. 3.108b and a polynomial of degree 'n' in r_1, r_2 and r_{12} the coefficients of which are also treated as variational parameters in addition to Z' and obtained a theoretical value of the ground state energy of He in excellent agreement with the experimental value.

3.9 More on Variational Methods

(a) Since much of the practical approaches to calculation of electronic structures of atoms and molecules rely on variational approximation, it is expedient that we discuss the method in greater detail before turning our attention to molecular quantum mechanics in the next chapter. So far, we have always chosen the trial wavefunctions in specific forms. Let us suppose that we do not restrict ourselves to specific forms of trial wavefunctions, but vary the forms as well under the restriction (the only restriction) that the trial functions remain normalized, and proceed to make $< E >$ stationary with respect to 'all possible variations' in the forms of trial wavefunctions. It turns out that the stationary points of the energy functional are the eigenvalues of Hamiltonian. Let $\hat{H}$ be the Hamiltonian and Ψ the trial wavefunction of a single particle system (the single particle restriction is only for the sake of simplicity). Then the mean value of energy

$$< E > = < H > = \int \Psi^* H \Psi dv \tag{3.110}$$

is restricted only by the normalization constraint

$$\int \Psi^* \Psi dv = 1 \tag{3.111}$$

Now by varying $< E >$ with respect to Ψ and keeping in mind the hermiticity of $\hat{H}$ we have, for stationarity of $< E >$

$$\delta < E > = \int \{\delta\Psi^* \hat{H}\Psi + \delta\Psi \hat{H}^* \Psi^*\}dv = 0 \tag{3.112}$$

In Eq. 3.112 we can not treat $\delta\Psi$ and $\delta\Psi^*$ as independent arbitrary variations as they are interconnected through the normalization condition (Eq. 3.111). In order to remove this interdependence and decouple the variations, we vary the condition expressed in Eq. 3.111 leading to the condition

$$\int \delta\Psi^* \Psi dv + \int \Psi^* \delta\Psi dv = 0 \tag{3.113}$$

which is multiplied by a Lagrange multiplier λ (λ has the dimension of energy) chosen in a way that makes $\delta\Psi$ and $\delta\Psi^*$ independent and added to Eq. 3.112 to get

$$\int \delta\Psi^*(H + \lambda)\Psi dv + \int \delta\Psi(H^* + \lambda)\Psi^* dv = 0 \tag{3.114}$$

The mutually independent and arbitrary nature of variations $\delta\Psi^*$ and $\delta\Psi$ then lead to the following equation for Ψ and Ψ^*:[14]

$$(H + \lambda)\Psi = 0 = (H - E)\Psi \tag{3.115a}$$

$$(H^* + \lambda)\Psi^* = 0 = (H^* - E)\Psi^* \tag{3.115b}$$

where we have identified λ with $-E$. Thus, we have arrived at the Schrödinger equation for Ψ and Ψ^*. Eq. 3.115a and b are merely complex conjugate of each other since E must be real ($E^* = E$) because of the self-adjoint (Hermitian) character of $\hat{H}$. Thus the second equation (Eq. 3.115b) does not provide any additional information over and above whatever information is conveyed to us by Eq. 3.115a.

Since we have arrived at the SEs for Ψ and Ψ^* by making $<E>$ stationary with respect to all possible variations, the eigenvalues (E) of $\hat{H}$ are stationary points of the energy functional and it is possible to show that they are the only stationary points.[14,15] The question that remains concerns the nature of the stationary points. It can be shown that the stationary point corresponding to the lowest eigenvalue (the ground state, E_0) is an absolute minimum of the functional. It also transpires that the stationary points corresponding to the lowest state of each symmetry are also minimal points of the energy functional, provided that completely unrestricted (all possible) variations have been carried out. For other states the stationary points are merely saddle points of the functional. Variational calculations on such states must be carried out keeping the trial function orthogonal to all the lower (exact) eigenstates of the same symmetry.

(b) Restricted variation: In practical calculation, for interacting n-particle system, one must always choose a trial wave-function of a particular form like, for example, a product of n-single particle wave-functions or a sum of such products and vary the forms of such functions so as to make $<E>$ stationary under orthonormality constraints imposed on one-electron functions being varied. That leads to effective single particle Schrödinger like equation that the 'optimal' single particle functions must satisfy. Often, these equations are solved iteratively because of their non-linearity, yielding the best one-electron wave-functions the product of which provides the best ground state wavefunction of the chosen form. Such variational calculations are manifestly 'restricted' and the calculated energy is an upper bound to the exact ground state energy. It is possible to improve the upper bound by choosing the trial wavefunction in more flexible forms with additional variational degrees of freedom built in.

(c) A variational recipe for the ground state of a two-electron atom (the Hartree method): The two-electron trial wavefunction for the system is chosen in a product form:

$$\Psi(\vec{r_1}, \vec{r_2}) = \Phi_1(\vec{r_1})\Phi_2(\vec{r_2}) \tag{3.116}$$

The mean value of energy is then

$$<E> = \int \Psi^*(r_1, r_2)\hat{H}(1,2)\Psi(r_1, r_2)dv_1 dv_2 \tag{3.117}$$

where

$$H(1,2) = H_1 + H_2 + \frac{e^2}{r_{12}}$$
$$H_1 = T_1 - \frac{Ze^2}{r_1}; \quad H_2 = T_2 - \frac{Ze^2}{r_2} \tag{3.118}$$

T_1, T_2 are the kinetic energy operators for electrons 1 and electron 2 respectively, Z being the nuclear charge carried by the atom. The target is to minimize $<E>$ subject to the normalization constraint on Φ_1 and Φ_2 separately for each particle, which leads (no orthogonality condition is imposed)

$$\int \Phi_1^* \Phi_1 dv_1 = \int \Phi_2^* \Phi_2 dv_2 = 1 \tag{3.119a}$$

or in combined form as

$$\int \Phi_1^* \Phi_2^* \Phi_1 \Phi_2 dv_1 dv_2 = 1 \tag{3.119b}$$

With trial function of the form specified in Eq. 3.116 $<E>$ now reads

$$<E> = \int \Phi_1^*(r_1)\Phi_2^*(r_2)\left(H_1 + H_2 + \frac{e^2}{r_{12}}\right)\Phi_1(r_1)\Phi_2(r_2)dv_1 dv_2$$

Making $<E>$ stationary with respect to all variations in Φ_1 and Φ_2, separately we have the condition

$$\int \left[(\Phi_2^*\delta\Phi_1^* + \Phi_1^*\delta\Phi_2^*)\left(\hat{H}_1 + \hat{H}_2 + \frac{e^2}{r_{12}}\right)\Phi_1\Phi_2 + \Phi_1^*\Phi_2^*\left(\hat{H}_1 + \hat{H}_2 + \frac{e^2}{r_{12}}\right)\right.$$
$$\left.(\Phi_2\delta\Phi_1 + \Phi_1\delta\Phi_2)\right]dv_1 dv_2 = 0 \tag{3.120a}$$

The normalization condition specified in Eq. 3.119b leads to the following condition expressing the connection between the variations:

$$\int \left\{ \delta\Phi_1^*\Phi_2^*\Phi_1\Phi_2 + \Phi_1^*\delta\Phi_2^*\Phi_1\Phi_2 + \Phi_1^*\Phi_2^*\delta\Phi_1\Phi_2 + \Phi_1^*\Phi_2^*\Phi_1\delta\Phi_2 \right\} dv_1 dv_2 = 0 \qquad (3.120b)$$

Multiplying Eq. 3.120b by the Lagrange multiplier λ and adding the resulting expression to Eq. 3.120a under the assumption that we have chosen 'λ' so as to make all the variations $\delta\Phi_1^*, \delta\Phi_2^*, \delta\Phi_1, \delta\Phi_2$ independent, we arrive at the Hartree equations that Φ_1 and Φ_2 satisfy:

$$\left\{ \hat{H}_1 + \int \Phi_2^*\hat{H}_2\Phi_2 dv_2 + \int \Phi_2^* \frac{e^2}{r_{12}}\Phi_2 dv_2 - E \right\}\Phi_1 = 0 \qquad (3.121a)$$

$$\left\{ \hat{H}_2 + \int \Phi_1^*\hat{H}_1\Phi_1 dv_1 + \int \Phi_1^* \frac{e^2}{r_{12}}\Phi_1 dv_1 - E \right\}\Phi_2 = 0 \qquad (3.121b)$$

and a similar pair of equation for Φ_1^* and Φ_2^* where we have set $\lambda = -E$, as before (see Eq. 3.114 and Eq. 3.115). The pair of equations for Φ_1 and Φ_2 are the Hartree equations. In Eq. 3.121a, $\hat{H}_1$ represents the kinetic energy (KE) and electron-nucleus interaction operator for electron 1, the second term is just the average kinetic plus nuclear attraction energy of electron 2 residing in Φ_2 while the third term is an integral operator representing Coulomb repulsion due to electron 2 that has a charge density equal to $e\Phi_2^*\Phi_2$. Similar interpretation can be given for terms in Eq. 3.120b. Clearly, Eq. 3.121a describes the motion of electron 1 in the attractive field of the bare nuclear charge (Ze) and the repulsive Coulomb field of the second electron while Eq. 3.121b describes the motion of electron 2 in the attractive field provided by the nuclear charge (Ze) screened by the repulsive field due to electron 1. Since the operators on the left hand side of Eq. 3.121 depend on the function to be determined (Φ_2 in Eq. 3.121a and Φ_1 in Eq. 3.121b), there is a built in non-linearity in these equations and these integro-differential equations can only be solved iteratively by invoking suitable numerical methods, starting with two guessed solutions Φ_1^0 and Φ_2^0 and iterating until there is no further changes in

$$\epsilon_1 = \int \Phi_1^*\hat{H}_1\Phi_1 dv_1$$

and

$$\epsilon_2 = \int \Phi_2^*\hat{H}_2\Phi_1 dv_2$$

or in the total energy E_0. That is, the solutions do not change further. They are said to be self-consistent as input and output orbitals become practically same.

We note, however, that the operators determining Φ_1 and Φ_2 are hermitian, but different and therefore the solutions will not be mutually orthogonal (no orthogonality constraint was imposed during variation). The total Hartree energy of the system can be easily calculated by multiplying Eq. 3.121a with Φ_1^* and Eq. 3.121b by Φ_2^* and integrating over the space of electron 1 in the first case and over the space of the second electron in the second case, and taking the half-sums of the resulting equations. That leads to

$$E = \sum_{i=1}^{2} \int \Phi_i^*\hat{H}_i\Phi_i dv_i + \frac{1}{2}\sum_{i=1}^{2}\sum_{j=1(j\neq i)}^{2} \int\int \Phi_i^*\Phi_j^* \frac{e^2}{r_{ij}}\Phi_i\Phi_j dv_i dv_j$$

The $j \neq i$ restriction has been imposed so that the total energy expression is not vitiated by the presence of self-interaction energy of an electron. Although we have taken the explicit instance of a two-electron atom, we can proceed exactly similarly in the case of n-electron atoms. There will now be n integro-differential equation to be solved for the n-Hartree orbitals $\Phi_1, \Phi_2, \cdots, \Phi_n$, which are mutually non-orthogonal but individually normalized. The Hartree wavefunctions is more localized in space and lack the symmetry that n-particle Fermionic wavefunction is required to have. V. A. Fock in 1930 further generalized the Hartree method by including spin in the trial wavefunction, which was restricted to be antisymmetric with respect to interchange (transposition) of the space spin coordinates of any pair of electrons. Thus grew the Hartree-Fock self-consistent field method. Even today, it provides the zeroth order description of an interacting *n*-electron system. In the chapter that follows we will briefly review the two-electron Hartree-Fock method and then proceed to generalize it further to handle *n*-electron systems, especially molecules.

REFERENCES

1. L. Pauling and E. B. Wilson, *Introduction to Quantum Mechanics: With Application to Chemistry*, McGraw-Hill, New York and London (1935).
2. L. I. Schiff, *Quantum Mechanics*, Third Edition, McGraw-Hill, New York and London (1968).
3. A. S. Kompaneyets, *A Course of Theoretical Physics*, Volume I, (Translated from the Russian by V. Talmy), MIR Publishers, Moscow (1978).
4. A. A. Sokolov, I. M. Ternov and V. Ch. Zhukovskii, *Quantum Mechanics*, (Translated from Russian by Ram. S-Wadhwa), MIR Publishers, Moscow (1984).
5. H. A. Kramers, *Quantum Mechanics*, (Translated from German by D. Ter Haar), Dover Publications, Mineola, New York (1957).
6. V. A. Fock, *Fundamentals of Quantum Mechanics*, MIR Publishers, Moscow (1986).
7. F. Schwebl, *Quantum Mechanics*, (Translated from German by Ronald Kates), Springer International Students Edition: First Narosa Publishing House Reprint, New Delhi (1992).
8. H. Eyring, J. Walter and G. E. Kimball, *Quantum Chemistry*, John Wiley, New York and London (1944).
9. D. R. Bes, *Quantum Mechanics: A Modern and Concise Introductory Course*, Springer, Second Revised Edition, Berlin, Germany (2004).
10. M. Tinkhan, *Group Theory and Quantum Mechanics*, Tata McGraw-Hill Publishing, New Delhi (1974).
11. W. Pauli, 'Exclusion principle and quantum mechanics', *Z. Phy.* (German), 31, 765–783 (1926); Nobel Lecture, 13 December (1946).
12. F. Hund, 'On the interpretation of complex spectra, in particular of elements from scandium to nickel' *Z. Phy.* (German) 33, 345–371 (1928).
13. G. T. Seaborg, 'Prospect of further considerable extension of the periodic table', *J. Chem. Edu.*, 46, 626–634 (1969).
14. A. Mesiah, *Quantum Mechanics*, Volumes 1 and 2, Dover Publication, Mineola, New York (1999).
15. S. T. Epstein, *The Variation Method in Quantum Chemistry*, Academic Press, New York (1974).

4

Molecular Quantum Mechanics

4.1 Introduction: Molecules as Building Blocks

If we classify the material world around us into living and non-living worlds, the building blocks of the living world are certainly molecules. At a fundamental level, the living world is shaped by carbohydrate, lipid, protein and nucleic acid molecules consisting mainly of carbon, hydrogen, nitrogen and oxygen atoms held together in stable configurations by what are known as covalent bonds. These large molecules, often called biomolecules, interact with each other via weak intermolecular interaction, which shape their functionalities, recognition and many other properties. In the non-living world also, molecules do serve as building blocks. A gaseous sample of air contains molecules of nitrogen (N_2), oxygen (O_2), carbon dioxide (CO_2) and traces of inert monatomic gases (He, Ne, for example). These molecules interact via van der Waals forces and under appropriate conditions can exists in solid or liquid phases. The seas, oceans and lakes are made of molecules of water (H_2O), which exists under atmospheric condition as a liquid material by virtue of a very special kind of weak force of interaction called hydrogen bonding. At low temperature ($0°C$) water freezes into a crystalline solid material called ice. The structure of ice is supported by a network of hydrogen bonds between individual water molecules. Only in high temperature steam at low enough pressure individual water molecules probably exist. The point we are trying to emphasize here is that molecules serve as building blocks of many materials, which can be thought of as huge collections of interacting molecules. To understand the properties of such materials we must understand the properties of individual molecules and the kind of interactions that exist among them. Individual molecules, be they small, medium or large, belong to the microscopic world and require quantum mechanics for a complete description and understanding. Since a molecule must have at least two or more interacting atoms, the problem becomes more complicated even in the simplest case where there is only one electron moving in the field of two or more nuclei (e.g. H_2^+ or H_3^{++}). At the simplest level, it is still a many particle problem. The cornerstone of molecular quantum mechanics is the realization that there are two kinds of microscopic particle in molecules – the fast moving light particles called electrons (-vely charged, mass m_e) and the slow moving massive nuclei (+vely charged, mass $\sim M$) – roughly 2000 times more massive compared to electrons. That huge difference in masses ($m_e/M << 1$) allows us to treat the nuclei to be virtually at rest relative to the electrons. In other words, the fast moving electron virtually sees a frozen nuclear framework defined by fixed nuclear coordinates $\{R_i\}$. The frozen nuclear coordinates $\{R_i\}$ define molecular geometry and shape or what may be called a structure. The quantum mechanical Hamiltonian operator of an electron in this frozen nuclear framework approximation depends parametrically on the frozen nuclear coordinates $\{R_i\}$ through the electron-nucleus Coulomb interaction terms ($V_{Ne}(r,R)$) and the internuclear repulsion-energy $V_{NN}(R)$. The problem therefore boils down to solving the time-independent Schrödinger equation for the molecular electronic Hamiltonian $\hat{H}_e(r,R)$ the energy eigenvalues $[E_k(R)]$ which parametrically depends on the frozen nuclear coordinates (R). Thus, a different frozen nuclear framework R' produces a different set of eigenvalues $[E_k(R')]$ of the molecular electronic Hamiltonian. The internuclear repulsion energy $V_{NN}(R)$ for each framework is a constant that depends on R and not on the electronic coordinates. $E_k(R)$ plus $V_{NN}(R)$ define the so called adiabatic molecular potential energy surfaces for the different electronic states. We will denote them by $U_k(R)$, $k = 0, 1, 2, \cdots$. The global minimum of $U_0(R)$, for example, defines the ground state equilibrium molecular structure. If the set of nuclear coordinates at this global minimum of $U_0(R)$ are $\{R_i^0\} \equiv \{R_1^0, R_2^0, \cdots, R_p^0\}$, they define the equilibrium nuclear framework or geometry of the molecule in the electronic ground state. Similarly, the global minimum of $U_1(R)$ leads to

DOI: 10.1201/9781003244882-4

the molecular structure in the first excited state and so on to the second, third, $\cdots$, nth excited electronic states. This adiabatic separation of the motion of the fast and slow particles (electron and the nuclei) is the first step in the celebrated Börn-Oppenheimer theory (BOT) of molecules.[1-2] The BOT paves the way to understand molecular structures, potential energy surfaces and has become the cornerstone of molecular spectroscopy.

In the second step of BOT, the nuclei are unfrozen and allowed to move on the potential energy surface $U_i(R)$ generated in the first step. The corresponding time-independent Schrödinger equation for the Hamiltonian

$$H_N^{(i)} = \sum_{I=1}^{N_1} -\frac{\hbar^2}{M_I} \nabla_I^2 + U_i(R) = T_N + U_i(R)$$

where T_N is the sum of nuclear kinetic energy (KE) operators when solved yields the total energy eigen-values and eigenfunctions for nuclear motion and leads to the generation and understanding of vibrational and ro-vibrational spectra of molecules.

The frozen-nucleii approximation thus simplifies the problem of understanding molecules, their structures and spectra. At the same time, it creates problems. The molecular electronic Hamiltonians are devoid of the full rotational symmetry of the atomic Hamiltonians (Chapter 2). Consequently, we can no longer use the spherical harmonics directly as the angular parts of the eigenfunctions of the molecular electronic Hamiltonians. However, whatever residual symmetry is still retained in $\hat{H}_e(r,R)$ can be suitably exploited through the use of group theory. Before developing the BO theory for molecules any further, it would be prudent to consider the simplest of all molecules – the hydrogen molecule ion (H_2^+). This one-electron two nuclei system illustrates many aspects of the problem of calculating and understanding molecular electronic structure and can serve as the prototypical species for understanding molecules and molecular states in much the same way hydrogen atom serves as the building block in the understanding of atoms and atomic states.

4.2 The Quantum States of Hydrogen Molecule Ion (H_2^+)

If we assume that the nuclei of two hydrogen atoms (two protons) are at rest, we are led to consider the motion of only a single electron moving in the attractive Coulomb potential produced by the two proton (A, B) at fixed positions in space marked $\vec{R}_A$ and $\vec{R}_B$. The essentiality amounts to setting the nuclear kinetic energy operator to zero ($\vec{T}_N = 0$). The electronic positional coordinate is marked by $\vec{r}$ (Figure 4.1).

The Hamiltonian (H) of an electron in H_2^+ in the fixed nuclei approximation takes the following form:

$$\hat{H}_e(r,R) = -\frac{\hbar^2}{2m} \nabla^2 - \frac{e^2}{|\vec{r} - \vec{R}_A|} - \frac{e^2}{|\vec{r} - \vec{R}_B|} + \frac{e^2}{|\vec{R}_A - \vec{R}_B|} \tag{4.1}$$

The first term in $\hat{H}_e$ is the kinetic energy operator for the electron, the second and third terms represent the potential energy of the electron in attractive field of nuclei (protons) A and B, respectively. The last term accounts for repulsive potential energy due to interaction between the two nuclei. This term depends

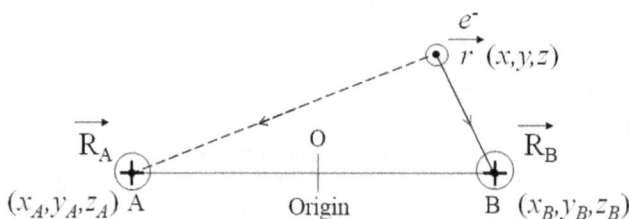

FIGURE 4.1 The framework of a hydrogen molecule ion.

only on the distance between the two fixed nuclei ($R = |\vec{R_A} - \vec{R_B}|$) and is independent of the electronic coordinates (x, y, z). The eigenvalue problem for $\hat{H}_e(r,R)$ is not separable in cartesian coordinates, but can be separated by transforming it to spherical coordinates and exactly solved. It is however, too complicated to be included here for our purpose. We will therefore try to construct approximate solutions (variational) to illustrate specific features of fixed nuclei solutions of the molecular electronic Schrödinger equation for H_2^+ ion.

Let us start by noting that at infinite separation, the H_2^+ ion will dissociate into $H_A + H_B^+(p)$ or $H_A^+(p)$ $+H_B$ as represented in the Figure 4.2. The electron in H_2^+ obviously and clearly does not see a spherically symmetric potential. The ion, however has reflection (inversion) symmetry about the origin located at the mid-point between the two nuclei, which is preserved whether 'R' is small or large or very large. That means, the eigenfunctions of $\hat{H}_e(r,R)$ must be either 'symmetric' (g) or 'antisymmetric' (u) under reflection through the origin whatever value the internuclear separation (R) may have. In the dissociation limit ($R \rightarrow \infty$), the situation I (electron attached to the hydrogen atom marked 'A', and a bare proton (p^+) at an infinitely large distance from 'A') is neither symmetric or antisymmetric with respect to inversion at the origin, and can not be an acceptable reality. On the same ground, we can reject the other situation (situation II) in which the electron is attached to the hydrogen atom marked B and a bare proton at infinite separation. Now the electron attached to the hydrogen atom H_A in situation I can be described by an eigenfunction of hydrogen atom with the origin shifted to the mid-point between the two-nucleii. For the ground state of H_2^+, this eigenfunction can be reasonably assumed to be the ground state energy eigenfunction of H_A (call it $\Phi_A(r)$) where

$$\Phi_A(r) = (\pi a_0^3)^{-\frac{1}{2}} \exp\left(\frac{-|\vec{r} - \vec{R_A}|}{a_0}\right) \tag{4.2a}$$

Similarly, in situation II, the electron attached to H_B can be described by the ground state wavefunction $\Phi_B(r)$ of H_B where

$$\Phi_B(r) = (\pi a_0^3)^{-\frac{1}{2}} \exp\left(\frac{-|\vec{r} - \vec{R_B}|}{a_0}\right) \tag{4.2b}$$

However, neither $\Phi_A(r)$ nor $\Phi_B(r)$ conforms to requirement of inversion symmetry that requires even approximate solutions to be either symmetric or antisymmetric with respect to reflection at the origin. The symmetry requirement can, however, be easily met by taking the following linear combinations of $\Phi_A(r)$ and $\Phi_B(r)$:

(i) $\Psi_+(\vec{r},R) = N_+(\Phi_A(\vec{r}) + \Phi_B(\vec{r}))$, N_+ being the normalization constant. Ψ_+ is symmetric under reflection through the origin; on reflection $\Phi_A(r)$ and $\Phi_B(r)$ merely interchange positions in the sum; thus, $\Psi_+(r)$ remains $\Psi_+(r)$.

(ii) $\Psi_-(\vec{r},R) = N_-(\Phi_A(\vec{r}) - \Phi_B(\vec{r}))$. On reflection at the origin $\Psi_-(r)$ becomes $N_-(\Phi_B(\vec{r}) - \Phi_A(\vec{r}))$, which is just $-\Psi_-(\vec{r},R)$. This function is antisymmetric under reflection or inversion at the origin.[2-4]

N_+ and N_- are normalization constants. The point to note is that $\Psi_+(\vec{r},R)$ remains symmetric under inversion whether the nuclei are close to each other or infinitely separated ($R \rightarrow \infty$) just as $\Psi_-(\vec{r},R)$ retains its antisymmetry for any value of R. The normalization constants can be determined from the condition:

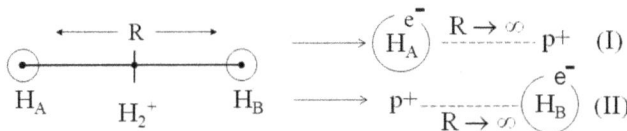

FIGURE 4.2 Two alternative dissociation modes for the H_2^+ ion.

$$\int |\Psi_+(\vec{r},R)|^2 dv = 1 = \int |\Psi_-(\vec{r},R)|^2 dv \qquad (4.3)$$

The first condition leads to the result

$$N_+ = \frac{1}{(2+2S(R))^{\frac{1}{2}}} = \frac{1}{\sqrt{2}(1+S(R))^{\frac{1}{2}}} \qquad (4.4a)$$

while the second leads to

$$N_- = \frac{1}{(2-2S(R))^{\frac{1}{2}}} = \frac{1}{\sqrt{2}(1-S(R))^{\frac{1}{2}}} \qquad (4.4b)$$

Thus, we have the following expressions for $\Psi_+(\vec{r})$ and $\Psi_-(\vec{r})$:

$$\Psi_\pm(\vec{r}) = \frac{1}{\sqrt{2(1+S)}}\{\Phi_A(\vec{r_1}) \pm \Phi_B(\vec{r_1})\} \qquad (4.4c)$$

$R = |\vec{R_A} - \vec{R_B}|$ measures the distance separating the two points. $S(R)$ is called the overlap integral defined as

$$S(R) = \int \Phi_A(\vec{r})\Phi_B(\vec{r})dv \qquad (4.5)$$

The integral in Eq. 4.5 can be evaluated easily giving

$$S(R) = \left(1 + \frac{R}{a_0} + \frac{R^2}{3a_0^2}\right)\exp\left(-\frac{R}{a_0}\right) \qquad (4.6)$$

Clearly $S(R) \to 0$ as $R \to \infty$, but remains positive under the
entire range of R from $0 \to \infty$. Speaking physically, S is a measure of the space shared by the two atom-centered function Φ_A and Φ_B. An electron located in the shared-space may be thought of as belonging to both the nuclei (A,B) simultaneously and contributing to process of bond formation (see later). The energy of H_2^+ ion in the state represented by $\Psi_+(\vec{r},R)$ or $\Psi_-(\vec{r},R)$ at the fixed value of the internuclear distance 'R' is given by the expectation value of the Hamiltonian $H(\vec{r},R)$ in the state concerned. Thus we have

$$U_+(R) = \int \Psi_+^*(\vec{r},R)\hat{H}_e(\vec{r},R)\Psi_+(\vec{r},R)dv \qquad (4.7a)$$

$$U_-(R) = \int \Psi_-^*(\vec{r},R)\hat{H}_e(\vec{r},R)\Psi_-(\vec{r},R)dv \qquad (4.7b)$$

$U_+(R)$ and $U_-(R)$ are the so called potential energy curves in the Ψ_+ and Ψ_- states of H_2^+. We will briefly sketch the steps leading to analytical expression for $U_+(R)$ and $U_-(R)$. The expression for expectation values, $U_+(R)$ and $U_-(R)$ (Eq. 4.7) can be written in expanded form as follows (using bra-ket notation)

$$U_\pm(R) = (2 \pm S)^{-1}[\langle\Phi_A|H_e|\Phi_A\rangle + \langle\Phi_B|H_e|\Phi_B\rangle \pm 2\langle\Phi_A|H_e|\Phi_B\rangle]$$
$$= (1+S)^{-1}[\langle\Phi_A|H_e|\Phi_A\rangle \pm \langle\Phi_A|H_e|\Phi_B\rangle] \qquad (4.8)$$

where,

$$\langle\Phi_A|H_e|\Phi_A\rangle = \langle\Phi_A|-\frac{\hbar^2}{2m}\nabla^2 - \frac{e^2}{|\vec{r}-\vec{R_A}|}|\Phi_A\rangle + \langle\Phi_A|-\frac{e^2}{|\vec{r}-\vec{R_B}|}|\Phi_A\rangle + \frac{e^2}{R}\langle\Phi_A|\Phi_A\rangle \qquad (4.9)$$

The first term in Eq. 4.9 is the expectation value of the Hamiltonian of the hydrogen atom 'A' in the state Φ_A, which has been assumed to be the $1s$ (ground state) of the hydrogen atom. So, it is just equal to the ground state energy of a hydrogen atom $E_0 = -1$ Ry. The second term represents the following integral

$$\langle\Phi_A|\frac{-e^2}{|\vec{r}-\vec{R_B}|}|\Phi_A\rangle = \int |\Phi_A|^2 \frac{-e^2}{|\vec{r}-\vec{R_B}|}dv$$

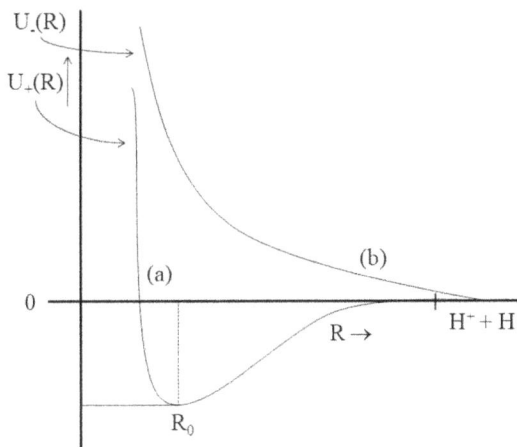

FIGURE 4.3 Ground and excited states of the H_2^+ ion.

which physically represents the attractive potential energy of the electron in the $1s$ state of the hydrogen atom A in the field of the bare atom B. The last term is simply $\frac{e^2}{R}$, the repulsive potential energy of the two protons separated by the distance R. The complete integral is evaluated readily and we have the result[2-4]

$$\langle \Phi_A | H_e(\vec{r},R) | \Phi_A \rangle = E_0 + \frac{e^2}{R}\left(1 + \frac{R}{R_0}\right)\exp\left(\frac{-2R}{a_0}\right) \tag{4.10}$$

where we have used the result that

$$\int \Psi_A^2(\vec{r}) \frac{e^2}{|\vec{r} - \vec{R_B}|} dv = \frac{e^2}{r}\left\{1 - \exp\left(\frac{-2R}{a_0}\right)\left(\frac{R}{R_0}+1\right)\right\} \tag{4.11}$$

The second term in Eq. 4.8 is more complex, but can be evaluated analytically and leads to

$$\langle \Phi_A(\vec{r}) | H_e(\vec{r},R) | \Phi_B(\vec{r}) \rangle = \left(E_0 + \frac{e^2}{r}\right)S(R) - \int \Phi_A(\vec{r})\Phi_B(\vec{r})dv + \frac{e^2}{|\vec{r} - \vec{R_B}|} \tag{4.12}$$

$$\langle \Phi_A(\vec{r}) | H_e(\vec{r},R) | \Phi_B(\vec{r}) \rangle = \left(E_0 + \frac{e^2}{r}\right)S(R) - \frac{e^2}{a_0}\left(1 + \frac{R}{a_0}\right)\exp\left(\frac{-2R}{a_0}\right) \tag{4.13}$$

The last integral on the right hand side of Eq. 4.12 is known as exchange integral $X(R)$ where

$$X(R) = \int \Phi_A(\vec{r})\Phi_B(\vec{r}) \frac{e^2}{|\vec{r} - \vec{R_B}|} dv$$

$$= \frac{e^2}{a_0}\left\{\left(1 + \frac{R}{a_0}\right)\exp\left(\frac{-2R}{a_0}\right)\right\} \tag{4.14}$$

Adding up, we have the analytical expressions for the potential energy curves:

$$U_\pm(R) = \left[1 \pm S(R)\right]^{-1}\left[E_0 + \frac{e^2}{R}\left(1 + \frac{R}{a_0}\right)\exp\left(\frac{-2R}{a_0}\right) \pm \left(E_0 + \frac{e^2}{R}\right)S(R) \mp \frac{e^2}{a_0}\left(1 + \frac{R}{a_0}\right)\exp\left(\frac{-R}{a_0}\right)\right] \tag{4.15}$$

It will be instructive now to examine $U_+(R)$ and $U_-(R)$ graphically as functions of R. The graphs are shown in Figure 4.3a-b. The plots reveal one striking feature:

$U_+(R)$ passes through a minimum at a certain value of 'R' $= R_0$ while $U_-(R)$ does not. Both the potential energy curves (PECs) however, approach E_0 as $R \to \infty$ indicating dissociation into a proton (H^+) and a hydrogen atom in its ground $1s$ state (-1 Ry $= \frac{e^2}{2a_0}$). At the minimum of $U_+(R)$, the energy of the H_2^+ ion is quite lower than E_0, the ground state energy of the H atom. Thus, the system state $\Psi_+(\vec{r},R)$ produces a

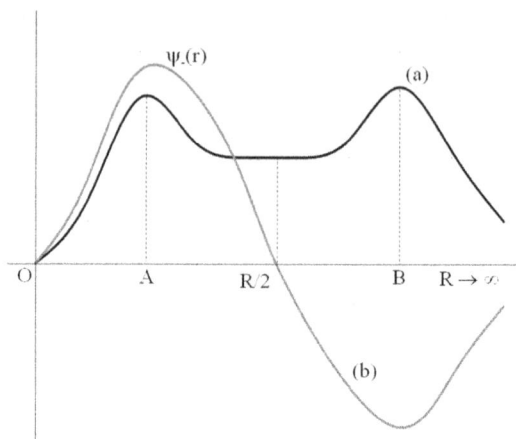

FIGURE 4.4　Plots of $\Psi_+(\vec{r})$ and $\Psi_-(\vec{r})$. Note that $\Psi_+(r)$ has zero slope at $R/2$ while $\Psi_-(r)$ changes sign at $R/2$.

bound H_2^+ ion while the antisymmetric $\Psi_-(\vec{r},R)$ fails to produce any such bound state. The reason is not far to seek. $\Psi_+(\vec{r},R)$ represents a constructive superposition of the two $1s$ functions (orbitals) centered on the atoms H_A and H_B, respectively, but in $\Psi_-(\vec{r},R)$, the two functions superpose destructively (they are in opposite phases). So, in the region between the two protons $\Psi_+(\vec{r},R)$ is larger than $\Psi_-(\vec{r},R)$, the latter vanishing on plane bisecting the line joining the two protons (A–B) as shown in Figure 4.4. $\Psi_+(\vec{r})$ is therefore said to describe a bonding one-electron state of H_2^+ (σ_g) while $\Psi_-(\vec{r})$ is called an antibonding one-electron state (σ_u) of H_2^+ ion. As we have already seen, the hydrogen molecule ion in the σ_u [$\Psi_-(\vec{r})$] state is not stable; it automatically dissociates into a proton (H^+) and a hydrogen atom in its ground state as indicated in the plot of $U_-(R)$ (see Figure 4.3). That is why it is called a repulsive potential. The minimum energy in the bonding state is predicted to occur at $R = R_0 = 1.32$ Å ($\frac{\partial U}{\partial R}|_{R=R_0} = 0$) and the binding energy $\Delta E_{binding}$ relative to the dissociated state $H^+ + H(1s)$ is estimated to be 1.76 eV. The exact values of the corresponding quantities obtained from the experiment or exact solution of the H_2^+ problem are $R_0 = 1.06$ Å and $\Delta E_{binding} = 2.79$ eV. The reason for the discrepancies are clear – the wavefunction $\Psi_+(\vec{r},R)$ is a crude approximation as it neglects the strong polarization of the $1s$ orbital on the hydrogen atom in the field of the proton has not been taken into account. In other words, the basis set of undistorted $1s$ orbitals of the H atom used to construct $\Psi_+(\vec{r},R)$ is physically inadequate to describe dissociation of the one-electron bonding state (σ_g molecular orbital) of the H_2^+.

4.3 The Quantum States of Hydrogen Molecule

The transition from H_2^+ to H_2 molecule appears to be conceptually simple. A second electron added to the H_2^+ ion produces the neutral hydrogen molecule at the given internuclear distance. The added electron can be accommodated in σ_g MO of H_2^+ leading to the orbital configuration σ_g^2 ($\sigma_g \downarrow\uparrow$). The presence of the second electron, in the same one-electron state of H_2^+, however, calls Pauli-exclusion principle into play and requires us to ensure that the antisymmetry of the total wavefunction under transposition of space as well as spin-coordinates of the two electrons is maintained. We make the following choice for the wavefunction for H_2 molecule in the ground state:

$$\Psi(\vec{r_1},\vec{r_2};S=0) = \Psi_+(\vec{r_1})\Psi_+(\vec{r_2})\eta_{spin}(s_1,s_2;S=0) \tag{4.16}$$

where $\Psi_+(\vec{r})$ represents the bonding one-electron state of H_2^+ ion at the given internuclear separation. Using the expression for $\Psi_+(\vec{r})$ in Eq. 4.4c we get the following form for $\Psi(\vec{r_1},\vec{r_2};S=0)$.

$$\Psi(\vec{r_1}, \vec{r_2}; S = 0; R) = \frac{1}{2(1+S)}\left\{(\Phi_A(\vec{r_1}) + \Phi_B(\vec{r_1}))(\Phi_A(\vec{r_2}) + \Phi_B(\vec{r_2}))\right\} \times$$

$$\eta_{spin}(s_1, s_2; S = 0) \tag{4.17}$$

$$\equiv \Psi_S(\vec{r_1}, \vec{r_2})$$

The two-electron spin function is necessarily the antisymmetric singlet spin eigenfunction of S^2 operator as two electrons occupy the same one-electron molecular state ($\Psi_+(r)$) forcing their spins to be antiparallel. Before proceeding further, let us have a look at Figure 4.5 showing how the electrons and nuclei are arranged at a given internuclear separation R at a certain instant of time, the lines joining the particles indicate interaction (single arrowhead denoting attraction, double arrowhead representing repulsion).

The nuclei A and B, both protons are frozen in space, their coordinate vectors being $\vec{R_A}$ and $\vec{R_B}$, respectively. The position vectors of electrons 1 and 2 from the common origin are $\vec{r_1}$ and $\vec{r_2}$, respectively. The Hamiltonian 'H' for the hydrogen molecule in the frozen nuclei approximation reads (nuclear kinetic energy terms are set equal to zero)

$$H = -\frac{\hbar^2}{2m}\nabla_1^2 - \frac{\hbar^2}{2m}\nabla_2^2 - \frac{e^2}{|\vec{r_1} - \vec{R_A}|} - \frac{e^2}{|\vec{r_1} - \vec{R_B}|} - \frac{e^2}{|\vec{r_2} - \vec{R_A}|} - \frac{e^2}{|\vec{r_2} - \vec{R_B}|} + \frac{e^2}{|\vec{r_1} - \vec{r_2}|} + \frac{e^2}{|\vec{R_A} - \vec{R_B}|} \tag{4.18}$$

Rearranging the terms on the right hand side of Eq. 4.18, we have

$$H = \left(-\frac{\hbar^2}{2m}\nabla_1^2 - \frac{e^2}{|\vec{r_1} - \vec{R_A}|} - \frac{e^2}{|\vec{r_1} - \vec{R_B}|}\right) + \left(-\frac{\hbar^2}{2m}\nabla_2^2 - \frac{e^2}{|\vec{r_2} - \vec{R_B}|} - \frac{e^2}{|\vec{r_2} - \vec{R_A}|}\right)$$

$$+ \frac{e^2}{|\vec{r_1} - \vec{r_2}|} + \frac{e^2}{|\vec{R_A} - \vec{R_B}|} \tag{4.19}$$

$$= (\hat{T}_1 + \hat{V}_{1A} + \hat{V}_{1B}) + (\hat{T}_2 + \hat{V}_{2A} + \hat{V}_{2B}) + \hat{V}_{12} + \hat{V}_{AB}$$

$$= \hat{H}_1 + \hat{H}_2 + \hat{V}_{12} + \hat{V}_{AB}$$

$\hat{H}_1$ represents the Hamiltonian of electron 1 in the attractive Coulomb fields of protons A and B while $\hat{H}_2$ represents the same entity for electron 2. $\hat{V}_{12}$ represents the operator for the mutual Coulomb repulsion between electrons e_1 and e_2 (Figure 4.5). $\hat{V}_{AB}$ does not depend on electronic conditions at all and represents the mutual repulsive potential energy of interaction between the two bare protons A and B. It depends only on the inter proton distance $R = |\vec{R_A} - \vec{R_B}|$ and therefore from the point of view of the electrons it can be regarded as a mere additive constant at each internuclear distance (R). It is now squarely possible to calculate the expectation value of energy by using singlet wavefunction $\Psi_S(\vec{r_1}, \vec{r_2})$ of Eq. 4.17 and the Hamiltonian $\hat{H}$ of Eq. 4.18. Instead of doing that, we will first examine the shortcomings of the ansatz proposed (this ansatz is called molecular orbital theory of hydrogen molecule since the space-part of the wavefunction is delocalized over the entire nuclear framework of the molecule). The first defect is that the wavefunction $\Psi_S(\vec{r_1}, \vec{r_2})$, at very short internuclear distances collapses into a product of 18 wavefunctions of hydrogen atom. But, the united atom limit of hydrogen molecule is clearly helium atom and we

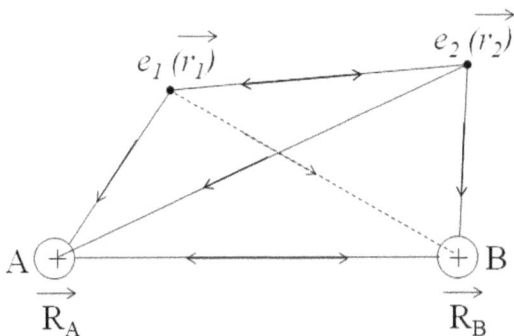

FIGURE 4.5 Interaction diagram for the H$_2$ molecule.

expect $\Psi_S(\vec{r_1},\vec{r_2})$ to turn over into products of $1s$ wavefunctions of the He atom. The shortcomings at the other extreme ($R \to \infty$) is even more glaring, and is revealed when we expand the space-part of the wavefunctions. Thus, expanding the product $\Psi_+(\vec{r_1}) \cdot \Psi_+(\vec{r_2})$, we get

$$\Psi_S(\vec{r_1},\vec{r_2}) \propto \left[\left\{ \Phi_A(\vec{r_1})\Phi_A(\vec{r_2}) + \Phi_B(\vec{r_1})\Phi_B(\vec{r_2}) \right\} + \left\{ \Phi_A(\vec{r_1})\Phi_B(\vec{r_2}) + \Phi_A(\vec{r_2})\Phi_B(\vec{r_1}) \right\} \right] \quad (4.20)$$

The terms inside the first pair of curly brackets reveal that both the electrons are resident in the $1s$ orbital of one hydrogen atom (A) or the other (B); so as $R \to \infty$, the H_2 molecule would produce $H^- + H^+$ as a product. The terms inside the second pair of curly brackets indicate dissociation into a pair of hydrogen atoms, each in its ground state ($H_2 \to H(1s) + H(1s)$ as $R \to \infty$). For a homopolar molecule like hydrogen, the second dissociation channel, i.e. dissociation into a pair of hydrogen atoms is energetically much lower than the dissociation channel producing $H^- + H^+$ ions; but the expression in Eq. 4.20 suggests that the ionic and atomic product will be produced equally as $R \to \infty$ – the weightage of the terms in the first and the second curly brackets being equal at all internuclear distances. The simple MO theory ansatz for describing the ground state energetics of the hydrogen molecule on the basis of one-electron states of H_2^+ as building blocks is not acceptable. An alternative formulation that ensures the correct homopolar dissociation as $R \to \infty$ was proposed by Heitler and London and goes by the name of valence bond theory of hydrogen molecule alias Heitler-London theory of covalent bond formation.[5]

4.4 Quantum Mechanics of Covalent Bond

The shortcomings of the simple MO theory in describing the behavior of the hydrogen molecule from the united atoms to the separated atoms limit as described in the preceeding section, calls for an alternative description that naturally grafts into itself the correct dissociation behavior. This alternative model is based on the one-electron states of the two hydrogen atoms as building blocks. Consider Figure 4.6: Electron 1 is in the ground state of the hydrogen atom marked 'A' and is resident in the orbital $\Phi_A(\vec{r})$, which is just the $1s$ orbital (call it $1s_A$). Similarly, electron 2 is resident in the orbital $\Phi_B(\vec{r})$ of the hydrogen atom marked B (call it $1s_B$). At infinite separation ($R \to \infty$) this distribution appears to be correct as the infinitely separated pair of electrons do not interact via Coulomb forces. Notice that we have neglected spin angular momentum of the electron at this point. However, the electrons are indistinguishable. Therefore, as alternative scenario in which electron 2 is resident on $\Phi_A(1s_A)$ with electron 1 in $\Phi_B(1s_B)$ is an equally valid description of the electron distribution. Energetically the two descriptions are degenerate. A valid spatial wavefunction for the ground state of hydrogen molecule must therefore be a superposition of the two degenerate descriptions: $\Phi_A(\vec{r_1}) \times \Phi_B(\vec{r_2})$ and $\Phi_A(\vec{r_2}) \times \Phi_B(\vec{r_1})$. Since they are energetically degenerate, they must appear in the superposition with equal amplitude, the phase being either $+1$ or -1. Thus choosing the phase of the superposition to be $+1$, we can write, for the space-part of the wavefunction for H_2:

$$\Psi_{space}(\vec{r_1},\vec{r_2}) = N_+ \left\{ \Phi_A(\vec{r_1})\Phi_B(\vec{r_2}) + \Phi_A(\vec{r_2})\Phi_B(\vec{r_1}) \right\} \quad (4.21)$$

N_+ being the normalization constant. Notice that the second product term within the curly brackets can be obtained from the first by merely exchanging the spatial coordinates of the two electrons and vice-versa. Hence $\Psi_{space}(\vec{r_1},\vec{r_2})$ is symmetric under $1 \leftrightarrow 2$ exchange. In order to conform to the requirements

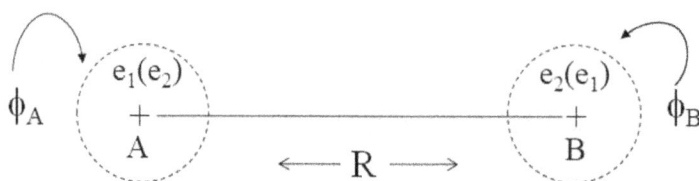

FIGURE 4.6   Electron exchange in the hydrogen molecule.

of the Pauli-exclusion principle $\Psi_{space}(\vec{r_1}, \vec{r_2})$ has to be multiplied with an antisymmetric two-electron spin eigenfunction of S^2 operator. That, as we have already seen in the previous chapter (cf. He ground state) is

$$\eta_{spin}^0(s_1, s_2) = \frac{1}{\sqrt{2}}\left\{\alpha(1)\beta(2) - \alpha(2)\beta(1)\right\}$$

$$\equiv \frac{1}{\sqrt{2}}\left\{|\uparrow\downarrow\rangle - |\downarrow\uparrow\rangle\right\}$$

$\eta_{spin}^0(s_1, s_2)$ as written corresponds to spin-singlet indicating spin paired states of the two electrons. Therefore, the correct (but approximate) choice for the ground state wavefunction for the H_2 molecule (correct in the sense that it dissociates into two hydrogen atoms in the $1s$ state) could be

$$\Psi_{HL}^0(1,2) = N_+\left\{\Phi_A(\vec{r_1})\Phi_B(\vec{r_2}) + \Phi_A(\vec{r_2})\Phi_B(\vec{r_1})\right\}\eta_{spin}^0(s_1, s_2) \tag{4.22}$$

Such a wavefunction for the H_2 molecule in its ground state, based on the products of atom centered or localized $1s$ orbital orbitals (they are non-orthogonal) of the two hydrogen atoms was first proposed and energetics of H_2 molecule formation tested by Heitler and London. It was hailed as the first successful theory of the 'covalent bond' that is supposed to hold the two hydrogen atoms together. Let us note however, that $\Psi_{HL}(1,2)$ suffers from the same drawback in the united atom limit ($R \rightarrow 0$) as the simple MO ansatz presented in section 4.3 does.

Another choice for the Heitler-London wavefunction could be to keep the space-part of the wavefunction antisymmetric under interchange of spatial coordinates of the two electrons (that corresponds to choosing the phase of superposition) while the spin wavefunction is chosen to remain symmetric under $1 \leftrightarrow 2$ interchange of spin coordinates. Such a spin-function would be a spin-triplet ($S = 1$), the $M_S = 0$ component of it being

$$\eta_{spin}^1(s_1, s_2) = \frac{1}{\sqrt{2}}\left\{\alpha(1)\beta(2) + \alpha(2)\beta(1)\right\}$$

$$\equiv \left\{|\uparrow\downarrow\rangle + |\downarrow\uparrow\rangle\right\}$$

Thus, the other choice for $\Psi_{HL}(1,2)$ would represent a triplet state of the H_2 molecule and we can readily write

$$\Psi_{HL}^1(1,2) = N_-\left\{\Phi_A(1)\Phi_B(2) - \Phi_A(2)\Phi_B(1)\right\} \times \eta_{spin}^1(s_1, s_2) \tag{4.23}$$

The question now is: Is H_2 molecule in its ground state a spin-singlet or a spin-triplet? The answer lies in the energetics, which we describe in what follows.

4.4.1 Energetics of Covalent Bond in H_2

Heitler-London used a perturbative method to calculate the energetics of bonding. What we are using is a variational recipe for the same problem. The variational parameter in our case has been fixed by symmetry. To understand the energetics of bond formation in the H_2 molecule we have to calculate the expectation values of the energy operator ($\hat{H}$ of Eq. 4.19) in the states represented by $\Psi_{HL}^0(1,2)$ and $\Psi_{HL}^1(1,2)$, respectively and examine the energy profiles ($E_+(R)$ and $E_-(R)$) as functions of internuclear distance just as we did in the case of H_2^+ ion, and find out which of them leads to a bound state with a deeper minimum or higher binding energy.

It is easy to find that

$$N_\pm = \frac{1}{\sqrt{2(1 \pm s^2)}} \tag{4.24}$$

where '+' sign refers to the Heitler-London singlet wavefunction $\Psi_{HL}^0(1,2)$ of Eq. 4.4 and the '-' sign to the HL triplet states wavefunction $\Psi_{HL}^1(1,2)$ of Eq. 4.23. The expectation value of '$\hat{H}$' in the two states are (using bra-ket notations)

$$\epsilon_+(R) = \langle \Psi_{HL}^0(1,2)| \hat{H} |\Psi_{HL}^0(1,2)\rangle \tag{4.25a}$$

$$\epsilon_-(R) = \langle \Psi_{HL}^1(1,2)| \hat{H} |\Psi_{HL}^1(1,2)\rangle \tag{4.25b}$$

where for the singlet state we have

$$\Psi_{HL}^S = \Psi_{HL}^0(1,2) = \frac{1}{\sqrt{2(1+S^2)}}\left\{\Phi_A(\vec{r_1})\Phi_B(\vec{r_2}) + \Phi_A(\vec{r_2})\Phi_B(\vec{r_1})\right\}\eta_{spin}^0(s_1,s_2) \tag{4.26a}$$

and for the triplet HL state

$$\Psi_{HL}^T = \Psi_{HL}^1(1,2) = \frac{1}{\sqrt{2(1-S^2)}}\left\{\Phi_A(\vec{r_1})\Phi_B(\vec{r_2}) - \Phi_A(\vec{r_2})\Phi_B(\vec{r_1})\right\}\eta_{spin}^1(s_1,s_2) \tag{4.26b}$$

the superscript '0' denoting singlet and '1' denoting triplet states.

The expectation value of $\hat{H}$ of Eq. 4.18-Eq. 4.19 in the two HL states then yields $\epsilon_+(R)$ and $\epsilon_-(R)$, which are upper bounds to the exact energy $E_{singlet}$ and $E_{triplet}$, respectively. Now replacing $\Phi_A(\vec{r})$ and $\Phi_B(\vec{r})$ by the ket vectors $|a\rangle$ and $|b\rangle$, respectively we can write Eqs. 4.25a and b as

$$\epsilon_\pm(R) = \frac{1}{1\pm S^2}\left\{\langle ab|\hat{H}|ab\rangle \pm \langle ab|\hat{H}|ba\rangle\right\} \tag{4.27}$$

where

$$\langle ab|\hat{H}|ab\rangle = \int\int dv_1 dv_2 \Phi_A(\vec{r_1})\Phi_B(\vec{r_2})H(\hat{1},2)\Phi_A(\vec{r_1})\Phi_B(\vec{r_2}) = \langle ab|\hat{H}|ba\rangle \tag{4.28a}$$

and

$$\langle ab|\hat{H}|ab\rangle = \int\int dv_1 dv_2 \Phi_A(\vec{r_1})\Phi_B(\vec{r_2})H(\hat{1},2)\Phi_A(\vec{r_2})\Phi_B(\vec{r_1}) = \langle ba|ab\rangle \tag{4.28b}$$

Now considering electron 1 or 2 resident in the $1S_a$ or $1S_b$ orbitals ($1S_a \equiv \Phi_A$, $1S_b \equiv \Phi_B$) we can write the following energy eigenvalue equation (time-dependent SE):

$$\left(-\frac{\hbar^2}{2m}\nabla_1^2 - \frac{e^2}{|\vec{r_1} - \vec{R_A}|}\right)\Phi_A(\vec{r_1}) = E_1\Phi_A(\vec{r_1}) \tag{4.29a}$$

$$\left(-\frac{\hbar^2}{2m}\nabla_1^2 - \frac{e^2}{|\vec{r_2} - \vec{R_B}|}\right)\Phi_B(\vec{r_2}) = E_1\Phi_B(\vec{r_2}) \tag{4.29b}$$

Then $\langle ab|\hat{H}|ab\rangle$ can be easily shown to be equal to the following expression: $\langle ab|\hat{H}|ab\rangle = 2E_1 + C$, where 'C' stands for what is known as the Coulomb energy and is expressed as the integral

$$Q = \int dv_1 \int dv_2 \Psi_A^2(\vec{r_1})\Psi_B^2(\vec{r_1}) \times \left\{\frac{e^2}{|\vec{r_1} - \vec{r_2}|} - \frac{e^2}{|\vec{r_1} - \vec{R_B}|} - \frac{e^2}{|\vec{r_2} - \vec{R_A}|} + \frac{e^2}{R}\right\} \tag{4.30}$$

Splitting into various terms, we have[2]

$$Q = -2\int dv_1 \frac{e^2}{|\vec{r_1} - \vec{R_B}|}\Psi_A^2(\vec{r_1}) + \int\int dv_1 dv_2 \Psi_A^2(\vec{r_1})\frac{e^2}{|\vec{r_1} - \vec{r_2}|}\Psi_B^2(\vec{r_2}) + \frac{e^2}{R} \tag{4.31}$$

The first term on the right hand side of Eq. 4.31 represents twice the Coulomb interaction energy of an electron localized in the orbital $\Phi_A(r)$ centered on the nucleus A, with the nucleus B, which by the basic symmetry of the problem must be equal to the Coulomb energy of the electron located in orbital Φ_B centered on the nucleus B (hence the factor 2). The second term clearly represents the repulsive Coulomb energy between the two electrons localized in the orbitals $\Phi_A(\vec{r_1})$ and $\Phi_B(\vec{r_2})$, respectively. The third term in Eq. 4.28 merely represents the Coulomb repulsion between the two fixed nuclei separated by a distance R.

Similarly it is easily found that

$$\langle ab| \hat{H} |ba\rangle = 2E_1 S^2 + K$$

where 'K' the so called exchange energy, which is defined as follows:

$$K = S^2 \frac{e^2}{R} + \int \int dv_1 dv_2 \Phi_A(\vec{r_1})\Phi_B(\vec{r_1})\Phi_A(\vec{r_2})\Phi_B(\vec{r_2}) \times \left\{ \frac{e^2}{|\vec{r_1} - \vec{r_2}|} - \frac{e^2}{|\vec{r_1} - \vec{r_B}|} - \frac{e^2}{|\vec{r_2} - \vec{r_A}|} \right\}$$

(4.32)

The first term on the right hand side of Eq. 4.32 is the nuclear repulsion energy weighted by the square of the overlap integral. The double integrals in the second term can be separated into electron repulsion and electron-nucleus attraction integrals leading to

$$K = \frac{S^2 e^2}{R} + \int \int \Phi_A(\vec{r_1})\Phi_B(\vec{r_1}) \frac{e^2}{|\vec{r_1} - \vec{r_2}|} \Phi_A(\vec{r_2})\Phi_B(\vec{r_1}) - 2S \int dv_1 \frac{e^2 \Phi_A(\vec{r_1})\Phi_B(\vec{r_1})}{|\vec{r_1} - \vec{r_A}|}$$

Note that the double integral physically represents the mutual repulsive energy of overlapping charge densities $e\Phi_A(\vec{r_1})\Phi_B(\vec{r_1})$ and $e\Phi_A(\vec{r_2})\Phi_B(\vec{r_2})$. Notice that the overlapping charge densities transform into each other under $1 \leftrightarrow 2$ exchange. The root of exchange energy lies in an interplay of Pauli's principle and Coulomb's interaction. It represents a purely quantum mechanical entity, arising from antisymmetry requirement.

We can now go back to the expressions for the singlet and triplet energy profiles $\epsilon_\pm(R)$ defined in Eq. 4.28 and get

$$\epsilon_\pm(R) = 2E_1 + \frac{Q \pm K}{1 \pm S^2}$$

(4.33)

the '+' sign referring to the singlet and '-' sign to the triplet state energies. The explicit analytical expressions for 'Q' and 'K' have been worked out (see, for example Y. Sugiura, Z. Phys., 45, 484, 1927). It turns out that both the Coulomb and exchange energy vary with the internuclear separation (R). Magnitude wise 'Q' is small remaining +ve (>0) everywhere whereas the exchange energy 'K' is negative (< 0) every where except for small values of R, exceeding the Coulomb term in magnitude. Thus the singlet state represents chemical binding ϵ_+ (R) being lower relative to the energy of the two hydrogen atoms ($2E_1$) at infinite separation. ϵ_+ (R) passes through a clearly defined minimum (Figure 4.7a) as a function of $\frac{R}{a_0}$. The triplet state energy profile (ϵ_- (R)) fails to produce any binding (Figure 4.7b). Thus, covalent bond formation in H_2 molecule is correctly described by $\Psi^0(1,2)$, which represents a spin-paired state, but not by $\Psi^+(1,2)$, which represents a state with two uncoupled spins. The notion therefore gained traction that mutual compensation of spins of the valence electrons takes place in the formation of the homopolar bond in the simplest diatomic entity – the hydrogen molecule. It is possible to generalize the results of Heitler-London theory of the homopolar H_2 molecule and suggest that the formation of homopolar molecules always taken place under the condition of mutual compensation (pairing) of spins of valence electrons. Indeed, such valencies are very often called spin-valencies.[6]

Before taking up general aspects of the theory homopolar or covalent bonds qualitatively, let us take a look into the quantitative aspects of the Heitler-London energetics of covalent bonding in the hydrogen molecule. The singlet state energy profile ($E_+(\frac{R}{a_0})$) in Figure 4.7a has a well defined minimum, which defines the stable bonded state of H_2. The depth of the minimum which occurs approximately at $R_0 = 1.52$ Å corresponding to a homopolar dissociation energy $D = 3.14$ eV ($H_2 \rightarrow H + H$ while the corresponding experimental quantities are $R_0^{exp} = 0.739$ Å and $D_0^{exp} = 4.48$ eV (without zero-point energy correction). Thus Heitler-London theory, though qualitatively successful in explaining the stability of the H_2 molecule is clearly inadequate when it comes to quantitative prediction of binding energy and equilibrium bond length. The inadequacy can be traced to the fact that the HL wavefunctions $\Psi_{HL}^{0,1}(1,2)$ are constructed from products of undistorted $1S$ orbitals on the interacting hydrogen atoms. In reality, the presence of another electron (e_2) on the neighbouring hydrogen atom (B) produces a screening the nuclear charge on atom 'A' that electron 1 feels and vice-versa. Thus, $1S$ orbitals used in HL description has to be motivated

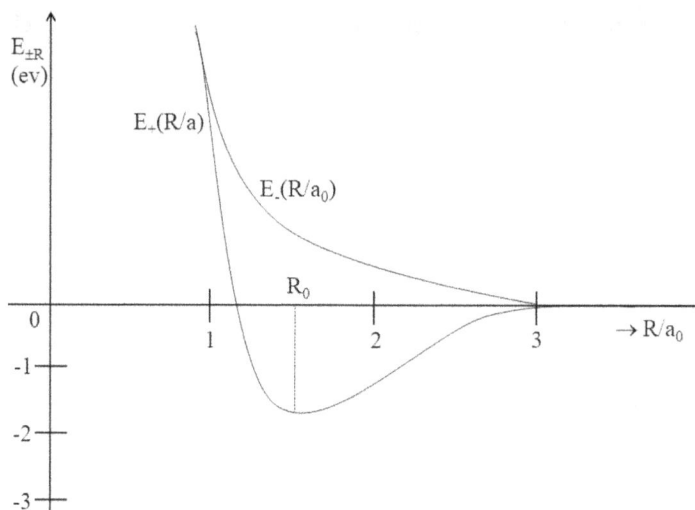

FIGURE 4.7   Ground and excited state potential energy curves of hydrogen molecule.

by incorporating the screening effect by introducing an optimizable effect nuclear charge (Z'). Let us call them $1S'$ orbitals where

$$\Phi'_{1S} = \left(\frac{Z'^3}{\pi a_o^3}\right)^{\frac{1}{2}} \exp\left(\frac{-Z'r}{a_0}\right) \tag{4.34}$$

$E_+(\frac{R}{a_0})$ now becomes a function of the effective nuclear charge, which is treated as a variational parameter. With the optimization of Z' ($Z' \neq 1$), R_0 is predicted to be 0.76 Å and D_0 to be 3.76 eV in much better agreement with the experiment results. With more flexible basis functions containing many variational parameters the results improve further as expected.

4.4.2 Electron Probability Density Distribution in Heitler-London States

Traditional wisdom describes covalent bonding as sharing of valence electron pairs between the interacting atoms. It is expedient now to probe whether HL wavefunction $\Psi^0_{HL}(1, 2)$ lives up to the traditional picture.

The probability density of electron distribution in the spin-paired (singlet) spatially symmetric ground state of a hydrogen molecule is given by

$$\rho^0_{HL} = (\Psi^0_{HL}(1,2))^2 = \frac{1}{2(1+S^2)}\left\{\Phi^2_A(1)\Phi^2_B(2) + \Phi^2_A(2)\Phi^2_B(1) + 2\Phi_A(1)\Phi_B(1)\Phi_A(2)\Phi_B(2)\right\} \tag{4.35a}$$

For comparison, we may also compute the probability density in the triplet state as well. It is given by the following expression

$$\rho^1_{HL} = (\Psi^1_{HL}(1,2))^2 = \frac{1}{2(1-S^2)}\left\{\Phi^2_A(1)\Phi^2_B(2) + \Phi^2_A(2)\Phi^2_B(1) - 2\Phi_A(1)\Phi_B(1)\Phi_A(2)\Phi_B(2)\right\} \tag{4.35b}$$

For the sake of simplicity we may consider the bonding and antibonding states of the H_2^+ and consider probability density distribution in the symmetric and antisymmetric states (spatial). If contours of constant electron density are plotted we find that the probability of an electron arriving at or around the mid-point of the line joining the two nuclei A and B is much higher in the spatially symmetric singlet state than in the spatially antisymmetric triplet state for which their probability practically vanishes (see Figure 4.8).

Since electrons located at or around the central point would bind the two nuclei rather strongly, the symmetric state is naturally expected to lead to the formation of a stable molecule. Moreover, the contour depicting distribution of electron around the two nuclei merge with one another as the nuclei approach

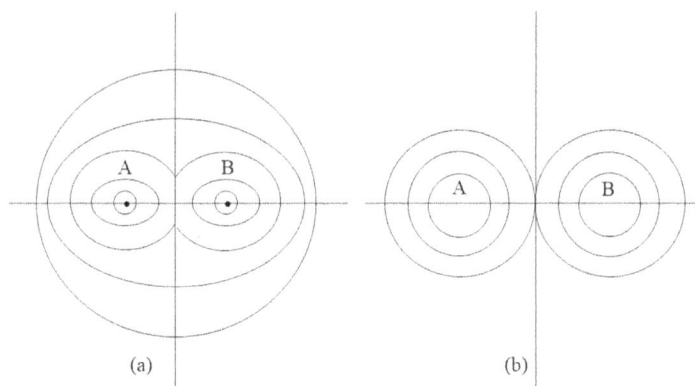

FIGURE 4.8   (a) Symmetric, (b) Antisymmetric states of H_2^+.

each other in the symmetric state. That conjures up a visual picture of electron sharing in a homopolar bond formation. It is a rather simple picture. Covalent binding between atoms is a cooperative process in which all electrons but not specific pairs, contribute (see Section 4.6).

4.4.3 Valency and Quantum Mechanics

Heitler-London's theory of the H_2 molecule and the homopolar bond paved the way to a better understanding of the intuitive concept of chemical valency, which represents the ability of the atoms of an element to combine with a certain number of atoms of another element and form molecules. The formation of heteropolar molecules was conceptually interpreted by Kossel in terms of redistribution or reorganization of electrons in the outer shells of chemically interacting atoms. The number of electrons donated by one atom to another in the process of heteropolar (ionic) bond formation was called the positive ionic valency of electrons while the number of electrons received by the other was termed the negative ionic valency. The process of molecule formation was supposed to the driven by the inherent tendency to redistribute the electrons in the outer subshells of interacting atoms so as to saturate their valencies. It was perhaps the first successful attempt to interpret chemical bonding in terms of electronic structure of atoms as elucidated by quantum mechanics. With the advent of Heitler-London (HL) theory, came the second success of quantum mechanics in the arena of valency when HL theory explained the making of homopolar (covalent) bond in H_2, the simplest of all molecules. In qualitative terms, the theory envisaged the formation of the covalent bond in H_2 by the mutual quenching or pairing of spins of valence ($1s$) electrons on the interacting pair of hydrogen atoms in a manner consistent with the requirements of Pauli's principle. The spin pairing, however, does not mean that HL theory takes into account magnetic interactions between spins or spin-orbit interactions. The Hamiltonian used in HL theory is free of spin, hence the spin variables in HL wavefunctions get integrated out, but the spatial parts of the wavefunctions $\Psi_{HL}^0(1,2)$ or $\Psi_{HL}^1(1,2)$ retains, so to say, a 'memory' of the spin wavefunctions in the symmetric or antisymmetric nature of the orbital products depending or whether one is dealing with the spin paired ground (singlet) or a spin uncoupled excited (triplet) state of the molecule. We have seen already that only the product of symmetric spatial wavefunction and spin-paired singlet (antisymmetric) spin eigenfunctions lead to stable H_2 molecule. The theory was easily generalized to measure the covalency of atoms by the number of electrons in the outershell with unpaired spins, the saturation of valency consisting of mutual compensation of spins of the valence electrons on the atom-pair involved in covalent bond formation.[6-7]

We will demonstrate the working of the idea in specific cases displayed in Figure 4.9 where the ground state electronic configuration of a number of elements are displayed. The one-electron states are represented by small boxes, the electrons in them by little arrows pointing up or down indicating orientations of their spins. Take the case of boron atom, for example. This group III element of the periodic table has the ground state electronic configuration $1s^2 2s^2 2p^1$, its spectroscopic term symbol being ^2P. With one

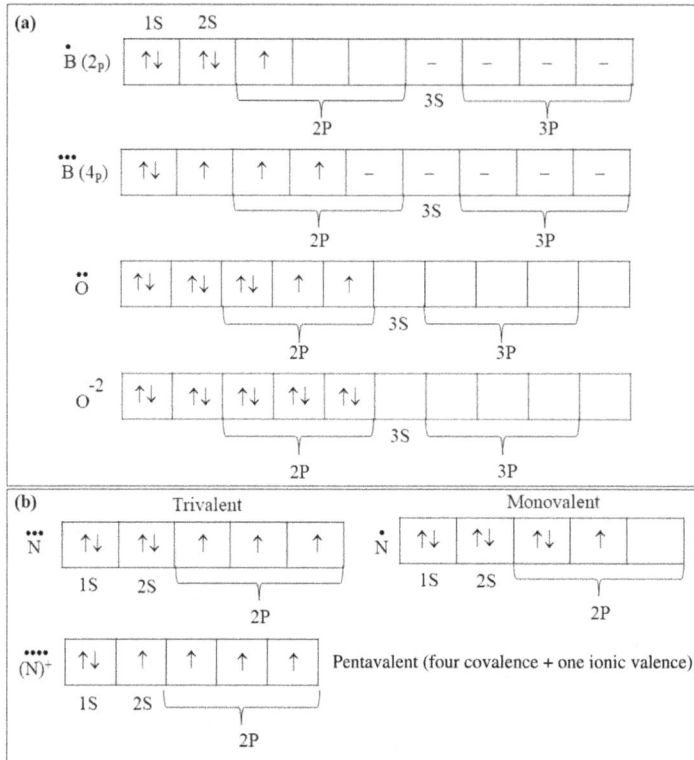

FIGURE 4.9   (a) Valency of boron and oxygen atoms. (b) Valency of nitrogen atom.

uncompensated spin boron atom (B in Figure 4.9) in ^2P state has a covalency of 1. Note, however that boron atom has a low lying quartet P state (^4P), which has three valence electrons unpaired ($\ddot{B}$ in Figure 4.9) generating a covalency equal to 3, which is manifested in many compounds of boron.

Oxygen atom ($\ddot{O}$ in Figure 4.9a), a group VI element, has two unpaired electrons in the $2p$ subshell (spectroscopic state ^3P). It displays a covalency of two uniformly. It can receive two electrons to produce O^{-2} anion producing the spectroscopic state $^1S_0$ (described by configuration $1s^22s^22p^6$). Oxygen has therefore an ionic valency of -2. Unlike other elements of this group, which display variable covalency of 2, 4 and 6, oxygen is restricted to a covalency of 2. This happens because oxygen atom can not easily create higher spin multiplicity states because promoting one or more electrons to the $n = 3$ shell is energetically demanding. In the heavier elements like sulphur of group VI, the $3d$ subshell of the valence shell is vacant and can be utilized for generating higher valence states with a small expenditure of energy. Proceeding in this manner, it can be shown that fluorine has a covalency of 1 as the ground state is a spin doublet (one unpaired spin, configuration $1s^22s^22p^5$). Also, it has an ionic valency of -1 as it can accept one electron in its valence shell quenching the only unpaired spin. In the heavier halogens like chlorine higher covalencies of 3, 5, 7 are also possible.

The case of nitrogen is interesting. The normal ground state of nitrogen atom is a spin quartet with three unquenched spins (Figure 4.9b) leading to a normal covalency of 3. It can also have a doublet valence state with one uncompensated spin and display a covalency equal to 1. In addition. it can have a pentavalent state by losing one of its $2s$ electron producing N^+ with four uncompensated spins (Figure 4.9b). At this trait we may consider the inert gases and their valencies. With singlet ground states with no uncompensated spins (the valence shell configuration being ns^2np^6) they are expected to be zero valent. Experimental data on heavier inert gases like Xe for example indicate that they can display covalencies of 2, 4, 6 and even 8. It appears that binding energies in such systems breaks spin-orbit coupling of electrons in the outer shell creating valence states with a number of uncompensated spins. It must be emphasized that the strictly binary classification of bonding into homopolar (covalent) and heteropolar (ionic) bonds is an idealization that does not describe reality. In practice there is a continuum of variation in bonding between the two extreme types. The bonding type depends on the distribution of electron density between the bonded pair. If the distribution is strongly asymmetric, we encounter ionic bonds. On the other hand, a completely symmetric distribution of electron density between the bonded atoms leads to pure covalency. In most of the cases, it is either covalent bonds with partial ionic character or ionic bonding with an admixture of covalent character.

4.5 Dynamics of Electron Exchange in Covalent Bond Formation

We have seen that exchange interaction plays a crucial role in imparting stability to the H_2 molecule in the ground state (spin-singlet). We will now examine the dynamics of exchange, if any. Let us consider the system of two widely separated hydrogen atoms, marked A and B as depicted in Figure 4.6 (note, A and B are identical atoms). The energy eigenstates and eigenvalues of the atoms A and B are assumed to be completely known and are solutions of the following eigenvalue equation:

$$\left[-\frac{\hbar^2}{2m}\nabla_i^2 + V(r_i)\right]\Phi_n(r_i) = \epsilon_n \Phi_n(r_i)$$

$$\text{or,} \quad H_i\Phi_n(r_i) = \epsilon_n \Phi_n(r_i)$$

(4.36a)

where $V(r_i)$ is the potential energy of electron 'i' in the Coulomb field provided by the protons located at A and B, respectively. We assume the set $\Phi_n(r_i)$ to be orthonormal. The two electrons interact with each other and the interaction potential is treated as a perturbation $H_{12} = V_{12} = H_{21} = V_{21}$. Our problem is to find the perturbed states first and determine how the unperturbed states evolve into the final perturbed states, in time. The unperturbed or zeroth order Hamiltonian of the problem is $H_0 = H_1 + H_2$ and the energy eigenvalue equation for H_0 is

$$\hat{H}_0\Psi_{01}(\vec{r_1}, \vec{r_2}) = (\hat{H}_1 + \hat{H}_2)\Psi_{01}(\vec{r_1}, \vec{r_2})$$

$$= E_0\Psi_{01}(\vec{r_1}, \vec{r_2})$$

(4.36b)

We can choose

$$\Psi_{01}(\vec{r_1}, \vec{r_2}) = \Phi_r(\vec{r_1})\Phi_s(\vec{r_2})$$

(4.37a)

assuming that the electron 1 is in the nth eigenstate of H_1 and electron 2 is in the sth eigenstate of H_2. E_0 in Eq. 4.36b is clearly the sum of ϵ_r and ϵ_s defined in Eq. 4.36a that determines the one-electron basis functions for the problem and their energy eigenvalues. Therefore, we can set $E_0 = \epsilon_r + \epsilon_s$. However, there is yet another product function that is consistent with the same unperturbed energy (E_0), which is

$$\Psi_{02}(\vec{r_1}, \vec{r_2}) = \Phi_r(\vec{r_2})\Phi_s(\vec{r_1})$$

(4.37b)

From Ψ_{01} and Ψ_{02} it is clear that, the two particles (electrons) have clearly changed their one-electron states between themselves and this exchange has not affected the unperturbed energy E_0 of the problem. We can therefore say that E_0 represent a two-fold degenerate unperturbed energy level. The solution, i.e. the eigenfunctions of the perturbed problem can therefore be chosen in the following form (a linear combination of the degenerate eigenfunctions)

$$\Psi(1,2) = a\Psi_{01}(1,2) + b\Psi_{02}(2,1)$$

(4.38)

If we substitute $\Psi(1,2)$ into the energy eigenvalue equation of the full Hamiltonian (H), we have

$$(\hat{H} - E)\Psi(1,2) = (\hat{H}_1 + \hat{H}_2 + \hat{H}_{12} - E)\Psi(1,2) = 0$$

(4.39)

Writing the perturbed energy levels E as $\epsilon_r + \epsilon_s + \lambda$, we get a very simple looking equation (λ is the energy correction due to the perturbation, $\epsilon_r + \epsilon_s$ is the unperturbed energy, E_0)

$$(H_{12} - \lambda)(a\Psi_{01} + b\Psi_{02}) = 0$$

(4.40)

Multiplying the equation above by Ψ_{01}^* and integrating over the coordinates of both the electrons we have

$$(D_{11} - \lambda)a + D_{12}b = 0 \qquad (4.41a)$$

Similarly, multiplying Eq. 4.40 by Ψ_{02}^* and integrating over the coordinates of both the electrons as before, we get

$$D_{21}a + (D_{22} - \lambda)b = 0 \qquad (4.41b)$$

The matrix elements $D_{11}, D_{12}, D_{22}, D_{21}$ appearing in Eqs. 4.41a and b are defined as follows:

$$D_{11} = \int_1 \int_2 \Psi_{01}^* \hat{H}_{12} \Psi_{01} dv_1 dv_2 = \int_1 \int_2 \Phi_r^*(\vec{r_1})\Phi_s^*(\vec{r_2})\hat{H}_{12}\Phi_r(\vec{r_1})\Phi_s(\vec{r_2})dv_1 dv_2 \qquad (4.42a)$$

$$D_{22} = \int_1 \int_2 \Psi_{02}^* \hat{H}_{12} \Psi_{02} dv_1 dv_2 = \int_1 \int_2 \Phi_r^*(\vec{r_2})\Phi_s^*(\vec{r_1})\hat{H}_{12}\Phi_r(\vec{r_2})\Phi_s(\vec{r_1})dv_1 dv_2 \qquad (4.42b)$$

D_{11} and D_{22} differ only in the labeling of the variables appearing in the integrals defining them and are therefore equal. We denote the value as (i.e. $D_{11} = D_{22} = Q$). The off-diagonal matrix elements D_{12} and D_{21} are similarly equal as $\hat{H}_{12} = \hat{H}_{21}$. Thus we can set $D_{12} = D_{21} = A$ (say) where

$$D_{12} = \int_1 \int_2 \Psi_{01}^* \hat{H}_{12} \Psi_{02} dv_1 dv_2$$

$$= D_{21}$$

$$= \int_1 \int_2 \Psi_{02}^* \hat{H}_{12} \Psi_{01} dv_1 dv_2$$

With these definitions we rewrite Eq. 4.41a and b in the following form:

$$(Q - \lambda)a + Ab = 0 \qquad (4.43a)$$

$$Aa + (Q - \lambda)b = 0 \qquad (4.43b)$$

Non-trivial solutions of Eq. 4.43 exist only if the determinant formed by the coefficients of a and b vanishes, leading to the condition,

$$(Q - \lambda)^2 - A^2 = 0$$

That means

$$Q - \lambda = \pm A$$

or

$$\lambda = Q \pm A$$

Of the two possibilities, $\lambda = Q + A$ demands $a = b$ and the corresponding perturbed energy level E' is given by

$$E' = E_0 + \lambda = \epsilon_r + \epsilon_s + Q + A$$

The wavefunction Ψ' for the energy level E'

$$\Psi' = a(\Psi_{01} + \Psi_{02})$$

$$= a\left\{ \Phi_r(\vec{r_1})\Phi_s(\vec{r_2}) + \Phi_r(\vec{r_2})\Phi_s(\vec{r_1}) \right\} \qquad (4.44a)$$

The other solution is $\lambda = Q - A$ and $a = -b$ which leads to the energy level (E'') and wavefunction Ψ'' where

$$E'' = \epsilon_r + \epsilon_s + Q - A$$

$$\Psi'' = a\left\{ \Phi_r(\vec{r_1})\Phi_s(\vec{r_2}) - \Phi_s(\vec{r_1})\Phi_r(\vec{r_2}) \right\} \qquad (4.44b)$$

The coefficient 'a' appearing in Eq. 4.44a,b can be fixed by the normalization condition:

$$\int_1 \int_2 \Psi_{01}^* \Psi_{01} dv_1 dv_2 = \int \int \Psi_{02}^* \Psi_{02} dv_1 dv_2$$

$$= 1$$

which leads to $a = \frac{1}{\sqrt{2}}$. We note here further that Ψ_{01} and Ψ_{02} are mutually orthogonal as well (can be easily checked by evaluating the integral $\int \int \Psi_{01}^* \Psi_{02} dv_1 dv_2$). So the spatial parts of the wavefunctions for the two perturbed energy levels E' and E'' are,

$$\Psi' = \frac{1}{\sqrt{2}}\Big[\Psi_{01} + \Psi_{02}\Big], \quad E' = \epsilon_r + \epsilon_s + Q + A \tag{4.45a}$$

$$\Psi'' = \frac{1}{\sqrt{2}}\Big[\Psi_{01} - \Psi_{02}\Big], \quad E'' = \epsilon_r + \epsilon_s + Q - A \tag{4.45b}$$

Since we are dealing with two electrons, each with half-integral spin quantum numbers, the complete two-electron wavefunctions must be multiplied the appropriate two-electron spin eigenfunctions so as to make the total wavefunction Φ antisymmetric with respect interchange of the space-spin coordinates of the two electrons. Since $\Psi'(1,2)$ is manifestly symmetric under the interchange of the spatial coordinates (r_1, r_2) it must be multiplied by the antisymmetric two electron spin eigenfunction $\chi_a(\sigma_1,\sigma_2)$ where

$$\chi_a(\sigma_1,\sigma_2) = \frac{1}{\sqrt{2}}\Big(\alpha(1)\beta(2) - \alpha(2)\beta(1)\Big)$$

which described a spin-singlet.

We therefore have

$$\Phi_a'(r_1,r_2,\sigma_1,\sigma_2) = \Psi'(r_1,r_2) \times \chi_a(\sigma_1,\sigma_2)$$

Similarly, for the other solution (E'', Ψ'') the complete antisymmetric wavefunction is

$$\Phi_a''(r_1,r_2,\sigma_1,\sigma_2) = \Psi''(r_1,r_2) \times \chi_s(\sigma_1,\sigma_2)$$

where the space-part $\Psi''(r_1,r_2)$ is antisymmetric under $1 \leftrightarrow 2$ interchange while the spin part $\chi_s(\sigma_1,\sigma_2)$ is symmetric. These are, three symmetric two-electron eigenfunctions of spin, which means Φ_a'' represents a spin-triplet, the three components being given by

$$\Phi_a'' \equiv \Psi'' \times \begin{cases} \frac{1}{\sqrt{2}}[\alpha(1)\beta(2) + \alpha(2)\beta(1)] \\ \alpha(1)\alpha(2) \\ \beta(1)\beta(2) \end{cases}$$

The states described by Φ' or Φ'' are energyeigen states of the hydrogen molecule expressed as linear combinations of the unperturbed eigenstates of H_0 (Ψ_{01} and Ψ_{02}). Let us suppose that initially, i.e. at $t = 0$, the system (two widely separated hydrogen atoms) was in the state Ψ_{01}. Then

$$\Psi(0) = \Psi(t = 0) = \Phi_r(\vec{r_1})\Phi_s(\vec{r_2}).$$

As the interaction term $\hat{H}_{12}$ is switched on Ψ evolves in time and we represent $\Psi(t)$ as the result of action of an operator $\hat{S}$ on $\Psi(0)$:

$$\Psi(t) = \hat{S}\Psi(0) \tag{4.46}$$

Using $\Psi(t)$ of Eq. 4.46 in the time-dependent Schrödinger equation we get

$$i\hbar\frac{\partial \hat{S}}{\partial t}\Psi(0) = H\hat{S}\Psi(0) \tag{4.47}$$

Integrating formally over t, we get

$$\hat{S} = \exp\left(\frac{-i\hat{H}t}{\hbar}\right)$$

$$\text{and} \quad \Psi(t) = \exp\left(\frac{-i\hat{H}t}{\hbar}\right)\Psi(0) \tag{4.48}$$

Let $\Psi(0)$ be expanded in a complete set of eigenstates of $\hat{H}$ (i.e. states satisfy the eigenvalue equation $\hat{H}\Psi_n = E_n\Psi_n$). Thus we get

$$\Psi(t) = \exp\left(\frac{-i\hat{H}t}{\hbar}\right)\sum_n c_n\Psi_n = \sum_n c_n\exp\left(\frac{-i\hat{E}_nt}{\hbar}\right)\Psi_n \tag{4.49}$$

where we have used the result

$$f(\hat{H})\Psi_n = f(E_n)\Psi_n \tag{4.50}$$

where $f(\hat{H})$ is an operator function of $\hat{H}$, and $\hat{H}\Psi_n = E_n\Psi_n$. It means that $\Psi(t)$ can be determined by expanding $\Psi(0)$ in a complete set of eigenstates of $\hat{H}$. In the problem at hand we have already found the two eigenstates of $\hat{H}$ in the zeroth order approximation:

$$\Psi' = \frac{1}{\sqrt{2}}\left(\Psi_{01} + \Psi_{02}\right)$$

$$\Psi'' = \frac{1}{\sqrt{2}}\left(\Psi_{01} - \Psi_{02}\right) \tag{4.51}$$

where Ψ' corresponds to the energy level $E' = \epsilon_r + \epsilon_s + Q + A$ and Ψ'' to the energy level

$$E'' = \epsilon_r + \epsilon_s + Q - A.$$

$\Psi_{01} = \Psi(0)$, accordingly can be expressed as

$$\Psi(0) = \frac{1}{\sqrt{2}}(\Psi' + \Psi'')$$

whereas

$$\Psi(t) = \frac{1}{\sqrt{2}}\left[\exp\left(\frac{-iE't}{\hbar}\right)\Psi' + \exp\left(\frac{-iE''t}{\hbar}\right)\Psi''\right] \tag{4.52}$$

Using the fact that

$$E' = \epsilon_r + \epsilon_s + Q + A$$

and

$$E'' = \epsilon_r + \epsilon_s + Q - A$$

and the expressions for Ψ' and Ψ'' in terms of Ψ_{01} and Ψ_{02} (Eq. 4.51) in Eq. 4.52, we get

$$\Psi(t) = \frac{1}{2}\exp\left(\frac{-i}{\hbar}(\epsilon_r + \epsilon_s + Q)t\right)\left[(\Psi_{01} + \Psi_{02})\exp\left(\frac{-iAt}{\hbar}\right) + (\Psi_{01} - \Psi_{02})\exp\left(\frac{iAt}{\hbar}\right)\right]$$

$$= \exp\left(\frac{-i}{\hbar}(\epsilon_r + \epsilon_s + Q)t\right)\left\{\Psi_{01}\frac{[\exp(\frac{iAt}{\hbar}) + \exp(\frac{-iAt}{\hbar})]}{2} + \Psi_{02}\frac{[\exp(\frac{iAt}{\hbar}) - \exp(\frac{-iAt}{\hbar})]}{2i}\right\} \tag{4.53}$$

$$= C_1(t)\Psi_{01} + C_2(t)\Psi_{02}$$

where

$$C_1(t) = \exp\left(\frac{-i}{\hbar}(\epsilon_r + \epsilon_s + K)t\right)\cos\frac{At}{\hbar}$$

$$C_2(t) = \exp\left(\frac{-i}{\hbar}(\epsilon_r + \epsilon_s + K)t\right)\sin\frac{At}{\hbar} \tag{4.54}$$

Therefore the probabilities of finding the two-H atom system in the state $\Psi_{01} = \Phi_r(r_1)\Phi_s(r_2)$ or in the state $\Psi_{02} = \Phi_r(r_2)\Phi_s(r_1)$ are respectively given by

$$P_{01} = |c_1|^2 = \cos^2 \frac{At}{\hbar}$$

$$P_{02} = |c_2|^2 = \sin^2 \frac{At}{\hbar}$$

(4.55)

At $t = 0$, $P_{01} = 1$ and $P_{02} = 0$. The time taken by the pair of electrons (1, 2) to exchange their states from $\Psi_{01} = \Phi_r(r_1)\Phi_s(r_2)$ to $\Psi_{02} = \Phi_r(r_2)\Phi_s(r_1)$ can now be calculated by finding the time 'τ' at which $C_1(\tau) = 0$ and $C_2(\tau) = 1$, i.e.

$$\cos^2 \frac{A\tau}{\hbar} = 0 = \cos \frac{A\tau}{\hbar}$$

The equality demands that

$$\frac{A\tau}{\hbar} = \frac{\pi}{2}$$

(4.56)

$$\text{or,} \quad \tau = \frac{\hbar\pi}{2A}$$

The larger the value of the exchange interaction 'A' smaller is the electron interchange time (τ), and more facile is the process of covalent bond formation.[8] 'A' is small when the two atoms are widely apart, the electron interchange time is large, making the bond formation impossible. As the pair of H atoms approach each other more closely and are within the standard bonding distances, the value of 'A' becomes large reducing the electron interchange time to a very small value. The exchange now becomes so rapid that the distinction between electron 1 and 2 vis-a-vis their one electron states are completely lost and the superposed state $\frac{1}{\sqrt{2}}(\Psi_{01} + \Psi_{02})\chi_s^{singlet}$ describing the covalently bonded state $\Psi'(1,2)$ in which the electron spins are so coupled as to produce the only singlet state possible. It is the time scale of exchange that gives the notion of electron-pair bonds in molecules a degree of validity. However, the electrons in an atom or a molecule are all indistinguishable and all the electrons in the interacting pair of atoms are expected to contribute cooperatively to the process of bonding leading to the emergence of a state equilibrium molecular structure. Such consideration has led to the idea of 'forces in molecules' – forces that are entirely electrostatic in origin. The delicate balancing of such forces originating from the electron-nuclear attraction and internuclear repulsion produces the equilibrium molecular structure as described in what follows.

4.6 Forces in Molecules, Bonding and Equilibrium Structures

Let us consider an n-electron and M-nuclei molecules, $Z_\beta e$ being the positive charge carried by the nuclei 'β', with β running from 1 to M. In the Börn-Oppenheimer (BO) model, the nuclei coordinates (X_β, Y_β, Z_β) are frozen and serve as parameters in the BO molecular electronic Hamiltonian $H_e(x,\underline{X})$ and the its eigenstates $\Psi_n^e(x,X)$. The corresponding energy eigenvalues ($E_n^e(x)$) also depend parametrically on the nuclear coordinates. When the internuclear repulsion energy $V_{NN}(X)$ is added to $E_n(x)$, we get the potential energy surface $U_n(X)$ on which the nuclei move in the nth electronic state. The molecular electronic Schrödinger equation, as we have seen already is

$$H_e(x,\underline{X})\Psi_n^e(x,\underline{X}) = E_n^e(x,\underline{X})\Psi_n^e(x,\underline{X})$$

(4.57)

where

$$H_e(x,\underline{X}) = \sum_{i=1}^{n} -\frac{\hbar^2}{2m_e}\nabla_i^2 + \sum_{i=1}^{n}\sum_{\beta=1}^{M} \frac{-Z_\beta' e^2}{r_{i\beta}} + \frac{1}{2}\sum_{i\neq j}\sum \frac{e^2}{r_{ij}}$$

$$= \hat{T}_e(x) + V_{eN}(x,X) + V_{ee}(x,x')$$

and therefore,

$$U_n(X) = E_n^e(X) + V_{NN}(X) \tag{4.58}$$

where

$$V_{NN}(X) = \frac{1}{2} \sum_{\alpha=1}^{M} \sum_{\beta=1(\alpha \neq \beta)}^{M} \frac{Z_\alpha' Z_\beta' e^2}{R_{\alpha\beta}}.$$

In view of the nature of the $V_{NN}(X)$ term, we can assume that $V_{NN}(X)$ has been added to $H_e(x,\underline{X})$ and a new molecule electronic Schrödinger equation including internuclear repulsion has been set up and solved (let us assume, exactly).

$$H(x,X)\Psi = \left\{ H_e(x,X) + V_{NN}(X) \right\} \Psi_n(x,X) = U_n(X)\Psi_n(x,X) \tag{4.59}$$

In the stationary state that $\Psi_n(x,X)$ describes the Hellmann-Feynman theorem holds and we can write

$$\frac{\partial U_n}{\partial X_\beta} = \int (\Psi_n^e)^* \frac{\partial H}{\partial X_\beta} \Psi_n^e \, dv_{el} \tag{4.60}$$

where $\hat{H} = \hat{T}_e(x) + V_{eN}(x,X) + V_{ee}(x,x') + V_{NN}(X)$.

Since $\hat{T}_e(x)$, $V_{ee}(x,x')$ do not depend on the nuclear coordinates X_β, Y_β, Z_β of any atom β in the molecule, we can write $\frac{\partial H}{\partial X_\beta}$ only in terms of the corresponding derivatives of $V_{eN}(x,X')$ and $V_{NN}(X,X')$. It is then straightforward to find that

$$\frac{\partial U_n}{\partial X_\beta} = -Z_\beta' e^2 \int |\Psi_n^e|^2 \sum_{i=1}^{n} \frac{x_i - X_\beta}{r_{i\beta}^3} dv_{el} + \sum_{1(\alpha \neq \beta)}^{M} Z_\beta' Z_\alpha' \frac{X_\alpha - X_\beta}{R_{\alpha\beta}^3} \tag{4.61}$$

where

$$r_{i\beta} = \sqrt{(x_i - X_\beta)^2 + (y_i - Y_\beta)^2 + (z_i - Z_\beta)^2}$$

and

$$R_{\alpha\beta} = \sqrt{(X_\alpha - X_\beta)^2 + (Y_\alpha - Y_\beta)^2 + (Z_\alpha - Z_\beta)^2}$$

The $3n$ dimension integral over the spatial coordinates of all the electrons on the right hand side of Eq. 4.61 can be replaced by a three-dimensional integral over the spatial coordinates (x, y, z) of a single electron (the discrete spin coordinates being summed up as usual) by introducing the spinless single electron probability density function $\rho_n(x,y,z)$ in the state Ψ_n leading to the following equation:

$$\frac{\partial U_n}{\partial X_\beta} = -Z_\beta' e^2 \int \int \int \rho_n(x,y,z) \frac{x - X_\beta}{r_\beta^3} dx dy dz + \sum_{\alpha \neq \beta} Z_\beta' Z_\alpha' e^2 \frac{X_\alpha - X_\beta}{R_{\alpha\beta}^3} \tag{4.62}$$

with $r_\beta = \sqrt{(x - X_\beta)^2 + (y - Y_\beta)^2 + (z - Z_\beta)^2}$ representing the Euclidean distance of the point (x, y, z) from the nucleus with charge $Z_\beta' e$ fixed in space at the point $(X_\beta, Y_\beta, Z_\beta)$. Let us now recall that $U_n(X)$ represents the potential function generating the potential energy surface on which the nuclei move in the nth electronic state. In fact, $U_n(X)$ in BO recipe defines the nuclear Schrödinger equation

$$[T_N + U_n(X)]\Psi_n(X) = E_n \Psi_n(X) \tag{4.63}$$

Therefore, $-\frac{\partial U_n}{\partial X_\beta}$ is the X-component of what may be called the effective force on the nucleus of the atom β (charge on the nucleus $+Z_\beta' e$) due to all the electrons and all other nuclei present in the molecule in the electronic state represented by $\Psi_n(x,X)$. Defining $-\frac{\partial U_n}{\partial Y_\beta}$ and $-\frac{\partial U_n}{\partial Z_\beta}$ in a similar fashion, we can write down the expression for the effective force $(\overrightarrow{F_\beta})$ acting on the nucleus β in the electronic state 'n' of the molecule using vector notation, we have,

$$\overrightarrow{F_{\beta(n)}} = -i\frac{-\partial U_n}{\partial X_\beta} - j\frac{-\partial U_n}{\partial Y_\beta} - k\frac{-\partial U_n}{\partial Z_\beta} = -Z_\beta' e^2 \int \int \int \rho_n(x,y,z) \frac{\overrightarrow{r_\beta}}{r_\beta^3} dx dy dz + e^2 \sum_{\alpha \neq \beta} Z_\beta' Z_\alpha' \frac{\overrightarrow{R_{\alpha\beta}}}{R_{\alpha\beta}^3} \tag{4.64}$$

where

$$\vec{r_\beta} = i(x_\beta - x) + j(y_\beta - y) + k(z_\beta - z)$$

and

$$\vec{R_{\alpha\beta}} = i(X_\beta - X_\alpha) + j(Y_\beta - Y_\alpha) + k(Z_\beta - Z_\alpha)$$

Eq. 4.64 has become known as electrostatic Hellmann-Feynman theorem – a special form of the generalized Hellmann-Feynman theorem applied to molecules under BO approximation.[9] The physical content of Eq. 4.64 is now quite clear: the first term on the right hand side of Eq. 4.64 represents the attractive force exerted on the nucleus 'β' carrying $Z'_\beta e$ units of positive charge by the entire electronic charge density $e\rho_n(x,y,z)$ while the second term measures the total repulsive force on the nucleus β due to all other nuclei present in the molecule. The first term is calculated by electrostatics of distributed electronic charge density and localized positive nuclear point charge. The second term uses purely point-charge electrostatics. At the equilibrium geometry of the molecule in the nth electronic state the net force on each nucleus must be zero $(\vec{F_{\beta(n)}} = 0)$. Thus, the emergence of an equlibrium stable nuclear framework of a molecule is entirely due to a complete balance between the attractive and repulsive electrostatic forces originating from electron-nucleus and internuclear interactions. There is thus no need to invoke special 'covalent forces' to explain the stability and geometrical configuration of a 'covalently bonded' molecule in the ground or an excited electronic state. Only classical electrostatics is enough to do the job. There is one rider, however: the exact electronic charge density $e\rho_n(x, y, z)$ in the given electronic state must be available for the force calculations, either from experiments or from theory. That, of course is a tall order.

The second point to emphasize here is the fact that $e\rho_n(x, y, z)$ is a delocalized quantity spread out all over the molecule and that all the n-electrons whether from core or valence regions contribute to the build-up of $\rho_n(x, y, z)$. The notion of localized electron-pair bonds envisaged in VB theory appears to be redundant in force-based elucidation of equilibrium structures of molecules. 'Bonding' appears to be a cooperative process in which all the electrons in the system contribute.

4.7 Bonding and Anti-bonding Region in a Molecule, Berlin Diagrams

It is possible to apply the force equation (Eq. 4.61) directly to a diatomic molecule and determine its equilibrium bond length assuming that $\rho_n(x, y, z)$ is known for every nuclear configuration. It is a straightforward application of the force equation (Eq. 4.61). We will omit the details here and focus instead, on the physical picture that emerges from such applications.

Let us consider a diatomic molecule $A - B$, $+Z'_A e$ being the nuclear charge carried by A while $+Z'_B e$ is the charge on atom 'B' separated by a distance R as shown in Figure 4.10.

$P(x, y, z)$ is an arbitrarily chosen point in the molecular space, where the electronic charge density (supposedly known) is $e\rho_n(x, y, z)$ and the amount of electronic charge in the infinitesimal volume Δv around the point P is $dv\rho_n(x, y, z)e = \Delta qe$ (says). The repulsive force on A due to B is $\frac{Z'_A Z'_B e^2}{R^2}$ while that

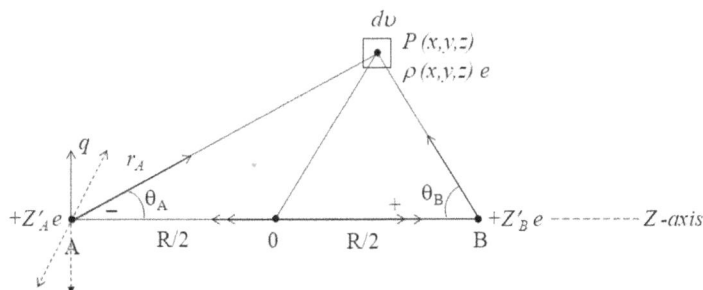

FIGURE 4.10 The diatomic molecule $A - B$ is lying on the Z-axis with the origin at O.

on B due to A is also $\frac{Z'_A Z'_B e^2}{R^2}$, but oppositely directed. They balance each other the attractive force exerted on the nucleus of A by the electronic charge enclosed in the volume element dv at P is $\frac{Z'_A \Delta q e^2}{r_A^2}$ and acts along the r_A. This force can be resolved into its components along the X, Y and Z axes. We note here that P is an arbitrary point and when we consider all such positions of P, the cylindrical symmetry of the system ensures that X and Y components of the force averaged over the entire space sum up to zero, leaving out only the Z-component of the electronic force on 'A' acting along the +ve direction of the Z axis (the internuclear axis), which can be clearly written as

$$\overrightarrow{f_A^Z} = Z_A e^2 \int \frac{\rho(x,y,z)}{r_A^2} \cos\theta_A \, dx\, dy\, dz \tag{4.65a}$$

Similar considerations lead to the expression for the Z-component of the electronic force on the nucleus of atom B, acting along the -ve direction of the Z-axis

$$\overrightarrow{f_B^Z} = (-)Z_B e^2 \int \frac{\rho(x,y,z)}{r_B^2} \cos\theta_B \, dx\, dy\, dz \tag{4.65b}$$

If point P is located in space such that

$$\frac{Z'_A e^2 \rho \, dx\, dy\, dz}{r_A^2} \cos\theta_A > -\frac{Z'_B e^2 \rho \, dx\, dy\, dz}{r_B^2} \cos\theta_B, \tag{4.66}$$

the electronic force due to the charge enclosed in the volume $dx\,dy\,dz$ around P will tend to push the nucleus A toward B (or B toward A) and thus be conducive to a 'bond formation' between the two atoms. The electronic probability density $\rho > 0$, everywhere, so that the inequality in Eq. 4.66 above representing the condition favoable for bonding can be written as (dividing both sides by ρ)

$$\frac{Z_A \cos\theta_A}{r_A^2} + \frac{Z_B \cos\theta_B}{r_B^2} > 0 \tag{4.67a}$$

Similarly, if the point $P(x,y,z)$ is such that the left hand side of the above equation is <0, i.e.

$$\frac{Z_A \cos\theta_A}{r_A^2} + \frac{Z_B \cos\theta_B}{r_B^2} < 0 \tag{4.67b}$$

the electronic force will tend to draw 'A' away from 'B', thereby destabilizing the molecule. The inequality in Eq. 4.67a can be interpreted as the equation of a surface that divides the entire molecular space into two regions namely the binding and the antibonding regions. Electronic charge density accumulated at points in the binding region (BR) satisfy the inequality Eq. 4.67a and tends to force the two atoms towards each other, thus contributing positively to the bond formation. Any electronic charge density located at points in the antibonding region satisfy Eq. 4.67b and tends to draw the atoms away from each other, thereby contributing negatively to bond formation. Thus we can visualize the bond formation as the process in which the entire electronic charge densities of the interacting atoms get distributed or delocalized all over the molecular space. Some of the electronic charge density concentrates in the binding region (Eq. 4.67a). A stable bonded diatomic (AB) emerges at equilibrium where the net force on each other of the bonded atoms is zero, the electronic 'pull' being completely neutralized by the 'nuclear push' experienced by each atom. For a homonuclear diatomic the binding and antibonding regions and their dividing surfaces are shown in Figure 4.11.

Diagrams of the type that maps out the binding and antibonding regions in a molecule on the basis of force equations like Eq. 4.67a,b are known as Berlin's diagrams.[11] The concept of such diagrams has been generalized and applied to polyatomic molecules also. The interested reader may consult the original work.[11] An excellent in-depth analysis of the force concept may be found in the following reference.[12]

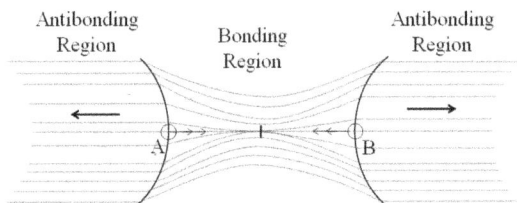

FIGURE 4.11 Berlin diagram for a homonuclear diatomic molecule ($A = B$).

4.8 Ionic Bonds and Ionic Solids

A strongly asymmetric electronic charge density distribution in a diatomic molecule 'AB' takes place when A for example, is a metal atom 'M' (like Na) and 'B' is a halogen atom X (like chlorine). Since the metal atom has a pronounced tendency to lose electrons from the outermost valence shell (it has low ionization potential and moreover, by losing one or two electrons, can attain closed-shell structure of its inert gas neighbor in the periodic table) and the halogen atom has a strong tendency to accept an electron in its outermost shell (it has high electron affinity and can attain the closed-shell electronic configuration of its neighboring inert gas atom in the process), it becomes energetically feasible or even favorable to form a chemically bonded ion-pair, $M^+ - X^-$ in which the bonded atoms display their ionic valencies (+1 and −1 in this case), respectively with complete quenching of all unpaired spins. Such molecules are called ionically bonded since the 'shared-pair' has shifted entirely to the more electronegative atom (X) resulting in the formation of a halide anion (X^-) and metal cation (M^+). The ions produced interact via attractive Coulomb forces. The overall energetics (ΔE) of formation of ionic bond in a metal halide like NaCl, for example looks simple, at the rudimentary level and can be represented as follows:

$$\Delta E = I_{Na} - A_{Cl} - \frac{e^2}{R} \tag{4.68}$$

where 'R' is the distance separating the ions of opposite charges. Since the cation (Na$^+$) has a small size and low polarizability and the anion (Cl$^-$) has a much larger size and higher polarizability, the cation will inevitably polarize the anion, leading to ion-induced dipole and still higher order interactions, further stabilizing the ionically bonded pair ($M^+ - X^-$) or what may be called an isolated ionic molecule. Several questions, however, crop up:

(i) Does the energetics represented by Eq. 4.68 imply formation of an isolated stable Na$^+$Cl$^-$ molecule with a well defined equilibrium interionic distance R_0? (ii) Do NaCl molecules even exist in isolation? In fact, an isolated Na$^+$Cl$^-$ molecule can exist only in gas phase at very high temperature and low pressure. At lower temperature and higher pressure, there is an inherent tendency of the Na$^+$Cl$^-$ unit to undergo polymerization and form larger units (see Figure 4.12). Let us note that once the ion pair states polymerizing, crystalline solid NaCl will eventually emerge with a crystal structure that occupies a well defined minimum on the free energy surface under the prevailing condition defined by temperature and pressure. In the isolated Na$^+$Cl$^-$, there is only a long range Coulomb attraction ($-\frac{e^2}{R}$), which supposedly overcomes the net energy expenditure implied by the first two terms on the right hand side of Eq. 4.68 at a certain value of 'R'. The Coulomb attraction brings the spherical Na$^+$ and Cl$^-$ ions in close proximity where they are held almost in 'touching-sphere' configuration. Eq. 4.68 shows that as R → 0 and ΔE → −∞, implying collapse. We must note that Coulomb attraction can not bring the ions any closer than the 'touching sphere configuration' as both are closed-shell ions and therefore can not share common space without violating Pauli-exclusion principle. Strong short range repulsive forces are called into play and a balance of Coulomb attraction and steeply rising forces of repulsion originating from Pauli-exclusion principle lead to the emergence an equilibrium distance (R_0) between the two ions. A minimum appears in ΔE-R profile at R_0. The formation of a stable NaCl molecule can therefore take place in gas phase. This short range repulsive force is usually modeled by one of the following forms of potential, there being no direct way to arrive at the exact distance dependence:

(a) $Na^+ \cdots Cl^- \cdots Na^+ \cdots Cl^- \cdots Na^+ \cdots Cl^-$ **1D**

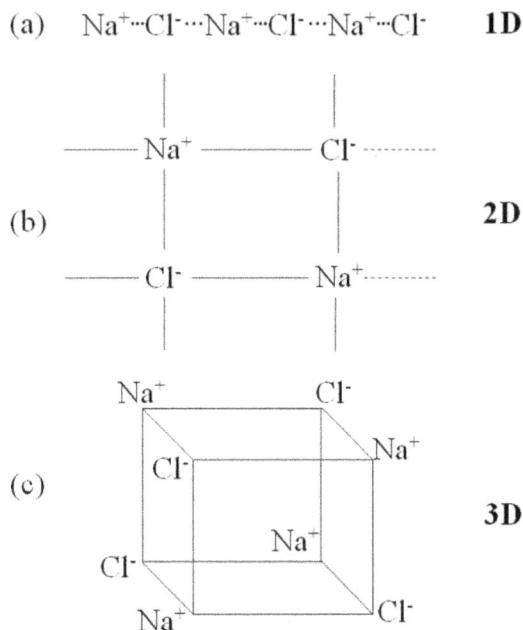

(b)   **2D**

(c)   **3D**

FIGURE 4.12   (a) Linear chain. (b) Square two-dimensional units extending in all direction. (c) Square blocks on top of each other forming a cube that extends in all directions.

$$U_{rep}(r) = C_0 . \exp\left(\frac{-R}{R_0}\right) \tag{4.69a}$$

and

$$U_{rep}(r) = \frac{C}{R^{12}} \tag{4.69b}$$

C_0, C are constants, R_0 being called a range parameter. Their values can be chosen so as to reproduce equilibrium bond length data available from experiments.

4.8.1 Cohesive Energy of Ionic Solids

Unlike the isolated Na^+Cl^- molecule in the gaseous phase, an ionic solid has many cations and anions arranged regularly in a lattice. There is thus both attractive and repulsive Coulomb forces operating simultaneously in the crystal. The cohesive energy (the energy that holds the collection of cations and anions together at equilibrium) of the crystal may perhaps be obtained by summing over the Coulomb interaction energies of all the ion pairs in the crystal (Na^+Cl^-, $Na^+ \cdots Na^+$, $Cl^- \cdots Cl^-$) located at various distances from each other. If the sum is finite and < 0, and attains a minimum, a stable cohesive ionic solid results. If the sum is < 0, but tends to $-\infty$ as more and more ion pairs are included in the sum and interionic distances are lowered, the structure becomes unstable and implodes. On the other hand, if the sum > 0, and tends to ∞, Coulomb explosion is a possibility and there is complete collapse of the structure. The evaluation of the interaction energies of all the ion pairs in a crystal is a challenging task requiring a special technique. The idea can be explained with reference to an infinite two-dimensional square lattice of oppositely charged ions, the shortest distance between the oppositely charged species being R_0 (say). A portion of the infinite regular (square) network of oppositely charged species is shown in Figure 4.13.

We can now proceed to calculate the total energy of an ion, say a cation, due to its interaction with all other ions in the lattice. To evaluate the sum, let us note that the chosen ion (A^+) has (see Figure 4.13) the following neighbors:

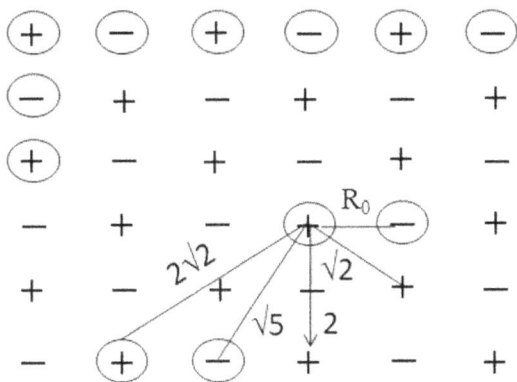

FIGURE 4.13 2D square lattice of oppositely charged ions.

(i) Four oppositely charged (B^-) neighbors at a distance of separation R_0;

(ii) four similarly (+ve ly) charged neighbors at a distance of separation $\sqrt{2}R_0$;

(iii) four similarly charged species (A^+) at the distance of $2R_0$;

(iv) eight oppositely charged ions at a distance $\sqrt{5}R_0$;

(v) four similarly charged ions (A^+) at a distance of separation of $2\sqrt{2}R_0$, and so on.

Before writing down the sum of all such interactions let us note that each term represents the interaction energy of an ion pair and only half of it can be apportioned to the ion of our interest. Thus, the expression for Coulomb energy per ion (U_C) in the network takes the following form

$$U_C = \frac{1}{2}\left[4 \times \frac{(-)e^2}{R_0} + \frac{4e^2}{\sqrt{2}R_0} + \frac{4e^2}{2R_0} + \frac{8(-)e^2}{\sqrt{5}R_0} + \frac{4e^2}{2\sqrt{2}R_0} + \cdots\right] \tag{4.70}$$

The expression has an infinite number of terms and can be written in the form of a series

$$U_C = \frac{e^2}{2R_0}\left\{-\frac{4}{1} + \frac{4}{\sqrt{2}} + \frac{4}{2} - \frac{8}{\sqrt{5}} + \frac{4}{2\sqrt{2}} + \cdots\right\} \tag{4.71}$$

Unfortunately, the infinite series within the curly brackets is very slowly convergent, many thousands of terms must be added to get meaningful results. Alternatively, special convergence acceleration techniques may be invoked to evaluate the sum of the series. In either case the converged Coulomb energy per ion turns out to be negative and finite[13]

$$U_C \sim \frac{-1.569e^2}{R_0} \tag{4.72}$$

indicating that the ions could remain bound to be network. However, $U_C \to -\infty$ as $R_0 \to 0$, implying the possibility of collapse of the network (implosion). The reason, as we noted earlier, is the absence of any short range repulsion term in the expression for U_C, which would have accounted for the strong Pauli repulsion that is brought into play as the two closed-shell entities (A^+ and B^-) begin to intrude into each others' space. The short range repulsion modified $U_C(R_0)$ could be expected to pass through a well defined minimum as a function of R_0 (the lattice constant) and lead to a more refined estimate of cohesive energy of the ionic network without any collapse of the structure. A similar approach can also be adopted for estimating the cohesive energy of an ionic 3D crystal of say NaCl, displayed in Figure 4.14.

The net Coulomb energy per ion (say Cl^-) in a crystal of sodium chloride can be expressed as an infinite series of terms, each representing the Coulomb interaction energy of the Na^+ ion with either similarly or oppositely charged ions present in the first nearest neighbor, second nearest neighbor, third nearest neighbor, etc. shells drawn around the Cl^- ion as a test case. As before, we can list the interaction as we move out from the Cl^- ion and take into account interactions with ions distributed in shells at progressively larger distances: A reference to the cubic lattice in Figure 4.14 helps us to list the interactions:

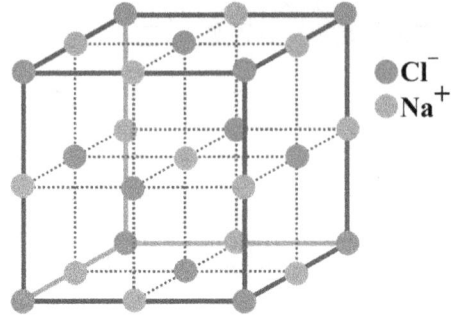

FIGURE 4.14 Ionic arrangement in a 3D crystal of solids like NaCl.

(i) Attraction interactions with six oppositely charged ions (Na^+) at a distance of R_0 (lattice constant).
(ii) Repulsive interaction with 12 similarly charged ions (Cl^-) in a shell of radius of $\sqrt{2}R_0$.
(iii) Attractive interaction with eight Na^+ ions in a shell of radius $\sqrt{3}R_0$.
(iv) Repulsive interactions with six similarly charged ions Cl^- located in a shell of radius $2R_0$ and so on.

These interactions, when added together produced an infinite slowly converging series half the sum of which (the Madelung sum) yields an estimate of the cohesive energy U_C per ion in the NaCl crystal. The expression for U_C takes the following form:

$$U_C = \frac{e^2}{2R_0}\left\{6 \times \frac{-1}{1} + \frac{12 \times 1}{\sqrt{2}R_0} + \frac{-8 \times 1}{\sqrt{3}R_0} + 6 \times \frac{1}{2R_0} - \cdots\right\} \tag{4.73}$$

In a NaCl type crystal, the converged sum turns out to be a finite negative quantity

$$U(R_0) = \frac{1}{2}\frac{e^2}{R_0}\left\{-1.748\right\} \tag{4.74}$$

Just as observed in the case of the two-dimensional network of oppositely charged ions, here too, $U(R) \to -\infty$ as $R_0 \to 0$, signifying possibility of collapsed. We can prevent the collapse by bringing in a short range Pauli-repulsion. If we represent the Pauli-repulsive energy term by the expression in Eq. 4.69b, and take into account the repulsive energy due to the six nearest neighbors only, the total cohesive energy per ion comes out to be

$$U(R_0) = (-)\frac{1.748e^2}{2R_0} + \frac{6C}{R_0^{12}} \tag{4.75}$$

Minimizing $U(R_0)$ with respect to R_0, we now obtain a value of $R_0 = R_0^{eq}$ at which the oppositely charged ions in the crystal can be held in equilibrium without collapse and that R_0 turns out to be

$$R_0^{eq} = \left[\frac{36C}{1.748e^2}\right]^{\frac{1}{11}} \tag{4.76}$$

The corresponding equilibrium cohesive energy then turns out to be

$$U_0 = \frac{-0.801e^2}{R_0^{eq}} \tag{4.77}$$

We can convert the expression for cohesive energy per ion into an estimate of cohesive energy of per mole by noting that NaCl is a two component solid (Na^+, Cl^-), which contains two ions per formula unit (NaCl). The per mole cohesive energy formula then becomes

$$U = 2NU_0 \tag{4.78}$$

'N' being Avogadro's number of course the estimate is an approximate one and can be refined further by taking into account higher order interactions among the oppositely charged entities (ion-induced dipole, induced dipole induced dipole etc.).

4.9 Weak-Binding

Two neutral atoms like argon or molecules like H_2 appear to experience a short ranged force of attraction when separated by a large distance (R), with $R >> a$, 'a' being the typical atomic or molecular dimension. It appears a little bit weird as the interacting atoms or molecules are charge neutral and their average dipole moments are zero. This force of attraction between neutral atoms or molecules turns out to be universal and plays an important role in shaping the behavior of many materials, specially molecular materials. It is therefore important to try to understand how such a force of attraction appears, its distance dependence and range, its strength, etc. We will do that with reference to a model problem, which can be handled analytically as well as perturbatively and then extend it to cover more general systems.

Let us suppose that we have two model hydrogen atoms (1 and 2) separated by a large distance $(R >> a_0)$. The model assumes that the lighter electron $(-e)$ is bound to the massive proton $(+e)$ by a harmonic force and the motion takes place in one dimension (x) only. Each of the two atoms is assumed to possess an instantaneous dipole moment $d = ex$. The instantaneous charge configuration of such a system is displayed in Figure 4.15.

We have thus reduced the problem of two hydrogen atoms at a large distance to a problem of two interacting charged linear oscillation. The potential energy of interaction of the system of charged oscillators is equal to

$$V(R, x_1, x_2) = \frac{e^2}{R} + \frac{e^2}{R + x_2 - x_1} - \frac{e^2}{R + x_2} - \frac{e^2}{R - x_1}$$

$$\approx \frac{-2e^2}{R^3}(x_1 x_2) \tag{4.79}$$

In the classical case, if the dipoles are not oscillating, i.e. $x_1 = x_2 = 0$ as it should be in the ground state, there is no interaction and $V = 0$ in the ground state. They do, however, interact when excited. In quantum mechanics, the oscillators have zero-point energy – i.e. even in the ground state, the oscillators do not cease to oscillate about the equilibrium position and $V \neq 0$, even in the ground state. Our next task is therefore to set up the Schrödinger equation for the pair of coupled linear oscillators $h_0(x_1)$ and $h_0(x_2)$ when they are interacting, the interaction Hamiltonian H_{int} (i.e. the perturbation) being given by[7]

$$V(R, x_1, x_2) = \frac{-2e^2}{R^3} x_1 x_2 \tag{4.80}$$

and find out the stationary state energies of the system. The equation reads

$$-\frac{\hbar^2}{2m}\frac{\partial^2 \Psi(x_1, x_2)}{\partial x_1^2} + \frac{1}{2}m\Omega_0^2 x_1^2 \Psi - \frac{\hbar^2}{2m}\frac{\partial^2 \Psi(x_1, x_2)}{\partial x_2^2} + \frac{1}{2}m\Omega_0^2 x_2^2 \Psi - \frac{2e^2}{R^3}x_1 x_2 \Psi$$

$$= E\Psi(x_1, x_2)$$

$$\text{or,} \quad \left[h_0(x_1) + h_0(x_2) + H_{int}(x_1, x_2) - E \right]\Psi(x_1, x_2) = 0$$

$$\text{or,} \quad \left[H_0(x_1, x_2) + H_{int}(x_1, x_2) \right]\Psi(x_1, x_2) = E\Psi(x_1, x_2) \tag{4.81}$$

where m is the reduced mass of the electron, Ω_0 is the unperturbed oscillator frequency and $\Psi(x_1, x_2)$ is the wavefunction of the coupled linear oscillators. If no interaction is present, i.e. $H_{int} = 0$, the equation

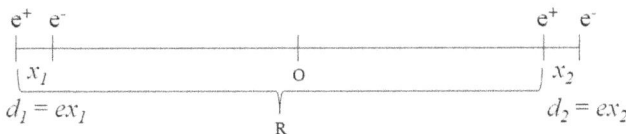

FIGURE 4.15 Interactions in a system of two charged oscillators at a distance R.

reduces to the Schrödinger equation for two non-interacting harmonic oscillator of frequency Ω_0, reduced mass 'm' and the stationary states $\Psi^0_{n_1,n_2}(x_1,x_2)$ of the system and their energies $E^0_{n_1,n_2}$ are given by

$$E^0_{n_1,n_2} = (n_1 + n_2 + 1)\hbar\Omega_0; \quad n_1, n_2 = 0, 1, 2, \cdots \tag{4.82a}$$

$$\Psi^0_{n_1,n_2}(x_1,x_2) = \Psi_{n_1}(x_1).\Psi_{n_2}(x_2) \tag{4.82b}$$

If both the oscillators are unexcited, $n_1 = n_2 = 0$ and the ground state energy of the coupled oscillators is given by

$$E_{0,0} = \frac{1}{2}\hbar\Omega_0 + \frac{1}{2}\hbar\Omega_0 = \hbar\Omega_0 \tag{4.83a}$$

$$\Psi_{0,0}(x_1,x_2) = \Psi_0(x_1) \cdot \Psi_0(x_2) \tag{4.83b}$$

where $\Psi_0(x_1)$ and $\Psi_0(x_2)$ are the ground state wavefunctions for the two identical non-interacting linear oscillators of frequency Ω_0. In order to find the eigenenergies of the system of two interacting oscillators with $H_{int} = -2e^2\frac{x_1x_2}{R^3}$, we make the following transformation or change of variables:

$$X_+ = \frac{1}{\sqrt{2}}(x_1 + x_2)$$
$$X_- = \frac{1}{\sqrt{2}}(x_1 - x_2) \tag{4.84}$$

The transformation converts the interacting linear oscillator Hamiltonian of Eq. 4.81 into sum of two new non-interacting oscillator Hamiltonians with frequencies Ω_+ and Ω_-:

$$H_0(x_1,x_2,\Omega_0) + H_{int}(x_1,x_2,R) \equiv H_0(x_+,\Omega_+) + H_0(x_-,\Omega_-)$$
$$\equiv -\frac{\hbar^2}{2m}\frac{\partial^2}{\partial x_+^2} - \frac{\hbar^2}{2m}\frac{\partial^2}{\partial x_-^2} + \frac{1}{2}m\Omega_+^2(x_+^2) + \frac{1}{2}m\Omega_-^2(x_-^2) \tag{4.85}$$

where

$$\Omega_+ = \Omega_0\left\{1 - \frac{2e^2}{\Omega_0^2 mR^3}\right\}^{\frac{1}{2}}$$
$$\Omega_- = \Omega_0\left\{1 + \frac{2e^2}{\Omega_0^2 mR^3}\right\}^{\frac{1}{2}} \tag{4.86}$$

Expanding bionomically we find that[7]

$$\Omega_+ \approx \Omega_0 - \frac{e^2}{m\Omega_0 R^3} - \frac{e^4}{2m^2\Omega_0^3 R^6}$$
$$\Omega_- \approx \Omega_0 + \frac{e^2}{m\Omega_0 R^3} - \frac{e^4}{2m^2\Omega_0^3 R^6} \tag{4.87}$$

The energy eigenvalue equation of the two dressed harmonic oscillators with frequencies Ω_+ and Ω_- is given by

$$\{H_0(X_+,\Omega_+) + H_0(X_-,\Omega_-)\}\Psi_n(X_+,X_-) = E_n\Psi_n(X_+,X_-) \tag{4.88a}$$

where

$$E_n \equiv E_{n_1,n_2} = \left(n_1 + \frac{1}{2}\right)\hbar\Omega_+ + \left(n_1 + \frac{1}{2}\right)\hbar\Omega_-$$
$$\text{and,} \quad \Psi_n(X_+,X_-) = \Psi^0_{n_1}(X_+,\Omega_+) \cdot \Psi^0_{n_2}(X_-,\Omega_-) \tag{4.88b}$$

The ground state energy $E_{0,0}$ of the system is

$$E_0 = E_{0,0} \equiv \frac{1}{2}\hbar\Omega_+ + \frac{1}{2}\hbar\Omega_- \tag{4.89}$$

which is just the zero-point energy of the two interacting oscillators. The shift in zero-point energy due to the interaction is therefore

$$\begin{aligned}
\Delta E_0(R) &= \frac{1}{2}\hbar\Omega_+ + \frac{1}{2}\hbar\Omega_- - 2 \cdot \frac{1}{2}\hbar\Omega_0 \\
&= -\frac{\hbar e^4}{2m^2\Omega_0^3} \cdot \frac{1}{R^6}
\end{aligned} \tag{4.90}$$

$\Delta E_0(R)$ may be identified as the potential energy of interaction between the two neutral model atomic oscillators at a large separation of R. We can therefore assume that the van der Waals interaction energy has the form:

$$V(R) = -\frac{C}{R^6}$$
$$\text{where} \quad C = \frac{\hbar e^4}{2m^2\Omega_0^3} \tag{4.91}$$

The form of $V(R)$ we have arrived at is based on one-dimensional harmonic model of hydrogen atom. Clearly, $C \to 0$ as $\hbar \to 0$. This result is entirely expected as the zero-point energy (ZPE) is a purely quantum phenomenon and $V(R)$ is linked to the shift in ZPE due to interactions.[2,7,14]

The result described above can also be arrived at perturbatively by partitioning the Hamiltonian into an unperturbed Hamiltonian (H_0) and a perturbation H_{int} where

$$H_0 = -\frac{\hbar^2}{2m}\frac{\partial^2}{\partial x_1^2} + \frac{1}{2}m\Omega_0^2 x_1^2 - \frac{\hbar^2}{2m}\frac{\partial^2}{\partial x_2^2} + \frac{1}{2}m\Omega_0^2 x_2^2$$

$\equiv$ sum of two identical linear oscillators of reduced mass 'm' and frequency Ω_0.

and $H_{int} = -\frac{2e^2}{R^3}x_1 x_2$.

The unperturbed eigenenergies and the corresponding eigenfunctions are easily found to be

$$H_0|n_1 n_2\rangle = E^0_{n_1 n_2}|n_1 n_2\rangle \tag{4.92}$$

$$E^0_{n_1 n_2} = (n_1 + n_2 + 1)\hbar\Omega_0 \tag{4.93}$$

The unperturbed ground state is $|00\rangle$ with energy $\hbar\Omega_0$. The first order energy correction obviously $\Delta E^{(1)}_{00} = 0$. The second order energy correction for the ground state is (using the standard Rayleigh Schrödinger perturbation theory (RSPT))

$$\begin{aligned}
\Delta E^{(2)}_{00} &= \sum_{n_1 n_2 \neq 0} \frac{|\langle 00| \frac{-2e^2 x_1 x_2}{R^3}|n_1 n_2\rangle|^2}{E_{00} - E_{n_1 n_2}} \\
&= \frac{4e^4}{R^6} \sum_{n_1 n_2 \neq 0} \frac{\langle 00|x_1 x_2|n_1 n_2\rangle \langle n_1 n_2|x_1 x_2|00\rangle}{E_{00} - E_{n_1 n_2}} \\
&= \frac{4e^4}{R^6} \sum_{n_1 n_2 \neq 0} \delta_{n_1 1}\delta_{n_2 1} \frac{\langle 00|x_1 x_2|n_1 n_2\rangle \langle n_1 n_2|x_1 x_2|00\rangle}{E_{00} - E_{n_1 n_2}}
\end{aligned} \tag{4.94}$$

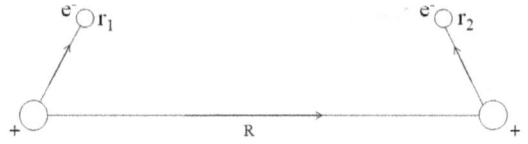

FIGURE 4.16   Two H atoms separated at a large distance R $(R >> r_1, r_2)$.

$$= \frac{4e^4}{R^6} \left\{ \frac{\langle 0|x_1|1\rangle^2 \langle 1|x_2|0\rangle^2}{-2\hbar\Omega_0} \right\}$$

$$= -\frac{1}{2} \frac{\hbar e^4}{m^2 \Omega_0^3 R^6}$$

$$= -\frac{C}{R^6}$$

Let us note that in arriving at the final result (Eq. 4.94) we have used the matrix element rules for the harmonic oscillator

$$\langle 0|x|n\rangle = \delta_{n1}\sqrt{\frac{\hbar}{2m\Omega_0}}$$

Thus the second order energy correction or shift due to the interactions between the model atomic oscillators also predict an expression for van der Waals potential energy already obtained by a more exact method.

Now let us consider the same problem in three dimensions. It is easy to find the interaction energy between the two instantaneous atomic dipoles $(d_1 = er_1, d_2 = er_2)$ in leading order to be[7]

$$H_{int}(r_1, r_2) = \frac{e^2}{R^3}\left(\vec{r_1} \cdot \vec{r_2} - 3(\vec{r_1} \cdot \vec{R})(\vec{r_2} \cdot \vec{R})\frac{1}{R^2}\right) \tag{4.95}$$

The unperturbed eigenstates of hydrogen atoms obey the eigenvalue equation (H_i^0 is the hydrogen atom Hamiltonian for i^{th} H atom, $i = 1, 2$)

$$H_1^0 |n_1\rangle = E_{n1}^0 |n_1\rangle$$
$$H_2^0 |n_2\rangle = E_{n2}^0 |n_2\rangle \tag{4.96}$$

The unperturbed ground state is $|00\rangle$. The perturbative correction to the ground state energy (upto second order) is

$$\Delta E(R) = \langle 00| H_{int} |00\rangle + \sum_{n_1 n_2}' \frac{|\langle 00| H_{int} |n_1 n_2\rangle|^2}{E_{00}^0 - E_{n_1 n_2}^0} \tag{4.97}$$

where $\sum_{n_1 n_2}'$ indicates that the summation is carried out over all the states ($|n_1 n_2\rangle$) except the ground state $|00\rangle$, and $E_{n_1 n_2}^0 = E_{n_1}^0 + E_{n_2}^0$.

The first order energy correction for H atom or atoms (like Ar, for example) with spherical ground state having zero dipole moment is zero by symmetry. The second order energy correction is always negative for the ground state. To estimate it for a pair of hydrogen atoms, we assume that $\vec{R}$ is directed along the Z-axis, which reduces to the interaction Hamiltonian to the form[2]

$$H_{int} = \frac{e^2}{R^3}\left(x_1 x_2 + y_1 y_2 - 2z_1 z_2\right) \tag{4.98}$$

The summation in Eq. 4.98 is over all the excited states. From symmetry considerations we can immediately set

$$\langle 00| H_{int} |0n_2\rangle = \langle 00| H_{int} |n_1 0\rangle = 0$$

which means that the singly excited states of H atoms do not contribute to the second order energy correction for the ground states. Therefore, $\Delta E^2(R)$ is given by

$$\Delta E^2(R) = -\frac{e^2}{R^6} \sum_{n_1 n_2}' \frac{|\langle 00| H_{int} |n_1 n_2\rangle|^2}{E_{n_1 n_2} - E_{00}} \tag{4.99}$$

where the summation is over all those states of the system in which both the atoms are excited. It is rather difficult to evaluate the sum accurately. However, we can estimate it fairly well by way of the closure approximation. Applying closure, we have

$$\Delta E^2(R) \approx -\frac{e^4}{R^6} \cdot \frac{1}{\Delta E} \left\{ \langle 00| v^2 |00\rangle - \langle 00| v |00\rangle^2 \right\} \tag{4.100}$$

where ΔE is an average energy denominator and $v = (x_1 x_2 + y_1 y_2 - 2z_1 z_2)$. The energy denominator $E_{n_1 n_2} - E_{00}$ varies between $\frac{3}{2}$Ry to 2Ry. This is because $E_{n_1 n_2} \approx 0$ if n_1 and n_2 are both large while $E_{00} = \frac{-e^2}{a_0}$ (both the H atoms are in the ground state) where the energy denominator becomes $E_{n_1 n_2} - E_{00} = \frac{e^2}{a_0} = 2$Ry. If both the hydrogen atoms are in the first excited state ($n_1 = n_2 = 2$), $E_{n_1 n_2} = -\frac{e^2}{2a_0}(\frac{1}{2^2} + \frac{1}{2^2}) = -\frac{e^2}{2a_0}(\frac{1}{2}) = -\frac{e^2}{4a_0}$, which determines $E_{n_1 n_2} - E_{00} = -\frac{e^2}{4a_0} + \frac{e^2}{a_0} = \frac{3e^2}{4a_0} = \frac{3}{2} \cdot \frac{e^2}{2a_0} = \frac{3}{2}$Ry. The average value of ΔE can therefore be approximated as

$$\Delta E_{av} = \frac{1}{2} \frac{5e^2}{2a_0} \approx \frac{e^2}{a_0}$$

Let us also note that $\langle 00| v |00\rangle$ is zero by symmetry for the given form of v. That simplifies the second order energy expression to

$$\begin{aligned} \Delta E^2(R) &= -\frac{a_0}{e^2}(\langle 00| v^2 |00\rangle) \cdot \frac{e^4}{R^6} \\ &\approx -\frac{6e^2 a_0^5}{R^6} \end{aligned} \tag{4.101}$$

The sum over states defined in Eq. 4.99 has been precisely evaluated and found to be[15]

$$\Delta E^2(R) = -\frac{6.47 e^2 a_0^5}{R^6} \tag{4.102}$$

so that the result obtained by applying the closure approximation for evaluating the sum is more or less acceptable. Eq. 4.101 and Eq. 4.102 therefore confirm the form of the potential of van der Waals force to be

$$V(R) = -\frac{C}{R^6} \tag{4.103}$$

with $C \approx 6e^2 a_0^5 - 6.47 e^2 a_0^5$.

Note that $a_0^3 \sim \alpha$ (polarizability of the hydrogen atom in the ground state) leads to the following form for van der Waals forces between atoms

$$V(R) = -\frac{e^2}{a_0} \cdot \frac{\alpha_1 \alpha_2}{R^6} \times A \quad (A \text{ is a constant}) \tag{4.104}$$

Thus van der Waals interaction energy would be higher between highly polarizable atoms. The calculations reported here are based on static Coulomb potential. We may point out that electromagnetic interactions occur between the interacting subsystems through exchange of photons. Photons, however travel with a finite velocity. Electromagnetic interactions are therefore not instantaneous, but travels with a finite speed – the speed of light (c). In our case, this travel time (τ) for photons (light) is

$$\tau_l = \frac{2R}{c}$$

where R is the distance of separation between the atoms. On the other hand, the characteristic time for one complete revolution of the electron in the atom is

$$\tau_e = \frac{1}{me^4}.$$

For small R, $\tau_l << \tau_e$ and the form of $V(R)$ defined on the basis of static model leads to $V(R) \sim -\frac{C}{R^6}$ as we have already seen. For very large values of $R(R >> \lambda)$, retardation effects become important. When retardation effects are included, $V(R)$ between a pair of atoms turns out to be[2]

$$V_{eff}(R) = -\frac{23}{4\pi} \hbar c \frac{\alpha_1 \alpha_2}{R^7} \qquad (4.105)$$

The inclusion of retardation effects makes $V_{eff}(R) \propto \frac{1}{R^7}$ as opposed to the $V_{static}(R) \propto \frac{1}{R^6}$. It appears that the energy of the atoms $(\frac{e^2}{a_0})$ appearing in Eq. 4.104 for the van der Waals energy with static case has been replaced by $\frac{\hbar c}{R}$ – the energy of photons of wavelength $\lambda = R$, when retardation effects are included. Two important facts about van der Waals force are (i) it is rather short ranged (vis-a-vis Coulomb force) (ii) it is much weaker than forces of covalent binding (approximately 10^3 times weaker). Thus solids formed by van der Waals interaction has much lower cohesive energy than covalent or ionic solids line. Properties of such solids are accordingly modified. The calculation of cohesive energy of van der Waals solids can be carried out in the same way as done in the case of ionic solids.

4.10 Weak-Binding: Hydrogen Bonds

Although much weaker compared to ionic or covalent bonds, hydrogen bonds are typically much stronger than van der Waals '*bonds*' and play an important role in providing stability especially to many organic materials including bio-materials. Water is one non-organic molecule, the bulk properties of which either in the liquid or in the solid (ice) phase are shaped by hydrogen bonding. It turns out that hydrogen bonds are formed by the cooperative interplay of both covalent and ionic '*forces*' and display the strong directionality of covalent bonds and polarity of ionic-binding. The formation of H bonds requires, in general, a strongly polar A-H bond and a highly electronegative '*acceptor*' atom (B). The A-H bond is '*covalent*' but the high asymmetry between the electron affinities of the atoms A and H produces a strongly polarized distribution of the shared electron-density. An atom with a complete shell of electrons has very low, almost zero, electron affinity. A strongly polarized covalent A-H bond essentially means that the valence shell of the H atom is no-longer spherically symmetric around the nucleus of H. This loss of spherical symmetry may be thought of as producing a '*descreening*' of valence shell so that the nuclear charge of 'H' can now be '*seen*' by electron on another atom in another A-H or B-X molecule, which are attracted to it leading to the formation of a hydrogen bond (Figure 4.17).

The interaction between $H^{\delta+} \cdots A^{\delta-}$ is ionic, but the strong directionality of the covalent A-H bond imparts directionality to the bond between $H^{\delta+} \cdots A^{\delta-}$ or the H bond. Thus H bonding is essentially an admixture of both covalent and ionic interactions. The different crystalline forms of water are classic examples of H bonded solids. They stand testimony to how H bonding could affect the properties of such solids. Calculation of cohesive energies of H bonded solids (molecular materials) or for that matter, of covalent solids and predicting equilibrium structure of such solids is far more difficult than the calculation of cohesive energies and other properties ionic or van der Waals solids. The difficulties arise mainly from the directionality of covalent as well as hydrogen bonds. The absence of directionality in ionic or van der

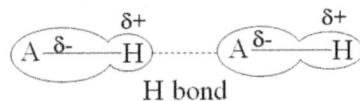

FIGURE 4.17 Directionality of hydrogen bond.

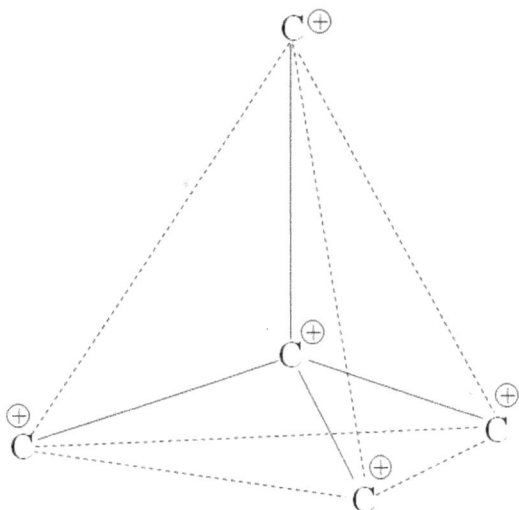

FIGURE 4.18 Directionality of covalent bonds in diamond.

Waals interactions means that the space around each atom or ion is homogeneous. These solids therefore try to pack as many ions of opposite charge as possible around a given ion (in ionic solids) and as many atoms as possible around a given atom (in molecular solids). In this way, the solid strives to attain the most stable or the minimum energy configuration (geometry). The problem is far more complex in the case of covalent solids (diamond, for example) or in hydrogen bonded solids (like ice, for example). Since the sp^3 hybrid orbitals on each carbon atom are tetrahedrally oriented around each carbon atom, two such carbon atoms can interact favorably only if one is at the center and the other is at an apex of the tetrahedron so that the sp^3-hybridized orbitals on them can overlap properly producing maximum bond strength and form a covalent bond (Figure 4.18).

In a covalent solid like diamond the packing of atoms is thus strongly dictated by the geometrical dispositions of sp^3-hybridized electronic orbitals of the carbon atoms in space. Arbitrarily adding one extra neighboring atom around a given atom without fulfilling the geometrical requirement for in-phase orbital overlap to take place, will not generally give rise to '*additional*' binding or stabilization. The covalent solids are therefore often seen to have rather open structure and are less dense. When covalently bonded solid materials melt into the liquid state, the open crystalline structure of the solid breaks down. This happens because the constantly and randomly changing configuration that thermal agitation brings in, makes it difficult to maintain correct orientation with respect to the neighboring particles. Such materials would tend to suffer volume contraction on melting raising the density. Silicon and germanium belong to this category of materials. In water, the directionality of the hydrogen bonds forces ice (solid) to assume an open crystal structure. Ice is therefore less dense than liquid water in which the solid crystalline structure of ice has partially collapsed, still retaining, however remnants of the ice structure. It turns out from advanced simulation studies, that only a few percent of H bonds gets bent in the liquid water, but are not broken by thermal agitation. The density of water (liquid) is therefore higher. We refrain from discussing metallic bonds and the metallic state here. The relevant issues will be addressed in Chapter 8 where we take up issues relating to the understanding of metallic materials.

4.11 Directed Valence and Chemical Binding

We have so far discussed issues concerning chemical valence within the framework of two types of approximate methods, namely the molecular orbital method and valence-bond method. In the MO description we considered one-electron states or eigenfunctions delocalized over the entire nuclear framework of the molecule (as advocated by Mulliken, for example). In the valence bond description the starting point has been to construct the so called valence orbitals that are localized in the region between the bonded pair of atoms (these orbitals are primarily bicentric, localized, accommodating the shared electron pair with opposite spins). We will use in what follows, the second approach to explain the directional properties of covalent bonds, which are so important in giving rise to molecules with varieties of shapes and materials with different types of structures. As we have already seen, chemical binding within the framework of localized valence orbital method (VB method) arises primarily from the so called '*interference*' terms present in the connected approximate wavefunctions. This '*interference*' leads to the build up of electronic charge in the region of low potential energy lying between the interacting atoms. A rough but useful measure of this build-up of electronic charge density is provided by the extent of overlap between the one-electron functions centered on the individual atoms participating in the bonding process. Linus Pauling exploited this feature to propose his overlap criterion of maximum bond strength. In what follows we use simple symmetry-based arguments (a more formal, group-theory-based approach can also be used) to find out how sets of atomic orbitals can be combined to form a set of strongly overlapping and directed valence orbitals on a given atom in a given chemical environment.

As an example, we take up the standard, much used example of tetrahedral bonding displayed by carbon in methane. A free carbon atom has the ground state electronic configuration $1s^2 2s^2 2p^2$. Since the $1s^2$ core (the K-shell) is not much perturbed by the bonding, we turn to work with the L-shell orbitals only (i.e. $2s$ and $2p$ functions). Disregarding the difference in the radial parts of the $2s$ and $2p$ eigenfunctions we consider only the angular parts of the valence shell orbitals of the carbon atomic wavefunctions (one-electron states) with polar angles θ and ϕ:

$$\Psi_{2s} \sim 1, \Psi_{2p_x} \sim \sqrt{3}\sin\theta\cos\phi,$$

$$\Psi_{2p_y} \sim \sqrt{3}\sin\theta\sin\phi,$$

$$\Psi_{2p_z} \sim \sqrt{3}\cos\theta.$$

Our task now is to try to form four completely equivalent orthogonal valence orbitals each of which is directed along one of the four regular tetrahedral directions. Symmetry-based arguments lead to the following directed valence orbitals:[16-17]

$$\Psi_{111} = \frac{1}{2}\left(\Psi_{2s} + \Psi_{2p_x} + \Psi_{2p_y} + \Psi_{2p_z}\right)$$

$$\Psi_{1-1-1} = \frac{1}{2}\left(\Psi_{2s} + \Psi_{2p_x} - \Psi_{2p_y} - \Psi_{2p_z}\right)$$

$$\Psi_{-11-1} = \frac{1}{2}\left(\Psi_{2s} - \Psi_{2p_x} + \Psi_{2p_y} - \Psi_{2p_z}\right)$$

$$\Psi_{-1-11} = \frac{1}{2}\left(\Psi_{2s} - \Psi_{2p_x} - \Psi_{2p_y} + \Psi_{2p_z}\right)$$

The subscripts on Ψ on the left allude to the direction cosines of the major lobes of the directed valence orbitals defined with respect to the axes of the cube in which the tetrahedron is sufficiently inscribed. The four orthogonal valence orbitals disposed along the four distinct tetrahedral directions are called sp^3-hybridized orbitals or simply the sp^3 hybrids, which are roughly sketched out in Figure 4.19.

We can now roughly measure the bond strength formed by each of the four equivalent tetrahedral valence orbitals in terms of the maximum numerical values of the angular valence function along each bond axes. As an example, we may take up the case of Ψ_{111}, which is directed along the (111) direction. Along this direction $\cos\theta = \frac{1}{\sqrt{3}}$, $\sin\theta = \sqrt{\frac{2}{3}}$ while $\sin\phi$ and $\cos\phi$ are each equal to $\frac{1}{\sqrt{2}}$. Using these values of θ and ϕ, it is easy to find that along the 111 direction $\Psi_{2p_x} = \Psi_{2p_y} = \Psi_{2p_z} = 1$. The Ψ_{2s} orbital is spherical allowing us to fix $\Psi_{2s} = 1$ along each one of the four tetrahedral bond axes. The maximum value of angular valence functions along the 111 axis is therefore $\frac{1}{2}(1+1+1+1) = 2$. Taking the three other bond axes and proceeding similarly it is possible to show that the maximum numerical value of the angular valence functions along the $(1, -1, -1)$, $(-1, 1, -1)$ and $(-1, -1, 1)$ bond directions is also two in each case. It has been shown by Linus Pauling that the tetrahedrally disposed valence orbitals produce the maximum

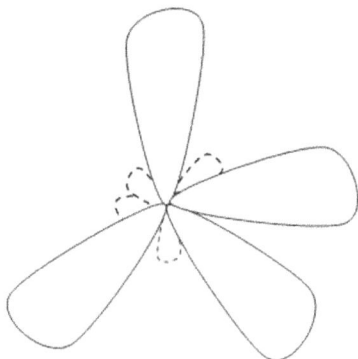

FIGURE 4.19 sp^3 hybrid orbitals on a carbon atom.

bond strength (overlap) out of all other hybrids of $2s$ and $2p$ functions. It seems natural then to expect that carbon atom will commonly form sp^3-hybridized tetrahedral bonds whenever its chemical environment makes it possible to form four equivalent bonds.

A minor question remains to be addressed. Since the three $2p$ one-electron functions are energetically degenerate, mixing them up for constructing directed valence functions has no cost in terms of energy. Mixing the $2s$ function with the $2p$ functions on the other hand ($2s$ orbital in carbon has lower energy than the $2p$ orbitals) involves an energy cost – the energy required to promote an electron from the $2s$ to a vacant $2p$ state. Fortunately, this promotion energy is overcompensated by the gain in bond strength, which is maximum for the sp^3 hybrids. The propensity for the carbon atom to form trigonal bonds (three equivalent in-plane bonds) is quite large. These bonds are separated from each other by an angle of $\frac{2\pi}{3}$ and are formed by mixing up the $2s$ and two $2p$ functions ($2p_x$, $2p_y$). They are known as sp^2-hybridized bonds or simply the sp^2 hybrids. These orthogonal and well directed valence orbitals can be written as

$$\Psi_1 = \frac{1}{\sqrt{3}}\Psi_{2s} + \sqrt{\frac{2}{3}}\Psi_{2p_x}$$

$$\Psi_{2+} = \frac{1}{\sqrt{3}}\Psi_{2s} - \frac{1}{\sqrt{6}}\Psi_{2p_x} + \frac{1}{\sqrt{2}}\Psi_{2p_y}$$

$$\Psi_{2-} = \frac{1}{\sqrt{3}}\Psi_{2s} - \frac{1}{\sqrt{6}}\Psi_{2p_x} - \frac{1}{\sqrt{2}}\Psi_{2p_y}$$

These directed valence orbitals are normalized to unity, mutually orthogonal and equivalent, the last property being confirmed by noting that Ψ_1, Ψ_{2+} and Ψ_{2-} get transformed into each other under the symmetry operations of the D_{3h} point group. The bond strength achieved with these sp^2 hybrid functions can be calculated by following the same principles followed for the tetrahedral bonds. Thus the bond strength achieved with Ψ_1 corresponds to the numerical value of the angular part of the one-electron valence function or orbital directed along the bond axis for which $\theta = \frac{\pi}{2}, \phi = 0$. The value turns out to be 1.99, which is just marginally lower than the bond strength achieved with the sp^3 hybrid valence orbitals. The Ψ_{2p_z} orbital remains unused in the trigonal scheme, and can be utilized for forming a much weaker π bond by lateral overlap only (strength = 1.73). The trigonal scheme is operative in the ethylene molecule, for example. Thus the directional nature of the covalent σ bonds are clearly explained by adopting an approximate scheme of hybridization like sp^3 for methane or diamond, sp^2 for ethylene and sp for acetylene.

In case the valence shell has d orbitals along with s and p, a variety of hybridization schemes can be constructed to explain the stereochemical disposition of the covalent bonds. However, when alternative schemes are there, the symmetry consideration or the group theoretical tools alone can not decide which scheme would lead to an energy minimum and the consequent ability to exist in nature. Along with symmetry-based reasoning, one must do detailed energy calculation (which are possible nowadays) or invoke approximate schemes based on Pauling's criterion of maximum bond strength.

4.12 Many Electron Systems

The approximate Hamiltonian of a many electron atom of atomic number Z reads (non-relativistic Schrödinger formulation)

$$H = \sum_{i=1}^{n} \left(-\frac{\hbar^2}{2m} \nabla_i^2 - \frac{Ze^2}{r_i} \right) + \frac{1}{2} \sum_{i,j(i\neq j)}^{n} \frac{1}{r_{ij}} \qquad (4.106)$$

It is quite difficult to deal with this Hamiltonian in a straightforward manner. It is possible to start with a rather simple spherically symmetric zeroth order Hamiltonian (H_0) by artificially introducing a central field potential $V_0(r_i)$ and write

$$H_0 = \sum_{i}^{n} \left[-\frac{\hbar^2}{2m} \nabla_i^2 + V_0(r_i) \right] \qquad (4.107)$$

and define a perturbation Hamiltonian H' as

$$H' = \sum_{i} \left[\frac{1}{2} \sum_{j\neq i} \frac{e^2}{r_{ij}} - V_0(r_i) - \frac{Ze^2}{r_i} \right]$$

This is done on the basis of a rather simple physical idea that each of the n individual electrons move in an effective central field potential $V_0(r_i)$ of the atomic nucleus with a reduced nuclear charge due to screening by other electrons. The advantage of introducing the spherical (a central field) potential is that the separation of angular and radial variable becomes possible and thus each electron can be assigned a well defined angular momentum value and the zeroth order state described by the normally used configuration labels.

What remains to be found is then the radial wavefunction of each individual electron and then the effects of the perturbation term H' needs to be calculated on the energy of various electronic states arising from the particular electronic configurations. A systematic exploration of the scheme that was fairly successful for the light atoms involved guessing a simple form for each radial function with adjustable parameters included in them, which were then variationally optimized by minimizing the energy with respect to the relevant parameters (Rayleigh-Ritz variational method). The use of more a systematic and elegant type of variational trial function was advocated and elaborated by Hartree. We will briefly address the method proposed by Hartree.

4.13 Hartree Method

The approximate wavefunction for the N-electron atom proposed by Hartree reads

$$\Psi(\vec{r_1}, \cdots, \vec{r_N}; \vec{s_1}, \cdots, \vec{s_N}) = u(\vec{r_1})\phi_1(s_1) \cdots u(\vec{r_N})\phi_1(s_N) \qquad (4.108)$$

where $u_i(\vec{r_i})$s are the one-electron functions or orbitals, $\chi_i(s_i)$s are the up ($\uparrow$) and down ($\downarrow$) spin eigenfunctions, respectively. The simple product wavefunction Ψ is clearly an approximate one because the probability distribution represented by $|u(\vec{r_1})|^2$ is clearly independent of the coordinates of the remaining $N-1$ electrons even though the Coulomb repulsion terms $\frac{e^2}{r_{ij}}$ would force the other electrons to avoid the region of space close to the instantaneous location of electron 1 at $\vec{r_1}$. This feature would have been present in an exact or more accurate wavefunction of the N-electron system. The case with all the other electrons is similar. Thus, the Ψ of Eq. 4.108 could be expected to yield an average value of energy for the ground state or the lowest state of each symmetry, which are only upper bounds to the corresponding values of exact or true energies of the N-electronic system. This would happen whatever choice is made for the one-electron functions $u_i(r_i)$ because of the operation of variational principle. However, we can do better, by requiring the $u_i(r_i)$s to be so chosen as to minimize the ground state energy of the system (the N-electron atom) subject to normalization condition. These optimized functions are called Hartree functions. Note that in this approach, no fixed analytical form is chosen a priori for the u_is – but they are kept flexible enough and shaped by the requirement of energy minimization and normalization. Thus we can expect Hartree method to work better than working with sets of analytical $u_i(r_i)$ functions of specific forms and only optimizing the parameters in them to arrive at the best N-electron product wavefunction of the Hartree type. However, the 'best' $u_i(r_i)$s can only be found by an iterative process because of the inherent non-linearity of the problem.

The Hartree procedure consists of the following steps:

1. One makes a choice of the initial set of the N functions $(u_1^0,....,u_N^0)$.

2. One then uses these functions for calculating the spherically averaged repulsive potential felt by each electron (i) due to the $N-1$ number of other electrons, which is added to the attractive Coulomb potential seen by each electron (i) due to interaction with the atomic nucleus of charge $+Ze$. The net potential $V_i(r_i)$ seen by the electron 'i' is thus found to be

$$V_i(r_i) = -\frac{Ze^2}{r_i} + \left\langle \sum_{j \neq i} \frac{e^2}{r_{ij}} \right\rangle_{\text{averaged spherically}} \qquad (4.109)$$

3. A better u_i function is then calculated by numerically integrating effective single electron Schrödinger equation of the following form:

$$\left[-\frac{\hbar^2}{2m}\nabla_i^2 + V_i(r_i) \right] u_i(r_i) = \epsilon_i u_i \qquad (4.110)$$

These solutions for u_is will in general be different from the solutions $u_i^0(r_i)$ one started with. Let us name the new functions (they are known as Hartree orbitals) as $u_1^{(1)}(r_1)$, $u_2^{(1)}(r_2)$, $u_3^{(1)}(r_3)$, $\cdots$, $u_N^{(1)}(r_N)$, the superscript (1) indicates that the functions have undergone one iteration.

4. One now goes back to the step '2' and the whole process is repeated with the modified Hartree orbitals $\{u_i^1(r_i)\}$ yielding the second iterates $\{u_i^2(r_i)\}$, etc.

5. These iterations are repeated over and over again until the solutions become acceptable-self-consistent, i.e., input and output orbitals do not differ by more than a preassigned value:

$$Max|u_i^{n+1}(r_i) - u_i^n(r_i)| < \epsilon.$$

It is possible to establish formally that the simple minded Hartree method described here is equivalent to self-consistently solving the integro differential equation that arise out of variational requirement of energy minimization keeping the $u_i(r_i)$s normalized. The Hartree wavefunctions and energies for the ground states of a large number of many electron atoms are available in tabulated form in a number of books. The Hartree approach was the first serious attempt to solve the many electron problem by introducing the concept of what is nowadays called mean-field approximation. Indeed, it was the first mean-field theory of many electron atoms.

The Hartree wavefunctions are fine as a first approximation to the many electron atomic wavefunctions. We note here that the self-consistent Hartree orbitals can be normalized (explicitly), but are not mutually orthogonal as the orthogonality requirement was not introduced in the entire procedure as a constraint. They tend to be more localized in space.

That apart, a more serious drawback of the Hartree wavefunctions lies in its lack of appropriate permutation symmetry. Thus one notices that all the electrons are dynamically equivalent entities and are indistinguishable from each other. An immediate consequence of this equivalence is that $\{u_i(r_i)u_j(r_j)\}$ is as good a wavefunction as $u_j(r_i)u_i(r_j)$ and both must have the same energy. Any linear combination of the two functions would also be equally acceptable. In general, this lack of symmetry affects all the simple minded product wavefunctions of the Hartree type. A way out seems to be to impose permutation symmetry on the simple product function by linearly combining all the energetically degenerate product functions produced by permutations of the order of the electrons in the simple product function so far used:

$$\Phi(1,2,\cdots,N) = u_1(r_1)u_2(r_2)\cdots u_i(r_i)\cdots u_N(r_N) \qquad (4.111)$$

Such a permutation-adjusted N-electron wavefunction could be

$$\Psi(1,2,\cdots,N) = \sum_p \alpha_p \hat{P}\left\{u_1(1)u_2(2)\cdots u_i(i)\cdots u_N(N)\right\} \tag{4.112}$$

where $u_i(i) = u_i(r_i)\phi_i(s_i)$ is spin-orbital and $\hat{P}$ permutes the order of the electrons appearing in the functions within the curly brackets $\{\}$, α_p is the weightage assigned to the pth permutation. Let us examine what happens to the expectation value of an operator $\hat{A}$ (a function of electronic coordinates) when an arbitrary exchange of j and k is effected through the permutation operator $\hat{P}_{jk}$. On physical grounds, the interchange must leave the expectation value

$$\langle A \rangle = \int \Psi_{jk}^* \hat{A} \Psi_{jk} dv_1 dv_2 \cdots dv_N$$

unchanged. That means $\hat{P}_{jk}\Psi$ must be equal to $\exp(i\delta)\Psi$, which keeps $|\Psi_{jk}|^2 = |\Psi|^2$. An additional restriction on Ψ appears from the requirement that two successive application of $\hat{P}_{jk}$ on Ψ must leave the wavefunction unchanged.[16] This is expected on the physical ground that the first application of $\hat{P}_{jk}$ interchanges j and k, which is simply restored by the second application of $\hat{P}_{jk}$. The two conditions can be satisfied if we can ensure that

$$\hat{P}_{jk}\Psi = (\pm 1)\Psi$$

That is, the wavefunction is either symmetric or antisymmetric with respect to the binary interchange of particle coordinates ($j \leftrightarrow k$). It turns out that the spin half particles like electrons – they are called Fermions – belong to the one-dimensional antisymmetric representation of the permutation group. The Hartree product function is not adapted to the requirement of the correct permutation symmetry. The required antisymmetry with respect to interchange of space (r_j) and spin (s_j) coordinates of any pair of particles appearing in Ψ can be built in if $\Psi(1,2,...N)$ is chosen as a determinant of the spin orbitals $u(r_i)\Phi(s_i)$ as shown below:

$$\Psi(1,2,...N) = A \begin{vmatrix} u_1(r_1s_1) & u_2(r_1s_1) & \cdots & u_N(r_1s_1) \\ u_1(r_2s_2) & u_2(r_2s_2) & \cdots & u_N(r_2s_2) \\ u_1(r_3s_3) & u_2(r_3s_3) & \cdots & u_N(r_3s_3) \\ \cdot & & & \\ \cdot & & & \\ \cdot & & & \\ u_1(r_Ns_N) & u_2(r_Ns_N) & \cdots & u_N(r_Ns_N) \end{vmatrix}$$ where A is the normalization constant and

the spin orbital $u_i(r_k,s_k) = u_i(r_k)\phi_i(s_k)$. It can be easily shown that for a set of N orthonormal spin orbitals $A = \frac{1}{\sqrt{N!}}$. These determinants are known as Slater determinants. Two important features of this determinantal wavefunction must be mentioned here:

(i) If two spin orbitals, say u_i and u_j are identical, the ith and jth columns of Ψ become identical and the determinant vanishes. Thus the single determinant wavefunctions conform to the requirement of Pauli's exclusion principle that two electrons can not occupy the same one-electron state.

(ii) Again if $(r_k, s_k) = (r_k, s_l)$, two rows of the determinant becomes identical and Ψ once again vanishes. The single determinant wavefunction therefore ensures that two electrons with parallel spins can not be found at the same place (r_k). This kinematic correlation arises from the antisymmetry property of N-Fermion wavefunction and is often called Fermi correlation.

(iii) Note, however, Ψ does not vanish if $s_k \neq s_l$, i.e. the electrons are at the same point r_k in space with their spins antiparallel. The electrons are identically charged particles and two electrons irrespective of their spins, would certainly tend to avoid coming too close to each other. That means the single determinant wavefunction does not take Coulomb correlation into account. A variational method can now be invoked and the expectation value of the ground state energy with respect to all possible variations of the one electron functions $u_i(r_i)$s (the space-part of the spin orbitals, the spin part remaining fixed) in the single determinant wavefunction under the constraint that they remain orthonormal. The variational procedure leads us to the Hartree-Fock equations for the N-electron system-atoms or molecules or even solids. Before introducing these equations, it is necessary to point out that the calculation of energy with determinantal wavefunctions and the many electron Hamiltonian of Eq. 4.106 requires us to calculate matrix elements of $\hat{H}$ with respect to determinantal wavefunctions built up from a set of N orthonormal spin orbitals. We will, in what follows only summarize the Slater-Condon rules for computing the different types matrix elements of one- and two-electron operators in H with Slater determinant wavefunctions. The rules can be derived easily.[5,16]

4.13.1 Slater Condon Rules

Let $\Psi(A)$ be an antisymmetrized product of N orthonormal spin orbitals a_i, which we write as single Slater determinant

$$\Psi(A) = \frac{1}{\sqrt{N}}\left\{|a_1(1)a_2(2)a_3(3)\cdots a_N(N)|\right\}$$

with $a_1(1)$, $a_2(2)$, $\cdots$, $a_N(N)$ arranged in a standard order based for example, on the quantum number labelling the orbitals.

Consider the operator $\hat{O}_1$, which we assume is a sum of one-electron operators each of which depends on the coordinates of only electron (e.g. $T = -\frac{\hbar^2}{2m}\nabla_i^2$, $V = \sum_i -\frac{Ze^2}{r_i}$, etc.),

In general,

$$\hat{O}_1 = \sum_{i=1}^{N} \hat{O}_i$$

Rule-1:

$$\langle \Psi(A)|\hat{O}_1|\Psi(B)\rangle = 0$$

if the determinants $\Psi(A)$ and $\Psi(B)$ differ in more than one pair of occupied spin orbitals.

Rule-2:

$$\langle \Psi(A)|\hat{O}_1|\Psi(A)\rangle = \sum_{i=1}^{N} \langle a_i|\hat{O}|a_i\rangle$$

$$= \sum_{i=1}^{N} \int a_i^*(1)O(1)a_i(1)dv$$

Rule-3:

$$\langle \Psi(A)|\hat{O}_1|\Psi(B)\rangle = \pm \langle a_k|\hat{O}|b_l\rangle$$

$$= \pm \int a_k^*(1)\hat{O}(1)b_l(1)dv_1 \tag{4.113}$$

when $\Psi(A)$ and $\Psi(B)$ differ by just one spin orbital (the spin orbital a_k in $\Psi(A)$ has been replaced by b_k in $\Psi(B)$, all other orbitals in $\Psi(A)$, $\Psi(B)$ being the same). The $\pm$ sign is fixed by the number of permutations that is required to bring b_l in $\Phi(B)$ into coincidence with a_k in $\Psi(A)$.

Rule-4: Let G be a sum of two electron operators $g(i,j)$ ($\sum$ indicates summation over both i and all j).

$$\hat{G} = \sum_{i>j} g(i,j) = \sum_{i>j} \frac{e^2}{r_{ij}}$$

Thus $\langle \Psi(A)|\hat{G}|\Psi(B)\rangle = 0$ if $\Psi(A)$ and $\Psi(B)$ differ in more than two pairs of spin orbitals.

Rule-5: If $\Psi(A)$ and $\Psi(B)$ differ in two pairs of spin orbital functions (a_k and a_l in $\Psi(A)$ have been replaced by b_m and b_n in $\Psi(B)$)

$$\langle\Psi(A)|\hat{G}|\Psi(B)\rangle = \pm\left\{\langle a_k a_l|g(1,2)|b_m b_n\rangle - \langle a_k a_l|g(1,2)|b_n b_m\rangle\right\}$$

$$= \pm\left[\int\int a_k^*(1)a_l^*(2)g(1,2)b_m(1)b_n(1)d\tau_1 d\tau_2 - \right. \qquad (4.114)$$

$$\left.\int\int a_k^*(1)a_l^*(2)g(1,2)b_n(1)b_m(2)d\tau_1 d\tau_2\right]$$

The $\pm$ sign depends upon the permutations required to bring $\Psi(B)$ into maximum coincidence with $\Psi(A)$.

Rule-6: If $\Psi(A)$ and $\Psi(B)$ differ in just one pair of spin orbitals (a_k in $\Psi(A)$ has been replaced by b_l in $\Psi(B)$) the matrix elements

$$\langle\Psi(A)|\hat{G}|\Psi(B)\rangle = \pm\sum_{r=1}^{n}(\langle a_k a_r|g|b_l a_r\rangle - \langle a_k a_r|g|a_r b_l\rangle)$$

$$= \pm\sum_{r=1}^{n}\int\int a_k^*(1)a_r^*(2)g(1,2)b_l(1)a_r(2)d\tau_1 d\tau_2 - \int\int a_k^*(1)a_r(2)g(1,2)a_r(1)b_l(2)d\tau_1 d\tau_2$$

$$(4.115)$$

The $\pm$ sign in the matrix element is fixed by the number of permutations required to bring the determinant B into maximum coincidence.

Rule-7: If $\Psi(A)$ and $\Psi(B)$ are identical, i.e. all the spin orbitals in $\Psi(A)$ and $\Psi(B)$ are the same ($\Psi(A)$ = $\Psi(B)$).

$$\langle\Psi(A)|\hat{G}|\Psi(A)\rangle = \sum_{k>r}\left\{\langle a_k a_r|g(1,2)|a_k a_r\rangle - \langle a_k a_r|g(1,2)|a_r a_k\rangle\right\}$$

$$= \sum_{k>r}\left\{\int\int a_k^*(1)a_r^*(2)g(1,2)a_k(1)a_r(2)d\tau_1 d\tau_2 - \right. \qquad (4.116)$$

$$\left.\int\int a_k^*(1)a_r^*(2)g(1,2)a_k(2)a_r(1)d\tau_1 d\tau_2\right\}$$

Note that $g(1,2) = \frac{e^2}{r_{12}}$ and the integration volume elements $d\tau_1, d\tau_2$ denote space-spin volume elements for electron 1 and electron 2, respectively wherever they appear in this section. The two-electron integrals in the expressions for the diagonal matrix elements, which carry a negative sign before them, are called exchange integrals where those without the -ve sign are called direct integrals. They are called exchange integral because the functions a_k and a_r are exchanged between the initial and final states in the corresponding integral. They appear solely due to the use of antisymmetric wavefunctions built up of for N orthonormal spin orbitals. A simple product of N one-electron functions (spin orbitals) would not produce proper exchange integrals. These integrals as we will see later in Chapter 10, play an important role in shaping magnetic properties of materials.

4.14 Hartree-Fock Method

Our objective is now to look for the 'best' single determinant wavefunction for the ground state of an N-electron system. Let us suppose that we are considering an N electron atom/ion with nuclear charge $+Ze$. Let N be an even number and let us suppose that we are looking for the singlet ground state in which all the one-electron functions or orbitals are occupied by a pair of electrons with opposite spins. The corresponding Slater determinant (SD) wavefunction $\Psi(s=0, M_s=0)$ then can be written as

$$\Psi(1,2,\cdots,N) = \frac{1}{\sqrt{N!}}|\Phi_1(1)\overline{\Phi_1}(2)\cdots\Phi_i\overline{\Phi_i}\cdots\Phi_n\overline{\Phi_n}| \qquad (4.117)$$

where the spin orbitals $\Phi_1(1)$, $\overline{\Phi_1}$, $\cdots$, Φ_n, $\overline{\Phi_n}$, are each a product of a space-dependent and spin part.

$$\Phi_i = u_i(r_i)\alpha(i) \equiv u_i(r_i)|\uparrow\rangle$$

$$\overline{\Phi_i} = u_i(r_i)\beta(i) \equiv u_i(r_i)|\downarrow\rangle$$

The spin function $\alpha(i)$ is an eigenfunction of $\hat{S}_{iz}$ with eigenvalue $m_s\hbar$ ($m_s = \frac{1}{2}$) and $\beta(i)$ an eigenfunction of S_{iz} with eigenvalue $-\frac{1}{2}\hbar$ ($m_s = -\frac{1}{2}$). Since each u_i function accommodates two electrons with opposite spins, the total $M_s = 0$ and the single determinant, a perfectly spin-paired wavefunction is an eigenfunction of both S^2 and S_z operators with $S = 0$ and $M_s = 0$ ($\hat{S}^2 = \sum_i S^2(i)$, $S_z = \sum_i S_{iz}$).

It is necessary now to spell out the Hamiltonian for our atomic many electron problem, which reads

$$H(1,2,\cdots,N) = \sum_{i=1}^{N}\left(-\frac{\hbar^2}{2m}\nabla_i^2 - \frac{Ze^2}{r_i}\right) + \sum_{i>j}\sum_j \frac{e^2}{r_{ij}}$$

$$= \sum_{i=1}^{N}\left(T_i + V(r_i)\right) + \sum_{i>j}\sum_j \frac{e^2}{r_{ij}} \qquad (4.118)$$

$$= \sum_{i=1}^{N} h(i) + \sum_{i>j}\sum_j \frac{e^2}{r_{ij}}$$

We mention here that the potential $V(r_i)$ in the atomic case is a one-center potential (center at the nucleus of charge Ze). However, we can replace it with a suitable many center potential to make the technique to be described here work even in the case of molecules as well as solids in general. Using the appropriate Slater-Condon rule, we can easily write down the expectation value of energy for the system described by 'H' of Eq. 4.118 and $\Psi(1,2,\cdots,N)$ of Eq. 4.117 (after integrating out the spin coordinates)

$$\langle E \rangle = \int \Psi^* H \Psi dv = \sum_{i=1}^{n} 2 \int u_i^*(1)h(1)u_i(1)dv_1 + \sum_{i>j}\sum_j 2 \int\int u_i^*(1)u_j^*(2)\frac{e^2}{r_{12}}u_j(1)u_j(2)dv_1 dv_2$$

$$- \sum_{i>j}\sum_j \int\int u_i^*(1)u_j^*(2)\frac{e^2}{r_{12}}u_j(1)u_j(2)dv_1 dv_2 \qquad (4.119)$$

In the first and second term, a factor of 2 appears because the summation runs over all space orbitals, which are doubly occupied with electrons of opposite spins. The factor of 2 does not appear in the third sum because the exchange integrals vanish because of the orthogonality of the eigenfunctions of $\hat{s}_z$ with antiparallel spins. In a shorthand (bra-ket) notation the energy expression takes the form

$$E = \sum_{i=1}^{n} 2\langle u_i|h(1)|u_i\rangle + \sum_{i>j}\sum_j 2\langle u_i(1)u_j(2)|\frac{e^2}{r_{12}}|u_i(1)u_j(2)\rangle - \sum_{i>j}\sum_j \langle u_i(1)u_j(2)|\frac{e^2}{r_{12}}|u_j(1)u_i(2)\rangle$$

$$(4.120)$$

If we now subject the space functions u_i to infinitesimal changes ($u_i \to u_i + \delta u_i$) the first order change in energy is

$$\delta E = \sum_i 2\langle \delta u_i|h(1)|u_i\rangle + \sum_i 2\langle u_i|h(1)|\delta u_i\rangle$$

$$+ \sum_i\sum_j 4\langle \delta u_i(1)u_j(2)|\frac{e^2}{r_{12}}|u_i(1)u_j(2)\rangle \qquad (4.121a)$$

$$-\sum_i \sum_j 2 \langle \delta u_i(1)u_j(2) | \frac{e^2}{r_{12}} | u_j(1)u_i(2) \rangle$$

$$-\sum_i \sum_j 2 \langle u_i(1)u_j(2) | \frac{e^2}{r_{12}} | \delta u_j(1)u_i(2) \rangle$$

Energy minimization demands $\delta E = 0$. The energy minimization must however be carried out under the constraint of preserving the orthonormality of the one-electron functions. To first order, this leads to the constraint condition

$$\sum_i \sum_j \left\{ \langle \delta u_i | u_j \rangle + \langle u_i | \delta u_j \rangle \right\} = 0 \qquad (4.121b)$$

Multiplying the expression on the left by the Lagrange multipliers $(-\epsilon_{ij})$ and adding it to the expression for δE leads to the constrained minimization condition, which must be satisfied for any arbitrary and independent variations in u_i. That leads to the following pair of equations (Euler-Lagrange equations), which the optimal Hartree-Fock orbitals must satisfy: for $i = 1, 2, \ldots, n$.

$$\left[\hat{h}(1) + \sum_j (2\hat{J}_j - \hat{K}_j) \right] | u_i \rangle = \sum_j | u_j \rangle \epsilon_{ij} \qquad (4.122a)$$

$$\langle u_i | \left[\hat{h}(1) + \sum_j (2\hat{J}_j - \hat{K}_j) \right] = \sum_j \langle u_j | \epsilon_{ij} \qquad (4.122b)$$

If we take the adjoint of the second equation above noting that $\hat{h}(1)$, $\hat{J}_j$ and $\hat{K}_j$ operators are all hermitian and substract the resulting equation from the first we get

$$\sum_j (\epsilon_{ij} - \epsilon_{ji}^*) | \phi_j \rangle = 0, i = 1, 2, \cdots, n \qquad (4.123)$$

Since $| \phi_j \rangle s$ are orthonormal one-electron functions they are linearly independent and this condition holds only if

$$\epsilon_{ij} = \epsilon_{ji}^*$$

That is, the matrix of Lagrange multipliers is a hermitian matrix, which can therefore be diagonalized by a unitary transformation of the orbitals (u_is). It is easy to show that the same unitary transformation leaves h, $\sum_j J_j$ and $\sum_j K_j$ invariant so that we can switch over to the new one-electron functions that diagonalize the ϵ matrix and arrive at the so called canonical Hartree-Fock equations for the optimal orbitals:

$$\hat{F} | u_i' \rangle = \epsilon_{ii}' | u_i' \rangle, i = 1, 2, \cdots, n \qquad (4.124)$$

where $\hat{F} = (\hat{h}(1) + \sum_j (2\hat{J}_j - \hat{K}_j))$ may be called the Hartree-Fock operators or simply Fock operator. To understand the exact nature of the equations in Eq. 4.124 we have to examine the Coulomb ($\hat{J}$) and exchange ($\hat{K}$) operators, which are defined as follows (ϕ is an arbitrary one-electron function)

$$\hat{J}_i(1)\phi(1) = \left(\int \frac{e^2}{r_{12}} | u_i(2) |^2 dv_2 \right) \phi(1) \qquad (4.125a)$$

$$\hat{K}_i(1)\phi(1) = \left(\int \frac{e^2}{r_{12}} u_i^*(2)\phi(2) dv_2 \right) u_i(1) \qquad (4.125b)$$

The Coulomb operator $J_i(1)$ represents the classical potential energy of electron 1 due to its interactions with a distributed electronic charge of electron 2 having the probability density $|u_i(2)|^2$. The exchange operator has no simple physical interpretation or a classical analog. It arises from the antisymmetry requirement of many electron wavefunctions. Note, however, that $\hat{K}_i$ operator is non-local in the sense that it depends on the function $\phi(i)$ on which it acts. ϵ'_{ii} in the equation is clearly the ith eigenvalue of the Fock operator the corresponding eigenfunction being u'_i. It turns out that $-\epsilon'_{ii}$ gives an approximation to the ionization energy of the electron from the orbital ϕ_i, also called Koopman's estimate of ionization potential. Let us note that the Fock operator $\hat{F}$ depends explicitly on the Hartree-Fock orbitals $u_i, i = 1, 2, ..., n$ (the prime on canonical HF orbital u_i is dropped from this point), which are not known to start with. Therefore, Eq. 4.124 only appears to be an eigenvalue equation while actually it is a pseudo eigenvalue equation with a built in non-linearity. These equations can only be solved iteratively. The second point to note is that J, K are integral operators and therefore the HF equation is an integrodifferential equation, which can be solved by suitable numerical integration techniques. The third point to note is that unlike Hartree orbitals all the Hartree-Fock orbitals are eigenfunctions of the same Fock operator $\hat{F}$ (Eq. 4.124) and are therefore mutually orthogonal. The fourth point to take note of is that the single determinant HF wavefunctions ψ contains the n-lowest energy HF orbitals $u_1, u_2, \cdots, u_n$, each being doubly occupied with electrons of opposite spins, while Eq. 4.124 can provide many more additional solutions $u_{n+1}, u_{n+2}, \cdots, u_{n+m}, \cdots$, orthogonal to the n-doubly occupied variationally optimal HF orbitals. However, they lack the physical meaning that the occupied HF orbitals have and are called virtual orbitals. They play an important role in constructing and describing excited states or improving the ground state description by means of linear variational calculations in a procedure that has come to be known as the method of configuration interaction (CI). We describe in what follows the essential aspects of CI methods after introducing LCAO-MO approximation to the actual Hartree-Fock self-consistent-field (HF-SCF) orbitals.

4.15 LCAO-MO-SCF-CI Calculations

The numerical integration scheme for the Hartree-Fock equations in non-canonical or canonical forms works well for the atomic problems for ground state calculations or calculations on the lowest states of each symmetry. For molecules, however, the scheme is just workable only for diatomic molecules. The basic difficulty stems from the many center potential seen by each electron due to its interaction with all the nuclei in the molecule and consequent loss of central field symmetry. The major breakthrough that made HF-SCF calculations on polyatomic molecules not only feasible, but also sufficiently accurately was made by CCJ Roothaan (1951) who proposed that the Hartree-Fock orbitals (the space-part) can be expanded in terms of a complete set of linearly independent one-electron functions or states, usually centered on different atoms. In practice, one is forced to use a large but fixed number M of one-electron atom centered basis functions (χ_i) and write[5]

$$u_i(\overrightarrow{r}) = \sum_{p=1}^{M} \chi_p(r)C_{pi}, \quad \text{for } i_0 = 1, 2, \cdots, N$$

where $\chi_p(\overrightarrow{r})$s are M number of fixed functions ($M >> N$) and C_{pi}s are unknown expansion coefficients, which are to be variationally optimized to produce the minimum energy ground state wavefunction subject to orthonormality conditions on u_is. The variational treatment converts the integrodifferential equation in Eq. 4.123 into a matrix eigenvalue problem

$$FC = SC\epsilon \qquad (4.126)$$

where F stands for the matrix form of Hartree-Fock operator $\hat{F}$ in the basis of atomic orbitals (χ_p). It is an $M \times M$ hermitian matrix where M is the number of one-electron basis functions used in the expansion of the ϕ_is (the MOs) in Eq. 4.126 ,'S' is the $M \times M$ matrix of overlap integrals ($S_{pq} = \int \chi_p^* \chi_q dv(1)$) and ϵ is the matrix (hermitian) of Lagrange multipliers, which can therefore be diagonalized by a unitary transformation. However, Eq. 4.126 is in the form of a generalized pseudo-eigenvalue problem due to the presence of the S matrix and the dependence of the Fock matrix elements (F_{pq}) in the atomic orbital basis on the expansion coefficients $\{C_{pi}\}$. So the equation has to be first reduced to the standard pseudo-eigenvalue form by noting that $S = S^{\frac{1}{2}}S^{\frac{1}{2}}$ and $S^{-\frac{1}{2}}S^{\frac{1}{2}} = 1$, so that we have

$$S^{-\frac{1}{2}} F S^{-\frac{1}{2}} (S^{\frac{1}{2}} C) = (S^{\frac{1}{2}} C)\epsilon$$

$$F'C' = C'\epsilon \tag{4.127}$$

where $C' = S^{\frac{1}{2}} C$, and $F' = S^{-\frac{1}{2}} F S^{-\frac{1}{2}}$ is the Fock matrix constructed in a symmetrically orthogonalized (Lowdin orthogonalized) atomic orbital basis with the expansion coefficients now being given by $C' = S^{\frac{1}{2}} C$, and $C = S^{-\frac{1}{2}} C'$. The pseudo-matrix-eigenvalue problem (Eq. 4.127) can be solved iteratively by repeated diagonalization of the Fock matrix (F') yielding M molecular orbitals out of which the $\frac{M}{2}$ lowest energy ones are doubly occupied by electrons of opposite spins. The coefficient matrix C' is transformed back to C and F matrix is constrained. The elements of the Fock matrix (F) depend upon the AO-MO expansion coefficients (C_{pi}s) in the following manner.[5]

$$F_{rs} = h_{rs} + \sum_{t,u=1}^{N} P_{tu} \left\{ \langle rs|tu \rangle - \frac{1}{2} \langle ru|ts \rangle \right\} \tag{4.128}$$

where the charge density bond order matrix 'P' has elements

$$P_{tu} = 2 \sum_{j=1}^{\frac{N}{2}} C_{tj}^* \cdot C_{uj}$$

The factor '2' appears as the orbitals are doubly occupied.

$$\langle rs|tu \rangle = \int \int \chi_r^*(1)\chi_s(1)\frac{e^2}{r_{12}}\chi_t^*(2)\chi_u(2)dv_1 dv_2 \tag{4.129}$$

and

$$h_{rs} = \langle \chi_r(1)| h(1) |\chi_s(1) \rangle$$

with $h(1) = -\frac{\hbar^2}{2m}\nabla_1^2 + \sum_A \frac{-Z_A e^2}{r_{1A}}$, A running over all the nuclei in the molecule. The one-electron operator $h(1)$ represents the kinetic energy of the electron and potential energy due to interaction with all the nuclei (A) with nuclear charges $Z_A e$. Note that F_{rs} are the Fock matrix elements in the basis of non-orthogonal atomic orbitals $\{\chi_p\}$. Before diagonalization, as noted already, F is transformed into the orthogonalized basis (expansion coefficients $\{C'_{pi}\}$). The formulation by Roothaan brought molecules small, medium and even large, under the fold of the Hartree-Fock method. The formulation has become known as the LCAO-MO-SCF method and is the starting point for any many electron calculation on atoms, molecules and solids. We note here that the diagonalization of the Hartree-Fock matrix 'F' on the basis of M atomic orbital basis function yields M orthonormal molecular orbitals out of which $\frac{N}{2}$ are doubly occupied with electrons of antiparallel spins. The remaining $(M - \frac{N}{2})$ number of orbitals are unoccupied and are called virtual orbitals. The single determinant Hartree-Fock singlet wavefunctions for the ground state is uniquely determined by the variationally optimized $\frac{N}{2}$ number of the occupied orbitals. The $(M - \frac{N}{2})$ virtual orbitals provide the orthogonal complement. Thus the entire configuration space has been uniquely partitioned by the variational process into an occupied subspace P and an unoccupied subspace Q with

$$P + Q = 1$$

$$\hat{P} = \sum_{i=1}^{\frac{N}{2}} |\phi_i\rangle \langle\phi_i| \tag{4.130}$$

$$\hat{Q} = \sum_{j=\frac{N}{2}+1}^{M} |\phi_j\rangle \langle\phi_j| = (1 - \hat{P})$$

$\hat{P}$ is thus the projector for the occupied subspace while $\hat{Q}$ is the orthogonal complement to $\hat{P}$ so that $\hat{P}\hat{Q} = \hat{Q}\hat{P} = 0$. The electrons can be excited in ones, twos, etc. from the occupied subspace into the unoccupied subspace generating new configuration functions, adapted to have (i) the same spin symmetry that the ground state ($S = 0$, $M_s = 0$) has and (ii) the same spatial symmetry that Ψ_0 has. It requires some extra effort, but makes the analysis much easier. These configuration functions provide a fixed N-electron anti-symmetric orthonormal basis functions for forming the molecular electronic Hamiltonian matrix, which may then be diagonalized to yield improved estimates of ground and excited state energies of the system each of which is a strict upperbound to the corresponding exact ground and excited state energies. The procedure is a linear variational recipe and is called the CI (configuration interaction) method.

The configuration functions Ψ_I in a CI calculation are classified as 'singly excited', 'doubly excited', 'triply excited' and 'quadruply excited', etc. depending on whether one, two, three or four electrons have been excited from the orbitals occupied in the Hartree-Fock ground state single determinant wavefunction (Ψ_0). The trial CI wavefunction takes the following form:[5]

$$\Psi = \sum_{I=0}^{M} C_I \Psi_I \tag{4.131}$$

If I runs over all possible configurations that can be generated, Ψ is called the full CI wavefunction and linear variational calculation leads to the following matrix eigenvalue problem:

$$HC = CE \tag{4.132}$$

The eigenvalues $E_1, E_2, \cdots, E_N$ are the full CI eigenvalues – the best that can be obtained from the atomic orbital basis set $\{\chi_p\}$ used in the Hartree-Fock calculation. However, the number of configurations in a full CI calculation increases astronomically as the number of electrons and the number of basis functions rise, the former determining the maximum level of excitations that can be included. Except for small molecules and small basis sets, one must therefore truncate the CI expansion leading to various restricted CI approximations. For a reasonable level of accuracy both in the computed energy values and one-electron properties, like the dipole moment, one must include all singly and doubly excited configurations in the CI calculations, the resulting level of approximation is designated as the SD-CI level of approximation. It is possible and often sensible to cut down the number of configurations further by freezing out the core-electrons and allowing only the valence electrons to be excited into the unoccupied one-electron levels as the core electrons are not expected so much to influence the commonly targeted energy quantities like excitation energies, dissociation energies, etc. For achieving still higher accuracy it is necessary to include at least up to quadruple excitations although the demand on computational labor and resources may prevent one from doing so. It is possible to compute the ground and only a few higher eigenvalues by special techniques, thereby cutting down the computing time.

Let us mention at this point an 'energy quantity' called electron-correlation energy, which is defined as the difference between the actual energy of the system (experimental) and the energy predicted by the exact Hartree-Fock method (single determinant). In the CI calculations, one tries to recover the correlation energy by bringing in many excited state determinants in the variational calculation although part of the effort is often wasted in accounting for the inadequacy of the basis set used for the calculation of Hartree-Fock wavefunction and energy. It is therefore necessary to use a large basis set so that the zeroth order energy is not vitiated by basis-inadequacy effects.

4.16 Perturbative Correction to HF Wavefunction and Energy

The CI method is variational in nature. It is time now to explore if we could also take a perturbative method to improve the HF description. We will briefly describe in what follows such a theory for closed shell ground states of atoms and molecules, which was proposed by Møller and Plesset and is even now the most widely adopted perturbative scheme for computing atomic and molecules properties.[18]

The zeroth order Hamiltonian in MP theory is taken as the sum of the effective one-electron Hamiltonians – the Fock-operators for all the electron in the system. Thus

$$H_0 = \sum_{k=1}^{n} \hat{f}(k) \tag{4.133}$$

where

$$\hat{f}(k) = -\frac{\hbar^2}{2m}\nabla_k^2 - \sum_A \frac{Z_A e^2}{r_{kA}} + \sum_{i=1}^{n}\left(J_i(k) - K_i(k)\right) \tag{4.134}$$

'A' sums over all the nuclei present in the molecule, $Z_A e$ being the nuclear charge carried by the Ath nucleus. $\hat{J}_i$ and $\hat{K}_i$ are respectively the Coulomb and exchange operators for the kth electron resident in the spin orbital u_i. The zeroth order wavefunction $\Psi_0^{(0)}$ is the Hartree-Fock single determinant wavefunction

$$\Psi_0 = \frac{1}{\sqrt{n!}}|u_1(1)u_2(2)\cdots u_n(n)|$$

which satisfies the eigenvalue equation

$$H_0\Psi_0^{(0)} = \left(\sum_{i=1}^{n}\epsilon_i\right)\Psi_0^{(0)} \tag{4.135}$$

ϵ_is being the Hartree-Fock orbital energies that appear as the eigenvalues of the Fock operator $\hat{f}_i$:

$$\hat{f}_i(k)u_i(k) = \epsilon_i u_i(k); \quad i = 1, 2, \cdots, n \tag{4.136}$$

$\Psi_0^{(0)}$ is just the single determinant wavefunction Ψ_0 obtained variationally in the HF theory and has $n!$ permutations of the orbital product $u_1(1)u_2(2)\cdots u_k(k)\cdots u_n(n)$ each such product being an eigenfunction of $H_0 = \sum_{i=1}^{n}\hat{f}_i$ with the eigenvalue $\epsilon = \sum_{i=1}^{n}\epsilon_i$; hence the determinant $\Psi_0^{(0)}$ also is an eigenfunction of H_0 with the same eigenvalue. The zeroth order energy in MP theory is thus

$$E_0^{(0)} = \sum_{i=1}^{n}\epsilon_i$$

The perturbation Hamiltonian (H') in Møller-Plesset theory is the difference between the actual many electron Hamiltonian $\hat{H}$ and the zeroth order Hamiltonian H_0. Noting that

$$H = \sum_{k=1}^{n}-\frac{\hbar^2}{2m}\nabla_k^2 - \sum_A\sum_k\frac{Z_A e^2}{r_{kA}} + \frac{1}{2}\sum_{i\neq j}\sum_j\frac{e^2}{r_{ij}}$$

and H_O is (viz. Eq. 4.133 and Eq. 4.134) given by

$$H_0 = \sum_{k=1}^{n}-\frac{\hbar^2}{2m}\nabla_k^2 - \sum_A\sum_k\frac{Z_A e^2}{r_{kA}} + \sum_{i=1}^{n}(\hat{J}_i(k) - \hat{K}_i(k))$$

H' becomes

$$H' = H - H_0 = \frac{1}{2}\sum_i\sum_{j(\neq i)}\frac{e^2}{r_{ij}} - \sum_{k=1}^{n}\sum_i(\hat{J}_i(k) - \hat{K}_i(k)) \tag{4.137}$$

Physically, it is the difference between the actual inter-electronic repulsion energy of n-electrons (the first term in H') and averaged about interelectronic interactions represented by $\sum_k\sum_i(\hat{J}_i(k) - \hat{K}_i(k))$ (the second set of terms in Eq. 4.137).

The first order correction to the ground state energy in the MP theory is the expectation value of H' in the unperturbed ground state $(\Psi_0^{(0)})$. Thus

$$E_0^{(1)} = \langle\Psi_0^{(0)}|H'|\Psi_0^{(0)}\rangle$$

$$= \langle\Psi_0^{HF}|H'|\Psi_0^{HF}\rangle$$

$E_0^{(1)}$ taken together with $E_0^{(0)}$ of MP theory gives us just the HF energy in the ground state

$$E_0^{(0)} + E_0^{(1)} = \langle \Psi_0^{(0)} | H_0 | \Psi_0^{(0)} \rangle + \langle \Psi_0^{(0)} | H' | \Psi_0^{(0)} \rangle$$
$$= \langle \Psi_0^{HF} | (H_0 + H') | \Psi_0^{HF} \rangle$$
$$= E_0^{HF}$$

The variationally obtained ground state Hartree-Fock energy E_0^{HF} is therefore equal to the Møller-Plesset perturbation theoretical estimate of the same energy upto the first order in the perturbation.

The second order correction to the ground state energy in Møller-Plesset theory can be obtained by writing down the second order expression for $E_0^{(2)}$ in the Rayleigh-Schrödinger perturbation theory (RSPT):

$$E_0^{(2)} = (-1) \sum_{k \neq 0} \frac{| \langle \Psi_k | H' | \Psi_0^{HF} \rangle |^2}{(E_k^0 - E_0^0)} \qquad (4.138)$$

where k sums over all the excited (singly, doubly, etc.) Slater determinant wavefunctions (spin-singlet) obtained by exciting 1, 2, $\cdots$ electrons from the spin orbitals $(i, j, \cdots)$ occupied in Ψ_0^{HF}, to virtual orbitals $(a, b, \cdots)$, which are empty. The singly excited determinants $\Psi_{i \to a}^{s=0}$ would not mix with Ψ_0^{HF} under the perturbation (by virtue of Brillouin's theorem) so that we are left to consider the doubly excited determinants $(i \to a, j \to b)$, the corresponding configuration function being $\Psi_0^k \equiv \Psi_{i \to a, j \to b}^0$ and the energy denominator in Eq. 4.138 being $\epsilon_a + \epsilon_b - \epsilon_i - \epsilon_j$. The second order Møller-Plesset energy expression then becomes (Slater-Condon rules, note that triply and higher excited states do not contribute as the corresponding matrix elements vanish)

$$E_0^{(2)} = (-1) \sum_{k(i \to a, j \to b)} \frac{| \langle ab | \frac{1}{r_{12}} | ij \rangle - \langle ab | ji \rangle |^2}{\epsilon_a + \epsilon_b - \epsilon_i - \epsilon_j} \qquad (4.139)$$

The ground state energy correct to second order then reads

$$E_0 = E_0^{HF} + E_0^{(2)} \qquad (4.140)$$

with the first order correction being absorbed in E_0^{HF} itself. Higher order MP theory has been developed and has been designated as MP_3, MP_4, etc. The cost of computation, specially in the context of geometry optimization of molecules, generally compels users to restrict calculations at the MP_2 level.

4.17 The Rise of Density Functional Theory

The many (N) electron wavefunction depends on $3N$ spatial coordinates (x, y, z, $\cdots$, x_N, y_N, z_N) and N spin coordinates (s_1, s_2, $\cdots$, s_N). Even if the exact many electron wavefunction of an atom, molecule or solid was available, the calculation of energy, for example, would formally require integration over only six spatial coordinates as the many body atomic/molecular Hamiltonians contain only one and two body (Coulombic) interactions. Thus the many electron $3N$ dimensional wavefunction appears to contain a lot of redundant information only a small part of which is formally required for practical applications. That apart, Ψ (x, y, z, $\cdots$, x_N, y_N, z_N) is not, in itself a physical entity. The connection with the normal physical world is established through $|\Psi|^2$, which represents probability density in a space of $3N$ dimensions, assuming that all the spin degrees of freedom have been integrated out. Thus a need for the search of a physical entity, a function of a much smaller number of coordinates or variables was felt right from the start. It turned out that one-electron probability density $\rho(x, y, z)$ function can be a convenient, physically meaningful and useful entity for the purpose at hand.[19]

In a seminal work P. Hohenberg and Walter Kohn (1964) proved that the ground state energy, wavefunction and electronic properties of atoms, molecules and materials (solids) is uniquely determined by the corresponding ground state electronic probability density function $\rho_0(x, y, z)$, the zero-subscript referring to the ground state.[20] Note that, the ρ_0 for an n-electron system is a function of three spatial coordinates as opposed to Ψ_0, which is a function of $3n$ coordinates, disregarding spin coordinate. That means, a huge contraction in the dimensionality of the working space is afforded by switching over to a density-based description and procedure. The question is how can one exploit the single electron density function?

Let us consider an n-electron molecule. The many electron Hamiltonian H, at a frozen nuclear configuration expressed in atomic system of units ($\hbar = m_e = e = 1$) reads

$$H = -\frac{1}{2}\sum_{i=1}^{n}\nabla_i^2 + \sum_{i=1}^{n}V(\vec{r_i}) + \frac{1}{2}\sum_{i\neq j}^{n}\sum_{j}^{n}\frac{1}{|\vec{r_i}-\vec{r_j}|}$$

$V(\vec{r_i})$ represents the interaction of the ith electron with the M number of frozen nuclei, each with charge $+Z_A$ and is called in the parlance of density functional theory the external potential acting on the system of n-electrons:

$$V_{ext}(\vec{r_i}) = -\sum_{A=1}^{n}\frac{Z_A}{r_{iA}} \tag{4.141}$$

The *identity* of a system of n-electrons is thus defined by the *external potential*. Once $V_{ext}(\vec{r_i})$ has been specified along with the frozen nuclear framework and the number of electrons in the system, we can set up the molecular electronic Schrödinger equation and in principle, solve it for the quantized molecular energy levels and the corresponding wavefunctions. That is the standard recipe, we have already described. Hohenberg and Kohn proved that for a non-degenerate ground state, the ground state electron density $\rho_0(\vec{r_i})$ determines the number of electrons (n) in the system as well as the external potential uniquely upto an arbitrary additive constant.[20] That also determines the ground state electronic wavefunction and energy. It is trivial to show that $\rho_0(\vec{r_i})$ determines n. Recall that $\rho_0(x,y,z)$ can be obtained from $|\Psi(\vec{r_1},\vec{r_2},\cdots,\vec{r_n};s_1,s_2,\cdots,s_n)|^2$ by integration over the spatial coordinates of all but one electron and summing or integrating over all the spin variable $s_1,\cdots,s_n$ so that

$$\rho_0(\vec{r}) = n\int|\Psi(\vec{r},\cdots,\vec{r_n})|^2 d\vec{r_2}\cdots d\vec{r_n}$$

The factor 'n' arises as $\vec{r}$ could be any one of the 'n' electronic coordinate. Since Ψ is normalized to unity it is trivial to prove that

$$\int\rho_0(\vec{r})d\vec{r} = n\int|\Psi|^2 d\vec{r_1}\cdots d\vec{r_n}$$

$$= n$$

To prove that $\rho_0(\vec{r})$ also fixes up the external potential $V_{ext}(\vec{r_i})$, we can following HK, assume that $\rho_0(\vec{r})$ does not fix $V_{ext}(\vec{r_i})$ uniquely and that two different external potentials $V_1 = \sum_i v_1(\vec{r_i})$ and $V_2 = \sum_i v_2(\vec{r_i})$ can produce the same ground state density $\rho_0(\vec{r})$. The two n-electron Hamiltonians defined by V_1 and V_2 may be called H_1 and H_2, which produce the ground state wavefunctions Ψ_{01} and Ψ_{02} (normalized), respectively with E_{01} and E_{02} as the corresponding ground state energies. Since the external potentials V_1 and V_2 have been assumed to be different, Ψ_{01} and Ψ_{02} must also be different functions as they are the ground state eigenfunctions of two different Hamiltonians. Suppose now, Ψ_{02} is used as a trial wavefunction for the Hamiltonian H_1. Variation theorem asserts that

$$\langle\Psi_{02}|H_1|\Psi_{02}\rangle > E_{01} \tag{4.142}$$

as Ψ_{02} is not the ground state wavefunction for H_1.

We can restate the contents of the inequality in Eq. 4.142 as follows:

$$E_{01} < \langle\Psi_{02}|H_1|\Psi_{02}\rangle = \langle\Psi_{02}|H_1+H_2-H_2|\Psi_{02}\rangle$$

$$= \langle\Psi_{02}|H_2|\Psi_{02}\rangle + \langle\Psi_{02}|H_1-H_2|\Psi_{02}\rangle \tag{4.143}$$

$$< E_{02} + \langle\Psi_{02}|V_1(\vec{r})-V_2(\vec{r})|\Psi_{02}\rangle$$

Since the kinetic energy and the interelectronic repulsive energy operators for any two n-electron molecules are the same, $H_1 - H_2$ represents the difference between the external potentials of the system.

The second term on the right hand side represents expectation value of one-electron operators and can be expressed in terms of the exact one-electron density function $\rho_0(\vec{r})$ so that we can write the inequality (Eq. 4.143) equivalently as

$$E_{01} < E_{02} + \int \rho_0(\vec{r})\Big(V_1(\vec{r}) - V_2(\vec{r})\Big) d\vec{r} \tag{4.144}$$

Now, one could have also started with the variational result that

$$E_{02} < \langle \Psi_{01} | H_2 | \Psi_{01} \rangle$$

and obtain, following the same line of reasoning

$$E_{02} < E_{01} + \int \rho_0(\vec{r})\Big(V_2(\vec{r}) - V_1(\vec{r})\Big) d\vec{r} \tag{4.145}$$

since the ground state density $\rho_0(\vec{r})$ has been assumed to be the same for both H_1 and H_2. Adding up Eq. 4.144 and Eq. 4.145 we get

$$E_{01} + E_{02} < E_{02} + E_{01} \tag{4.146}$$

which is impossible to hold true. This absurd result has been obtained since our starting assumption that the two different n-electron systems represented by Hamiltonians H_1 and H_2 both have the same ground state density $\rho_0(\vec{r})$ must be false. We may conclude therefore that the ground state electron density determines the external potential uniquely to within an additive constant, which can only shift the zero of energy. As we have already seen, $\rho_0(\vec{r})$ also determines the number of electrons 'n', which in turn determines the kinetic energy and the inter electronic repulsion energy parts of the n-electron of H $(=\sum_{i=1}^{n} -\frac{\hbar^2}{2m}\nabla_i^2 +$

$\frac{1}{2}\sum_{i\neq j}\sum_{j} \frac{e^2}{r_{ij}})$. Therefore, $\rho_0(\vec{r})$ determines the molecular electronic Hamiltonian $H(\vec{r})$ completely, and thereby determines the ground state energy and the wavefunction. Thus for a given external potential 'V $(\vec{r})$' the ground state energy E_0 is a functional of $\rho_0(\vec{r})$. How does one map the 3D density function into energy?

The ground state energy of a molecule (with frozen nuclear coordinates) is the sum of quantum mechanical expectation values of the kinetic energy operator ($<T_e>$) of the electrons, electron nucleus attraction energy operator $<V_{ne}>$ and the expectation value of interelectronic repulsion energy operator $<V_{ee}>$, each of which is determined by the many electron ground state n-electron wavefunction Ψ_0 through the standard rule of quantum mechanics that

$$< A(r) > = \int \Psi^* \hat{A} \Psi d\tau_1 d\tau_2 \cdots d\tau_n$$

which we have already seen to be determined completely by ρ_0. Therefore, all the expectation values appearing in $<H>$ must be a functional of the density function $\rho_0(\vec{r})$ – a function in the 3D space. Thus the ground state energy is

$$E_0 = E_v[\rho_0] \equiv T_e[\rho_0] + V_{ne}[\rho_0] + V_{ee}[\rho_0] \tag{4.147}$$

Of the three terms on the right hand side of Eq. 4.147 the functional $V_{ne}[\rho_0]$ is given by

$$V_{ne}[\rho_0] = \int \rho_0(\vec{r}) v(\vec{r}) d\vec{r}$$

where $v(\vec{r})$ is the electron-nucleus attraction potential felt by an electron at the point $\vec{r}$ in 3D space. The other two functionals, $T_e[\rho_0]$ and $V_{ee}[\rho_0]$ are universal (do not depend on v, and are defined by the number of electrons), but unknown. We can therefore write

$$E_0 = E_v[\rho_0] = \int \rho_0(\vec{r}) v(\vec{r}) d\vec{r} + T_e[\rho_0] + V_{ee}[\rho_0]$$

$$= \int \rho_0(\vec{r}) v(\vec{r}) d\vec{r} + F[\rho_0] \tag{4.148}$$

where $F[\rho_0]$ is the elusive universal functional of ground state electron density ρ_0. The existence of $F[\rho_0]$ is proved by Hohenberg-Kohn theorem. The theorem does not give any clue to the form of the functional. In traditional quantum mechanical formulation, one must solve the many electron Schrödinger equation for the unknown Ψ_0, which is then integrated to generate the ground state density. Hohenberg-Kohn (HK) theorem turns the problem around and the search for Ψ_0 is replaced by a search for the unknown functional $F[\rho_0]$. Even if the functional form of F were known, the problem of directly determining $\rho_0(\overrightarrow{r})$ would still be there for which Hohenberg and Kohn suggested a variational recipe. It may be identified as the second theorem of Hohenberg and Kohn: for any trial density function $\tilde{\rho}_0(\overrightarrow{r})$ satisfying the constraints $\rho_0(\overrightarrow{r}) \geq 0$ for every $\overrightarrow{r}$ and $\int \rho_0(\overrightarrow{r})dr = n$ (number of electrons) the following inequality is satisfied:

$$E_v[\tilde{\rho}] \geq E_0 = E_v[\rho_0]$$

$$\text{where} \quad E_v[\tilde{\rho}] = \int \tilde{\rho}(\overrightarrow{r})v(\overrightarrow{r})dr + F[\tilde{\rho}] \tag{4.149}$$

The physical content of the inequality in Eq. 4.149 is that the true ground state density $\rho_0(\overrightarrow{r})$ minimizes the energy functional $E_v[\tilde{\rho}]$ in the absolute sense. For any other density function, the functional value is only an upper bound to $E_v[\rho_0]$. The proof of the second theorem is straightforward.[19−20]

Let $\tilde{\Psi}$ be the n-electron antisymmetric wavefunction corresponding to the trial density $\tilde{\rho}(\overrightarrow{r})$ satisfying the necessary constraints. Using $\tilde{\Psi}$ as the trial wavefunction with the n-electron molecular electronic Hamiltonian $\hat{H}$, we have

$$\langle \tilde{\Psi} | \hat{H} | \tilde{\Psi} \rangle \geq E_0 = E_v(\rho_0)$$

$$\text{or,} \quad \langle T \rangle + \langle V_{ee} \rangle + \langle V_{ne} \rangle \geq E_v(\rho_0)$$

Since the expectation values of $\hat{T}$ and $\hat{V}_{ne}$ are universal functionals of electron density, we have

$$T[\tilde{\rho}] + V_{ee}[\tilde{\rho}] + \int \tilde{\rho}(\overrightarrow{r})v(\overrightarrow{r})dr \geq E_v[\rho_0]$$

which proves the existence of a variational principle for the density function for the ground state in much the same way as it exists for the ground state wavefunction of an n-electron molecule. However, the problem of guessing a form for trial density without alluding to trial wavefunctions for the ground state still remained unsolved impeding large scale applications of the density functional theory propounded by Hohenberg and Kohn. Needless to mention again that the form of the universal functional $F[\rho]$ remains elusive even today and will perhaps remain so forever.

4.17.1 The Kohn-Sham Method

The most important step that transformed density functional theory into a practicable tool for computing ground state molecular electronic structure and properties without the help of information from some wavefunction was taken by W. Kohn and L. J. Sham.[21] In the Kohn-Sham (KS) formalism one starts by assuming that there is a fictitious system of *n-non-interacting* electrons moving under the actual external potential $v(\overrightarrow{r})$ and this system has exactly the same ground state density as the real system to which the fictitious system refers to. The existence of such a system has never been proved.

The reference system of *n-non-interacting* electron enables us to transform the *n*-electron system into a sum of n one-electron systems. The corresponding n-electron wavefunction of the system can therefore be represented by a Slater determinant of n orthonormal orbitals – the Kohn-Sham orbitals, which are solutions of an effective one-electron Schrödinger equation. The Hamiltonian of the reference system(s) can be written as (in atomic system of units) follows:

$$\hat{H}_s = \sum_{i=1}^{n} \left(-\frac{1}{2}\nabla_i^2 + v_s(r_i) \right)$$

$$= \sum_{i=1}^{n} h_i^{KS}$$

where $h_i^{KS} = (-\frac{1}{2}\nabla_i^2 + v_s(r_i))$, and $v_s(r_i)$ is the external potential felt by each of n non-interacting electron and has been so 'chosen' as to generate the ground state electron density $\rho_s(\vec{r})$ of the reference system that exactly matches with the true ground state density $\rho_0(\vec{r})$ of the actual molecule or system we are working with h_i^{KS} are the one-electron Kohn-Sham Hamiltonians, the eigenfunctions are the Kohn-Sham orbitals (space-part)

$$h_i^{KS}\phi_i^{KS} = \epsilon_i^{KS}\phi_i^{KS}, \quad i = 1,2,\cdots,n$$

The Kohn-Sham spin-orbitals u_i^{KS} are then defined as the product of Kohn-Sham space orbitals ϕ_i^{KS} and spin eigenfunction α or β of $\hat{S}_z$

$$u_i^{KS} = \phi_i^{KS}(r_i) \times \alpha(i) \text{ or } \beta(i)$$

To make further progress with the task of determining the ground state electron density function $\rho_0(r)$ of the actual system we must now be able to model the energy functional $E_v[\rho]$ of Hohenberg and Kohn. Kohn-Sham theory works with the following model for $E_v[\rho]$:

$$E_v[\rho] = E_{ext}[\rho] + T_s[\rho] + E_c[\rho] + \Delta T[\rho] + \Delta V_{ee}[\rho] \tag{4.150}$$

where $E_{ext}[\rho] = \int \rho(\vec{r})v(\vec{r})d\vec{r}$ is the potential energy of the non-interacting electrons in the external potential (electron-nucleus interaction energy), $T_s[\rho]$ is the kinetic energy functional of the non-interacting electrons, $E_c[\rho]$ is the classical Coulomb interaction energy given by

$$E_c[\rho] = \frac{1}{2}\int \frac{\rho(\vec{r})\rho(\vec{r'})d\vec{r}\,d\vec{r'}}{|r-r'|}$$

$\Delta T[\rho]$ is the difference between the true or a actual kinetic energy functional of n interacting electrons ($T[\rho]$) and the kinetic energy functional $T_s[\rho]$ of the non-interacting electrons of the reference system ($\Delta T[\rho] = T[\rho] - T_s[\rho]$) with the same density ρ. $\Delta V_{ee}[\rho]$ stands for the difference between actual interaction energy calculated with correlated electron density (taking both Coulomb and Fermi correlations into account) and the $E_c[\rho]$.

We can now define the key quantity of DFT – the exchange correlation energy functional $E_{XC}[\rho] = \Delta T[\rho] + \Delta V_{ee}[\rho]$ and write down the energy functional for variational calculation of ρ_0 as

$$E_v[\rho] = \int \rho(\vec{r})v(\vec{r})d\vec{r} + T_s[\rho] + \frac{1}{2}\int \rho(\vec{r_1})\frac{1}{r_{12}}\rho(\vec{r_1})d\vec{r_1}\,d\vec{r_2} + E_{xc}[\rho] \tag{4.151}$$

For any N-representable density, the first three integrals on the right hand side of Eq. 4.151 can be easily evaluated as soon as ρ is specified. But the critical quantity is the 4th term, which has understandably small magnitude but important effects on molecular properties. The exact form of the exchange correlation potential is still elusive and needs to be appropriately modeled to give reasonably accurate molecular properties. For the reference system $\rho = \rho_s$ is obtained easily from the Slater determinant formed by the Kohn-Sham spin orbitals and reads

$$\rho = \rho_s = \sum_{i=1}^{n}|\phi_i^{KS}|^2 \tag{4.152}$$

The kinetic energy of the reference electrons $T_s[\rho]$ can be easily expressed in terms of KE integrals involving KS orbitals by invoking Slater-Condon rules and reads (in atomic units):

$$T_s[\rho] = -\frac{1}{2}\sum_{i=1}\langle \phi_i^{KS}(1)|\nabla_1^2|\phi_i^{KS}(1)\rangle \tag{4.153}$$

$E_{ext}[\rho]$ is easily evaluated as

$$E_{ext}[\rho] = -\sum_A Z_A \int \frac{\rho(\vec{r_1})d\vec{r_1}}{r_{1A}}d\vec{r_1}$$

once $\rho = \rho_s$ has been calculated from Eq. 4.152. The ground state energy $E_v[\rho] = E_0[\rho]$ becomes

$$E_v[\rho] = E_0[\rho] = -\sum_A Z_A \int \frac{\rho(\vec{r_1})d\vec{r_1}}{r_{1A}}d\vec{r_1} - \frac{1}{2}\langle \phi_i^{KS}(1)|\nabla_1^2|\phi_i^{KS}(1)\rangle + \frac{1}{2}\int \frac{\rho(\vec{r_1})\rho(\vec{r_2})d\vec{r_1}\,d\vec{r_2}}{|r_1-r_2|} + E_{xc}[\rho] \tag{4.154}$$

Thus, $E_v[\rho]$ is known if we know the KS orbitals (that defines $\rho \equiv \rho_s$) and the form of $E_{xc}[\rho]$.

The KS orbitals are found by variationally minimizing $E_v[\rho]$ by varying the KS orbitals under the usual orthonormality constraints. The variational process leads to an integrodifferential equation that the Kohn-Sham orbitals must satisfy and reads

$$\left[-\frac{1}{2}\nabla_1^2 - \sum_A \frac{Z_A}{r_{1A}} + \int \frac{\rho(\vec{r_2})d\vec{r_2}}{r_{12}} + v_{xc}(1) \right] \phi_i^{KS}(1) = \epsilon_i^{KS}\phi_i^{KS}(1) \qquad (4.155)$$

The $v_{xc}(1)$ term in the Kohn-Sham equation is the exchange correlation potential, which is obtained as the functional derivative of the exchange correlation energy $E_{xc}[\rho]$:

$$v_{xc}(\vec{r}) = \frac{\partial E_{xc}[\rho(\vec{r})]}{\partial \rho(\vec{r})}$$

Once $E_{xc}[\rho]$ is analytically specified, the calculation of $v_{xc}(\vec{r})$ is straightforward. We repeat once again that the exact form or expression for $E_{xc}[\rho]$ is hitherto unknown and must be approximated. A wide-variety of approximate forms for $E_{xc}[\rho]$ is available, but there is yet no well defined procedure for progressively improving $E_{xc}[\rho]$ following mathematical and physical reasoning. Nevertheless, the Kohn-Sham method has become an integral part of DFT protocol for molecular electronic structure calculations.

We must mention that Eq. 4.155 that KS orbitals satisfy looks very similar to the Hartree-Fock equation determining HF orbitals. Since v_{xc} in KS equation is supposed to contain Fermi, as well as Coulomb, correlation contribution, KS-orbitals, which belong to the reference system of n-non-interacting electrons have a very different physical meaning, if any. Similarly, the KS eigenvalues ϵ_i^{KS} do not have the same meaning as the Hartree-Fock orbital energies have, even though ϵ_i^{KS} are often used just as if they are the HF orbital energies.

The exchange correlation energy functional in the KS theory needs closer scrutiny. It contains

  (i) kinetic correlation energy term ($\Delta T[\rho]$) that measures the difference in the average kinetic energy of the actual system and the reference KS system containing an equal number of non-interacting electrons and having the same density as the real system;

  (ii) the exchange energy arising from the antisymmetry requirement;

  (iii) Coulombic correlation energy component emanating from the interelectronic repulsion between any pair of electrons, which rises rapidly as $r_{12} \rightarrow 0$;

  (iv) self-energy correction term, which is present in the classical Coulomb energy part of $E_v[\rho]$.

Several approximation to $E_{xc}[\rho]$ has been proposed from time to time, with progressively increasing accuracy and predictive power. Initially, the Coulomb correlation was neglected [$E_c[\rho] \approx 0$] and the exchange energy $E_x[\rho]$ was simply approximated by J. C. Slater's $\rho^{\frac{1}{3}}$ formula used in the X_α method. In 1980, Vosko *et al.*[22] developed a solution to the electron correlation problem in a homogeneous electron gas and constructed the corresponding Coulomb correlation potential. The resulting method including the exchange energy part also, is nowadays called the local density approximation (LDA). The method has been rather successful in handling metals. The failure of the method in describing molecules and ionic solids came as a rude shock.

The KS formalism was initially proposed for handling closed-shell systems only. Later the method was extended to cover open-shell systems under the guise of spin unrestricted Kohn-Sham formalism. This method is very much similar to the unrestricted Hartree-Fock formalism of wavefunction-based approach and suffers from the same drawbacks as the KS – single determinant wavefunction is not an eigenfunction of S^2 operator. It represents a mixture of all the possible spin multiplets. This open-shell version of LDA is usually called the local spin density approximation (LSDA). The LDA and LSDA methods succeeded in predicting molecular geometry and vibrational frequencies of surprisingly good quality, their defects not withstanding. They failed, however to predict binding energies of acceptable accuracy.

The DFT methods that go beyond the LDA can be broadly classified into two categories:

(i) The gradient corrected (GC) DFT.

(ii) The hybrid DFT.

In methods belonging to set (i) the calculation of exchange correlation functional ($E_{xc}[\rho]$) involves ρ as well as $\nabla\rho$ – that is electron density function is not assumed to be homogeneous, recognizing the presence of shell-structure. The number of GC-functionals is steadily growing. The Becke exchange energy functional (B) and the Perdew-Wang (PW) exchange correlation functional are the most-used ones under this category. The PW is commonly referred to as a generalized gradient approximation or GGA.[23]

A much used and popular gradient corrected (GC) correlation functional was proposed by Lee Yang and R. G. Parr, which is often used with Becke's exchange energy functional, giving rise to a class of methods, the so called BLYP methods.[24] Becke also proposed a class of hybrid methods in which the exchange energy is estimated by Fock energy calculation and is mixed with the local and GC correlation functional, the weightage of each being obtained from a fit to experimental heats of formation data of a large number of different types of molecule.[25] The most well-known and popular hybrid method is the so called B3LYP method in which the number '3' stands for the number of parameters used in the fitting of the experimental data. DFT methods are still under development and a lot of improvement can be expected to take place in the near future. We can foresee possible directions that the improvement may come from:

1. Developement of accurate interpolation formulae for exchange/correlation functionals that are accurately known at very low and very high densities and use of such formulae for calculation to be done at intermediate densities.

2. Extensive use of artificial intelligence methods specially genetic computing for constructing such interpolation formulae.

As of now, we expect hybrid DFT methods to be used for large scale calculation of electronic structure of large molecules and different classes of materials (solids) both for predicting properties and for designing new materials.

4.18 The Basis Sets for Molecular Calculation

For molecules, except perhaps for the diatomic molecules, the Hartree-Fock equations can not be solved accurately by numerical integration. One is therefore forced to adopt the LCAO-MO-SCF route proposed by CCJ Roothaan (see Section 4.13). At the heart of Roothaan's procedure is a set of atomic orbitals or atom centered orbitals or one-electron functions in terms of which the HF orbitals are expanded:

$$\phi_i(HF) = \sum_{p=1}^{N} C_{pi}\chi_p, \quad i = 1, 2, \cdots, m \tag{4.156}$$

$\{\chi_p\}$ being the set of atomic one-electron functions (orbitals) present in the basis set chosen.

For diatomic molecules the one-electron functions in the basis set are Slater-type orbitals (STOs) centered on each of the pair of atoms a, b

$$\chi_a = \underbrace{N r_a^{n-1} e^{-\zeta r_a}}_{\text{radial part}} \underbrace{Y_{lm}(\theta_a, \phi_a)}_{\text{angular part}} \tag{4.157}$$

For non-linear molecules, the spherical harmonics ($Y_{lm}(\theta_a, \phi_a)$) are replaced by linear combinations, which are real:

$$Y_{lm}^{\pm}(\theta_a, \phi_a) = \frac{1}{\sqrt{2}}\left(Y_{lm}^*(\theta_a, \phi_a) \pm Y_{lm}(\theta_a, \phi_a)\right)$$

Thus, the molecular Hartree-Fock orbitals can be expressed as linear combinations of a number of STOs, centered on different atoms, and belonging to different shells of the atoms. For polyatomic molecules, one is required to calculate many four-centered two-electron integrals, which are complicated and difficult to evaluate. The time required to evaluate the integrals can be quite large.

Boys proposed in 1950, that the use of Gaussian-type functions in place of STOs could reduce the computational time consumed in the calculation of two-electron four-centered orbitals.

A normalized Cartesian Gaussian-type function 'g' centered on an atom marked 'a' takes the following form:

$$g_{ijk}^a = N \cdot x_a^i y_a^j z_a^k e^{-\xi_a r_a^2} \qquad (4.158)$$

i, j, k are non-negative integers, N being the normalization constant. ξ_a is the exponent of the Gaussian. The normalization constant is readily evaluated and is given by

$$N = \left(\frac{2\xi_a}{\pi}\right)^{\frac{3}{4}} \frac{(8\xi_a)^{i+j+k} i! j! k!}{(2i)!(2j)!(2k)!} \qquad (4.159)$$

If $i = j = k = 0$, the Gaussian function is called an s-type Gaussian while $i + j + k = 1$ defines p-type Gaussians and $i + j + k = 2$, similarly define d-type Gaussian functions and so on.

The fact that product of two Gaussians centered at two different points (say atoms) can be set equal to a single Gaussian centered at a third point immediately reduces the four-centered two-electron integrals to two-centered two-electron integrals, thereby simplifying evaluation of two-electron integrals both in terms of computational labor and complexity.

Let us first consider the nomenclature relating to STO basis sets. A basis set comprising of one STO for each valence and inner-shell atomic orbitals of each atom is known as a minimal basis set. Thus, H_2O at the minimal basis set level, is described by five STOs $(1s, 2s, 2p_x, 2p_y, 2p_z)$ on the oxygen atom and two STOs $(1s)$, one each H atom making a total of seven basis functions. A better STO basis set is obtained by replacing each of STOs of the minimal set by two STOs with different exponents (ξ_s) and is commonly called double-zeta basis set. Similarly, one can construct a triple-zeta basis. If one makes a distinction between the core and valence shells and uses one STO for each inner shell, but more than one STO for the valence shell AOs on each atom, we have a split-valence basis set to deal with (like valence double zeta, valence triple zeta, etc.). AOs on each atom may get distorted or polarized by electron demands of other atoms present in the molecule. Such polarization can be taken care of by adding extra STOs to the basis set of each atom. These STOs correspond to angular momentum quantum number $l' > l_{max}$ of the valence shell of that atom. They are called polarization functions. Basis sets so generated are called polarized basis sets, a commonly used one being the double zeta plus polarization basis set, acronymed DZP basis or DZ + P basis. One commonly adds five $3d$ AOs as polarization for the atoms of the first two rows of the periodic table and three $2p$ AOs for each H atom as polarization functions.

The same nomenclature finds use also in Gaussian basis set. The trend has been not to use the primitive Gaussian basis set elements $(g_u s)$ directly. Instead, few normalized Gaussian primitives $(g_u s)$ are linearly combined to produce what is called a contracted Gaussian-type of function (CGTF). Thus primitives g_1, $g_2, g_3, \cdots, g_n$ are used to produce a single contracted Gaussian χ_p where

$$\{\chi_p\} = \sum_{k=1}^{n} d_{pk} g_k$$

The contraction coefficients d_{pk} are not allowed to change, but held fixed while performing a variational calculation. $g_k s$ are centered on specific atoms (a), the primitive Gaussians on atom a having the same i, j, k values, but different values of ξ_a. That saves a lot of computing requirements and time. The STO basis set terminologies can be extended to CGTF basis set easily.

Thus, a minimal basis set of contracted Gaussian function comprises of one contracted Gaussian function for each AO of the inner shells as well as for each valence shell AO while a DZ basis consists of two contracted Gaussian for each AO. A DZP set extends the DZ basis by adding to it contracted Gaussian with higher 'l' values, 'l' being equal to $i + j + k$ (see Eq. 4.158). How does one form the contracted Gaussian functions? For minimal CGTFs, it is invariably done by least squares fitting with the corresponding STOs of minimal set by varying the contraction coefficients $\{d_{pk}\}$ and the exponents ξ_a of the Gaussians. Alternatively, one can use uncontracted Gaussian functions as basis for doing atomic SCF calculations and then form the contracted sets.

Huzinaga used a $(9S\ 5P)$ basis set of uncontracted Gaussian functions for performing SCF calculations on the ground states of all atoms from lithium to neon and then formed the contracted GTFs for representing the AOs in molecular calculation.[26] Dunning's DZ (4S 2P) and Dunning and Hay's split valence [3S 2P] contractions of Huzinaga's (9S 5P) atomic orbitals of the first row atoms have found frequent use in molecular SCF calculations.[27–28] The other popular split valence sets are the 3-21G and the 6-31G sets. The 3-21G set has all inner-shell AOs represented by a single CGTF that is a linear combination of three primitive Gaussians while each valence shell AO is represented by a CGTF that is a linear combination of two primitive Gaussians and another single Gaussian with a rather small exponent (ξ_a), called a diffuse function as it decays very slowly as r increases. The 6-31G set uses six Gaussian primitives for each inner-shell AO while each valence shell AO is represented one CGTF with three primitive and a single Gaussian with one primitive. The contraction coefficients and orbital exponents are optimized with reference to atomic SCF calculations. Some of the exponents thus determined are often scaled further to make them more suitable for molecular calculations. The 6-31G^* basis set is obtained by adding 6-d-type cartesian Gaussian polarization function to the 6-31G basis for all atoms from Li-Ca. Similarly, 6-31G^{**} basis set is formed by adding to the 6-31G^* basis set a set of 3-p-type Gaussian polarization function on H and He atoms. There are many other basis sets available and every year new basis sets are being added to the databases.

The designing and choice of basis sets continue to be a difficult optimization problem. In fact it is a difficult multiobjective optimization problem as the set must be linearly independent, complete and capable of producing correct values of all the properties of a molecule one is interested in. The practice so far has been to design transferable atom centered basis sets **but with the revolution in digital and artificial intelligence-based computing taking place, it would soon be possible to allow each molecule to design its own optimal and universal basis set.**

REFERENCES

1. A. Mesiah, 'The variational method and associated problems', In *Quantum Mechanics* (pp. 762–781), Dover Publications, Mineola, New York (1999).
2. F. Schwebl, *Quantum Mechanics*, (Translated from German by Ronald Kates), Narosa Publishing House Reprint, New Delhi (1992).
3. L. Pauling and E. B. Wilson Jr., *Introduction to Quantum Mechanics*, McGraw-Hill, New York and London (1935); Dover Publications, Mineola, New York (1985).
4. H. Eyring, J. Walter and G. E. Kimball, *Quantum Chemistry*, John Wiley, New York and London (1944).
5. R. G. Parr, *Quantum Theory of Molecular Electronic Structure*, W. A. Benjamin, New York (1963).
6. J. N. Murrell, S. F. A. Kettle and J. M. Tedder, *Valence Theory*, Second Edition, John Wiley, New York and London (1970).
7. A. A. Sokolov, I. M. Ternov and V. Ch. Zhukovski, *Quantum Mechanics*, (Translated from Russian by Ram. S. Wadhwa), MIR Publishers, Moscow (1984).
8. J. C. Slater, *Introduction to Chemical Physics*, McGraw-Hill, New York and London (1939).
9. J. D. Bjorken, 'Relativistic quantum fields', *Am. J. Phys.*, 34, 367 (1966).
10. I. N. Levine, *Quantum Chemistry*, Fifth Edition, Pearson Education, London (2000).
11. T. Berlin, 'Binding regions in diatomic molecules', *J. Chem. Phys.*, 19, 208 (1951).
12. B. M. Deb, Ed., *The Force Concept in Chemistry*, Van Nostrand Reinhold, New York (1981).
13. M. de Podesta, *Understanding the Properties of Matter*, Second Edition, Taylor and Francis, London and New York (2002).
14. A. S. Davydov, *Quantum Mechanics*, Second Edition, Pergamon, Oxford (2013).
15. J. C. Slater, *Quantum Theory of Molecules and Solids, Vol:1, Electronic Structure of Molecules*, McGraw Hill, New York and London (1963).
16. M. Tinkham, *Group Theory and Quantum Mechanics*, TMH Edition. Reprinted in India by arrangement with McGraw Hill, New York (1974).
17. L. Pauling, *The Nature of the Chemical Bond and the Structure of Molecules and Crystals: An Introduction to Modern Structural Chemistry*, Cornell University Press, New York (1960).
18. C. Møller and S. Milton Plesset, 'Note on an approximation treatment of many electron system', *Phys. Rev.*, 46, 618–622 (1934).
19. R. G. Parr and W. Yang, *Density Functional Theory of Atoms and Molecules*, Oxford University Press, Oxford (1989).

20. P. Hohenberg and W. Kohn, 'Density functional theory (DFT)', *Phys. Rev. B*, 136, 864 (1964).

21. W. Kohn and L. Sham, 'Self-consistent equations including exchange and correlation effects', *Phys. Rev. A*, 140, 1133 (1965).

22. S. H. Vosko, L. Wilk and M. Nusair, 'Accurate spin-dependent electron liquid correlation energies for local spin density calculations: A critical analysis', *Can. J. Phys*, 58, 1200 (1980).

23. J. P. Perdew and Y. Wang, 'Erratum: Accurate and simple analytic representation of the electron-gas correlation energy', *Phys. Rev. B*, 45, 13244 (1992).

24. C. Lee, W. Yang and R. G. Parr, 'Development of the Colle-Salvetti correlation-energy formula into a functional of the electron density', *Phys. Rev. B*, 37, 785 (1988).

25. A. D. Becke, 'Density-functional exchange-energy approximation with correct asymptotic behavior', *Phys. Rev. A*, 38, 3098 (1998).

26. S. Huzinaga, 'Gaussian-Type functions for polyatomic systems. I', *J. Chem. Phys.*, 42, 1293 (1965).

27. T. H. Dunning Jr., 'Gaussian basis functions for use in molecular calculations. I. Contraction of (9s5p) atomic basis sets for the first-row atoms', *J. Chem. Phys.*, 53, 2823 (1970).

28. T. H. Dunning Jr. and P. J. Hay, In *Methods of Electronic Structure Theory*, Volume 3, H. Schaefer III, Ed., Springer Science and Business Media, Berlin (2013).

5

Quantum States of Solids

5.1 Introduction

As mentioned in Chapter 1, many of the technologically important materials are solids, either crystalline or amorphous. The most important property of crystalline solids is their structural periodicity, which gives rise to the notion of a crystal lattice. In a crystalline atomic solid, the lattice structure implies the existence of an arrangement of atoms, which can be generated by a mere repitition of an elementary cell. A crystal is therefore said to have a translationally invariant structure. The invariance ensures that the structure of the entire crystal can be obtained from the knowledge of the structure of just one cell – the unit cell. Let $\vec{a_1}$, $\vec{a_2}$ and $\vec{a_3}$ be three non-coplanar unit basis vectors and n_1, n_2 and n_3 be a group of three integers. A lattice vector $\vec{R_n}$ for the crystal can then be defined as $\vec{R_n} = n_1\vec{a_1} + n_2\vec{a_2} + n_3\vec{a_3}$. The translational invariance of its structure implies invariance with respect to displacements by the vector $\vec{R_n}$ for any choice of the integers n_1, n_2 and n_3. The question that immediately crops up concerning understanding the quantum states of electrons moving in such a periodic structure (potential) and how such states shape important properties of solids. Of special interest to us will be those properties that could not be properly explained within the framework of classical physics. It turns out that quantum mechanics enables us to understand both qualitatively and quantitatively the most important properties of solids that emanate from the unique structural properties of solids and their influence on the motion of electrons. The electrons in a solid move in the attractive electric field that the atomic nuclei produce together with the repulsive field produced by the interactions with all other electrons. The resulting many electron problem is a complex one and is too difficult to solve satisfactorily. However, a number of approximation schemes exist and can be invoked to understand the basic nature of the quantum states that appear in solids both qualitatively and at least, semiquantitatively as well. One such approximation scheme that has been quite successful is the one-electron approximation method. The method starts by replacing the motion of many electrons by the motion of a single electron moving in an effective potential (V_{eff}) that takes into account the field produced by the massive nuclei at fixed positions in the lattice plus the average field that all other electrons produce. As we will see in what follows, the quantum states of electrons in the solid in such an effective potential is profoundly shaped by the structural periodicity or the translational invariance of the crystal. Such states are called Bloch states.[1-3]

5.2 One-Electron Approximation, Translational Symmetry, Bloch States and Brillouin Zone

The Hamiltonian operator for the relevant one-electron problem is $\hat{H}$ where

$$\hat{H} = -\frac{\hbar^2}{2m}\nabla^2 + V_{eff}(r) \tag{5.1}$$

and the wavefunction $\Psi(r)$ that defines a quantum state of the electron in the crystal, satisfies the eigenvalue equation

$$H\Psi(r) = E\Psi(r) \tag{5.2}$$

DOI: 10.1201/9781003244882-5

From the nature of the problem, it is clear that $V_{eff}(r)$ must have the translational symmetry of the crystal, which means that $V_{eff}(r)$ must be a spatially periodic function with the same lattice period as that of the crystal. That imposes the following condition on $V_{eff}(r)$

$$V_{eff}(\vec{r_e} + \vec{R_n}) = V_{eff}(\vec{r}),$$ (5.3a)

which in turn means,

$$\hat{T}_n H(\vec{r}) = H(\vec{r})$$ (5.3b)

$\hat{T}_n$ representing the operator for translation (see later). The crystal, we assume, has infinite dimension. We can therefore impose cyclic boundary conditions on $\Psi(\vec{r})$. Before attempting a solution of the eigen value Eq. 5.2, let us enquire about the specific properties of the energy eigenfunctions determined by the periodic nature of the crystal structure or the potential in which the electron moves. $\hat{T}_n$ is the translational operator, which acts on the wavefunction $\Psi(r)$ to produce a displacement of the coordinate by the lattice period as written below:

$$\hat{T}_n \Psi(\vec{r}) = \Psi(\vec{r} + \vec{R_n})$$ (5.4)

Since $\hat{T}_n$ leaves $\hat{H}(\vec{r})$ or $V_{eff}(\vec{r})$ invariant (Eq. 5.3), we can immediately conclude that $[\hat{T}_n, H] = 0$ ($\hat{T}_n$ commutes with the Hamiltonian) so that $\hat{T}_n$ and H must have simultaneous eigenfunctions:

$$\hat{H}\Psi = E\Psi$$
$$\hat{T}_n\Psi = t_n\Psi$$ (5.5)

where E is the energy eigenvalue and 't_n' is eigenvalue of the translational operator. From the definition of the operator $\hat{T}_n$, and translational invariance of H we have,[1-4]

$$\hat{T}_n \Psi(\vec{r}) = \Psi(\vec{r} + \vec{R_n}) = t_n \Psi(\vec{r}) \equiv \Psi(r)$$ (5.6)

Therefore, the eigenvalues 't_n' must have modulus equal to unity. That demands

$$t_n = e^{i\vec{k}\cdot\vec{R_n}}$$ (5.7)

where $\vec{k}$ is called the Bloch vector of the wave and $\hbar\vec{k}$ is called the quasi-momentum or crystal momentum, the nature of which will become clear as we proceed further. Suffice it to mention at this point that if $V_{eff}(\vec{r}) = 0$ everywhere, we have the free electron problem and $\hbar\vec{k}$ then represents the actual or the true momentum. In the context of the motion of an electron in crystal, let us note that the values of the quasi-momentum $\hbar k$ characterize the quantum states of the electrons in the crystal. Accordingly, k can be regarded as a good quantum number for labeling the quantum states of an electron in the crystal. Generally, therefore, the eigenstates of $\hat{T}_n$ will satisfy the following conditions:

$$\hat{T}_n \Psi_{k,\lambda}(\vec{r}) = \Psi_{k,\lambda}(\vec{r} + \vec{R_n}) = e^{i\vec{k}\cdot\vec{R_n}}\Psi_{k,\lambda}(\vec{r})$$ (5.8)

where apart from the quasi-momentum or crystal momentum quantum number k, we have used 'λ', a collection of all other good quantum numbers, to label the eigenfunction of the Hamiltonian. It is more covenient and physically more descriptive now to switch over to a new form of notation for the crystal wavefunction and write

$$\Psi_{k,\lambda}(\vec{r}) = e^{i\vec{k}\cdot\vec{r}} U_{k,\lambda}(\vec{r})$$ (5.9)

These functions are called Bloch functions. They are modulated plane waves with amplitudes of modulation determined by the form of the periodic potential, $V_{eff}(\vec{r})$ and the magnitude of the quasi-momentum ($\hbar k$). The functions $U_{k,\lambda}(\vec{r})$ are periodic as can be seen by substituting the Bloch function in Eq. 5.8 whence we get

$$e^{i\vec{k}\cdot(\vec{r}+\vec{R_n})}U_{k,\lambda}(\vec{r} + \vec{R_n}) = e^{i\vec{R}\cdot\vec{R_n}} e^{i\vec{k}\cdot\vec{r}} U_{k,\lambda}(\vec{r})$$

The function $U_{k,\lambda}(\vec{r})$ is thus required to be periodic with a period equal to the lattice period:

$$U_{k,\lambda}(\vec{r}+\vec{R_n}) = U_{k,\lambda}(\vec{r}) \tag{5.10}$$

Physically, it means that the function $U_{k,\lambda}(\vec{r})$ that determines the amplitude of modulation of the plane wave has the exact periodicity of the lattice. We may now look for the equation that the periodic function $U_{k,\lambda}(\vec{r})$ would satisfy. To this end, we substitute the Bloch functions (Eq. 5.9) in the Schrödinger equation (Eq. 5.2) and after some simple steps, arrive at the following equation for $U_{k,\lambda}(\vec{r})$

$$\left[\frac{\hbar^2}{2m}(\nabla + i\,\vec{k})^2 + E_\lambda(\vec{k}) - V_{eff}(r)\right]U_{k,\lambda}(\vec{r}) = 0 \tag{5.11}$$

This equation has to be solved for one elementary cell only. The cyclic boundary conditions then allow us to continue the solution to the next cell with the same periodicity and then to the next to next cell, and so on.

Let us now focus on the wave vector $\vec{k}$ and the associated quasi-momentum $\hbar\,\vec{k}$ appearing in the Bloch functions. As noted earlier, the quasi-momentum turns out to be the true momentum when the motion of the electron is free which is possible when $V_{eff}(\vec{r}) = 0$. When the free motion condition is not met ($V_{eff}(\vec{r}) \neq 0$), the Bloch functions are not eigenfunctions of the momentum operator $\hat{p} = \frac{\hbar}{i}\nabla$ and therefore $\hbar\,\vec{k}$ can not be taken to represent the true momentum of the electrons in a periodic potential or a crystal. Not only are $\hbar\,\vec{k}$ s not the eigenvalues of the momentum operator $\hat{P} = -i\hbar\nabla$, the periodicity of the crystal potential has the additional consequence that the wave vector $\vec{k}$ is non-unique. This becomes clear when we consider the effect of the following transformation on $\vec{k}$:[4]

$$\vec{k'} = \vec{k} + \vec{G_m}$$
$$\vec{G_m} = 2\pi\,\vec{\tau} \tag{5.12}$$
$$\text{and} \quad \vec{\tau} = m_1\vec{b_1} + m_2\vec{b_2} + m_3\vec{b_3}$$

In the transformation defined above, $m_i (i = 1,2,3)$ are integers, $\vec{b_1}$, $\vec{b_2}$ and $\vec{b_3}$ are basis vectors and $\vec{\tau}$ is the lattice vector of the so called inverse (reciprocal) lattice, and $\vec{G_m}$ is manifestly a vector of the reciprocal lattice. The $\vec{b_i}$ vectors are related to the basis vectors $\vec{a_1}$, $\vec{a_2}$ and $\vec{a_3}$ through the following relations:

$$\vec{b_1} = \frac{\vec{a_2} \times \vec{a_3}}{A_0}$$
$$\vec{b_2} = \frac{\vec{a_3} \times \vec{a_1}}{A_0} \tag{5.13}$$
$$\vec{b_3} = \frac{\vec{a_1} \times \vec{a_2}}{A_0}$$

where A_0 represents the volume of an elementary cell and is equal to $|(\vec{a_1} \cdot (\vec{a_2} \times \vec{a_3}))|$. The definition of b vectors imply that the following conditions hold:

$$\vec{a_i} \cdot \vec{b_j} = \delta_{ij} \tag{5.14}$$

Expressing $\vec{R_n}$ and $\vec{\tau_m}$ in relevant basis vectors, we have

$$\vec{R_n} = \sum_i n_i \vec{a_i}$$

and

$$\vec{\tau_m} = \sum_j m_j \vec{b_j}$$

and thus,

$$e^{i\vec{k'}\cdot\vec{R_n}} = e^{i\vec{k}\cdot\vec{R_n}} \cdot e^{i\vec{G_m}\cdot\vec{R_n}}$$

Note that, from the definition of $\vec{G}$ we have

$$\vec{G_m}\cdot\vec{R_n} = 2\pi\,\vec{\tau_m}\cdot\vec{R_n}$$

$$= 2\pi\sum_i\sum_j m_j n_i\,\vec{b_j}\cdot\vec{a_i}$$

$$= 2\pi\sum_i\sum_j m_j n_i\delta_{ij}$$

$$= 2\pi\cdot q$$

where q is an integer given by the sum of the products of integers $(q=\sum_j m_j n_j)$.

Let us now turn our attention to the Bloch function $\Psi_{k,\lambda}(\vec{r})$ where

$$\Psi_{k,\lambda}(\vec{r}) = e^{i\vec{k}\cdot\vec{r}}\,U_{k,\lambda}(\vec{r}) \tag{5.15}$$

In view of Eq. 5.12, we can write Eq. 5.15 as $\Psi_{k,\lambda}(\vec{r}) = e^{i(\vec{k}+\vec{G_m})\cdot\vec{r}}\,U_{k+G_m}(\vec{r})$ whence it follows that

$$U_{k+G_m}(\vec{r}) = e^{-i\vec{G_m}\cdot\vec{r}}\,U_{k,\lambda}(\vec{r}) \tag{5.16}$$

That means, the function $U_{k+G_m}(\vec{r})$ has the periodicity of the **direct lattice**. We may therefore conclude that the wavefunction $\Psi_{k,\lambda}(\vec{r})$ and $\Psi_{k+G_m,\lambda}(\vec{r})$ correspond to the same energy state. The energy eigenvalues of an electron in a periodic potential are therefore periodic in the **inverse lattice**:

$$E(\vec{k}) = E(\vec{k}+\vec{G_m}) \tag{5.17}$$

Thus it is not possible to distinguish between $\vec{k}$ and $\vec{k}+\vec{G_m}$ and we have to consider them as essentially identical k values.

How does one uniquely fix the quasi-momentum? One usually chooses its lowest value, which means $\vec{k}$ is considered only within the first cell of the inverse lattice with its linear dimension multiplied by 2π. The cell thus produced is called the 'Brillouin zone'. We may restrict ourselves only to $\vec{k}$-vectors lying inside a finite zone of the reciprocal space or the $\vec{k}$-space because any k-vector lying outside the zone can be considered to be one lying inside the zone, by adding a suitable $\vec{G_m}$ to it.

It is expedient now to analyse and understand physical meaning of quasi-momentum ($\hbar k$). The physical meaning becomes clear by considering the motion of an electron in a periodic potential under external forcing. Let $\vec{F}$ be the external force (electric field, say) acting on the electron. The resulting motion of a localized particle can be elegantly described by forming a wave packet with Bloch functions in the interval $K_0-\Delta k$ and $K_0+\Delta k$ of wave numbers:[3]

$$\Psi(\vec{r},t) = \int_{\Delta k} U_{k,\lambda}(\vec{r})e^{(i\vec{k}\cdot\vec{r}-\frac{iEt}{\hbar})}d^3k \tag{5.18}$$

where $E = E(\vec{k})$.

The center of gravity of such a packet is known to move with a group velocity, $\vec{v} = \frac{1}{\hbar}\nabla_k E(\vec{k})$, $\vec{v}$ being identical with the velocity of the particle. If the spread $\Delta\vec{k}$ is very small, say, $\Delta k << |k_0|$, the amplitude of $U_{k,\lambda}(\vec{r})$ can be assumed to be constant over Δk for all practical purposes, making it possible for us to make use of all the general results on the motion of a wave packet.

Let us note now that the external force $(\vec{F})$ does work and changes the energy of the particle and we get[4]

$$\frac{dE(\vec{k})}{dt} = \nabla_k E(k)\frac{d\vec{k}}{dt} = \vec{v} \cdot \vec{F} \tag{5.19}$$

But in the present case,

$$\frac{1}{\hbar}\nabla_k E(\vec{k}) \cdot \vec{F} = \nabla_k E(\vec{k})\frac{d\vec{k}}{dt}$$

and

$$\vec{F} = \hbar\frac{d\vec{k}}{dt} = \frac{d}{dt}(\hbar k)$$

The equation above clearly expresses Newton's law of motion with momentum replaced by quasi-momentum. Thus, free particle momentum has been naturally replaced by quasi-momentum for an electron moving in a periodic field.[4]

5.3 Formation of Energy Bands

An important feature of the motion of an electron in a periodic field is the band structure of the energy spectrum. The energy $E(\vec{k})$ is no longer a continuous function of the momentum as observed in free motion of an electron $(E(\vec{k}) = \frac{\hbar^2 k^2}{2m_0})$. In a periodic field, $E(\vec{k})$ breaks up into a number of energy bands. The energy now remains continuous over wide intervals of $\hbar\vec{k}$ (reminiscent of free motion) but shows up discontinuities at certain values of $\vec{k}$, thereby dividing the entire spectrum of energy into '*strips*' or '*bands*' – i.e. regions of allowed energy values separated by gaps corresponding to what may be called forbidden values of energy. It is time now to consider a simple model of an electron in periodic field that illustrates the appearance of band structure of the energy spectrum. One such model is the nearly free electron model, which will now be considered.

5.3.1 Nearly Free Electron Model of Band Structure

The Schrödinger equation in a periodic potential $V(\vec{r})$ has energy eigenfunction $\Psi_{k,\lambda}(\vec{r})$ and energy eigenvalues $E_\lambda(k)$:

$$\left\{ -\frac{\hbar^2}{2m}\nabla^2 + V(\vec{r}) - E_\lambda(\vec{k}) \right\}\Psi_{k,\lambda}(\vec{r}) = 0 \tag{5.20}$$

If $V(\vec{r})$ is a constant potential, the Bloch function must turn over into an ordinary plane wave:

$$\Psi_{k,\lambda}(\vec{r}) = e^{i\vec{k}\cdot\vec{r}} U_{k,\lambda}(\vec{r}) = Ce^{i\vec{k}\cdot\vec{r}}$$

Here $U_{k,\lambda}(\vec{r})$ is a constant and the momentum $\hbar k$ defines the energy $E(\vec{k}) = \frac{\hbar^2 k^2}{2m}$, as expected for free motion of the electron.

Let us now suppose that $V(\vec{r})$ is not constant, but is so weak that its effects can be treated as a perturbation on the free motion. For simplicity, we will consider a one-dimensional potential $V(x)$ with lattice period equal to 'a' so that $V(x + a) = V(x)$, for any x. In that case, the Schrödinger equation takes the following form:

$$\left\{ -\frac{\hbar^2}{2m}\frac{d^2}{dx^2} + V(x) - E(k) \right\}\Psi_k(x) = 0 \tag{5.21}$$

The zeroth order or the unperturbed solution of the above equation are just the plane waves $\Psi_k^0(x)$ where

$$\Psi_k^0(x) = \frac{1}{\sqrt{Na}}e^{i\vec{k}\cdot x} \tag{5.22}$$

'*Na*' is the normalization-box length (L) that corresponds to the dimension of the crystal. The unperturbed energy of the electron given by

$$E^0(k) = \frac{\hbar^2 k^2}{2m} \tag{5.23}$$

forms a continuous energy spectrum. The perturbed wavefunction (upto first order) and energy (upto second order) are, on the basis of standard Rayleigh-Schrödinger perturbation theory, are respectively given by:

$$\Psi_k(x) = \Psi_k^0(x) + \sum_{k' \neq k} \frac{V_{k'k}(x)}{(\epsilon_k^0 - \epsilon_{k'}^0)} \Psi_{k'}^0(x) \tag{5.24a}$$

$$E(k) = \epsilon^0(k) + V_{kk} + \sum_{k' \neq k} \frac{|V_{kk'}|^2}{\epsilon_k^0 - \epsilon_{k'}^0} \tag{5.24b}$$

The matrix elements of the potential energy operator $V_{k'k}$ represents the integral:

$$V_{k'k} = \int \Psi_{k'}^{0*}(x) V(x) \Psi_k^0(x) dx$$

$$= \frac{1}{L} \int_0^L e^{i(k-k')x} V(x) dx \tag{5.25}$$

The diagonal element V_{kk} in the second order energy is reduced to a constant (V_0) free from any dependence on k. It produces just a constant shift in the energy spectrum and can be set equal to zero. The off-diagonal elements $V_{k'k}(k \neq k')$, however, contains a periodic function $V(x)$ in the integrand with period equal to 'a'. The exponent of the exponential term in the integrand must have the same periodicity if $V_{k'k}$ has to be non-zero. The following equality must therefore hold:

$$e^{i(k-k')\cdot(x+a)} = e^{i(k-k')x}$$

$$\text{i.e.} \quad e^{i(k-k')a} = 1 \tag{5.26}$$

Thus $V_{k'k} \neq 0$, if and only if

$$(k - k') = \frac{2\pi m}{a} = G_m \quad \text{with } m = (\pm 1, \pm 2, \dots) \tag{5.27}$$

The vector $\vec{G}$, we may recall, is connected with the inverse lattice vector. The condition stipulated in Eq. 5.27 suggests that $V_{k'k} \neq 0$ only if $k' = k - \frac{2\pi m}{a}$. This restriction, when used in the second order energy expression in Eq. 5.24b leads to the following expression for the energy $E(k)$:

$$E(k) = \epsilon^0(k) + \sum_{m \neq 0} \frac{|V_{(k-\frac{2\pi m}{a}),k}|^2}{\frac{\hbar^2}{2m}\left[k^2 - (k - \frac{2\pi m}{a})^2\right]} \tag{5.28}$$

Let us note that the terms in Eq. 5.28 above, for which the energy denominators are close to zero make the largest contribution to the second order energy expression. The perturbation theory used for $E(k)$ breaks down completely whenever the denominator is exactly zero; that is whenever[4]

$$k^2 - \left(k - \frac{2\pi m}{a}\right)^2 = 0$$

$$\text{or,} \quad k = \frac{\pi m}{a} = \frac{G_m}{2}, \ (m = \pm 1, \pm 2, \dots) \tag{5.29}$$

These values of k ($k = \frac{\pi m}{a}$) define the Brillouin zone boundaries and mark the points of discontinuity of the energy function $E(k)$. Note that the second order formula for $E(k)$ remains valid at large distances from the Brillouin zone boundaries.

In order to get perturbative results that remain valid even in the proximity of the points of discontinuity that is at points near the Brillouin zone boundaries, let us start by considering the ambiguity in the definition of quasi-momentum (Eq. 5.12). This, in our case, makes the problem degenerate in the zeroth order approximation as $\Psi_k^0(x)$ and $\Psi_{k-G}^0(x)$ belong to the same energy value, E_k^0 where

$$E^0(k) = \frac{\hbar^2 k^2}{2m_0}$$

The zeroth order wavefunction, for a fixed value of 'm' is doubly degenerate forcing us to choose

$$\Psi_k^0(x) = A\Psi_k(x) + B\Psi_{k-G}(x)$$

where A and B are undetermined coefficients and $\Psi_k(x)$ are the plane waves:

$$\Psi_k(x) = \frac{1}{\sqrt{L}} e^{ik \cdot x} \tag{5.30}$$

Invoking the standard formulation of degenerate perturbation theory, it is possible to obtain the following expressions for the perturbed energy and wavefunction in the first approximation:

$$E(k) = E_k^0 \pm \sqrt{|V_{k,k-G}|^2}$$
$$= E_k^0 \pm \sqrt{|V_{\frac{\pi m}{a}, \frac{-\pi m}{a}}|^2} \tag{5.31}$$

and

$$A = \pm B \tag{5.32}$$

These results suggest that the energy $E(k)$ passes through a discontinuity of a finite magnitude (ΔE) at the Brillouin zone boundary where

$$\Delta E = 2|V_{\frac{\pi m}{a}, \frac{-\pi m}{a}}| \tag{5.33}$$

The corresponding perturbed wavefunctions take the form

$$\Psi_k(x) = \frac{1}{\sqrt{2L}} \left\{ e^{\frac{i\pi m x}{a}} \mp e^{\frac{-i\pi m x}{a}} \right\} \tag{5.34}$$

The typical form of k-dependence of energy $E(k)$ (Eq. 5.31) for the free electron model has been displayed in Figure 5.1.

As can be seen clearly from Figure 5.1 the energy is not a continuous function of quasi-momentum. Because of discontinuities in $E(k)$ at certain values of quasi-momentum ($\hbar k$), $E(k)$ splits into *strips* or

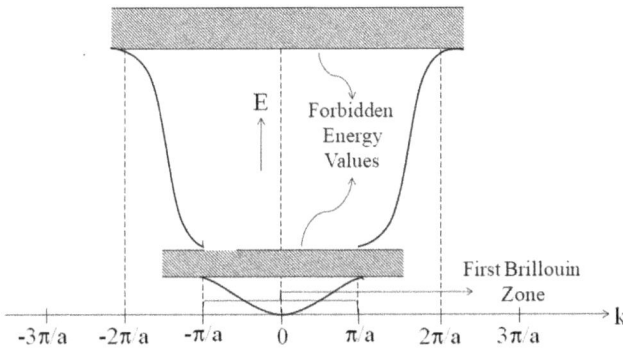

FIGURE 5.1 k-dependence of energy ($E(k)$) in the nearly free electron model of crystalline solids. Note the formation of energy bands.

bands. The most striking feature of the energy spectrum is the appearance of regions with *forbidden* values of energy. Within the allowed bands, energy still remains a continuous function of k (Figure 5.1). We note that the energy bands have their origin in the periodic structure or translational symmetry of crystalline solids. The band structure is a very fundamental feature of the electronic structure of solids – the discrete energy levels of the constituent atoms of the solids, as if merge producing allowed and forbidden energy bands. We will consider another simple but more realistic model of electrons in a one-dimensional periodic field. It is the celebrated Kronig-Penny (KP) problem. The band structure of the energy spectrum that the KP model leads to once again has allowed and forbidden regions.

5.3.2 Kronig-Penny Problem and Structure of Energy Bands

As before, we assume that the crystal is made up of a regular arrangement of N positive ions forming a periodic structure (lattice) in space and electrons moving in the electric field generated by the ions. In one dimension, the potential $V(x)$ that an electron effectively feels has the periodicity $V(x + a) = V(x)$, 'a' being the lattice period. Such a potential is displayed in Figure 5.2. In a classical model, an electron moving in the potential of Figure 5.2 would have been bound to a single ion, a situation that could also be realized in quantum mechanics if the ions forming the array were separated by a large distance, i.e. if 'a' had a very large value. In such a situation the N states in which the electron is bound to an atom of the array would have formed an N-fold degenerate orthogonal set of one-electron states. As the distance separating the ions is reduced, these states begin to overlap and the degeneracy is expected to be lifted. The new energy eigenvalues are expected to be distributed within a band. The problem here is to find out how exactly the eigenvalues are distributed when the more realistic potential is replaced by a periodic array of square well potential $V(x)$. We expect this idealization to work as the main effect of the potential lies in its translational symmetry and not so much in its detailed features like shape, etc. The model has been known as the Kronig-Penny model and the potential $V(x)$ as the Kronig-Penny potential.

The solutions of Schrödinger equation in KP potential takes different forms depending on whether the electron is in the region with zero potential energy or in the barrier region with potential energy V_0. In the region with zero potential energy $\Psi(x)$ take the form

$$\Psi_1(x) = Ae^{i\alpha x} + Be^{-i\alpha x}, \quad \alpha = \frac{\sqrt{2mE}}{\hbar} \tag{5.35a}$$

while in the barrier region it can be chosen in the form

$$\Psi_2(x) = Ce^{\beta x} + De^{-\beta x}, \quad \beta = \frac{\sqrt{2m(V_0 - E)}}{\hbar} \tag{5.35b}$$

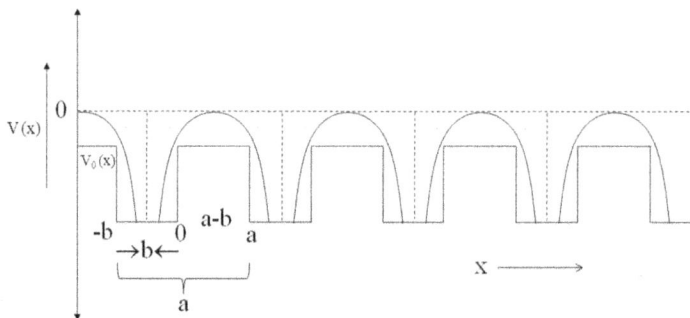

FIGURE 5.2 Sketch of Kronig-Penny potential.

The continuity of $\Psi(x)$ and $\Psi'(x)$ across boundaries together with the fact that $\Psi(x)$ conforms to the periodic form demanded by Bloch's theorem leads to the following equalities:

(i) $\Psi_2(0) = \Psi_1(0)$, $\Psi_2'(0) = \Psi_1'(0)$ These arise from matching Ψ_1, Ψ_2 (5.36a)

and their derivatives at $x = 0$

$$\begin{cases} (ii) \quad \Psi_2(-b) = e^{-i\gamma a}\Psi_1(a-b) \\ \qquad \Psi_2'(-b) = e^{-i\gamma a}\Psi_1'(a-b) \end{cases} \text{These follow from Bloch's theorem} \tag{5.36b}$$

where γ is a real quantity. If $\Psi_1(x)$ and $\Psi_2(x)$ as defined in Eqs. 5.35a,b are used in the matching condition in Eqs. 5.36a,b, we get the following equation for the determination of the five parameters, viz, A, B, C, D and γ:

$$C + D = A + B \tag{5.37a}$$

$$C - D = i\left(\frac{\alpha}{\beta}\right)(A - B) \tag{5.37b}$$

$$Ce^{-\beta b} + De^{\beta b} = e^{-i\gamma a}\left[A \cdot e^{i\alpha(a-b)} + B \cdot e^{-i\alpha(a-b)}\right] \tag{5.37c}$$

$$Ce^{-\beta b} - De^{\beta b} = i\left(\frac{\alpha}{\beta}\right)e^{-i\gamma a}\left[A \cdot e^{i\alpha(a-b)} - B \cdot e^{-i\alpha(a-b)}\right] \tag{5.37d}$$

Eqs. 5.37a–d can be combined into the following pair of equations:

$$(A + B)\left[\cosh\beta b - e^{-\gamma a}\cos\alpha(a-b)\right] = i(A - B)\left[\frac{\alpha}{\beta}\sinh\beta b + e^{-i\gamma a}\sin\alpha(a-b)\right] \tag{5.38a}$$

$$(A + B)\left[\sinh\beta b - \frac{\alpha}{\beta}e^{-\gamma a}\sin\alpha(a-b)\right] = i(A - B)\left[\frac{\alpha}{\beta}\cosh\beta b - \frac{\alpha}{\beta}e^{-i\gamma a}\cos\alpha(a-b)\right] \tag{5.38b}$$

In order for meaningful solutions to exist, the determinant formed by the coefficients of $A + B$ and $A - B$ must vanish, leading to the condition:[4]

$$\cos\gamma a = \frac{\beta^2 - \alpha^2}{2\alpha\beta}\sinh\beta b\sin\alpha(a-b) + \cosh\beta b\cos\alpha(a-b) \tag{5.39}$$

Eq. 5.39 can be solved graphically or numerically to construct the energy spectrum under the constraint that the modulus of the right hand side of Eq. 5.39 must not exceed unity. A clearer and simpler picture emerges if we replace the train of square well potential with chain of 'Delta-function' potential under the assumption that

$$V_0 \to \infty \text{ as } b \to 0$$

but the quantity $\frac{mV_0}{\hbar^2}ab$, which is proportional to the barrier area remains finite, i.e.

$$\frac{mV_0}{\hbar^2}ab = \lambda \quad \text{(finite)} \tag{5.40}$$

In this case, Eq. 5.38 is reduced to a new form:[3]

$$\cos\gamma a = \lambda\frac{\sin\alpha a}{\alpha a} + \cos\alpha a \tag{5.41}$$

To arrive at Eq. 5.41 we have assumed, under the approximations made, $\sinh\beta b \approx \beta b$ and $\cosh\beta b \approx 1$. In Eq. 5.41, γ is real so that the left hand side varies from -1 to $+1$. That means the equation can be satisfied only if the right hand side changes from -1 to $+1$. The situation is displayed in Figure 5.3.

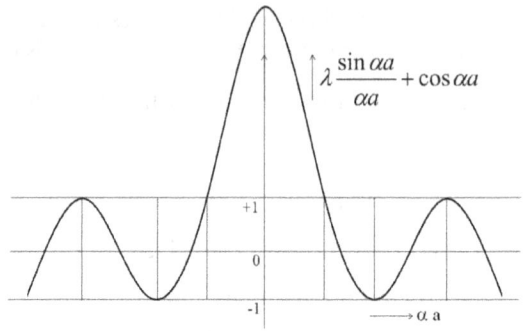

$\lambda \dfrac{\sin \alpha a}{\alpha a} + \cos \alpha a$

FIGURE 5.3 Allowed and forbidden energy band structure in the Kronig-Penny model.

The energy spectrum, in the KP model, just as in the nearly free electron model, splits into a band structure with alternating regions of allowed and forbidden values of energy (αa in Figure 5.3). It turns out that this feature of the energy spectrum splitting into allowed and forbidden bands of energy is quite general, irrespective of the details of the model used to represent the periodic field and appears to be a consequence of the translational symmetry of the potential. However, the detailed structure of the bands may be significantly more complicated in specific cases, with allowed energy bands even overlapping in some cases. The detailed calculation of the energy band structure of a specific crystalline solid is a complex and an arduous task. Our purpose here has been to reveal emergence of the band structure on the basis of simple models. It will not be wrong to conclude that the band structure has origin in the translational invariance of the crystal, which in turn is responsible for the creation of Bloch states.

5.3.3 The Tight Binding Model of Periodic Solids

Free or nearly free electron models described in the preceeding sections along with the Kronig-Penny model explain many important features of energy bands produced in periodic solids. Nevertheless, there is a need to approach the problem from another extreme point of view in which electrons are not free but are tightly bound. Let us begin by looking into the forms of energy band wavefunctions in a one-dimensional crystal, for different values of k. In the 1D case, the Bloch theorem asserts that[5]

$$\Psi_k(x) = U_k(x)e^{ikx} \tag{5.42a}$$

$$U_k(x+a) = U_k(x) \tag{5.42b}$$

a being the lattice constant. The only problem to focus on here is the determination of $U_k(x)$ within one unit cell. The periodicity then determines it over the rest of the crystal. In the free electron approximation (the so called empty lattice case or $V(x) = 0$ case) $U_k(x)$ can be set to be equal to 1 for all values of k. However, if k is restricted to have values $\leq \frac{\pi}{a}$, the choice $U_k(x) = 1$ is meaningful only for the lowest band. For the second band the choice that is consistent with the required periodicity is[4]

$$U_k(x) = e^{\frac{i2\pi x}{a}}, \quad \frac{-\pi}{a} \leq k \leq 0$$
$$U_k(x) = e^{-\frac{i2\pi x}{a}}, \quad 0 \leq k \leq \frac{\pi}{a} \tag{5.43}$$

so that we can recover the free electron wavefunctions of the same energy. For the third or still higher bands, we must choose

$$U_k(x) = e^{\pm \frac{i2\pi n x}{a}} \quad (n \text{ an integer}) \tag{5.44}$$

In the reduced zone scheme, energy levels may cross and at the points of crossing, the periodic potential can display remarkable effects.[5] Thus, choosing $k = \frac{\pi}{a}$, the wavefunctions for the first and the second

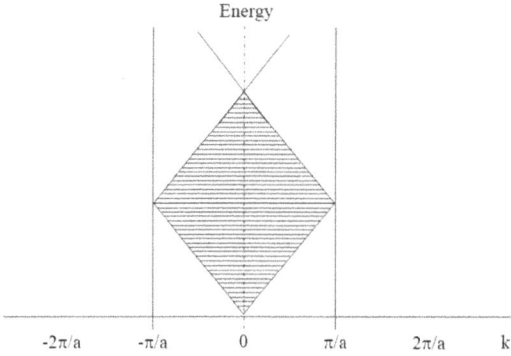

FIGURE 5.4 Energy $[E(k)]$-k plot in the 1D problem.

bands are

$$\Psi_1 = 1 \cdot e^{\frac{i\pi x}{a}}$$

$$\psi_2 = e^{-\frac{i2\pi x}{a}} \cdot e^{\frac{i\pi x}{a}} = e^{\frac{-i\pi x}{a}}$$

(5.45)

Clearly, Ψ_1 and Ψ_2 are degenerate. If the potential is such that $\langle \Psi_1 | V(x) | \Psi_2 \rangle \neq 0$, Ψ_1 and Ψ_2 combine to produce the proper eigenfunction Ψ_+ and Ψ_- where

$$\Psi_+ \sim \Psi_1 + \Psi_2 \sim \cos \frac{\pi x}{a}$$

$$\Psi_- \sim \Psi_1 - \Psi_2 \sim \sin \frac{\pi x}{a}$$

(5.46)

It is evident from Eq.5.46 that Ψ_+ and Ψ_- differ only in a phase shift of $\frac{a}{2}$ with respect to each other along the x-axis. The two proper eigenfunctions have different energies, leading to the appearance of a band gap (Figure 5.4). The gap has a width equal to $2 \langle \Psi_1 | V | \Psi_2 \rangle$.

Let us now assume that the periodic potential is *strong*. The free electron wavefunctions are no-longer adequate or appropriate zeroth order functions based on which a rapidly convergent perturbation scheme can be built up. A way out could be to appeal to the other extreme case in which an electron remains tightly bound to one center (or atomic site) along the chain with the translational symmetry just superimposed on the chain. Thus, if $\phi(x)$ is the wavefunction of the electron bound to one such site, a periodic function $U_k(x)$ can be generated by superposing $\phi(x)$ from all such sites (x_i) yielding

$$U_k(x) \sim \sum_i \phi(x - x_i)$$

(5.47)

When this $U_k(x)$ (Eq. 5.47) is ploughed back in Eq. 5.42 we get what has become known as the tight-binding approximation to the wavefunction for electrons in an energy band in one dimension. If we are dealing with energy bands in a 3-d atomic solid, the functions $\phi(x)$ represent the atomic wavefunctions ($1s$, $2s$, $2p$, $3p$, $3d$ etc.) so that bands produced can be referred to as $2s$ band or $3p$ or $3d$ band. Even when the binding is not that strong or tight, the band nomenclature based on atomic states does not entirely lose significance.

Let us consider for the sake of simplicity the band in our 1D model that corresponds to the $1s$ band in a solid. $\phi(x)$, in this case, is a nodeless function having a maximum at the origin with a smooth fall-off. Then, $U_k(x)$ as defined in Eq. 5.47 will also be a nodeless function with a maximum in each unit cell at the location of the atom as displayed in Figure 5.5a.

Notice that in this $k = 0$ state, the wavefunction is not just a constant as in the free electron case. It has a clear maximum in each unit cell. If $k \neq 0$, $e^{ikx} \neq 1$ for all x, and therefore it modulates the periodic function $U_k(x)$, the Bloch function, as we go from one unit cell to the next, and from the next to the next and so on.

A typical example of this modulation is displayed in Figure 5.5b for $k = \frac{\pi}{3} \cdot \frac{1}{a}$. We have plotted only the real part of $\Psi(x)$. The imaginary part of $\Psi(x)$ has a modulating envelope that is shifted by $\frac{3a}{2}$. The consequence is that the distribution of total charge density remains the same in each unit cell.[5]

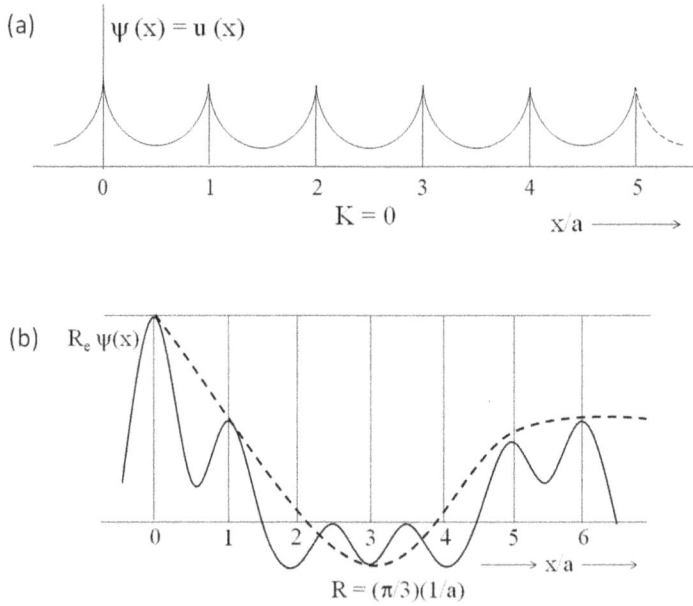

FIGURE 5.5   (a) The Bloch function for $k = 0$ in the 1D periodic lattice. (b) The real part of the wavefunction $\Psi(x)$.

Let us now consider a generalization of the tight-binding (TB) model and work with an elemental solid, which has just one atom in each unit cell. Each atom is assumed to have only one valence orbital $\phi(\vec{r})$. The Bloch-form of the wavefunction can be written as

$$\Psi_k = \frac{1}{\sqrt{N}} \sum_m e^{i\vec{k} \cdot \vec{R_m}} \phi(\vec{r} - \vec{R_m}) \tag{5.48}$$

$\vec{R_m}$ being a lattice vector. Let $\vec{t}$ be a translational vector. Then

$$
\begin{aligned}
\Psi_k(\vec{r} + \vec{t}) &= \frac{1}{\sqrt{N}} \sum_m e^{i\vec{k} \cdot \vec{R_m}} \phi(\vec{r} - (\vec{R_m} - \vec{t})) \\
&= \frac{1}{\sqrt{N}} \sum_m e^{i\vec{k} \cdot \vec{t}} \cdot e^{i\vec{k} \cdot (\vec{R_m} - \vec{t})} \phi(\vec{r} - (\vec{R_m} - \vec{t})) \\
&= e^{i\vec{k} \cdot \vec{t}} \frac{1}{\sqrt{N}} \sum_m e^{i\vec{k} \cdot (\vec{R_m} - \vec{t})} \phi(\vec{r} - (\vec{R_m} - \vec{t})) \\
&= e^{i\vec{k} \cdot \vec{t}} \cdot \Psi_k(\vec{r})
\end{aligned}
\tag{5.49}
$$

The last step utilizes the fact that $\vec{R_m} - \vec{t}$ is a lattice vector, since $\vec{R_m}$ is. The energy expectation value corresponding to the wavefunction $\Psi_k(\vec{r})$ is

$$E_k = \langle \Psi_k(\vec{r}) | H | \Psi_k(\vec{r}) \rangle$$

where $\hat{H}$ is the Hamiltonian of the system. It is straightforward to find

$$E_k = \sum_m \sum_n e^{i\vec{k} \cdot (\vec{R_m} - \vec{R_n})} \langle \phi_m | H | \phi_n \rangle \tag{5.50}$$

In the tight-binding model, $\phi_m(\vec{r})$, $\phi_n(\vec{r})$ are tightly bound to the atomic sites (m, n). The matrix element 'H_{mn}' is therefore expected to be negligibly small, unless $m = n$ (the diagonal element of the H-matrix) when the interaction is on-site only, or m and n are the nearest neighbor sites ($|m - n| = 1$) when the interaction is limited to occur between atomic states on the immediate neighboring sites. Beyond the nearest neighbor the interaction falls off rapidly with increasing intersite distance and can be neglected. We can therefore adopt the following parametrization on the basis of physical consideration already mentioned:

$$H_{mm} = H_{nn} = -\alpha, \quad \text{for all } m, n$$
$$H_{mn} = H_{nm} = -\gamma, \quad \text{for } |m - n| = 1 \tag{5.51}$$
$$= 0, \quad \text{otherwise}$$

With these definitions used in Eq. 5.50 we have

$$E_k = -\alpha - \gamma \sum_m e^{i \vec{k} \cdot \vec{R}_m} \tag{5.52}$$

Let us now apply the idea to a square lattice with lattice vectors $\{R_m\} = \{(a,0), (-a,0), (0,a), (0,-a)\}$ and wave vector $k = (k_x, k_y)$. Eq. 5.52 then yields

$$E_k = -\alpha - 2\gamma \{\cos k_x a + \cos k_y a\} \tag{5.53}$$

Since the value of the cosine terms on the right hand side of Eq. 5.53 range between -1 and 1, the energy E_k has the range from $-\alpha - 4\gamma$ to $-\alpha + 4\gamma$, thus producing an energy band with equal to 8γ. At $k = 0$ ($k_x = k_y = 0$), $E_0 = -\alpha - 4\gamma$. In the neighborhood of $k = 0$, k_x, k_y are small, allowing us to approximate the cosine terms in Eq.5.53 as

$$\cos k_x a \approx 1 - \frac{1}{2} k_x^2 a^2$$
$$\cos k_y a \approx 1 - \frac{1}{2} k_y^2 a^2 \tag{5.54}$$

which when used in Eq. 5.53 leads to

$$E_k \approx -\alpha - 4\gamma + \gamma (k_x^2 + k_y^2) a^2 \tag{5.55}$$

At or near the center of the 'Brillouin zone' ($k \approx 0$), E_k is thus seen to be described by circular constant energy surface. Let us now move out from the center to the zone-corners. If $k_x = k_y = \frac{\pi}{a}$, we may set, for example,

$$k_x = \frac{\pi}{a} - \Delta x$$
$$k_y = \frac{\pi}{a} - \Delta y$$

and get from Eq. 5.55

$$E_k = -\alpha - 2\gamma \left\{ \cos \left(\frac{\pi}{a} - \Delta x \right) a + \cos \left(\frac{\pi}{a} - \Delta y \right) a \right\}$$
$$= -\alpha - 4\gamma - \gamma \left\{ \Delta x^2 + \Delta y^2 \right\} a^2 \tag{5.56}$$

E_k in Eq. 5.56 implies circular constant energy surface. On the other hand, near the middle of the band we have,

$$\cos k_x a + \cos k_y a = 0 \tag{5.57}$$

which has solutions

$$k_x a = \pi - k_y a$$

That means the solutions of Eq. 5.57 are straight lines. These straight lines represent the constant energy surface at the middle of the band (see Figure 5.6). Proceeding in this manner and considering all the results, we can construct the constant energy surface in a 2D square lattice.

It is possible to extend the model to 3D periodic structures (solids) as well. We will not pursue it further here. The tight-binding Hamiltonian as described in this section is reminiscent of Huckel's model[6] Hamiltonian for molecules with translational symmetry superimposed on Huckel's description.

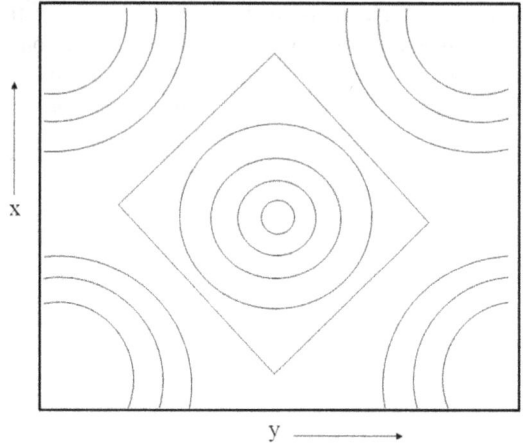

FIGURE 5.6 Contact energy surface in a 2D square lattice: tight-binding model.

5.4 The Idea of Band Gap and Electrical Transport in Solids

The band structure of the energy spectrum in solids, especially, the creation of forbidden zones or band gaps have an important bearing on the electrical conductivity of solids. The order of filling of the energy bands by electrons, the presence of the band gaps and vacant bands dominantly determine the type of electrical conductivity or response displayed by a given solid.

The filling of the energy bands by electrons can be assumed to be normally dictated by the tendency to keep the lowest available one-electron states (orbitals) occupied to produce the atomic ground state. As we will see in Chapter 7, at 0 K the electrons in solids will occupy all the energy bands upto the Fermi level (the upper limiting level of energy). Thus, all the states lying within the confines of a certain surface in the space of the wave vectors $\vec{k}$ will be occupied in the ground states. The confining surface is known as the Fermi-surface and the limiting energy measured from the bottom of the energy band is called the Fermi-energy (ϵ_F). In the free or nearly free electron models the Fermi-surface is a sphere defined by the relation $\frac{\hbar^2 k^2}{2m_0} \leq \epsilon_F$. We can classify solids from the point of view of the structure of their energy bands and position of the Fermi level which largely determine their gross electrical transport properties.

5.4.1 Electrical Conductors: Partially Filled Valence Band

The most important feature of conducting solids (metals, for example) is the presence of partially filled energy bands in the ground state (see Figure 5.7a). Why are such solids characterized by an ability to conduct electricity even when a small electric field is applied? A qualitative explanation may be offered at this point in the following manner:

Let us assume that the electrons in the partially filled band form pairs of oppositely moving electrons. In each pair, the two electrons move with the same speed but the direction of the motion is opposite. The average current in such a solid will be equal to zero as the current flowing in opposite directions quench each other completely. This statistical equilibrium may be easily broken in solids with partially filled energy bands, for example, by the application of a weak electric field. One of the free electrons may then go over into the nearest unoccupied energy level. The velocity compensation is now lost (average velocity is no-longer zero) leading to the appearance of electric current. It turns out that near the Fermi level the energy levels are rather closely spaced and application of even the smallest electric field produces electrical current. The metals are characterized by the pattern of filling of energy bands described here and are therefore electrically conducting. The situation becomes quite different if the sold is characterized by the presence of a completely filled principal or the valence band (band below the energy gap).

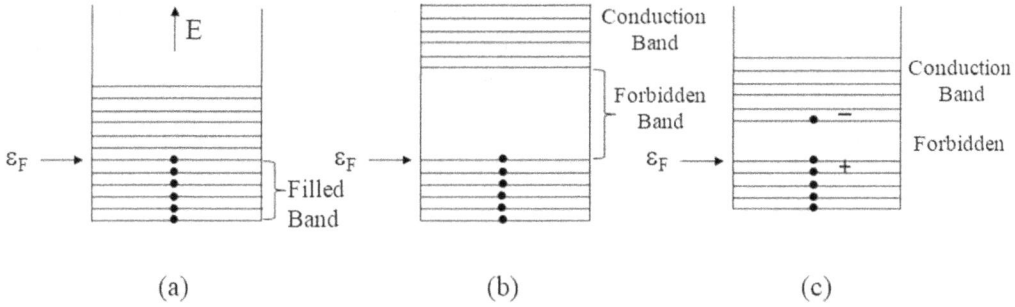

FIGURE 5.7 (a) Metallic energy band. (b) Energy band in insulator. (c) Semiconducting energy band.

5.4.2 Insulators: Completely Filled Valence Band

Solids having a completely filled valence band present a very different situation. Since the unoccupied states of the next band is separated from the valence or the principal band by a sizable forbidden energy gap, only the application of very large electric field can provide the energy required to overcome the gap energy, and produce a current. Such solids are called insulators (Figure 5.7b) and constitute a class of technologically important materials. Although the band gap varies widely from one solid to another each with a completely filled valence band, they all behave as insulators at 0 K. At room temperature, some of these solids with low band gap may conduct electricity. Such solids are called semiconductors.

5.4.3 Semiconductors

The size of the band gap is an important factor in determining whether the solid will behave like an insulator or a semiconductor. A completely covalently bonded solid like 'diamond' has a completely filled valence band as well as a large band gap (≈ 6.5 eV). Naturally, diamond responds like an insulator not just at 0 K, but also at room temperature or even at higher temperatures. In contrast, germanium has a completely filled valence band, but only a low band gap (≈ 0.7 eV) separates the free bands from the filled band. Thermal fluctuations can therefore push a sizable number of electrons into the free bands even at room temperature and germanium crystals behave as conductors at room temperature. At 0 K, however, thermal fluctuations die out completely, and germanium responds like an insulator. All semiconductors share this feature – they have zero electrical conductivity at $T = 0$ K, which rises significantly as the temperature increases.

The emergence of conductivity of semiconductors is, however, a more complex phenomenon and requires further elaboration. It turns out that conductivity of a semiconductor can be '*intrinsic*' – a property of the pure material. Presence of traces of suitable impurities can profoundly affect the response of a semiconductor to electric field. The dramatic rise in conductivity of semiconductors due to the presence of traces of certain impurities has made semiconductors technologically very important. This impurity effect has become known as 'impurity conductivity' as opposed to the 'intrinsic conductivity' of the semiconductor.[1-4]

The intrinsic conductivity of semiconductors arises as a result of excitation of an electron from the filled-lower band across the band gap, into the free higher band (see Figure 5.7c). The excitation of an electron leaves a '*vacancy*' in the filled band, which can be interpreted as a '*hole*' (carrying positive charge), the '*hole*' motion being interpreted as the motion of a positively charged particle. The intrinsic conductivity of a pure semiconductor may therefore be thought of as involving the motion of electrons in the upper band producing electron-conductivity and motion of the holes in the lower almost filled band giving rise to what has become known as the hole conductivity. The impurity atoms injected into a pure semiconductors can work in one of the two ways depending on its nature, to enhance conductivity of the semiconductors. The impurity atoms (if electron rich) can donate their valence-shell electron to the free or unoccupied states in the conduction band. Such electron-rich impurity atoms are called donors.

Semiconductors doped with donor impurity atoms are called n-type semiconductors. It is the movement of electrons in the unfilled conduction band that is responsible for the electrical conductivity of n-type semiconductors. On the other hand, if the impurity atom is electron-deficient, it can accept electrons from the lower filled band, creating '*holes*' in the nearly filled lower energy band. Such impurity atoms are called acceptors and the semiconductors doped with acceptors are called p-type semiconductors. The motion of holes in the nearly filled valence band shapes the electrical conductivity of the p-type semiconductors.

5.4.4 Effective Mass (m^*)

The motion of an electron in the conduction band leads naturally to the concept of effective mass (m^*). The motion of an electron in a crystal is associated with a total energy that can not be expressed in terms of its potential and kinetic components as simply as we do it for the free motion of an electron. We may try to understand the issue with reference to electron motion in a one-dimensional crystal. Expanding the energy $E(k)$ in a Taylor's series in the neighborhood of the point $k = k_0$, where $E(k)$ has an extremum we have

$$E(k) = E(k_0) + \frac{1}{2}(k - k_0)^2 \frac{\partial^2 E}{\partial k_0^2} + \cdots$$

$$= E(k_0) + \frac{\hbar^2}{2m^*}(k - k_0)^2 + \cdots \tag{5.58}$$

where m^*, the effective mass is equal to

$$m^* = \frac{\hbar^2}{\frac{\partial^2 E}{\partial k_0^2}} \tag{5.59}$$

At the present level of approximation, the electrons in the allowed energy bands therefore appear to move as particles of effective mass m^* given in Eq. 5.59. The effective mass of the electron may not only differ from its actual mass in magnitude, but also in sign.[4]

If $\frac{\partial^2 E}{\partial k_0^2}$ is positive, as is the case when there are only a few electrons in the band and k_0 is the point of minimum of $E(k)$, the electron will reside in states in the vicinity of the bottom of the band. The electronic conductivity in this case is characterized by positive effective mass of the electron. If the energy bands are nearly filled completely with electrons, k_0 will correspond to the maximum energy and $\frac{\partial^2 E}{\partial k_0^2}$ will be negative. In the vicinity of the upper band edge, the conducting electron will therefore behave as a charged particle with a negative effective mass. It can be shown that it is equivalent to the motion of a +vely charged particle with effective mass

$$m^*_{hole} = -m^* > 0 \tag{5.60}$$

The conductivity due to the motion of electrons with negative effective mass is therefore characterized as hole conductivity.

Let us recall here that when an electron from the filled band goes over into the conduction band, it leaves a vacancy, called a 'hole' in the lower energy band with a positive charge. The excited electron interacts with the 'hole' forming an electron-hole pair in which the oppositely charged particles may be assumed to be rotating about each other. Such a coupled electron-hole pair is known as an **exciton**.[1-4]

5.4.5 Lattice Vibrations, Phonons and Electrical Conductivity

While considering the motion of electrons in solids, we have so far adopted the simple picture in which electrons move in a periodic potential set up by the regular arrangement of atoms or ions in a lattice. The atoms or ions forming the crystal have been assumed to be at rest. In practice, the atoms or ions forming the lattice actually vibrate about their equilibrium positions. These vibrations play a significant role in shaping many physical properties of the solids like electrical resistance, specific heat, etc. and must be incorporated in the theoretical description of such properties. It is difficult to describe the motion of atoms in the crystal in detail without a detailed knowledge of the structural peculiarities of the crystal. A simplified model that assumes the motion of the atoms to be harmonic vibration about their equilibrium positions and considers only the long wavelength (low frequency) vibrations may, however be quite useful and can be developed quite easily. For this purpose, let us assume that the long wavelength vibrations are acoustic vibrations of a solid, which arise from the collective motion of the atoms of which the solid is made. Such motions can be treated as a strain-wave propagating in the solid. There is thus no need to treat the motion of each individual atom. Before proceeding further, two additional simplifications can be made:

(i) Each unit cell has only one atom.
(ii) Only one-dimensional motion is possible.

The energy of lattice vibrations can now be presented in the form

$$H = \sum_n \frac{M}{2} \dot{D}_n^2 + \frac{1}{2} \sum_n \sum_m C_m D_n D_{n+m} \tag{5.61}$$

where D_n represents the **displacement** of an atom in the nth cell from its equilibrium position, $\dot{D}_n$ is the velocity and M is the atomic mass. Note that the coefficients are required to satisfy the condition[4]

$$C_m = C_{-m} \tag{5.62}$$

and the second term in H represents interaction between atoms. The forms of interaction between atoms in our model depends only on the distance between the unit cells in which the pair of interacting atoms reside. That is why the condition in Eq. 5.62 has been imposed.

The **classical equation** of motion for the nth cell that takes into account the condition in Eq. 5.62 and the energy expression in Eq. 5.61 reads

$$M\ddot{D}_n = -\sum_m C_m D_{n+m} \tag{5.63}$$

The solution of the equation above can be written out in the form of a Fourier series:[3]

$$D_n = \frac{1}{\sqrt{N}} \sum_q (D_q e^{i\vec{q}\cdot\vec{n}} + D_q^* e^{-i\vec{q}\cdot\vec{n}}) \tag{5.64}$$

In Eq. 5.64, N is the total number of unit cells in the crystal under consideration and the summation runs over all values of the wave vector $\vec{q}$ lying in the interval $-\pi \le \vec{q}\cdot\vec{a}_i \le \pi$, $i = 1,2,3$. Clearly then, q lies within the cell of the inverse lattice (see Eq. 5.13). The expansion coefficients D_qs depend on time as described below:

$$D_q(t) = D_q^0 e^{-i\omega_q t} \tag{5.65}$$

where ω_qs are the vibrational frequencies that can be obtained from the relation

$$M\omega_q^2 = \sum_m C_m e^{i\vec{q}\cdot\vec{m}} \tag{5.66}$$
$$= C_q$$

which results from the use of Eq. 5.65 in Eq. 5.63, $\{C_q\}$ being the Fourier transformation of the coefficient $\{C_m\}$.

The kinetic energy term of the crystal, T can be easily expressed in the following form by transforming the $\frac{1}{2}\sum_m M\dot{D}_n^2$ term in the expression of energy in Eq. 5.61 with the help of equations Eq. 5.64 and Eq. 5.65 and the relation $\sum_n e^{i(q+q')n} = N\delta_{q,-q'}$

$$T = \sum_q \frac{1}{2} M\omega_q^2 \left(D_q D_q^* + D_q^* D_q - D_q D_{-q} - D_q^* D_{-q}^* \right) \tag{5.67}$$

The potential energy 'v' can also be similarly reduced to the form[3-4]

$$V = \sum_q \frac{1}{2} M\omega_q^2 \left(D_q D_q^* + D_q^* D_q + D_q D_{-q} + D_q^* D_{-q}^* \right) \tag{5.68}$$

The Hamiltonian for the vibrational energy then takes the simple quadratic form

$$H = T + V = \sum_q \frac{1}{2} M\omega_q^2 \left(D_q D_q^* + D_q^* D_q \right) \tag{5.69}$$

Now we can work out a quantum description of the crystal vibration[4] by replacing H of Eq. 5.69 by its quantum analog – the Hamiltonian operator in which D_q and D_q^* have been replaced by $\hat{D}_q$ and $\hat{D}_q^\dagger$. The quantum mechanical equation of motion can then be written in the Heisenberg form

$$\frac{dD_q}{dt} = \frac{i}{\hbar}[\hat{H},\hat{D}_q] = -i\omega_q D_q \tag{5.70}$$

For Eq. 5.70 to hold the operators $\hat{D}_q$ and $\hat{D}_q^\dagger$ must obey the following commutation relations:[3-4]

$$[\hat{D}_q, \hat{D}_{q'}^\dagger] = \frac{\hbar}{M\omega_q} \delta_{q',q}$$

$$[\hat{D}_q, \hat{D}_{q'}] = [\hat{D}_q^\dagger, \hat{D}_{q'}^\dagger] = 0 \tag{5.71}$$

Before proceeding further, we introduce two new operators replacing $\hat{D}_q$ and $\hat{D}_q^\dagger$ by $\hat{a}_q$ and $\hat{a}_q^\dagger$, respectively where

$$\hat{a}_q = \sqrt{\frac{M\omega_q}{\hbar}} \hat{D}_q$$

$$\hat{a}_q^\dagger = \sqrt{\frac{M\omega_q}{\hbar}} \hat{D}_q^\dagger \tag{5.72}$$

These new operators obey the following commutating relation:

$$[\hat{a}_q, \hat{a}_{q'}^\dagger] = \delta_{q,q'} \tag{5.73}$$

Using the definitions for a_q and $a_q^\dagger$ and their commutation properties (Eq. 5.72, Eq. 5.73), the Hamiltonian of our system can now be expressed easily in the following form

$$\hat{H} = \sum_q \left(a_q^\dagger a_q + \frac{1}{2} \right) \hbar\omega_q \tag{5.74}$$

It can be easily shown that the quadratic operator combinations, $a_q^\dagger a_q$ are diagonal with eigenvalues $n_q = 0, 1, 2, \cdots$. The operator therefore behaves as the number operator ($\hat{N} = a_q^\dagger a_q$). The energy eigenvalues of the system therefore turns out to be

$$E = \sum_q \left(a_q^\dagger a_q + \frac{1}{2} \right) \hbar\omega_q$$

$$= \sum_q \left(n_q + \frac{1}{2} \right) \hbar\omega_q, \quad n_q = 0, 1, 2, \ldots \tag{5.75}$$

The number n_q can be interpreted as the number of elementary excitations of the crystal or quasi-particles each having energy equal to $\hbar\omega_q$. They are called 'phonons' – the quasi-particles associated with acoustic vibrations of the crystal. The sum over q in Eq. 5.75 indicates that the vibrational energy of the crystal can be expressed as the total energy of all 'phonons' occupying states with energy $\hbar\omega_q$. The operator

$a_q^\dagger$ increases the number of phonons of energy $\hbar\omega_q$ and quasi-momentum $\hbar q$ from n_q to $n_q + 1$ while a_q decreases the number from n_q to $n_q - 1$. $a_q^\dagger$ and a_q can therefore interpreted as the creation and destruction operators for phonons, respectively. We can extend the present model for quantization of crystal vibrational energy to cover cases where more than one atoms are present in the unit cell and the atoms can vibrate in three mutually orthogonal directions (x, y, z). If the number of atoms in each cell is m, the vibrational degrees of freedom become $3m$. The total vibrational energy of the crystal can be written as (by extending the expression of Eq. 5.75)

$$E = \sum_q \sum_\alpha \left(n_{q,\alpha} + \frac{1}{2}\right)\hbar\omega_{q,\alpha} \tag{5.76}$$

where the summation now is not only over q but also over α with α ranging from 1 to $3m$. Phonons play an important role in the emergence of electrical resistance of materials like metals and quite paradoxically, in the emergence of electrical superconductivity or disappearance of electrical resistance of metals at very low temperatures. These phenomena appear to originate from the interaction of the conduction electrons with lattice vibrations, which is described as electron-phonon interactions in the quantum theory of conductivity and super conductivity, we will take up these aspects in greater detail in Chapter 8 where properties of metals and semiconductors will be examined within the framework of quantum and statistical mechanics.

5.5 Symmetry and Splitting of Bands

Let us now go back to Eq. 5.11 for considering the effects of the ($\vec{k}$) vector on the symmetry of the wave vectors in crystalline solids. After some rearrangements, we get the following pseudo Schrödinger equation for the radial part of the wavefunction.

$$-\frac{\hbar^2}{2m}\nabla^2 u + \left[V(\vec{r}) + \frac{\hbar}{m}\vec{k}\cdot\vec{p}\right] = E'u \tag{5.77}$$

With $E' = E - \frac{k^2\hbar^2}{2m}$. The change from E to E' means that energy is now being measured relative to the kinetic energy of a free electron with the same '$\vec{k}$' value. The Hamiltonian of the pseudo-Schrödinger equation contains the usual kinetic energy component $(\nabla^2 u)$, the periodic potential $V(\vec{r})$. However, it contains an additional term proportional to $\vec{k}\cdot\vec{p}$, which may be viewed as providing a momentum-dependent potential. The periodic potential $V(\vec{r})$ is invariant under all operations of the point group. The $\vec{k}\cdot\vec{p}$ term, on the other hand is invariant only under the operations of the group of wave vector at $\vec{k}$. Thus the relevant symmetry group for analyzing the symmetry of the eigenfunctions of the pseudo-Hamiltonian is that part of the point group, which is also present in the group of the wave vector at $\vec{k}$.[5]

In a simple cubic crystal, the point group is O_h and at $\vec{k} = 0$, we can make use of the full O_h group to classify the Bloch states according to symmetry. It is possible as all the operations under O_h will leave $\vec{k} = 0$ unchanged. However, if we take a non-zero wave vector along k_x, for example $(k_x \neq 0)$, the symmetry is reduced to C_{4v} (the tetragonal group). Note that k_x is left invariant under all the operations of the C_{4v} group. This reduction of symmetry also reduces the degeneracies of the eigenfunctions. Thus, C_{4v} has four 1-d and two 2-d irreducible representations, thereby restricting the maximum level of degeneracy to two. The three-fold degenerate p-like states (T_{1u}) at $\vec{k} = 0$ under O_h therefore splits into an A_1 and E representations, just as what happens in the case the local or the site symmetry of an atom or an ion in a crystalline solid is lowered from an octahedral to a tetragonal point symmetry. This feature is formally described by saying that the symmetry labels A_1 and E of the C_{4v} group for wavefunctions with $\vec{k}$ lying on the k_x axis is compatible with the symmetry label T_{1u} of the O_h group for a wavefunction at $k = 0$. The idea of compatibility and reduction in symmetry is the same as encountered in crystal field calculation where successive reduction in degeneracy is carried out by adding fields of progressively lower and lower symmetry to a Hamiltonian of higher symmetry. The lower symmetry fields act as perturbation to the higher symmetry Hamiltonian. In such cases, the original high symmetry eigenfunctions still serve as the basis functions for representing the new smaller group. We may say that the representation Γ_k of the smaller group (sub group) is compatible with the representation Γ_l of the larger (original) group if the basis for Γ_k is also included in the basis for Γ_l. We can pursue analysis along these lines and work out the mutually compatible symmetry labels of the wavefunction at all special symmetry points in the Brillouin zone, say in the case of cubic crystalline solids or crystalline solids of lower symmetry.[5]

5.6 Amorphous Solids and Localized Electronic States

Amorphous or non-crystalline solids are an important class of materials that find important applications in many walks of life. The amorphous solids are formally defined as solids having only short range, but no long range order or having disorder in the longer ranges. By 'order' we mean to convey both long range translational and orientational order. These solids therefore lack the periodicity of the crystalline structure. The consequence is that it is not possible to generate the structure or predicting the location of atoms or molecules making up the amorphous solid by means of a lattice vector $\vec{R_n} = n_1 \vec{a_1} + n_2 \vec{a_2} + n_3 \vec{a_3}$ with integer n_1, n_2, n_3; $\vec{a_1}, \vec{a_2}, \vec{a_3}$ being three no-coplanar lattice vectors as we could do for crystalline solids. The ideas of unit cells and translational invariance that were so useful in the development of theories for understanding electronic states in crystalline solids largely lose significance when it comes to understanding the nature of electronic states in amorphous solids or materials and describing their properties in terms of these electronic states. This is one way of viewing amorphous solids: they are solids with built in disorder at certain length scales and some residual order at much shorter length scales. Now liquids too, have only short range order, but no order in the longer ranges and it seems natural to try to understand amorphous solids starting from a model of a liquid. An important attribute of a liquid is its ability to flow. The actual process by which an atom or molecule moves through a liquid is a rather complex cooperative phenomenon in which several neighboring molecules take part – their chance motions constructively combining to allow a molecule to move past other molecules. The cell model of a liquid proceed by drastically simplifying the picture and producing a near universal model of liquids. The simplification involves assuming that each molecule in a liquid is confined by its neighbors into a 'cell' the physical dimension of which is of the order of atomic dimension. Within the confinement of its cell a cage-like space is created by the neighboring molecules within which each molecule oscillates. The physical transport of a molecule through the mass of the liquid then implies movement from one cell to another; but that requires an energy expenditure, which can only be provided by the thermal energy of neighboring molecule in a complex coordinated process involving chance motions. The actual potential energy surface on which the movement occurs is a very complex dynamic surface and the motion involves many body interactions and correlated movement. The cell model avoids these complexities[7] and assumes that only one molecule – a representative average molecule – moves on an averaged out potential energy surface seen by it due to all other molecules (Figure 5.8).

The simplification is no doubt drastic; but the model is still able to capture essential features of the process – the molecules vibrating about their mean positions for a while before moving away from their neighbors. The probability of that happening may be quite low. In fact, the probability of escape from the cage depends on temperature, molecular shape and how strong the interactions with neighbors are. Let us imagine now that the molecules are all completely trapped in their respective 'cells' in which they vibrate

FIGURE 5.8   The cell model potential.

about fixed positions and ΔE (Figure 5.8) is > kT. The model describes a solid that lacks crystalline order. It rather resembles a highly viscous liquid at a low temperature. The model also describes amorphous solids. Thus we can try to understand amorphous materials either in terms of a crystalline solid with disorder or as a highly viscous liquid without dynamic disorder. We must note here that many amorphous solids do not have the corresponding crystalline form with respect to which the disorder can be defined. Metals normally form crystalline solids, so also do some ceramic materials. Others like inorganic glasses are amorphous. Although the term 'glassy materials' has the same structural meaning that amorphous materials have, one usually assumes in addition that glassy materials display a characteristic glass transition temperature. Plastics or polymers are generally regarded as completely non-crystalline, although they can also be semi-crystalline. The formation of a crystalline solid on cooling a liquid is normally dictated by the relative ease of forming an ordered state from a disordered one. If the molecules are large and have rather complex structures, the transition to the ordered state meets with some difficulty as a lot of rearrangement may be required. Thus rapid cooling through the freezing temperature generally favors formation of amorphous or semi-crystalline materials, as not enough time is available for the molecules or atoms to organize themselves into a perfectly or near perfectly ordered state. That leads to the formation of a solid with some 'quenched in randomness or disorder'. A question of great practical importance comes up at this point: how do we experimentally determine whether a given material is amorphous or not? Several possibilities are there:

  (i) use calorimetry to determine the glass transition temperature and the associated thermodynamic quantities;
  (ii) check if X-ray studies give rise to the so-called 'amorphous halo';
  (iii) determine bond correlation from NMR data.

Even with these advanced analytical techniques, it is difficult to confirm if a substance is microcrystalline or amorphous. Amorphous forms that exhibit different microstructures and stabilities are not generally considered to be polymorphic. The crystal structure refers to a thermodynamic state and any microstructure is regarded as a kinetic modification. It can display different physical and chemical properties (e.g. solubility) but is not a new polymorph. The amorphous solids lack periodicity that characterizes crystalline solids and their characteristic geometry. The absence of long range order leads to the absence of elasticity, sharp melting point and other properties of crystalline materials. The properties of an amorphous substance are identical along all axes, and such solids are not characterized by a sharp melting point. They break up to form curved or irregular shapes unlike crystalline solids, which produce flat faces on being cleaved or sheared. The crystalline and amorphous states of matter therefore define two extreme on a continuum. Amorphous solids do however, retain some short range order in their structures and include both natural and man-made materials. Metallic glass is the most widely known amorphous inorganic materials that find wide use. Plastics, polymers, rubber, gel etc. are some of the well-known and useful organic amorphous materials. A question that arises naturally at this point concerns the nature of electronic states in amorphous solids. The presence of 'disorder' in amorphous solids critically affects the nature of electronic states, which may not be extended as expected but are actually localized.

5.6.1 Localization in Disordered Solids

P. W. Anderson theoretically considered the following scenario: suppose that a conducting crystalline solid (metal) is doped with some impurity atoms randomly. With the introduction of impurity centers, the regular periodic structure of the crystal potential is disrupted producing disorder, and more the disorder is, the more are the electronic states affected. The question that comes up now is: does a level of disorder exist where a spectacular change in the metallic properties takes place? If it does, what could be the reason.

In a seminal paper in 1958, P. W. Anderson conjectured that if the disorder is strong enough,[8] localization of states will occur no matter what the dimensionality of the solid is and once that happens the material will undergo metal-insulator phase transition as the critical strength of disorder is reached. It turned out that the abrupt loss of conductance is due to the localization of the electronic wavefunction, which is no longer extended all over the lattice, but is spatially localized in specific regions of the lattice. Anderson won the Nobel Prize in Physics in 1977 for this remarkable discovery that became known as Anderson localization. The phenomenon appeared to be quite intriguing at that time.

Anderson's localization phenomenon can be interpreted as the outcome of an interference effect among matter waves. Thus, in the tight-binding model used by Anderson, the electrons are able to tunnel between adjacent or neighboring sites on the lattice until the disorder becomes strong enough to produce large scale cancellation among the amplitudes of 'tunneling paths' thereby producing localization. An alternative model suggests that the incoming wave is scattered off the potential that disorder generates. At a critical level of disorder, the scattered waves interfere destructively forcing the wave to decay exponentially.[9] Hence the state becomes localized in a specific region of the lattice. It continues to remain an area of active interest of researchers.[10]

Let us consider a tight-binding model Hamiltonian with nearest-neighbor hopping (t = hopping strength) and randomly distributed on-site energies $\{\epsilon_n\}$ with a distribution width equal to 'W': ($-\frac{1}{2} \leq \epsilon_n \leq \frac{1}{2}$). The Hamiltonian reads ($<nm>$ indicates nearest neighbors and $h \cdot c$ hermitian conjugate)

$$\hat{H} = W \sum_n \epsilon_n a_n^\dagger a_n + t \left(\sum_{<nm>} a_n^\dagger a_m + h \cdot c \right) \qquad (5.78)$$

Let us set $t = 1$, measure all energies in unit of 't', and consider two specific cases:

Case 1: In this case, a chain of length 'L', with lattice spacing equal to 'a' is considered. In the limit $W = 0$, there is complete order and the eigenfunctions of H are the Bloch states with energy dispersion given by $E(k) = 2t \cos ka$. They are the normal delocalized states.

Case 2: In this case, $W \neq 0$ and finite. It can be shown that the solution of the Schrödinger equation $H|\psi\rangle = E|\psi\rangle$ with $|\psi\rangle = \sum_n \psi(n) a_n^\dagger |0\rangle$ can be obtained from the following recursion relation (Eq. 5.79):

$$E\psi(n) = \epsilon_n \psi(n) + \psi(n-1) + \psi(n+1) \qquad (5.79)$$

$\psi(n)$ being the probability amplitude at the nth lattice site. The problem can also be solved by transfer matrix method. Either way it can be and has been established that $\psi(n)$ decays exponentially ($\psi(n) \to 0$ as $n \to \infty$). The shape of the wavefunctions has the form $\psi(n) = \exp(-n/\xi)$, where ξ is a measure of the localization length in units of the lattice spacing 'a'. The eigenstates for $W \neq 0$ are therefore not extended over the whole lattice space, but are exponentially localized over specific regions of the lattice. The most important finding is: for non-zero values of 'W', all states in 1D are exponentially localized irrespective of the actual magnitude of 'W' – i.e. even for the weakest strength of disorder in the on-site energy distribution, localization takes place in 1D materials. Thus, a 1D material will lose its conduction or diffusion properties if the system size 'L' $>> \xi$ (localization length). Numerical experiments tend to predict emphatically that $\xi \propto W^{-2}$ where W is the width of the on-site energy distribution. The analysis has been extended to two-dimensional systems by invoking scaling argument and renormalization group approach. It turns out that the states in two dimensions are also localized in the presence of disorder.

The same idea has been pursued in 3D as well, albeit numerically. The 3D Hamiltonian has been formed on a cubic lattice of N^3 points and diagonalized exactly. It turns out, only for $d \geq 3$, a second order (continuous) phase transition from metal to insulator takes place once a critical strength of disorder is reached. Anderson localization has also been observed experimentally in a variety of classical and quantum disordered materials leading to modulation of light, acoustic and finally matter waves. Recent observation of Anderson localization in matter waves of 1D Bose-Einstein condensate of rubidium atoms confirms experimentally the occurrence of the phenomenon of Anderson localization.[11–12] There is no doubt that it occurs in two and three dimensions as well; but the direct experimental observation of the localization of the wavefunction has not so far been possible in 2D or 3D materials.

REFERENCES

1. C. Kittel, *Introduction to Solid State Physics*, Third Edition, John Wiley, New York (1966).
2. W. Harrison, *Solid State Theory*, New York (1970).
3. J. Ziman, *Principles of Theory of Solids*, Second Edition, Cambridge (1972).
4. A. A. Sokolov, I. M. Ternov and V. Ch. Zhukovskii, *Quantum Mechanics*, (Translated from Russian by Ram S. Wadhwa), MIR Publishers, Moscow (1984).
5. M. Tinkham, *Group Theory and Quantum Mechanics*, TMH Edition, New York (1974).
6. E. Huckel, 'Quantentheoretische Beitrage zum Benzolproblem', *Z. Phys.*, 70, 204 (1931).
7. (a) B. J. Adair, 'The cell theory of liquids', *Proc. Roy. Soc. London A*, 230, 390–398 (1951). (b) O. J. Heilnann et al., 'A cell model for liquids', *J. Phys. C: Solid State Physics*, 6 (22) , L403 (1973).
8. P. W. Anderson, 'Absence of diffusion in certain random lattices', *Phys. Rev.*, 109, 1492 (1958).
9. A. Aspect and M. Inguscio, 'Anderson localization of ultracold atoms', *Phys. Today*, 62, 30 (2009).
10. P. W. Anderson, In *50 Years of Anderson Localization*, Elihu Abraham, Ed., World Scientific Publishing, Singapore (2010).
11. D. Clement et al., 'Suppression of transport of an interacting elongated Bose-Einstein condensate in a random potential', *Phys. Rev. Lett.*, 95, 170409 (2005).
12. C. Fort et al., 'Effect of optical disorder and single defects on the expansion of a Bose-Einstein condensate in a one-dimensional waveguide', *Phys. Rev. Lett.*, 95, 170410 (2005).

6

Classical Statistical Mechanics

6.1 Introduction

So far, we have considered our systems be the atoms, molecules and solids to be at zero kelvin. That means, we have disregarded thermal effects on our systems. Whether the electrons are distributed among discrete energy levels or energy bands, we have assumed that these distributions are 'optimal', minimizing the system's own energy to the maximum possible extent in each case without considering thermal effects on these distributions. In real life applications, we are forced to bring in a thermal factor in the form of temperature 'T' when a large number of particles constitute our systems in contact with the environment – which can deliver thermal energy to the system or withdraw the same from the system across the system-environment boundaries. The question then crops up: what would be the equilibrium distribution of particles among the available and accessible states in a large many particle system? We can address the issues within the framework of classical physics wherever possible (systems at high temperature and density, and continuum of energy states) or invoke quantum mechanics when the systems are at low temperatures and densities and are characterized by discrete energy levels. In the former case, we are led into the branch of physics called classical statistical mechanics while in the latter we move into the regime of quantum statistical mechanics. In either case, the pivotal role is played by a statistical concept – the probability of a certain event occuring or the mathematical expectation of an event and the associated quantity called 'information' or information entropy. We will take up first the issues of classical statistical mechanics and its formulation.[1-3]

6.2 Types of Probability Distributions

Let us consider N-identical molecules of an ideal gas enclosed in a volume V at a temperature 'T'. The volume V is approximately equal to the molar volume and N is of the order of Avogardro's number. Being ideal the molecules do not interact with each other, or with the vessel walls. There is also no external field of force acting on the molecules. We now mentally mark out a tiny volume 'ΔV' inside the vessel (Figure 6.1) and try to figure out how many molecules could possibly be inside 'ΔV' on average. The molecules are in incessant chaotic motion and move throughout the volume of the vessel. The position of a molecule is in fact a random variable.[1] For large N and V, the question is best addressed by bringing in the concept of probability and rephrasing the question as 'what is the probability that 'n' out of N molecules are found inside ΔV at a certain time?' To answer the question, let us assume that there is only one molecule 'A' inside the vessel. With respect to A and the tiny volume element ΔV, there are only two mutually exclusive events:

   (i) A is inside ΔV;
   (ii) A is outside ΔV.

Let p be the probability of A getting inside ΔV; then $1 - p$ measures the probability that A is outside ΔV. Thus (i) and (ii) constitute a complete set of mutually exclusive events and the sum of their probabilities is $p + (1 - p) = 1$. If we bring in a second molecule the number of possible events increase and are:

DOI: 10.1201/9781003244882-6

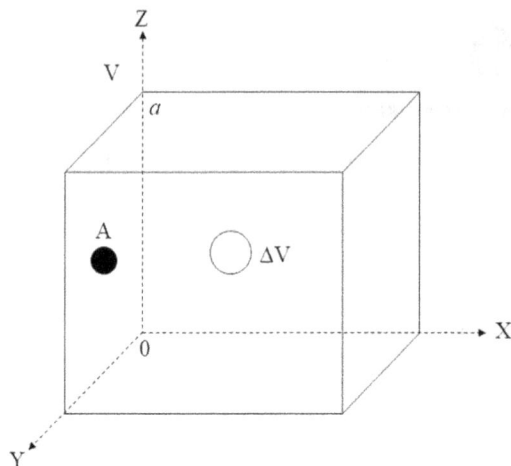

FIGURE 6.1   A small volume element ΔV carved out in a large vessel of volume $V = a^3$. The molecule marked 'A' is free to move inside the vessel.

(i) Both are inside ΔV,

(ii) one (say A) is inside ΔV and the other (say B) is outside and vice-versa;

(iii) both are outside ΔV. The probability of event (i) is p^2, of event (ii) is $2p(1-p)$ and of event (iii) is $(1-p)^2$. The sum of the probabilities of these complete set of mutually exclusive events is once again

$$p^2 + 2p(1-p) + (1-p)^2 = \{p + (1-p)\}^2 = 1$$

For three molecules in the vessel, the number of events increases further:

(i) All the three inside (probability $= p^3$)

(ii) Two inside and one outside (probability $= 3p^2(1-p)$)

(iii) One inside and two outside (probability $= 3p(1-p)^2$)

(iv) Three outside (probability $= (1-p)^3$).

The sum of probabilities of the mutually exclusive set of events (i-iv) is (writing $3 = 3!/2!$)

$$p^3 + \frac{3!p^2(1-p)}{2!1!} + \frac{3!p(1-p)^2}{2!1!} + (1-p)^3 = \{p + (1-p)\}^3 = 1$$

Note that the individual terms on the left belong to the binomial series for $\{p + (1-p)\}^3$. Now we can turn to the question we started with:

The probability that 'n' molecules out of N are found inside the volume ΔV is clearly given by

$$p(n) = \frac{N!}{n!(N-n)!} p^n (1-p)^{N-n} \tag{6.1}$$

which is just the nth term in the binomial expression of $\{p + (1-p)\}^N$. Eq. 6.1 represents what is known as binomial distribution of probabilities. Before using Eq. 6.1, let us consider what could be the value assigned to p? Since each molecule moves inside the volume 'V' randomly and independently, there is no external field of force influencing their motion, and the number density $\rho = N/V$ is the same everywhere (the density is translationally invariant) the probability 'p' of a molecule being inside the volume element ΔV in V must be $p = \Delta V/V$ (if $\Delta V = V$, the probability is clearly $V/V = 1$, as expected).[1]

Let us now take a closer look at $p(n)$ in Eq. 6.1. The argument n of $p(n)$ is not a static or a fixed value; it can take on all possible values of $n = 0, 1, 2, \cdots, N-1, N$ randomly at different instances. n can in fact be treated as a discrete random variable. The average value of n $(=<n>)$ in the volume element ΔV is therefore given by

$$<n>=\sum_{n_j=0}^{N}n_jp(n_j)=\sum_{n_j=0}^{N}n_j\frac{N!}{(N-n_j)!n_j!}p^{n_j}(1-p)^{n_j} \tag{6.2}$$

Let us note that $p=\frac{\Delta V}{V}$, and do some simple manipulations leading to the expression

$$
\begin{aligned}
<n> &= N\frac{\Delta V}{V}\sum_{k=0}^{N-1}\frac{(N-1)!}{k!(N-1-k)!}\left(\frac{\Delta V}{V}\right)^k\left(1-\frac{\Delta V}{V}\right)^{N-1}\\
&= N\frac{\Delta V}{V}\left\{\frac{\Delta V}{V}+1-\frac{\Delta V}{V}\right\}^{N-1}\\
&= N\frac{\Delta V}{V}=N\rho
\end{aligned} \tag{6.3}
$$

Thus the average number of molecules inside the designated volume ΔV is just the product of the total number of molecules in V and the ratio of the designated volume and the total volume (V). If we wish to calculate the mean value of a continuous random variable, we have to replace the summation in Eq. 6.1 by integration. Thus, the z-coordinate of a molecule in our reaction vessel of volume V can take on any value between 0 and a randomly, with probability $p(z)=\frac{1}{a}$. The average value of z is therefore calculated as

$$\int_0^a\frac{1}{a}zdz=\frac{1}{2}a \tag{6.4}$$

The binomial distribution of probability of random events as given in Eq. 6.1 is very general and can be applied to widely different types of problems. It is, however, not convenient to use as factorial of large numbers ($N!$, $N-n!$) must be calculated. That forces us to invoke Stirling's approximation, according to which $lnN!=NlnN-N$. From Eq. 6.3 we have the result that $<n>=\rho N$, while from Eq. 6.1 we have

$$P(n)=\frac{N!}{(N-n)!n!}p^n(1-p)^{N-n}$$

If we allow N to grow infinitely,

$$
\begin{aligned}
\lim_{N\to\infty}P(n) &= \lim_{N\to\infty}\frac{N!}{(N-n)!n!}p^n(1-p)^{N-n}\\
&= \lim_{N\to\infty}\frac{N^n}{n!}p^n(1-p)^N\\
&= \lim_{N\to\infty}\frac{(pN)^n}{n!}\lim_{N\to\infty}\left(1-\frac{Np}{N}\right)^N\\
&= \frac{<n>^n}{n!}\lim_{N\to\infty}\left(1-\frac{<n>}{N}\right)^N\\
&= \frac{<n>^n}{n!}\exp(-<n>).
\end{aligned}
$$

Thus in the limit of very large N and $N-n$, the binomial distribution law passes over into a new distribution law, called the Poisson distribution law:[1]

$$P_n=\frac{<n>^n}{n!}\exp(-<n>) \tag{6.5}$$

which is more convenient to use. Poisson distribution determines the probability of the volume ΔV containing n molecules when mean number of molecules in ΔV is equal to $<n>$ and is correct if $N>><n>^2$.

If we assume now that n is large enough to merit Stirling approximation for $n!$ to be used, we get $ln\,n! = n\,ln\,n - n$. Now taking logarithm of both sides of Eq. 6.5, we get, with the help of Stirling's formula

$$lnP_n = n\,ln <n> - <n> - ln\,n!$$

$$= -\left(<n> + \Delta n\right)ln\left\{1 + \frac{\Delta n}{<n>}\right\} + \Delta n \tag{6.6}$$

where $\Delta n = n - <n>$ is the deviation of n from the mean keeping in mind that $\Delta n <<< n>$, we get from Eq. 6.6 that $lnP_n \approx -\frac{(\Delta n)^2}{2<n>}$, correct to the second order in $\frac{\Delta n}{n}$. Thus, we arrive at a new distribution law – the law of Gaussian distribution for which

$$P_n = A.\exp\left(-\frac{(n - <n>)^2}{2<n>}\right) \tag{6.7}$$

The constant A can be fixed by the normalization condition $\int_{-\infty}^{+\infty} P_n dn = 1$, which yields

$$A = \frac{1}{\sqrt{2\pi <n>}}$$

This new distribution is called the Gaussian distribution law.

Poisson distribution is the limiting case of a binomial distribution when $\langle n \rangle << N$, the dispersion 'D' of a random variable obeying the Poisson distribution can be easily shown to be equal to the mean value of the number n:

$$D = \langle n \rangle \tag{6.8}$$

Using this fact, in Eq. 6.7, we get the standard formula for Gaussian distribution (note $P(n)$ has been replaced by $W(n)$ for convenience later).

$$W(n) = \frac{1}{\sqrt{2\pi D}}e^{-\frac{(n-<n>)^2}{2D}} \tag{6.9}$$

The Gaussian distribution can be generally used whenever the number of molecules in ΔV is quite large. It is important to note that in practice, we are always more interested in knowing what is the probability δW that the number of particles/molecules confined within the interval $n \leftrightarrow n + \delta n$, which is given by the product of $w(n)$ and δn

$$\delta W = w(n) \cdot \delta n$$

$$= \left\{\frac{1}{\sqrt{2\pi D}}e^{-\frac{(n-<n>)^2}{2D}}\right\}\delta n \tag{6.10}$$

In Eq. 6.10, $w(n)$ is the probability density. It tells us that the random variable 'n' (n is very large) can assume a value in the interval $n \leftrightarrow n + \delta n$ with the probability $w(n)$. It would have been meaningless to state that the probability of the random variable assuming the value 'n' is $w(n)$ as all the values of a random variable are equally probable.

Sometimes we may want to determine the probability W that n is found in the interval from n_1 to n_2, which is not small compared to $\sqrt{D}$. In that case $w(n)$ may change in the interval concerned and W is then found by integration (treating n as a continuous random variable)

$$W = \int_{n_1}^{n_2} w(n)dn = \int_{n_1}^{n_2} \frac{1}{\sqrt{2\pi D}}e^{-\frac{(n-<n>)^2}{2D}}dn \tag{6.11}$$

The same procedure can be adopted for determining the mean-value of a function $f(n)$ of the random variable:

$$\langle f(n) \rangle = \int_{l_1}^{l_2} f(n)w(n)dn \tag{6.12}$$

As to the lower and the upper limits l_1 and l_2, let us note the number of particles n can not be less than zero and greater than 'N' by conventional wisdom. However, it turns out that large deviations of 'n' from its mean value $\langle n \rangle$ is highly improbable and therefore we can replace l_1 and l_2 by $-\infty$ and $+\infty$, respectively without affecting the accuracy of the calculated value. That means we can compute $\langle f(n) \rangle$ as

$$\langle f(n) \rangle = \frac{1}{\sqrt{2\pi D}} \int_{-\infty}^{+\infty} f(n) e^{-\frac{(n-\langle n \rangle)^2}{2D}} \, dn \tag{6.13}$$

Let us emphasize here that Gaussian distribution (GD) is obeyed by many continuous random variables. The reason is that for a large number of trials GD is the limiting form for an entire set of distribution. There is the 'central limit theorem' of the probability theory[2] that lays down quite general sufficiency conditions for a limiting distribution to be Gaussian or normal.

Thus, if a continuous random variable (z) obeys a normal distribution, the probability $dw(z)$ that the values of the random variable z is confined within an interval $z \leftrightarrow z + dz$ is

$$dw(z) = \left[\frac{1}{\sqrt{2\pi D}} e^{-\frac{(z-\langle z \rangle)^2}{2D}} \right] dz \tag{6.14}$$

where $\langle z \rangle$ stands for the mean value and D for the dispersion of the continuous random variable (z).

6.2.1 Probability and Unexpectedness: The Entropy

Probability of a random event can be measured by its unexpectedness on a quite general ground. If the probability is low, the event rarely takes place and may be categorized as an unexpected event. On the other hand, if the probability of an event is close to unity (very high) the event is an expected one. The information theory defines the unexpectedness of an event as the natural logarithm of its probability of occurrence w taken with a negative (-) sign, i.e.

$$\text{unexpectedness} = -\ln w \tag{6.15}$$

Unexpectedness is thus a positive quantity as $0 < w \leq 1$. If $w = 1$, the event is sure to happen and the unexpectedness is $\ln 1$, which is zero as expected. If an event consists of two independent but simple events, the unexpectedness of the complex event is the sum of unexpectedness of the two simple events that the complex event can be broken into. This happens because the probability w of the complex event is just the product of the probability w_1 and w_2 of the two simple independent events:

$$\text{unexpectedness} = -\ln w = -\ln (w_1 w_2) = -\ln w_1 - \ln w_2$$

If an experiment confirms that an event has taken place, the unexpectedness of the event becomes zero while before the confirmatory experiment was done the unexpectedness was $-\ln w$, w being the probability of the event taking place. Thus the experiment has reduced the unexpectedness by the amount $-\ln w$, which can therefore be also used as a measure of information obtained in an experiment confirming the event.

Let us now consider a system characterized by a complete set of mutually exclusive events. If each of these events has their own unexpectedness, could there be a single quantity, a common characteristic for the system as a whole based on unexpectedness of individual events? It turns out that the mean value of the unexpectedness of the whole system can be such a quantity. Let w_i be the probability of the ith event. Then $-\ln w_i$ is the unexpectedness of the ith event. The mean value of the unexpectedness that characterize the entire system is therefore given by $\sum_{i=1}^{N} w_i(-\ln w_i)$ and is called the information entropy or simply entropy (S) of the system. Thus the mathematical expression for entropy (dimensionless)

$$S = -\sum_{i=1}^{N} w_i \ln w_i \tag{6.16}$$

where the sum extends over all the N members of the complete set of mutually exclusive events. The entropy 'S', as defined in Eq. 6.16 measures the average unexpectedness or the indeterminacy or the randomness of events occuring in system under the given situation. To understand this information theoretic idea of entropy, let us turn back to the system in Figure 6.1. If the designated volume ΔV is very small compared to the total volume V, the molecule 'A' will almost certainly be found outside ΔV making the indeterminacy very small and 'S' defined in Eq. 6.16 precisely conveys this property. Thus the probability w_1 that 'A' is inside ΔV is $w_1 = \frac{\Delta V}{V}$ and the probability w_2 that 'A' is outside ΔV is $w_2 = 1 - w_1 = 1 - \frac{\Delta V}{V}$. The entropy '$S$' is therefore given by the following expression:

$$S = -\frac{\Delta V}{V} \ln \frac{\Delta V}{V} - \left(1 - \frac{\Delta V}{V}\right) \ln\left(1 - \frac{\Delta V}{V}\right)$$

$$= -w_1 \ln w_1 - w_2 \ln w_2$$

(6.17)

where w_1 is the probability of 'A' getting into ΔV and w_2 is the probability of 'A', not getting into ΔV.

If $\frac{\Delta V}{V}$ is very small, the first factor in the first term in Eq. 6.17 tends to 0 while the second factor in the same term tends to $(-)\infty$. Note however, that $-\ln\frac{\Delta V}{V}$ grow very slowly as $\frac{\Delta V}{V}$ decreases so that the product tends to zero as $\frac{\Delta V}{V}$ is reduced to zero. This fact can be rigorously proven by invoking L' Hospital's rule. The second term in Eq. 6.17 behaves similarly as the first factor tends to become unity, and the second factor approaches zero as $\frac{\Delta V}{V}$ diminishes and becomes zero. Therefore, at a small value of $\frac{\Delta V}{V}$, the value of entropy defined in Eq. 6.16, would certainly be small and practically close to zero. If ΔV is large and is close to V, the molecule is almost certain to be inside the designated volume ΔV, leading to a very small indeterminacy in the system under consideration and therefore to a rather small value of entropy.

Since the entropy of the system is found to be small for both when $\frac{\Delta V}{V}$ is small (≈ 0) and $\frac{\Delta V}{V}$ is large ($\approx$ 1) the question arises: when does 'S' attain a maximum value as $\frac{\Delta V}{V}$ varies from zero to 1. The answer can be provided by differentiating both sides of Eq. 6.17 with respect to $\frac{\Delta V}{V} = w_1$ and setting the result equal to zero. The procedure leads to the result that S is maximum when $w_1 = \frac{\Delta V}{V} = \frac{1}{2}$. We conclude therefore that if the designated volume ΔV is half the total volume, the system is maximally indeterminate and the information entropy is also maximum under the given conditions. We will make use of these ideas later in our efforts to find the equilibrium state and equilibrium distribution functions.

6.3 The Equilibrium State and Distribution Functions

We have seen that the average number of molecules in the volume element ΔV in an enclosed mass of gas (N = number of molecules) at a given temperature T and volume V is

$$\langle n \rangle = N \frac{\Delta V}{V}$$

which does not depend on where ΔV is located inside the volume 'V'. $\langle n \rangle$ is therefore an invariant quantity under the condition that no external force acts on the molecules, which themselves are also mutually non-interacting. In other words, the gas molecules have the same number density everywhere within the volume V. They are uniformly distributed in 'space'. Let us note, however that constancy of $\langle n \rangle$ throughout the body of the gas does not mean that the volume element ΔV has the same number of molecules at any and every instant. Similarly, in any other non-coincident but equal volume element $\Delta V'$, the actual number of molecules at any time may be different from $\langle n \rangle$ – but these fluctuations – or deviations from the mean can be shown to be rather small and disregarded. The molecules or particles can thus be assumed to be distributed uniformly in space when the mass of gas is in a state of equilibrium as evidenced by the non-changing states of its macroscopic attributes like volume and temperature (see later). The molecules in the gas move with diverse velocities ($\vec{v_i} = \frac{\vec{p_i}}{m}$), and the question that naturally comes up concerns the probability of molecules having different values of it. The answer can be sought by treating the momentum of molecules inside the vessel as a continuous random variable characterized by an appropriate momentum probability density ω_p. To make further progress we will bring in the notion of momentum space. Let us assume that the momentum components of a molecule are laid out along the axes of the cartesian coordinate system (see Figure 6.2).

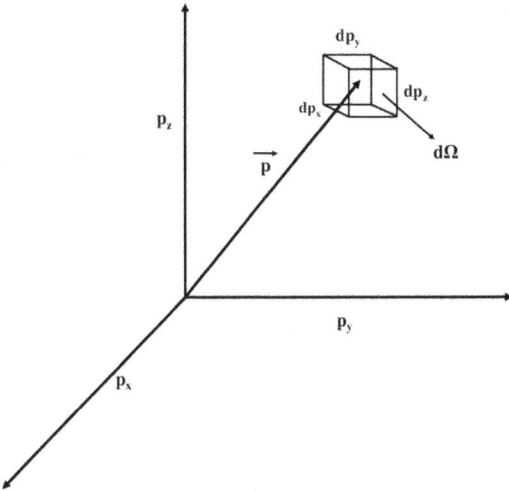

FIGURE 6.2 A small volume element ($d\Omega$) in the momentum space with components of momentum of a molecule laid out along the axes of a cartesian system of coordinates.

6.3.1 Maxwell's Distribution

The probability that the three components of momentum of a particular molecule are confined within the ranges from $p_x \leftrightarrow p_x + dp_x$, $p_y \leftrightarrow p_y + dp_y$ and $p_z \leftrightarrow p_z + dp_z$, can be formulated graphically with the aid of our Figure 6.2. The restrictions on the momentum components imposed by us mean that the tip of the momentum vector ($\vec{p}$) drawn from the origin of the coordinate system must be inside the elementary rectangular volume $d\Omega$ in the momentum space (Figure 6.2). That elementary volume has sides dp_x, dp_y and dp_z, which means

$$d\Omega = dp_x dp_y dp_z$$
$$\text{and,} \quad dW = \omega_p d\Omega \tag{6.18}$$

ω_p being the probability density of momentum that we have set out to calculate.

Maxwell showed that the form of ω_p can be established easily by assuming very reasonably that

  (i) the space is isotropic, i.e. the directions in space are all equivalent and therefore all the directions of $\vec{p}$ are equally probable. The probability density will therefore be isotropic.
 (ii) The motion along the three mutually orthogonal directions (x, y, z) are independent.

Therefore, the value of the x-component of the momentum (p_x) does not depend on the values of p_y, p_z.

However, this condition's independence of momentum components can be challenged at high molecular velocities close to the velocity of light. The probability dW_{p_x} that the x-component of the momentum p_x lies in the range p_x to $p_x + dp_x$ can be written in the following form:

$$dW_{p_x} = \omega_{p_x} dp_x \tag{6.19}$$

where ω_{p_x} represents the probability density for p_x. Since the probability of the values of x-component of p being $+p_x$ and $-p_x$ is identical as demanded by isotropy of p, probability density ω_{p_x} must be an even function of p_x:

$$\omega_{p_x} = \Phi(p_x^2) \tag{6.20a}$$

The isotropic nature of w further suggests that similar relations must also hold for ω_{p_y} and ω_{p_z}:

$$\omega_{p_y} = \Phi(p_y^2) \tag{6.20b}$$

$$\omega_{p_z} = \Phi(p_z^2) \tag{6.20c}$$

Note that the same function Φ (unknown at this point) appears in Eq. 6.20a-c. Since v_x, v_y and v_z and hence the x, y and z components of p have been assumed to be independent we can express ω_p in terms of ω_{p_x}, ω_{p_y} and ω_{p_z} as

$$\omega_p = \omega_{p_x}\omega_{p_y}\omega_{p_z} \tag{6.21}$$

The probability dW that the tip of the momentum vector $\overrightarrow{p}$ is within the volume element $d\Omega$ (Figure 6.2) in the momentum space is then obtained as

$$dW = \omega_p d\Omega \tag{6.22}$$

Eq. 6.21 also conveys the fact that the momentum components are within the ranges $p_x \leftrightarrow p_x + dp_x$, $p_y \leftrightarrow p_y + dp_y$ and $p_z \leftrightarrow p_z + dp_z$, respectively. We can therefore rewrite Eq. 6.22 as

$$\begin{aligned} dW = \omega_p d\Omega &= \omega_{p_x} dp_x \omega_{p_y} dp_y \omega_{p_z} dp_z \\ &= \Phi(p_x^2)\Phi(p_y^2)\Phi(p_z^2) dp_x dp_y dp_z \end{aligned} \tag{6.23}$$

which leads to the identification

$$\omega_p = \Phi(p_x^2)\Phi(p_y^2)\Phi(p_z^2) \tag{6.24}$$

ω_p, we have stressed earlier is isotropic in character. Hence ω_p must depend only on the magnitude of $\overrightarrow{p}$, and be independent of its direction. This leads us to demand that

$$\omega_p \equiv \Psi(p^2) \tag{6.25}$$

where $p^2 = p_x^2 + p_y^2 + p_z^2$, and Ψ is an unknown function as yet. Comparing Eq. 6.24 and Eq. 6.25 we obtain a relation between Ψ and Φ :

$$\Psi(p^2) = \Phi(p_x^2)\Phi(p_y^2)\Phi(p_z^2) \tag{6.26}$$

In order to establish the forms of the functions Ψ and Φ, let us proceed by taking logarithm of both sides of Eq. 6.26, yielding the identity

$$ln\Psi(p_x^2 + p_y^2 + p_z^2) = ln\Phi(p_x^2) + ln\Phi(p_y^2) + ln\Phi(p_z^2) \tag{6.27}$$

Differentiating both sides of the identity with respect to p_x, p_y and p_z, respectively we get

$$\frac{\Psi'(p^2)}{\Psi(p^2)} = \frac{\Phi'(p_x^2)}{\Phi(p_x^2)} \tag{6.28a}$$

$$\frac{\Psi'(p^2)}{\Psi(p^2)} = \frac{\Phi'(p_y^2)}{\Phi(p_y^2)} \tag{6.28b}$$

$$\frac{\Psi'(p^2)}{\Psi(p^2)} = \frac{\Phi'(p_z^2)}{\Phi(p_z^2)} \tag{6.28c}$$

which leads us to conclude that

$$\frac{\Phi'(p_x^2)}{\Phi(p_x^2)} = \frac{\Phi'(p_y^2)}{\Phi(p_y^2)} = \frac{\Phi'(p_z^2)}{\Phi(p_z^2)} \tag{6.29}$$

The first term in Eq. 6.29 above depends only on p_x while the second and third terms depend only on p_y and p_z, respectively, suggesting that each ratio must be equal to the same constant, say $-\beta$. Thus the differential equations that determines the form of the function Φ is

$$\frac{\Phi'(p_x^2)}{\Phi(p_x^2)} = \frac{\Phi'(p_y^2)}{\Phi(p_y^2)} = \frac{\Phi'(p_z^2)}{\Phi(p_z^2)} = -\beta \tag{6.30}$$

which are solved by

$$\Phi(p_x^2) = A.e^{-\beta p_x^2} = \omega_{p_x}$$

$$\Phi(p_y^2) = A.e^{-\beta p_y^2} = \omega_{p_y} \qquad (6.31)$$

$$\Phi(p_z^2) = A.e^{-\beta p_z^2} = \omega_{p_z}$$

where A and β are hitherto unknown constants. Once we know the form of Φ, we can exploit Eq. 6.26 to find Ψ. Let us stop here for a moment and reflect on what we have achieved so far: starting with the conditions of isotropy and independence of motion along mutually orthogonal axes (p_x, p_y, p_z) we have found the probability dW_{p_x} that the x-component of p is found in the interval between p_x and $p_x + dp_x$ is given by

$$dW_{p_x} = \omega_{p_x} dp_x = A.e^{-\beta p_x^2} dp_x \qquad (6.32)$$

and similar relations hold for p_y and p_z components as well. The probability dW that the tip of the momentum vector $\overrightarrow{p}$ lies within the volume element $d\Omega = dp_x dp_y dp_z$ is given by the product of dW_{p_x}, dW_{p_y} and dW_{p_z}, that is

$$dW = A^3 e^{-\beta(p_x^2 + p_y^2 + p_z^2)} dp_x dp_y dp_z$$

$$= A^3 e^{-\beta p^2} d\Omega \qquad (6.33)$$

Once we can fix the constants A and β, presumably from other conditions, the problem we set out to solve has been solved. Note that the constant A can be fixed from the normalization condition

$$\int_{-\infty}^{+\infty} \omega_{p_x} dp_x = 1 = A^2 \int e^{-\beta p_x^2} dp_x \qquad (6.34)$$

which requires β to be positive and yields upon integration $A = (\frac{\beta}{\pi})^{\frac{1}{2}}$. The determination of β requires us to refer to the properties of an ideal monatomic gas that kinetic theory and thermodynamics provide us with.

The kinetic energy of an individual molecule is $\epsilon = \frac{p^2}{2m_0}$ where m_0 is the mass of a single molecule. The mean value of energy $\langle \epsilon \rangle$ is then calculated as

$$\langle \epsilon \rangle = \frac{\langle p^2 \rangle}{2m_0} = \frac{\langle p_x^2 \rangle + \langle p_y^2 \rangle + \langle p_z^2 \rangle}{2m_0} \qquad (6.35)$$

The isotropic nature of the space demands $\langle p_x^2 \rangle = \langle p_y^2 \rangle = \langle p_z^2 \rangle$ so that

$$\langle \epsilon \rangle = \frac{3}{2} \cdot \frac{\langle p_x^2 \rangle}{m_0} \qquad (6.36)$$

The $\langle p_x^2 \rangle$ can be calculated by using the probability density distribution function for the momentum values as follows

$$\langle p_x^2 \rangle = A \cdot \int_{-\infty}^{+\infty} e^{-\beta p_x^2} p_x^2 dp_x$$

$$= \left(\frac{\beta}{\pi}\right)^{\frac{1}{2}} \cdot \frac{\sqrt{\pi}}{2} \cdot \beta^{\frac{-3}{2}} = \frac{1}{2\beta} \qquad (6.37)$$

The mean value of energy of a monatomic molecule is then

$$\langle \epsilon \rangle = \frac{3}{2} \cdot \frac{\langle p_x^2 \rangle}{m_0} = \frac{3}{2} \cdot \frac{1}{m_0} \cdot \frac{1}{2\beta} \qquad (6.38)$$

If 'g' is the mass of a gas with molar mass M, the number of moles in 'g' is g/M and the total number of molecules present is $\frac{g}{M} \times N_A = N$, N_A being the Avogadro's number. The internal energy of the given mass of gas (monatomic + ideal) is then

$$U = N \times \langle \epsilon \rangle = \frac{g}{M} \cdot N_A \times \frac{3}{2} \cdot \frac{1}{m_0} \cdot \frac{1}{2\beta} \tag{6.39}$$

The internal energy of an ideal monatomic gas at temperature 'T' is known to be

$$U = \frac{g}{M} \cdot \frac{3}{2} RT = \frac{g}{M} \cdot \frac{3}{2} (kT) \times N_A \tag{6.40}$$

k being the Boltzmann constant or gas constant per molecule. Comparing Eq. 6.39 and Eq. 6.40, we find that

$$\beta = (2m_0 kT)^{-1}$$

The probability density of momentum values at temperature T is now completely defined and is given by

$$\omega_{p_i} = \frac{1}{\sqrt{2\pi\, m_0 kT}} e^{\frac{-p_i^2}{2m_0 kT}}; \quad i = x, y, z \tag{6.41}$$

while the probability itself is given by (in component form)

$$\omega_{p_i} dp_i = \frac{1}{\sqrt{2\pi\, m_0 kT}} e^{\frac{-p_i^2}{2m_0 kT}} dp_i; \quad i = x, y, z \tag{6.42}$$

Similarly,

$$\omega_p d\Omega = dW = \frac{1}{(2\pi\, m_0 kT)^{\frac{3}{2}}} e^{\frac{-p^2}{2m_0 kT}} d\Omega \tag{6.43}$$

Eq. 6.41 defines the Maxwellian probability densities for the x, y and z components of momenta (p) while the Maxwellian probability that the momentum has value $p(= \sqrt{p_x^2 + p_y^2 + p_z^2})$ and the tip of $\overrightarrow{p}$ vector lies within the volume $d\Omega = dp_x dp_y dp_z$ is given by Eq. 6.43.

Turning our attention to the x-component of the momentum $\overrightarrow{p}$,

$$\langle p_x \rangle = \int_{-\infty}^{+\infty} \omega_{p_x} p_x dp_x = 0$$

and $\quad \langle p_x^2 \rangle = \int_{-\infty}^{+\infty} \omega_{p_x} p_x^2 dp_x = m_0 kT$

We find that the fluctuation in the x-component of momentum is $D = \langle p_x^2 \rangle - \langle p_x \rangle^2 = m_0 kT$, which means that ω_{p_x} takes the form

$$\omega_{p_x} = \frac{1}{\sqrt{2\pi D}} e^{-\frac{(p_x - \langle p_x \rangle)^2}{2D}} \tag{6.44}$$

It is indeed just what we had obtained for a continuous random variable. Similar results apply for ω_{p_y} and ω_{p_z} also. We can therefore say that the equilibrium momentum distribution law for an ideal gas at a temperature T is a Gaussian or a normal distribution.[2-3] The Maxwellian probability density can be immediately exploited to calculate the mean number of particles having definite values of momentum as follows:

Let dv be a volume element in the vessel of volume V enclosing the mass of gas at temperature T, the number density (number of molecules/unit volume) being 'n'. We have already seen that 'n' is uniform everywhere. The mean number of molecules inside the volume element is therefore $n\, dV$. Now the probability that the momentum of a molecule in the vessel belongs the momentum space volume element $d\Omega$ is $\omega_p d\Omega$. Therefore, the average number of molecules (dN) in the volume element dV having the given value of momentum (p) is

$$dN = n \cdot dV \times \omega_p d\Omega = f \times dv d\Omega \tag{6.45}$$

The function $f(=n\omega_p)$ is called Maxwell's distribution function, which when multiplied with $dV \cdot d\Omega$ gives the number of molecules in the volume element dV having their momentum lying in the specified range $(p_x \leftrightarrow p_x + dp_x, p_y \leftrightarrow p_y + dp_y$ and $p_z \leftrightarrow p_z + dp_z, p = \sqrt{p_x^2 + p_y^2 + p_z^2})$. Thus, for Maxwell's distribution, we have

$$dN = \frac{n}{(2\pi m_0 kT)^{\frac{3}{2}}} e^{\frac{-p^2}{2m_0 kT}} dV d\Omega \tag{6.46}$$

From Eq. 6.42 and Eq. 6.43, it is straightforward to obtain Maxwell's probability density ω_v for a random velocity vector $\overrightarrow{v}$:

$$\omega_v dv_x dv_y dv_z = \left(\frac{m_0}{2\pi kT}\right)^{\frac{3}{2}} e^{\frac{-mv^2}{2kT}} dv_x dv_y dv_z \tag{6.47}$$

Eq. 6.43 can be similarly translated into a probability distribution law for energy in an ideal gas by noting that the energy ϵ of a molecule is equal to $\frac{p^2}{2m_0}$ (only kinetic). Thus, we have

$$\omega_\epsilon d\epsilon = \frac{2}{\sqrt{\pi}} \left(\frac{1}{kT}\right)^{\frac{3}{2}} e^{-\frac{\epsilon}{kT}} \sqrt{\epsilon} d\epsilon \tag{6.48}$$

Notice that the energy-dependent part of ω_ϵ is $e^{-\frac{\epsilon}{kT}} \sqrt{\epsilon}$, which is also dependent on absolute temperature. Eq. 6.48 is Maxwell's law for the probability that the energy of an ideal gas molecule is confined in the interval $\epsilon \leftrightarrow \epsilon + d\epsilon$ at a temperature T.

6.3.2 The Equilibrium State and Boltzmann Distribution

The uniform distribution of particles in space is no-longer ensured when an external field is present and the particles (molecules) interact with it. The particles acquire a potential energy component. For example, in the gravitational field a particle of mass 'm' at a height 'h' from the bottom surface of the vessel has a potential energy equal to mgh, which is manifestly dependent on the positional coordinate of the particle. The task now is to find out the equilibrium distribution law for the particles in space when they are moving inside a potential well and the potential energy depends on the position of the particle in space. Let us first note that the incessant random movement of the molecules continue even in the presence of an external field of force or a position-dependent potential. So, there is a continuous exchange of particles between any two regions of space inside the well by virtue of the molecules possessing kinetic energy. In the state of equilibrium when macroscopic properties become invariant in time, the dynamic exchange of particles between any two regions continue unabated at the microscopic level in a balanced manner, the 'balance' being reflected in the time-invariance of macroscopic properties. The equilibrium is therefore a dynamic process at the microscopic level. Formally, this fact is expressed in the form of the principle of detailed balance, which asserts the following:

'In true equilibrium, with each flux, there is an associated opposite flux so that the equilibrium occurs not only as a whole, but also in detail, for each pair of opposite processes'. In other words, each flux across a surface inside the system is annulled by an opposite flux of equal magnitude. The macro level equilibrium can therefore be considered as a consequence of the principle of detailed balance.

Let us consider a system of gas molecules enclosed in a vessel, which is represented by a potential well (Figure 6.3) in which there are two regions. Region 1 has lower potential energy than region 2 where the potential energy is higher. We can conceptualize the system as a combination of two communicating vessels 1 and 2 (Figure 6.3), which communicate through an opening on the surface of the separating wall. Let the area of the opening be 'ds' and the x-axis be perpendicular to the separating wall. Molecules in each vessel obey Maxwell's distribution of momenta corresponding to some concentrations n_1, n_2 (number densities) and temperatures T_1, T_2, the subscript '1' referring to vessel 1 and '2' to vessel 2.

The number of molecules with their momenta lying in the interval $d\Omega$ migrating from vessel 1 to vessel 2 through the opening of area 'ds' is $ds dj_x$ where dj_x is the flux density across 'ds' from vessel 1 to 2.

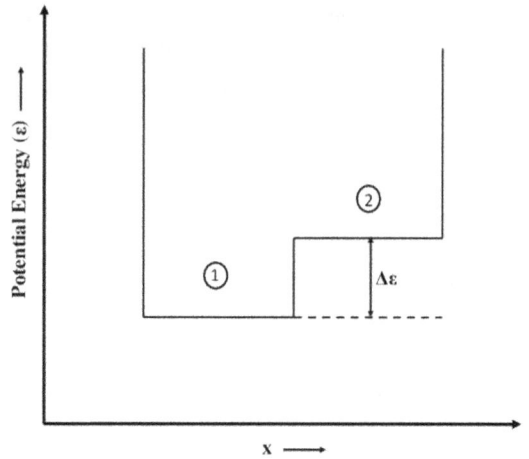

FIGURE 6.3 A potential energy well with a stepped bottom. Note the difference in potential energy in the two regions (1) and (2).

Invoking Maxwell's law of distribution of molecules by momenta we have

$$dsdj_x = \frac{p_x}{m_0} \cdot \frac{n_1}{(2\pi m_0 kT_1)^{\frac{3}{2}}} \cdot e^{\frac{-p^2}{2m_0 kT_1}} \, d\Omega ds \tag{6.49}$$

where p_x is the component of momentum perpendicular to the opening. T_1 is the temperature and n_1 is the number density of molecules in vessel 1. In calculating $dsdj_x$, we have, of course, assumed that p_x is directed from vessel 1 to vessel 2 and is large enough to negotiate the energy barrier on its path from vessel 1 to vessel 2. A molecule passing through ds along x-axis is acted upon by a force acting in the opposite direction along the x-axis. p_x is therefore reduced to, say p_x' while p_y and p_z components remain unaltered as they are perpendicular to the direction of force experienced by the molecule. The energy conservation law then demands that

$$\frac{p_x^2}{2m_0} = \frac{(p_x')^2}{2m_0} + \Delta\epsilon_p \tag{6.50}$$

$\Delta\epsilon_p$ is the rise in potential energy after the passage of the molecule through the opening from vessel 1 to vessel 2. The p_y and p_z components of momentum remain unchanged upon migration. The principle of detailed balance requires the flux from vessel 1 to vessel 2 across the area 'ds' must be exactly compensated by the reverse flux (j_x') across 'ds' for which we can write,

$$dsdj_x' = \frac{p_x'}{m_0} \frac{n_2}{(2\pi m_0 kT_2)^{\frac{3}{2}}} e^{\frac{-p'^2}{2m_0 kT_2}} \, d\Omega' ds \tag{6.51}$$

Here, n_2 is the number density, T_2 is the temperature of molecules in vessel and p' is their momentum distributed in accordance with Maxwell's law. Invoking equality of fluxes across ds in opposite directions, we have

$$\frac{n_1 p_x}{m_0 (2\pi m_0 kT_1)^{\frac{3}{2}}} e^{\frac{-p^2}{2m_0 kT_1}} \, dp_x dp_y dp_z = \frac{n_2}{m_0} \frac{p_x'}{(2\pi m_0 kT_2)^{\frac{3}{2}}} e^{\frac{-p'^2}{2m_0 kT_2}} \, dp_x' dp_y' dp_z' \tag{6.52}$$

We have mentioned already that p_y and p_z components are conserved upon migration so that,

$$dp_y = dp_y'; \quad dp_z = dp_z' \tag{6.53}$$

From the energy conservation condition (Eq. 6.50) we have

$$\frac{p_x}{m_0} = d\left(\frac{p_x^2}{2m_0}\right) = d\left(\frac{p_x'^2}{2m_0}\right) = \frac{p_x'}{m_0} \tag{6.54}$$

as the change $\Delta\epsilon_p$ in potential energy on migration from vessel 1 to vessel 2 is the same for all molecules irrespective of their momentum values. Using Eq. 6.74. Eq. 6.73 in Eq. 6.72 we arrive at the condition

$$\frac{n_1}{(2\pi m_0 kT_1)^{\frac{3}{2}}}e^{-\frac{(p_x^2+p_y^2+p_z^2)}{2m_0kT_1}} = \frac{n_2}{(2\pi m_0 kT_2)^{\frac{3}{2}}}e^{-\frac{(p_x'^2+p_y'^2+p_z'^2)}{2m_0kT_2}} \tag{6.55}$$

But we have seen before that the migration through the opening leaves p_y and p_z components unaltered so that we can enforce the equalities $p_y^2 = p_y'^2$ and $p_z^2 = p_z'^2$ in Eq. 6.55 leading to slightly modified condition for detailed balance:

$$\frac{n_1}{(2\pi m_0 kT_1)^{\frac{3}{2}}}e^{-\frac{(p_x^2+p_y^2+p_z^2)}{2m_0kT_1}} = \frac{n_2}{(2\pi m_0 kT_2)^{\frac{3}{2}}}e^{-\frac{(p_x'^2+p_y^2+p_z^2)}{2m_0kT_2}} \tag{6.56}$$

A reference to Eq. 6.70, enables us to express the kinetic energy $\frac{p_x'^2}{2m_0}$ of the molecule in vessel 2 in terms of the kinetic energy $\frac{p_x^2}{2m_0}$ in vessel 1 and the rise in potential energy on account of $1 \leftrightarrow 2$ transfer. Thus, we have

$$\frac{p_x'^2}{2m_0} = \frac{p_x^2}{2m_0} - \Delta\epsilon_p \tag{6.57}$$

which when ploughed back into Eq. 6.56 yields

$$\frac{n_1}{(2\pi m_0 kT_1)^{\frac{3}{2}}}e^{-\frac{(p_x^2+p_y^2+p_z^2)}{2m_0kT_1}} = \frac{n_2}{(2\pi m_0 kT_2)^{\frac{3}{2}}}e^{+\frac{\Delta\epsilon_p}{kT_2}}e^{-\frac{(p_x^2+p_y^2+p_z^2)}{2m_0kT_2}} \tag{6.58}$$

Since $p_x^2 + p_y^2 + p_z^2 = p^2$, we have the detailed balancing condition in the final form

$$\frac{n_1}{(2\pi m_0 kT_1)^{\frac{3}{2}}}e^{-\frac{p^2}{2m_0kT_1}} = \frac{n_2}{(2\pi m_0 kT_2)^{\frac{3}{2}}}e^{\frac{\Delta\epsilon_p}{kT_2}}e^{-\frac{p^2}{2m_0kT_2}} \tag{6.59}$$

This condition (Eq. 6.59) must hold not only for any specific value of 'p' but also for any value of p. In other words, Eq. 6.59 must be an identity independent of p. That is possible only if $T_1 = T_2$ whence we arrive at the distribution law

$$n_2 = n_1 e^{-\frac{\Delta\epsilon_p}{kT}} \tag{6.60}$$

with vessel 1 as the reference point where potential energy is lower. Eq. 6.60 is the celebrated equilibrium distribution law for particles moving in a potential that changes with position and is known as *Boltzmann distribution law or simply Boltzmann law*. For establishment of the equilibrium, it is essential that the temperature 'T' must be the same at every part of system. Only then does Boltzmann's law of distribution of particles according to their potential energy take over. Boltzmann distribution is thus the signature of the underlying equilibrium, which is a direct consequence of the condition of detailed balancing principle holding at the microscopic level. If we set the potential energy in vessel 1 as zero, Boltzmann distribution takes the form

$$n_2 = n_1 e^{-\frac{\epsilon_2}{kT}} \tag{6.61}$$

where we have used $\Delta\epsilon_p = \epsilon_2 - \epsilon_1 = \epsilon_2 - 0 = \epsilon_2$. If ϵ is a continuous function of $\vec{r}$ (the positional coordinates) of a particle and we choose a reference point $\vec{r_0}$ where the potential energy is $\epsilon_p(r_0)$. Boltzmann distribution can be written in a more generalized fashion as

$$n(\vec{r}) = n(\vec{r_0})e^{-\frac{(\epsilon_p(\vec{r})-\epsilon_p(\vec{r_0}))}{kT}} \tag{6.62a}$$

where $n(\vec{r_0})(= n_0)$ is the number of particles with potential energy $\epsilon_p(r_0)$ and $n(\vec{r})(= n_r)$ is the number at $\vec{r}$ when equilibrium has set in at the temperature T. More simply and commonly, one writes

$$n_r = n_0 e^{-\frac{\epsilon_r}{kT}} \tag{6.62b}$$

where ϵ_r stands for $\epsilon_p(r) - \epsilon_p(r_0)$.

6.3.3 Maxwell-Boltzmann Distribution

We have already seen how to go about calculating the mean number of particles having definite values of momentum, velocity or energy, from our knowledge of Maxwellian probability density in the absence of any external field of force. Let us recall that the distribution function 'f' (Eq. 6.46) determines, in the absence of external field, the number (dN) of particles in the volume element $d\Omega$ of the momenta space and in the volume element dV of the configuration or the conventional space:

$$dN = f dv d\Omega = \frac{n}{(2\pi m_0 kT)^{\frac{3}{2}}} e^{-\frac{p^2}{(2m_0 kT)}} dv d\Omega$$

n being the mean number of molecules per unit volume.

What would be the form of equilibrium distribution function when field of force is present? We have seen in the previous section that the number density now becomes position-dependent and is given by $n(r)$ (see Eq. 6.82a-b). The momentum distribution continues to remain Maxwellian as it was in the absence of the field of force. Therefore the probability (dw) that the momentum of a molecule is confined in the volume element $d\Omega$ of the momentum space is given by

$$dw = \frac{1}{(2\pi m_0 kT)^{\frac{3}{2}}} e^{-\frac{p^2}{2m_0 kT}} d\Omega \tag{6.63}$$

The number of molecules dN in the volume dV with momentum values restricted within the momentum space volume $d\Omega$ is

$$n(r) dv \times dw = \frac{n(r)}{(2\pi m_0 kT)^{\frac{3}{2}}} e^{-\frac{p^2}{2m_0 kT}} dv d\Omega \tag{6.64}$$

From the Boltzmann distribution we can write

$$n(r) = n(r_0) e^{-\frac{\epsilon_p(r) - \epsilon_p(r_0)}{kT}} \tag{6.65}$$

where $\epsilon_p(r_0)$ is the potential energy of a molecule at the reference point r_0. Therefore,

$$dN = n(r_0) e^{+\frac{\epsilon_p(r_0)}{kT}} \frac{1}{(2\pi m_0 kT)^{\frac{3}{2}}} e^{-\frac{\epsilon_p(r)}{kT}} \times e^{-\frac{p^2}{2m_0 kT}} dv d\Omega$$

$$= C_N e^{-\frac{\left(\epsilon_p(r) + \frac{p^2}{2m_0}\right)}{kT}} dv d\Omega \tag{6.66}$$

where C_N is a constant factor that does not depend on the instantaneous coordinate $\vec{r}$ of the gas molecule and is given by

$$C_N = n(r_0) e^{-\frac{\epsilon_p(r_0)}{kT}} \times \frac{1}{(2\pi m_0 kT)^{\frac{3}{2}}} \tag{6.67}$$

If we integrate Eq. 6.66 over all the values of the coordinates and momenta, we find

$$\int dN = N = C_N \int e^{-\frac{\left(\epsilon_p(r) + \frac{p^2}{2m_0}\right)}{kT}} dv d\Omega \tag{6.68}$$

which yields

$$C_N = \frac{N}{\int e^{-\frac{\left(\epsilon_p(r) + \frac{p^2}{2m_0}\right)}{kT}} dv d\Omega} \tag{6.69}$$

The normalization condition (Eq. 6.68) directly leads to the evaluation of the constant C_N. Eq. 6.66 above represents the equilibrium distribution function in the presence of external fields of force with C_N determined by Eq. 6.69, i.e. the normalization condition. The specific distribution function combines Maxwell's law of distribution of momenta (in Ω space) and Boltzmann, distribution law of particles in the conventional configuration space.[1-3] The combination has become known as the 'Maxwell-Boltzmann distribution law' – a classical equilibrium distribution function when 'energy' varies continuously.

6.4 Gibbs Distribution

6.4.1 Maxwell-Boltzmann Probability Density

We have so far been frequently using the idea of molecules (or particles) having their coordinates confined within the elementary volume $dv = dxdydz$ and momenta restricted within the interval $d\Omega = dp_x dp_y dp_z$. It would be advisable to switch over to another description based on a six-dimensional phase space or the so called mu-space for the sake of clarity and economy. In this space let us imagine that the coordinates of a particle are laid out along three axes (x, y, z) and the components of momenta are also laid down along the three other coordinates (p_x, p_y, p_z) of the six-dimensional space. The position of a point in this six-dimensional phase space is therefore completely determined by all six coordinates. Thus when we say that a molecule or a particle is at a point in the phase space, we mean that the spatial coordinates (three) of the molecule are set or given along with the three components of momentum. An elementary volume in the phase space (or mu-space), say $d\gamma$ is then set by the product of differentials of all six coordinates $(x, y, z; p_x, p_y, p_z)$ and we have

$$d\gamma = dxdydzdp_x dp_y dp_z = dvd\Omega \tag{6.70}$$

In this newly introduced phase space Eq. 6.66 becomes

$$dN = C_N e^{-\frac{\left(\epsilon_p(r) + \frac{p^2}{2m_0}\right)}{kT}} d\gamma \tag{6.71}$$

Similarly, Eq. 6.68 now reads

$$N = C_N \int_r e^{-\frac{\left(\epsilon_p(r) + \frac{p^2}{2m_0}\right)}{kT}} d\gamma \tag{6.72}$$

Instead of saying that a molecule is in the volume element dv of the traditional space with its momentum confined to the region Ω of the momentum space as we have been doing so far, it suffices to state now that a molecule is in the region $d\gamma$ of the newly introduced phase space without losing or missing out any information.

Let us now return to Eq. 6.71 where the mean number of molecules in the designated phase-space volume $d\gamma$ is dN. The probability that one molecule is found in the phase-space volume element $d\gamma$ is

$$dW(\overrightarrow{r}, \overrightarrow{p}) = \frac{dN}{N} = \frac{C_N}{N} e^{-\frac{\left(\epsilon_p(r) + \frac{p^2}{2m_0}\right)}{kT}} d\gamma \tag{6.73}$$

$$\text{or, } dW = a_w e^{-\frac{\epsilon}{kT}} d\gamma = \omega(\overrightarrow{r}, \overrightarrow{p}) d\gamma$$

where $a_w = \frac{C_N}{N}$, and $\epsilon = \epsilon_p(r) + \frac{p^2}{2m_0}$ is the total energy possessed by the molecule at the designated region of the phase space, and $\omega(\overrightarrow{r}, \overrightarrow{p}) = a_w e^{-\frac{\epsilon}{kT}}$. $a_w e^{-\frac{\epsilon}{kT}}$ is therefore the probability density $\omega(\overrightarrow{r}, \overrightarrow{p})$ that a molecule is found at the point $(\overrightarrow{r}, \overrightarrow{p})$ of the phase space with total energy ϵ. The constant a_w appearing in $\omega(\overrightarrow{r}, \overrightarrow{p})$ can be fixed by the normalization condition

$$\int dw = 1 = a_w \int_r e^{-\frac{\epsilon}{kT}} d\gamma$$

$$\text{i.e.} \quad a_w = \frac{1}{\int_r e^{-\frac{\epsilon}{kT}} d\gamma} \tag{6.74}$$

Eq. 6.73 describes what may be called the Maxwell-Boltzmann probability density with a_w fixed by normalization condition leading to Eq. 6.74.

6.4.2 The Gibbs Distribution: Probability of an Equilibrium State

We have seen how Maxwell-Boltzmann distribution is arrived at and the probability density $\omega(\overrightarrow{r}, \overrightarrow{p})$ of a molecule to be at the point $(\overrightarrow{r}, \overrightarrow{p})$ in phase space is calculated. The probability of finding a molecule in the phase-space volume $d\gamma$ at the designated point in phase space is then

$$dW(\vec{r}, \vec{p}) = \omega(\vec{r}, \vec{p}) d\gamma$$

Let us now ask a broader question: what is the probability $dW(\vec{r_1}, \vec{p_1}; \vec{r_2}, \vec{p_2})$ that molecule 1 gets into the phase-space volume element $d\gamma_1$ when molecule 2 is in the element $d\gamma_2$? If the interactions between 1 and 2 are small enough to be negligible, the two events can be regarded as independent. The laws of probability then suggest that (Eq. 6.73)

$$dW(\vec{r_1}, \vec{p_1}; \vec{r_2}, \vec{p_2}) = dW(\vec{r_1}, \vec{p_1}) \times dW(\vec{r_2}, \vec{p_2})$$

$$= a_w e^{-\frac{\epsilon_1}{kT}} d\gamma_1 \times a_w e^{-\frac{\epsilon_2}{kT}} d\gamma_2 \tag{6.75}$$

$$= (a_w)^2 e^{-\frac{(\epsilon_1 + \epsilon_2)}{kT}} \times d\gamma_1 d\gamma_2$$

We can further broaden the question and ask: what would be the probability that out of the N molecules in a gas sample at temperature T, molecule 1 is in the phase-space volume element $d\gamma_1$, molecule 2 is in volume element $d\gamma_2$, molecule 3 is in $d\gamma_3$ and continuing further the Nth molecule is in the volume element $d\gamma_N$? We can rephrase the question and ask what is the probability that such a state is indeed realized in a sample of an ideal gas consisting of N molecules enclosed in a volume V at temperature T? If the temperature T is high, and the number density (N/V) of the molecules is low, i.e. the gas is in a sufficiently rarefied state we can extend Eq. 6.75 and write

$$dW(\vec{r_1}, \vec{p_1}; \vec{r_2}, \vec{p_2}; \cdots; \vec{r_N}, \vec{p_N}) = (a_w)^N e^{-\frac{(\epsilon_1 + \epsilon_2 + \cdots + \epsilon_N)}{kT}} d\gamma_1 d\gamma_2 \cdots d\gamma_N \tag{6.76}$$

Now the sum $\sum_{i=1}^{N} \epsilon_i$ defines a state with the total energy ε so that we can write

$$\varepsilon = \varepsilon(\vec{r_1}, \vec{p_1}; \vec{r_2}, \vec{p_2}; \cdots; \vec{r_N}, \vec{p_N}) = \sum_{i=1}^{N} \epsilon_i \tag{6.77}$$

The arguments in the total energy ε define the dynamical state that the assembly of N-non-interacting molecules are in. Thus, even when we suppress these arguments and write ε, we must keep in mind that the total energy ε refers to a state of the N molecules defined by the $3N$ positional coordinates $(x_1, y_1, z_1; x_2, y_2, z_2; \cdots; x_N, y_N, z_N)$ and $3N$ projections of momenta of all the N particles of the system $(p_{x_1}, p_{y_1}, p_{z_1}; p_{x_2}, p_{y_2}, p_{z_2}; \cdots; p_{x_N}, p_{y_N}, p_{z_N})$. The total phase space Γ of the entire system of N-molecules/particles defines a $6N$-dimensional space along the axes of which $3N$ positional coordinates and $3N$ projections of momenta of the N-particles are laid out. A point in the $6N$-dimensional Γ-space determines the coordinates and momenta of all the N-particles and thereby completely defines the dynamical state of the system. It would be useful to define a volume element $d\Gamma$ of the total phase space of the system, which is

$$d\Gamma = dx_1 dy_1 dz_1 dp_{x_1} dp_{y_1} dp_{z_1} \cdots dx_N dy_N dz_N dp_{x_N} dp_{y_N} dp_{z_N}$$

$$= (dv_1 d\Omega_1)(dv_2 d\Omega_2) \cdots (dv_N d\Omega_N) \tag{6.78}$$

$$= d\gamma_1 d\gamma_2 \cdots d\gamma_N$$

Using the expression for $d\Gamma$ in Eq. 6.76 we have

$$dW(\vec{r_1}, \vec{p_1}; \vec{r_2}, \vec{p_2}; \cdots; \vec{r_N}, \vec{p_N}) = a_w^N e^{-\frac{\epsilon}{kT}} d\Gamma \tag{6.79}$$

a_w^N is a function of the temperature T. Let us now take following Gibbs, the important step and write

$$a_w^N = e^{\frac{F}{kT}} \tag{6.80}$$

so that,

$$F = kT ln\left(a_w^N\right) \tag{6.81}$$

'*F*' does not depend on the position and momentum coordinates of individual molecules, but depends on collective coordinates like T and V. Clearly, F has the dimension of energy, but it requires to be further interpreted. With the change in notation Eq. 6.79 turns into an expression for the total probability:

$$dW(\vec{r_1}, \vec{p_1}; \vec{r_2}, \vec{p_2}; \cdots; \vec{r_N}, \vec{p_N}) = e^{\frac{F-\epsilon}{kT}} d\Gamma \qquad (6.82)$$

The equation above has become known as the now famous **Gibbs' distribution**, which determines the probability of realizing a particular state of a mass of gas consisting of N-molecules as a whole. The Maxwell-Boltzmann distribution, on the other hand, determines the probability of the occurence of a given state for one molecule of an ideal gas.

The Gibbs distribution is simple in appearance but embodies lot of physics concealed in its simple form, so much so, that it has been given the status of the 'Central relation of the statistical physics of equilibrium states' – not only of an ideal gas but also of any other system. Gibbs distribution is therefore a profoundly basic equation of equilibrium statistical mechanics from which all the other properties of systems in equilibrium, most notably, the thermodynamic properties can be worked out correctly. The importance of the central dogma of classical statistical mechanics of equilibrium states can hardly be overemphasized. Now, Eq. 6.82, representing Gibbs distribution determines the probability of a system in equilibrium at temperature to be in one of the states belonging to the phase-space volume element $d\Gamma$ and possessing energy ϵ. When integrated over the entire phase space, Eq. 6.82 leads to

$$1 = \int dW = \int e^{\frac{F-\epsilon}{kT}} d\Gamma$$

$$= e^{\frac{F}{kT}} \left(\int e^{\frac{-\epsilon}{kT}} d\Gamma \right) \qquad (6.83)$$

$$= e^{\frac{F}{kT}} (Z) = 1$$

where

$$Z = \int e^{\frac{-\epsilon}{kT}} d\Gamma \qquad (6.84)$$

is called the state integral or the partition function. Taking logarithm of both sides of Eq. 6.83, we arrive at the definition of F,

$$F = -kTlnZ \qquad (6.85)$$

'*F*' as defined in Eq. 6.85 is called the free energy of the system (see later). Before applying Gibbs' distribution law to a specific problem let us emphasize that Gibbs law fixes the probability of realizing a state of the system from among the states described by points in the phase space volume $d\Gamma$, at temperature T.

6.5 Classical Statistical Mechanics

In an ideal gas, molecules do not interact with each other. The total energy ϵ appearing in the Gibbs distribution (see Eq. 6.82) is just the sum of energies ϵ_i of N-individual molecules:

$$\epsilon = \sum_{i=1}^{N} \epsilon_i$$

Energy of each individual molecule (ϵ_i) is a sum of its kinetic and potential energies $\epsilon_i = \frac{p_i^2}{2m_0} + \epsilon_{p,i}(\vec{r})$ indicating the dependence of ϵ_i on the position and momentum coordinates of individual molecules. For an ideal gas molecule $\epsilon_{p,i}(\vec{r})$ is zero within the confining walls and infinite outside. Thus we have

$$\epsilon = \sum_{i=1}^{N} \left[\frac{p_i^2}{2m_0} + \epsilon_{p,i}(\vec{r}) \right] \qquad (6.86)$$

Once ε in Eq. 6.82 is defined, we need to fix the value of 'F' in order to determine Gibbs' distribution completely for the problem at hand. Since $F = -kTlnZ$, the task boils down to calculating the state integral Z. In view of the formula for ε given in Eq. 6.86, we have

$$Z = \int e^{\frac{-\varepsilon}{kT}} d\Gamma \tag{6.87}$$

$$Z = \int_1 \cdots \int_N e^{\frac{\sum\limits_{i=1}^{N} -\epsilon_i}{kT}} d\gamma_1 d\gamma_2 \ldots d\gamma_N$$
$$= \int e^{\frac{-\epsilon_1}{kT}} d\gamma_1 \int e^{\frac{-\epsilon_2}{kT}} d\gamma_2 \cdots \int e^{\frac{-\epsilon_N}{kT}} d\gamma_N \tag{6.88}$$

Thus we have expressed the $6N$-dimensional integral in Eq. 6.87 as a product of N six-dimensional integrals. A little reflection will indicate that each of the N six-dimensional integrals is of the same form, differing only in the dummy integration variable, which allows us to write

$$Z = \left[\int e^{-\frac{\epsilon_1}{kT}} \right]^N \tag{6.89}$$

$$Z = z^N \tag{6.90}$$

where

$$z = \int e^{-\frac{\epsilon_1}{kT}} d\gamma_1$$
$$= \int e^{-(\frac{p_i^2}{2m_0} + \epsilon_{p1})} dv_1 d\Omega_1 \tag{6.91}$$
$$= \int e^{-\frac{p_i^2}{2m_0}} d\Omega_1 \int e^{-\frac{\epsilon_{p1}}{kT}} dv$$

z is thus a product of two three-dimensional integrals. The first integral was encountered in Maxwellian distribution law and is easily found to be $(2\pi m_0 kT)^{\frac{3}{2}}$. The value of the second integral can be found by noting that ϵ_{p1} is infinite outside the boundaries of the confining vessel or potential well so that the integral is zero. Inside the vessel ϵ_{p1} is zero, making the integral 'unity' inside the vessel. Hence

$$\int e^{-\frac{\epsilon_{p1}}{kT}} dv = \int dv = V \tag{6.92}$$

where 'V' is the volume of the vessel confining the gas. Thus, 'z' is calculated to be

$$z = (2\pi m_0 kT)^{\frac{3}{2}} \times V \tag{6.93}$$

whence the state integral Z turns out to be

$$Z = (z)^N = (2\pi m_0 kT)^{\frac{3N}{2}} V^N \tag{6.94}$$

Now we are in a position to evaluate the free energy 'F' of an ideal gas confined to a volume 'V' at temperature T completely as

$$F = -kTlnZ$$
$$= -kTN \left\{ lnV + \frac{3}{2}lnT + \frac{3}{2}ln(2\pi m_0 k) \right\} \tag{6.95}$$

If we, as claimed earlier, accept Gibbs' distribution as the very fundamental postulate of statistical physics or mechanics of equilibrium states, we should be able to recover the Maxwellian law of distribution of momenta or the Boltzmann law of distribution of coordinates directly from Gibbs' law. With this purpose in mind, let us consider a system consisting of a single molecule of an ideal gas. The total phase space Γ is just the phase space of a single molecule, which is γ. That is we can replace Γ by γ. The Gibbs formula then predicts the probability of realizing a given state of our system of a single molecule to be (cf. Eq. 6.82)

$$dW = e^{\frac{F-\varepsilon}{kT}} d\gamma \qquad (6.96)$$

where $\varepsilon = \frac{p^2}{2m_0} + \epsilon_p(r)$ is the total energy of the molecule (kinetic plus potential). We can easily rearrange the right hand side of Eq. 6.96 to make it coincident with the Maxwell-Boltzmann distribution. The rearrangement leads to

$$dW = e^{\frac{F}{kT}} \cdot e^{-\frac{\epsilon}{kT}} dv d\Omega$$
$$= a_w e^{-\frac{p^2}{2m_0 kT}} e^{-\frac{\epsilon_p(r)}{kT}} dv d\Omega \qquad (6.97)$$

where $a_w = e^{\frac{F}{kT}}$ is the normalization constant.

Carrying out the normalization

$$\int dW = a_w \int e^{-\frac{p^2}{2m_0 kT}} d\Omega \int e^{-\frac{\epsilon_p(r)}{kT}} dv$$
$$= 1$$

That means,

$$a_w = \left[\int e^{-\frac{p^2}{2m_0 kT}} d\Omega \right]^{-1} \left[e^{-\frac{\epsilon_p(r)}{kT}} dv \right]^{-1}$$
$$= (2\pi m_0 kT)^{\frac{-3}{2}} \cdot \frac{1}{V} \qquad (6.98)$$

assuming that we are dealing with an ideal gas.

Thus Maxwell-Boltzmann distribution is fully recovered from Gibbs' distribution, demonstrating the fundamental importance of Gibbs' formulation of statistical mechanics of equilibrium states. It is not that Gibbs' formulation works only for ideal gases. One can choose many different types of interaction potential $\epsilon_p(r)$. The evaluation of the second integral appearing on the right hand side of Eq. 6.98 may turn out to be a crucial step in handling more. We refrain from pursuing these problems in this work.

6.6 Classical Statistical Mechanics and Macroscopic Properties

Macroscopic systems and their properties, especially those associated with the description of thermal phenomena have been widely and traditionally studied by the methods of thermodynamics. The key to thermodynamics lies in a set of laws – the laws of thermodynamics, which are empirical, but infallible in nature. The statistical mechanics formulated by Gibbs targets the same systems and their properties on the basis of a distribution law for the dynamic quantities like position ($\vec{r_i}$) and momentum ($\vec{p_i}$) of the microscopic constituents – the atoms and molecules of the system.

In this section, we begin by analyzing how we can use Gibbs' distribution law in constructing the equation of state of an ideal gas, which expresses the dependence of the pressure 'P' of the gas on its volume (V) and temperature (T) like $P = f(V, T)$. The equation of state is an important thermo-mechanical relation and is well-established for an ideal gas. This equation of state reads

$$P = \frac{m}{M} \frac{RT}{V} \qquad (6.99)$$

where 'm' is the mass of the gas of molecular weight M enclosed in a volume 'V' at the temperature T (in Kelvin). Our endeavor will be to derive a generalized equation of state that holds for all systems including

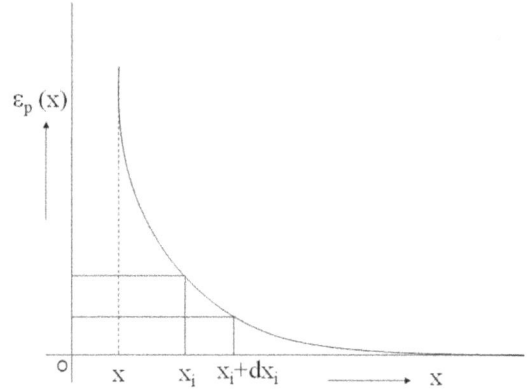

FIGURE 6.4 The variation of potential energy (ϵ_p (x)) of a particle as a function of distance (x_i) from the wall at x.

an ideal gas. The key in this approach will be the quantity we have called free energy (F), which appeared naturally in the Gibbs distribution. The starting point in this enterprise is a system enclosed in a vessel. The constituent part of the system (may be the ideal gas, non-ideal gas or perhaps, a liquid) are continuously impinging on the walls of the vessel and repulsed. Let us consider one of the walls perpendicular to the axis 'OX' (Figure 6.4). The coordinate of this wall has a value equal to x. The potential energy (ϵ_p) of one of the constituent atoms/molecules (say i) of the system as it approaches the wall and bounces off varies sharply with the distance x_i of the ith molecule from the wall and has the following profile (Figure 6.4).

The ith molecule is acted upon by a repulsive force f_i' from the wall, which rises sharply as the ith molecule comes closer to the wall and dies out quickly as it moves away from the wall. The work done in the process appears as the potential energy ϵ_p, which depends only on the distance between the molecule and the wall so that we can formally write ϵ_p as a function of $(x - x_i)$:

$$\epsilon_p = \epsilon_p(x - x_i) \tag{6.100}$$

The work done by the repulsive force f_i' exerted by the wall on the ith molecule in moving it from $x_i \to x_i + dx_i$

$$dA = f_i' dx_i \tag{6.101}$$

and the potential energy gets lowered (Figure 6.4). The magnitude of change is

$$d\epsilon_p = \epsilon_p(x_i) - \epsilon_p(x_i + dx_i)$$
$$= -\frac{d\epsilon_p}{dx_i} \tag{6.102}$$

Comparing Eq. 6.101 and Eq. 6.102, we are led to an expression for the force exerted by the wall on the ith molecule

$$f_i' = -\frac{d\epsilon_p}{dx_i} \tag{6.103}$$

However, we are trying to calculate the pressure exerted by the gas molecules on the wall and hence must find the force f_i that the ith molecule acts on the wall with. By Newton's third law of motion, f_i must be equal to f_i' (Eq. 6.103) and oppositely directed. Therefore, we find

$$f_i = -f_i' = \frac{d\epsilon_p}{dx_i} \tag{6.104}$$

Using Eq. 6.100 for ϵ_p, we can immediately see that f_i' can be expressed as derivative with respect to the wall coordinate (x);

$$f_i = -\frac{d\epsilon_p(x - x_i)}{dx_i} \tag{6.105}$$

The total force exerted by all the 'N' molecules of the gas on the wall is clearly obtained by summing up both sides of Eq. 6.105 over 'i' yielding

$$F = \sum_{i=1}^{N} f_i$$

$$= -\frac{d}{dx} \sum_{i=1}^{N} \epsilon_p(x - x_i)$$

The Gibbs distribution can now be invoked to find the average force $\langle F_1 \rangle$ on the wall, which reads

$$\langle F_1 \rangle = \int \left\{ -\frac{d}{dx} \sum_{i=1}^{N} \epsilon_p(x - x_i) \right\} e^{\frac{F-\epsilon}{kT}} d\Gamma \equiv -\int \frac{d\epsilon}{dx} e^{\frac{F-\epsilon}{kT}} d\Gamma$$

where we have used $\epsilon = \epsilon_p + \epsilon_k + \epsilon_p^{int} + \epsilon_p^{ext} = \sum_i (\epsilon_p(x - x_i) + \frac{p_i^2}{2m_i})$, p_is are in no way dependent on the wall-coordinate 'x', and the gas molecules are ideal (so, no potential energy term appears in ε due to interparticle interactions, ($\epsilon_p^{int} = 0$), neither is there any external force acting on the molecules ($\epsilon_p^{ext} = 0$)). Note, however, that even if $\epsilon_p^{int} \neq 0$, $\epsilon_p^{ext} \neq 0$, these energy components, like the kinetic energy component (ϵ_k) would not depend on the coordinate 'x' of the wall. So, in any case we can write

$$\frac{d\epsilon}{dx} = \frac{d}{dx} \sum_{i=1}^{N} \epsilon_p(x - x_i) \tag{6.106}$$

The pressure (P) on the wall is just the average force $\langle F_1 \rangle$ per unit area (area = a), which then becomes

$$P = \frac{\langle F \rangle}{a} = -\int \frac{1}{a} \frac{d\epsilon}{dx} e^{\frac{F-\epsilon}{kT}} d\Gamma \tag{6.107}$$

where $\frac{d\epsilon}{dx}$ is given by Eq. 6.106, and a is the surface area of the wall. adx in the above equation can be identified with the change in volume 'dV' of the system when the wall of the vessel is displaced by dx. With that Eq. 6.107 becomes

$$P = -\int \frac{d\epsilon}{dV} e^{\frac{F-\epsilon}{kT}} d\Gamma \tag{6.108}$$

The integral above is easily evaluated by taking an indirect route that starts from the normalization condition (cf. Eq. 6.83)

$$\int e^{\frac{F-\epsilon}{kT}} d\Gamma = 1 \tag{6.109}$$

Differentiating the equation above (Eq. 6.109) with respect to volume (V) we have

$$\int e^{\frac{F-\epsilon}{kT}} \cdot \frac{1}{kT} \left(\frac{\partial F}{\partial V} - \frac{\partial \epsilon}{\partial V} \right) d\Gamma = 0 \tag{6.110}$$

In arriving at Eq. 6.110 we have made use of the fact that both free energy (F) and total energy (ε) depend on the wall-coordinate (x) and therefore on volume $V (= a \cdot x)$.

Eq. 6.110 leads to the condition

$$\int \frac{\partial F}{\partial V} e^{\frac{F-\epsilon}{kT}} d\Gamma = \int \frac{\partial \epsilon}{\partial V} e^{\frac{F-\epsilon}{kT}} d\Gamma \tag{6.111}$$

Note that F does not depend on the integration variables in $d\Gamma$. $\frac{\partial F}{\partial V}$ can therefore be taken outside the integral sign allowing us to write

$$\int \frac{\partial \epsilon}{\partial V} e^{\frac{F-\epsilon}{kT}} d\Gamma = \frac{\partial F}{\partial V} \int e^{\frac{F-\epsilon}{kT}} d\Gamma = \frac{\partial F}{\partial V} \tag{6.112}$$

Using Eq. 6.108 in Eq. 6.112 we get

$$-P = \frac{\partial F}{\partial V}$$

$$\text{or,} \quad P = -\frac{\partial F}{\partial V} \tag{6.113}$$

Earlier, in section 6.5, we arrived at an expression for the free energy F of an ideal monatomic gas (Eq. 6.95). Using it in Eq. 6.113, we find

$$P = \frac{kNT}{V} \tag{6.114a}$$

If $N = N_A$ (Avogardro's number), we have $kN_A = R$, whence we get the molar equation of state

$$P = \frac{RT}{V} \tag{6.114b}$$

If we have a mass 'm' of the gas with molecular weight M, N in Eq. 6.114a can be replaced by $N = (m/M) \cdot N_A$ and we have the ideal gas equation in a more general form ($n = m/M$ = number of moles)

$$P = \frac{nRT}{V} \tag{6.114c}$$

where 'n' is the number of moles of the gas in the vessel of volume 'V' at a temperature T. The Gibbs method is perfectly general and can be applied to any system in a similar manner be it an ideal or non-ideal gas or a liquid to arrive at the corresponding equation of state. The only requirement is that we must first workout an analytical expression for 'F' – the free energy of the system. We will not pursue these more involved problems here, but refer the readers to more specialized texts and point out that for specific systems, the toughest task is to construct an appropriate yet workable potential energy function $\epsilon_p(r)$ for the particles in the system and apply Gibbs' formulation diligently.

We have seen how the free energy 'F' in Gibbs' distribution is computed for an ideal gas without ascribing any physical meaning to it. Before leaving this section, we will briefly examine the physical meaning that 'F' has.

Let us imagine an isothermal process taking place in our ideal gas system that leads to a change in volume at constant temperature. The free energy F is a function of both volume and temperature, hence we have

$$dF = \frac{\partial F}{\partial V} dV + \frac{\partial F}{\partial T} dT \tag{6.115}$$

$dT = 0$ for an isothermal process. Therefore,

$$dF = \frac{\partial F}{\partial V} dV = -PdV \tag{6.116}$$

where we have made use of Eq. 6.113 derived from classical statistical mechanics formulated by Gibbs. PdV, as we know, is the mechanical work 'dA' done by the system in the course of an infinitesimal volume expansion (dV) leading to the identification

$$dF = -dA = -PdV$$

For a finite isothermal expansion, we can arrive at (by integration of the relevant equation)

$$\Delta F = -\Delta A$$

As we have stated earlier, 'F' depends on both volume (V) and temperature (T), but not on the microscopic variables p_i, r_i of individual particles, i.e. the Γ (phase) space variables. F is thus a state function. The change ΔF of which in an isothermal process is equal to the mechanical work ΔA, albeit with opposite sign. This completes the thermodynamic interpretation of F – the central quantity in Gibbs' formulation as free energy.

6.6.1 Gibbs-Helmholtz Equation from Classical Statistical Mechanics: Internal Energy (U)

Let us start with the normalization condition

$$\int e^{\frac{F-\varepsilon}{kT}} d\Gamma = 1$$

Differentiating both sides with respect to 'T' (the temperature of the system at equlibrium) we arrive at the condition

$$\int \frac{\partial}{\partial T}\left(e^{\frac{F-\varepsilon}{kT}}\right) d\Gamma = 0$$

$$or, \int e^{\frac{F-\varepsilon}{kT}}\left\{\frac{1}{kT}\frac{\partial F}{\partial T} - \frac{F-\varepsilon}{kT^2}\right\} d\Gamma = 0$$

$$or, \int e^{\frac{F-\varepsilon}{kT}}\left\{T\frac{\partial F}{\partial T} - F + \varepsilon\right\} d\Gamma = 0 \tag{6.117}$$

$$or, \int \left(T\frac{\partial F}{\partial T} - F\right)e^{\frac{F-\varepsilon}{kT}} d\Gamma + \int \varepsilon e^{\frac{F-\varepsilon}{kT}} d\Gamma = 0$$

$$or, \left(T\frac{\partial F}{\partial T} - F\right)\int e^{\frac{F-\varepsilon}{kT}} d\Gamma + \int \varepsilon e^{\frac{F-\varepsilon}{kT}} d\Gamma = 0$$

In the last step we have exploited the independence of F as well as $\frac{\partial F}{\partial T}$ on the phase-space variables, i.e. the integration variables of our problem ($d\Gamma$). We observe now

(i) $\int e^{\frac{F-\varepsilon}{kT}} d\Gamma = 1$,

(ii) $\int \varepsilon e^{\frac{F-\varepsilon}{kT}} d\Gamma = \langle \varepsilon \rangle = U(\text{say})$

where the average total energy at temperature T has been identified as the thermodynamic quantity called internal energy U.

$$T\frac{\partial F}{\partial T} - F + U = 0$$

$$\text{i.e. } U = F - T\frac{\partial F}{\partial T} \tag{6.118}$$

which is the Gibbs-Helmholtz equation we encounter in thermodynamics. 'U' – the internal energy of the system can thus be interpreted as the average total energy of the system at temperature T defined with the aid of Gibbs' distribution and its value can be computed in terms of the free energy F and its temperature coefficient $\frac{\partial F}{\partial T}$ with the aid of equation Eq. 6.117, which is a general relation and holds for any system. Thus, we can model the thermodynamic quantity U (internal energy) of any system by invoking Gibbs' formulation of statistical mechanics to calculate the free energy 'F' appearing in Gibbs' distribution function and then using the Gibbs-Helmholtz equation (Eq. 6.117). As an example, we can consider one mole of an ideal monatomic has and make use of F already calculated in the previous section (cf. Eq. 6.95) and find,

$$U = F - T\frac{\partial F}{\partial T} = \frac{3}{2}\left(kN_A\right)T = \frac{3}{2}RT$$

$$\text{and, } \left(\frac{\partial U}{\partial T}\right)_V = C_V = \frac{3}{2}R$$

6.6.2 Entropy: Statistical Mechanical and Thermodynamic Interpretation

The free energy F is a state function. So, its derivatives or other quantities expressed entirely in terms of F and its derivatives are also a state function. Two such quantities already introduced are pressure (P)

and internal energy U. The temperature derivative of F appearing on the right hand side of Eq. 6.118 is presumably another state function and we may call the state function 'S' and define

$$S = -\frac{\partial F}{\partial T} \qquad (6.119)$$

In thermodynamics 'S' is called entropy, a collective 'coordinate' of the system. Its conjugate coordinate is the temperature T so that TS represents the 'work done' due to changes in this collective coordinate. The 'work done' is actually the amount of 'heat' that is delivered to or withdrawn from the system across the system surrounding boundary. In thermodynamics, one shows by analyzing a reversible process with the aid of the Gibbs-Helmholtz equation that

$$dS = \frac{dU + dA}{T} \qquad (6.120)$$

The first law of thermodynamics, as we know, asserts the law of energy conservation taking thermal changes into account. The numerator in Eq. 6.120 represents the sum of the changes in internal energy and the work done by the system. It must also represent the amount of the heat supplied to the system (dQ), thereby leading to an expression

$$dS = \frac{dQ}{T} \qquad (6.121)$$

Thus the 'collective coordinate', 'S' is a state function, an infinitesimal change in which in a reversible process is equal to the ratio of the infinitely small amount of heat added to the system (dQ) and the temperature T at which the process is taking place. If the process is a finite reversible process, the change in the state function 'S' can be calculated by integration:

$$S_2 - S_1 = \Delta S = \int_1^2 ds = \int_1^2 \frac{dQ}{T} \qquad (6.122)$$

The integral above is independent of the shape of the contour around which the calculation is being carried out – a rather remarkable property ensured by a state function. Gibbs' formulation of statistical mechanics provides a direct route to the entropy 'S' of a system through the free energy 'F' appearing in the Gibbs distribution function. Thus, for a system comprising of N atoms of a monatomic ideal gas with atomic mass m, as we have already seen (Eq. 6.95) is

$$F = -kTN\left[lnV + \frac{3}{2}lnT + \frac{3}{2}ln(2\pi m_0 k)\right]$$

Differentiating F with respect to the temperature T we have

$$S = -\frac{\partial F}{\partial T} = kN\left[lnV + \frac{3}{2}lnT + ln(2\pi m_0 k)^{\frac{3}{2}}\right] + \frac{3}{2}kN$$

Assuming $N = N_A$ (Avogadro's number), we arrive at the following expression for molar entropy of a monatomic ideal gas ($kN_A = R$):

$$S = R\left[lnV + \frac{3}{2}lnT + ln(2\pi m_0 k)^{\frac{3}{2}}\right] \qquad (6.123)$$

where 'e' is the base of natural algorithm. Thus we have been able to provide a means for quantitative evaluation of the newly introduced thermodynamic state function called entropy by invoking the principles of classical statistical mechanics enunciated by Gibbs. More importantly, as we will see now, statistical mechanics provides a clear physical (but statistical) meaning to entropy in a natural manner and allows us to understand why entropy plays such an important role in thermodynamics and statistical physics. Thus from Gibbs-Helmholtz equation the entropy function 'S' is given by

$$S = \frac{U - F}{T} \qquad (6.124)$$

We have already seen that the internal energy is the average of total energy

$$U = \langle \varepsilon \rangle = \int \varepsilon e^{\frac{F-\varepsilon}{kT}} d\Gamma \qquad (6.125)$$

while we can write formally

$$F = \int F e^{\frac{F-\varepsilon}{kT}} d\Gamma \qquad (6.126)$$

Knowing that F does not depend on the integration variables in $d\Gamma$ and that $\int e^{\frac{F-\varepsilon}{kT}} d\Gamma = 1$, we can write combining Eqs. 6.124-6.126,

$$S = -k \frac{F - \langle \varepsilon \rangle}{kT} \qquad (6.127)$$

$$S = -k \int \frac{F - \varepsilon}{kT} e^{\frac{F-\varepsilon}{kT}} d\Gamma \qquad (6.128)$$

As we know, $e^{\frac{F-\varepsilon}{kT}}$ in Gibbs' formulation is the probability density that a state, representing a point in the phase-space volume element $d\Gamma$ is actually realized. Note that $\frac{F-\varepsilon}{kT}$ is just $ln e^{\frac{F-\varepsilon}{kT}} = ln\omega$ so that Eq. 6.128 becomes

$$S = -k \int \omega ln\omega \, d\Gamma = -k \langle ln\omega \rangle \qquad (6.129)$$

Eq. 6.129 is just the information theoretical defination of entropy as a measure of randomness we introduced in section 6.2 except that the discrete sum has been replaced by integration as the 'events' are no longer countable here, and that the appearance of the Boltzmann constant in Eq. 6.129 adds a dimension to 'S' while it was treated as a dimensionless entity in Section 6.2.

The statistical mechanical definition of entropy and its interpretation as a measure of randomness in the system are important concepts, which will play a key role in setting up the entire apparatus of quantum statistical mechanics of equilibrium states. Before leaving the topic, let us note that Eq. 6.129 has an important shortcoming, which we will take up in the next chapter and show that it disappears when a proper quantum equivalent of Eq. 6.129 is used.

6.7 Statistical Mechanics and Numerical Simulation

The science of numerical simulation is a rapidly evolving and a fast growing branch of theoretical science and has become an essential tool for analyzing, interpreting and understanding the behavior of condensed matter systems and properties of materials. In this section we will briefly review the two basic approaches of simulating a system and predicting its properties in the context of statistical mechanics.

The advent of highly powerful digital computers has made a strong impact on statistical mechanics by enabling the scientists to solve problems requiring rather complicated calculations and in the process has made significant contributions to our understanding of the basic concepts of statistical mechanics and ensembles. Of the two computational schemes, the first is the method that works by solving the classical equations of motion of all the particles in the system numerically and calculating the positions and velocities (or momenta) of all the particles (atoms, molecules, ions, etc.) as functions of time. This approach involving the calculation of molecular motion can be viewed as a numerical experiment that involves calculating the trajectory of motion explicitly and then predicting equilibrium properties of the system by taking appropriate trajectory averages of the quantities of interest. One pertinent question that arises here concerns whether equilibrium would be reached and the properties would indeed be the equilibrium properties. It is now firmly believed that equilibrium states would indeed emerge provided that there are a large enough number of molecules (particles) in the system and the trajectories have been calculated for a long enough time for a set of initial conditions. The collection of equations of a molecular motion-based approach is called classical molecular dynamics simulation and nowadays is regarded as a branch of statistical mechanics.

As already mentioned, the calculation of the kind described represent a type of experiment in which the positions and momenta of all the molecules in the system are calculated for a long time. Any information (thermodynamic, structural, etc.) about the system during the period of calculation can be obtained from the trajectory and each equilibrium property of the system can be obtained by averaging. What are the advantages of the procedure? Most importantly, these calculations do not employ approximations (except numerical approximations). Once a model is proposed and initial conditions are set, the outcome of the numerical experiment is fixed. One might even argue that these schemes are more powerful than analytical treatments employing assumption and approximations and the results are therefore directly comparable to those obtained by real experiments. Such comparisons are often carried out, wherever possible, to justify the model. The second approach is Monte Carlo simulations of the system. The states generated by a deterministic equation of motion (Newton equation) is now replaced by generating a random sequence of dynamical states and equilibrium properties are calculated using these states by appropriate averaging of the desired quantities. The Monte Carlo simulations may appear to have done away with trajectories. In effect they also calculate trajectories, but instead of Newton's law, makes use of a model of random (stochastic) motion. This approach also can be regarded as a numerical experiment, just as the molecular dynamics (MD) simulations.

6.7.1 MD Simulations (Basic Idea)

From the introductory discussion in the previous section, we know that the systems to be dealt with here are such that all the degrees of freedom are explicitly taken care of without introducing any stochastic elements, for example, the interaction of the system with a heat bath. The MD simulations start with a well defined microscopic description of the physical system, usually a many body system. This description may be based on a Hamiltonian, Lagrangian or expressed directly in the form of Newton's equation of motion. We will only deal with the third option and proceed to solve Newton's equation of motion for the system of particles numerically on a computer.

This can be done by approximating the EOM by suitable schemes that can be easily evaluated numerically on a computer. That inevitably involves the transition from a description based on continuous variables with differential operators to a description with discrete variables and finite difference representation of operators. That entails errors, which, in principle, can be made as small as possible, depending on the capabilities of the hardware at our disposal, namely the speed and memory of the computer.

By definition MD simulation methods compute the phase-space trajectories of a collection of molecules, which individually obey the classical laws of motion. The system need not be just a collection of point particles but may be a collection of particles with subunits or internal structures with or without internal constraints. In the early days, MD simulations were carried out for systems where the energy would be a constant of motion (a conserved quantity). Accordingly, for property calculations, one used averaging over the microcanonical ensembles for which the particle number (N), the volume (V) and the energy (E) are constants. If we are interested in the behavior of the system at a constant temperature (T), the appropriate ensemble for averaging would be the canonical ensemble. The MD simulations technology now available has advanced to a stage where computations can be carried out with the ensembles that are not necessarily microcanonical.

In general, we will have to deal with equations of the following form:

$$\frac{dy(t)}{dt} = k\Big(y(t), t\Big) \tag{6.130}$$

where $y(t)$ is the unknown quantity, say a velocity, position, an angle, etc.; 'k' is a known operator and 't' is time. We will restrict ourselves only to situations where 'k' does not involve stochastic elements and initial conditions are precisely known.

Let us now consider, for the sake of simplicity a monatomic system of N units in which the atoms interact with pair-wise additive central force potential. This choice of the system and interactions guarantee that molecular interactions are orientation-independent. However, one can generalize the approach to include orientation-dependent interactions and constraints due to connectivity (as in a polymer chain). The systems of our interest will be described in general, by the Hamiltonian (classical) H where

$$H = \frac{1}{2} \sum_i \frac{P_i^2}{m_i} + \sum_{i<j} u(r_{ij}) \qquad (6.131)$$

r_{ij} being the distance between the particles (atoms) i and j. The configurational interaction energy (the potential energy) can be represented in an abbreviated manner as

$$U(r) = \sum_{ij} u(r_{ij}) \qquad (6.132)$$

where $u(r_{ij})$s are additive pair potentials.

Since we will focus on properties of the N particle system at a specific density ρ, we have to specify the volume of the so called MD cell so as to create a bulk at a specified density. If our system is in thermal equilibrium the shape of the MD cell is immaterial for liquids and gases in the limit of large volume. For crystalline solids, the shape of the MD cell does make a difference and has to be chosen with care.

For liquids and gases, a cubic MD cell may be chosen with volume $V = L^3$, L being the linear size of the cell. The consequence of introducing a cubic cell with a finite volume introduces six unwanted surfaces. If the number of particles (N) in the system is small, any computed property will be vitiated by strong contribution from these surface. To eliminate or at least to reduce the effect of the surfaces, it is convenient to impose periodic boundary conditions. Physically that means, the basic MD cell is repeated an infinite number of times. Mathematically, this amounts to imposing on an observable O the following conditions for all integer values of (n_1, n_2, n_3):[1–3]

$$O(\vec{X}) \equiv O(\vec{X} + \vec{n} L) \qquad (6.133)$$

where $\vec{n} = (n_1, n_2, n_3)$ is an integer vector in 3D space. In practice it amounts to ensuring computations that a particle (molecule) crossing a surface of the basic MD cell, it must re-enter through the opposite surface (wall) with unchanged velocity. Thus, with the imposition of periodic boundary conditions, we have been able to eliminate the unwanted surfaces and create a quasi-infinite volume that simulates the macroscopic system much more closely. In this representation, each component of the position vector will be a number between zero and L. If a particle (molecule) is at the location $\vec{r_i}$, there is supposed to be a set of image particles at locations $\vec{r_i} + \vec{n} L$. The imposition of periodic boundary conditions thus brings in a host of image particles and that affects the potential energy U as follows:

$$U(\vec{r_1}, \vec{r_2}, \ldots, \vec{r_N},) = \sum_{i<j} u(r_{ij}) + \sum_{\vec{n}} \sum_{i<j} u(r_i - r_j) \qquad (6.134)$$

The second term has an infinite number of terms. $\vec{n}$ is an integer vector in three-dimensional space. The problem of evaluating an infinite sum is avoided by adopting a convention (called the minimum image convention) that determines how the interparticle distances are to be computed. Stated explicitly, the minimum image convention proposes that the distance r_{ij} between the particle 'i' at $\vec{r_i}$ and 'j' at $\vec{r_j}$ is to be fixed by setting

$$r_{ij} = min\left(|\vec{r_i} - \vec{r_j} + \vec{n} L|\right) \text{ over all integer vectors } \vec{n} \qquad (6.135)$$

Condition Eq. 6.135 ensures that every particle in the MD cell interacts only with the $(n-1)$ other particles in the cell or their nearest images.[6–7] The 'minimum image convention' effectively cuts off the potential by the condition $r_c < L/2$. That introduces some error in the calculated quantities. It would have been better to evaluate the interaction of each particle with all the image particles by adopting the elegant technique of Ewald (Ewald sum). To the best of our knowledge, the exact extent by which the computed properties are affected by the adoption of minimum image convention vis-a-vis properties computed by adopting Ewald's method are not fully understood yet. We will therefore advocate the use of the minimum image convention with 'L' large enough to ensure that interactions that would have taken place between particles separated by distances larger than $L/2$ would be negligibly small.

6.7.2 Calculation of Thermodynamic Properties

In MD simulation, the equations of motion, upon integration generate phase-space trajectories, and ensemble average has to be replaced by time-average over the trajectories. In a traditional MD simulation, the number of particles (N) is fixed along with the volume V. We can regard the total linear momentum also as a conserved quantity and set its value to zero in order to avoid any movement of the system as a whole. Let the trajectory generated by an MD algorithm (see later) corresponding to a given initial position ($\vec{r^N}(0)$) and momentum ($\vec{p^N}(0)$) be ($\vec{r^N}(t), \vec{p^N}(t)$). If the energy is a conserved quantity and if the trajectories spend equal time in equal volume with the same energy, then the trajectory average of a thermodynamic quantity $A(\vec{r^N}(t), \vec{p^N}(t), V(t))$ is given by[4]

$$\bar{A} = \lim_{t' \to \infty} \frac{1}{(t' - t_0)} \int_{t_0}^{t'} dt A(\vec{r^N}(t), \vec{p^N}(t)) \tag{6.136}$$

The ergodic theorem asserts that under the stipulated conditions the trajectory average becomes equal to the microcanonical ensemble average $\langle A \rangle_{NVE}$ whereas the subscripts specifying quantities are conserved. In what follows, we will describe in brief details how a microcanonical ensemble molecular dynamics is set up and implemented.

6.7.3 Microcanonical Ensemble Molecular Dynamics

As already mentioned, our starting point is the definition of the Hamiltonian of the system and constraints, if any. For a two body potential with spherical symmetry ($u(r_{ij})$), the N particle Hamiltonian reads (all the particles are assumed to have mass m)

$$H = \frac{1}{2} \sum_i \frac{p_i^2}{m} + \sum_{i<j}^{N} u(r_{ij}) \tag{6.137}$$

The particle number N and energy E are conserved quantities of the system along with the total linear momentum which is set to be equal to zero. From the point of view of numerics, the MD method is an initial value problem for which many algorithms have been developed. In the present context many of these algorithms are not suitable for computational reasons. Let us consider the equation of motion in Eq. 6.130 in the context of the Hamiltonian defined in Eq. 6.137. The equations of motion are

$$\frac{d\vec{r_i}}{dt} = \frac{\vec{p_i}}{m}$$

$$\frac{d\vec{p_i}}{dt} = \sum_{i<j}^{N} \vec{F_i}(r_{ij}), \ i = 1, 2, \cdots, N \tag{6.138}$$

It turns out that each evaluation of the right hand sides of Eq. 6.138 for the N particles tasks $\frac{N(N-1)}{2}$ time consuming operations, which makes the scheme unworkable. One has to adopt therefore simpler integration schemes without sacrificing the required accuracy and conservation properties. The commonly used technique involves discretization that converts the differential equation of motion (Eq. 6.138) into equations involving finite difference schemes from which suitable recursion relations for positions and momenta (velocities) of the particles are constructed. The recursion relations must be such that efficient evaluation can be made and the time propagation scheme remains numerically stable. To illustrate the construction of a practicable scheme for integration of the relevant equations of motion let us consider the equations in Eq. 6.138 combined into the corresponding second order differential equation – Newton's equation of motion:

$$\frac{d^2 r_i(t)}{dt^2} = \frac{1}{m} \sum_{i<j} F_i(r_{ij}) \tag{6.139}$$

To discretize, let us adopt the explicit central difference formula for the second order differential operator on the left hand side of Eq. 6.139 at time 't', which yields

$$\frac{d^2 r_i}{dt^2} = \frac{1}{h^2}\left\{r_i(t+h) - 2r_i(t) + r_i(t-h)\right\}$$
$$= \frac{1}{m}F_i(t); \quad i = 1,2,\ldots,N \tag{6.140}$$

We can rearrange Eq. 6.140 into the following algorithmic form:

$$r_i(t+h) = 2r_i(t) - r_i(t-h) + \frac{F_i(t)h^2}{m} \tag{6.141}$$

It is now clear that if the force acting on the ith particle of mass m is known at time t, we can predict the position of the particle at the next instant, i.e. at $t = t + h$, provided we know the particle positions at the two immediately preceeding time steps, i.e. $r_i(t)$ and $r_i(t-1)$. For computer implementation it is convenient to introduce the following change in symbols and let

$$t_n = nh$$
$$r_i^n = r_i(t_n) \tag{6.142}$$
$$F_i^n = F_i(t_n)$$

and recast Eq. 6.141 in the following form:

$$r_i^{n+1} = 2r_i^n - r_i^{n-1} + \frac{h^2}{m}F_i^n; \quad i = 1,2,\ldots,N \tag{6.143}$$

Thus we may start with two positions for each particle, r_i^0 and r_i^1 and use the recursion formula in Eq. 6.143 to determine the positions at subsequent times. Notice that we also need the velocities of the particles at every time step for computing the kinetic energy of the system of particles. To obtain the velocities, let us note that $v_i^t = \frac{dr_i^t}{dt}$ and make use of the finite difference formula for evaluating the first derivative of r_i^t, to obtain

$$v_i^n = \frac{r_i^{n+1} - r_i^{n-1}}{2h}; \quad i = 1,2,\ldots,N \tag{6.144}$$

The formula above immediately reveals that at the $(n+1)$th time step, the velocities available are for the previous time step only. That means, the kinetic energy will lag one step behind the potential energy. Eq. 6.143 and Eq. 6.144 together constitute the very popular Verlet algorithm for performing MD simulation at constant particle number (N), volume (V) and energy (E). We will summarize now the algorithmic steps for *NVE* molecular dynamics:

1. Provide positions r_i^0 and r_i^1 for all the particles in the MD cell.
2. Compute the forces at time step n on each particle (F_i^n).
3. Update the positions at time step $(n+1)$ using the formula in Eq. 6.141.
4. Compute the velocities at the nth time step using the formula in Eq. 6.144.
5. Compute kinetic and potential energies from updated positions and velocities.

The form of Verlet algorithm described above suffers from being non-self-starting. It requires, as we have seen, not one set of the initial positions of each particle, but also another set of positions for the algorithm to start. However, if initial positions (r_i^0) and also initial velocities (v_i^0) are given, we can reformulate the procedure to update position at the next time step by writing

$$r_i^1 = r_i^0 + hv_i^0 + \frac{1}{2m}h^2 F_i^0$$

and the set (r_i^0, r_i^1) is fed back to the Verlet algorithm, which then proceeds normally as described already.

In a reformulated form, the Verlet algorithm has been designed to predict the positions and velocities for the same time step with enhanced overall numerical stability. The reformulated scheme has become known as the Verlet algorithm in velocity form. We outline in what follows the steps executed in an *NVE* MD simulation in velocity form:

1. Provide the initial positions: r_i^1; $i = 1,2, \ldots, N$.
2. Specify the initial velocities: v_i^1; $i = 1,2, \ldots, N$.
3. Compute the positions at $(n + 1)$th time step by the recursion formula $r_i^{n+1} = r_i^n + h v_i^n + \frac{1}{2} \left(\frac{F_i^n}{m} \right) h^2$; $i = 1,2,\ldots,N$.
4. Compute the velocities at the $(n + 1)$th time step by the following formula

$$v_i^{n+1} = v_i^n + h \frac{1}{2} \frac{(F_i^{n+1} + F_i^n)}{m}$$

The velocity-Verlet algorithm possesses enhanced numerical stability and many other advantages. In general the MD simulations can be broadly divided into three parts:

1. Initialization.
2. Equilibration.
3. Production.

In the initialization step, initial conditions like positions and velocities of the particles are specified at $t = 0$. In general, precise initial conditions are unknown. Several choices are therefore possible. Most commonly, one assigns initial positions on a lattice and the initial velocities are drawn from a Boltzmann distribution. One need not worry too much about the initial conditions as the system will ultimately lose all memory of the initial state. However, the system set up following the prescription of initialization described already will not have the desired energy and almost certainly the state will not correspond to an equilibrium state.

In the equilibration phase, the system is nudged to reach the equilibrium state by either removing or adding energy until the energy of the system has attained the desired value. The energy removal is achieved by scaling down the velocities (KE is reduced) while energy addition is realized by scaling the velocities up (KE increases) and in either case, the system is allowed to relax into an equilibrium state by integrating the equations of motion forward for a number of time steps. When the mean values of the kinetic and potential energies have settled down, the system signals attainment of equilibrium. Now the system is ready for calculation of the desired quantities. The attainment of equilibrium may require a long time (many integration steps) if the relaxation time of the system (τ) is large and the simulation time steps are small. It is often possible to avoid the problem by approximate scaling of variables. A second problem likely to be faced concerns the system getting trapped in a long lived metastable state. Once the equilibration phase is over, the actual calculation of quantities, thermodynamics or otherwise is carried out, using the trajectory of the system in phase space. We must remember that the results obtained with a given set of initial positions and velocities is only one particular realization out of a host of possible ones. Had we started with another set of initial positions and velocities, the system would have evolved along another trajectory or path on the constant energy surface. However, even if a single path is followed over an infinitely long duration so that the system spends equal time in equal volumes of the phase space, we may invoke the ergodic hypothesis and accept the quantities calculated as time averages over a very long time along a single trajectory as the 'ensemble averaged quantities'. But following a path for infinite time on a computer is not feasible. One must therefore compute the quantities by following the trajectory for different lengths of time and ascertain if the asymptotic behavior has already set in.

Before closing the section, let us consider the computational complexity of the algorithm and time-saving instruments that can be invoked. We note that the integration steps require approximately N operations. Assuming that the two body additive central field potentials are used, the computation of forces require us to evaluate $\frac{1}{2}N(N - 1)$ terms at each steps. One can, however, reduce the theoretical complexity by taking advantage of the fact that the potential used has a cut off at $r = r_c$ so that we can avoid calculation of forces and set them equal to zero whenever $r > r_c$.

Higher order algorithms for integration of equation of motion are often used in simulation of large systems for long durations. Such algorithms belong to the multi-step predictor – corrector class developed for example, by Gear,[7] Beeman[8] and Toxvaerd.[9] They require far more storage space compared to one-and two-step methods. Hardware limitation often forces the users to avoid predictor-corrector algorithms. To move from NVE-MD to NVT-MD, we have to modify the equations of motion so far used, which allow us to propagate only on a constant energy surface (NVE-MD). For NVT-MD, we must be able to investigate the dynamics along an isotherm rather than along a line of constant energy. The necessary modification can be introduced by assuming that the system under investigation is coupled to a heat bath, which causes the 'system-energy' to fluctuate. The properties of the system can then be defined as averages over an appropriate ensemble. The ensemble appropriate for describing the systems-heat bath equilibrium is a canonical ensemble where the number of particles (N), volume (V) and the temperature (T) are fixed. The total energy (E) is not a constant of motion at constant temperature and fluctuates. The average fluctuations for a constant T can be achieved by integrating the equations of motion subject to an appropriate constraint. In isokinetic MD, the equation of constraint is

$$\frac{1}{2} \sum_i m v_i^2 = constant \tag{6.145}$$

with the constant depending on the temperature.

To constrain the kinetic energy of the system to a given value, and allow the system to equilibrate energy can be added to or withdrawn from the system by scaling the velocities. Once the targetted temperature is reached, the system is left undisturbed. This is the basic idea of the NVT-MD, which executes the following steps.

1. Define the initial positions (r_i^0, $i = 1, 2, \ldots, N$).
2. Specify the initial velocities (v_i^0, $i = 1, 2, \ldots, N$).
3. Calculate the positions at time step $n + 1$: $r_i^{n+1} = r_i^n + h v_i^n + \frac{h^2}{2m} F_i^n$.
4. Compute the velocities at $(n + 1)$th time step as $v_i^{n+1} = v_i^n + \frac{h}{m} \frac{(F_i^{n+1} + F_i^n)}{2}$.
5. Compute $\frac{1}{2} \sum_i m (v_i^{n+1})^2$ and calculate the scaling factor λ from the constraint equation.
6. Scale all velocities: v_i^{n+1} (scaled) $= \lambda v_i^{n+1}$.

There are various other ways of realizing NVT-MD.[4] The numerical experiments that the MD simulations perform are often much simpler to organize and manage than real experiments and has the added advantage that observations can be made on any scale, both for equilibrium and non-equilibrium systems. There are shortcomings though vis-a-vis real experiments.

1. For one, even a collection of one thousand molecules is much much smaller than a typical macroscopic system or an object. The simulation time, even if it allows a million collisions or so, is still too short compared to macroscopic scales of time.
2. The simulations described are entirely classical while atoms and molecules are quantum mechanical objects. Therefore all quantum features are lost.

However, deficiency (1) is not that serious. Particularly, when we consider equilibrium properties of systems for which the computer model of the system used must be much larger than the correlation length and the simulation time is much longer than the correlation time. If the correlation length is small, we need not consider too many molecules, $N = 1000$ being sufficient. If the correlation time is small, we need not use a very long observation (simulation) time. In the neighborhood of a critical point, however, both correlation length and time become very large and much larger computer models must be used. That can bring the MD simulations to a very harsh test.

The deficiency (2) is fundamentally more serious and raises critical questions. There are many important problems that are inherently quantum mechanical. The behavior of electrons in metals can not be properly described or understood avoiding quantum mechanics (see Chapter 7). So is the case with the properties of graphene or likes of graphene (see Chapter 11). Full quantum MD simulation of many interacting atoms or molecules or electrons on a computer, is still far beyond the scope of traditional computing hardware and software. In spite of such limitations, the computer simulation has still a fairly wide range of applicability. This happens because many interesting phenomena like melting, crystallization, certain collective properties of heavy atoms/molecules are not essentially or intrinsically quantum in nature. However, with the advent of quantum computer and quantum computing exploiting quantum parallelism it would soon become possible to simulate intrinsically quantum systems or quantum computers.

6.7.4 Monte-Carlo Simulations

So far, we have used the classical equation of motion to construct paths in the phase-space and evaluated the quantities of interest along this path (Eq. 6.152). The MC method takes a different route. Here one starts with a description of the system in terms of a Hamiltonian and an appropriate ensemble for the problem is chosen. All the properties or observables are then computed using the associated distribution function and the partition function. The basic idea of the approach is to sample the main contributions to arrive at an approximate estimate of the observable of interest. The general problem that the MC method addresses is the following:

Let N be the number of particles. The set of dynamical variables associated with the ith particle is q_i, which essentially represents a degree of freedom and the set $(q_1, q_2, \ldots, q_N)$ defines the phase space (Ω). Let $\overrightarrow{X}$ denote a point in the Ω space and let us further assume that $\mathcal{H}(X)$ is the Hamiltonian from which the KE term has been dropped. It creates no problem as the effects of the KE term can always be taken care of analytically when required. Our task is to compute the value of the observable '**O**' of the system. Then

$$\langle \mathbf{O} \rangle = Z^{-1} \int_{\Omega} \mathbf{O}(X) f(\mathcal{H}(X)) dX \tag{6.146}$$

where $f(\mathcal{H}(X))$ is the energy distribution function for the chosen ensemble, say the canonical ensemble, and the partition function Z is defined as

$$Z = \int_{\Omega} f(\mathcal{H}(X)) dX \tag{6.147}$$

The omission of the KE term in the Hamiltonian means that the MC method will yield configurational properties of the system as opposed to the MD method which provide the actual dynamics of evolution of the system. The calculation of $\langle \mathbf{O} \rangle$ involves the evaluation of an integral of high dimensionality. In the MC method, for evaluation of the integral (Eqs. 6.146, 6.147) we proceed by assuming that the phase space is discrete, but without making any additional approximation. Let us suppose that we are working with a canonical ensemble for which the distribution function $f(\mathcal{H}(X))$ is simply proportional to the Boltzmann distribution, i.e.,

$$f(\mathcal{H}(X)) \propto e^{-\frac{\mathcal{H}(X)}{kT}} \tag{6.148}$$

it tells us immediately that a state X with large energy will make a small contribution and only a certain number of states will make a large contribution to the integral Eq. 6.146. Let X_i be a large set of states selected from the phase space randomly with equal probability.

$$\langle \mathbf{O} \rangle = \sum_{i=1}^{n} \mathbf{O}(X_i) f(\mathcal{H}(X_i)) \bigg/ \sum_{i=1}^{n} f(\mathcal{H}(X_i)) \tag{6.149}$$

We can do better by introducing the idea of importance sampling and generate states X_i with a probability $P(X_i)$. Now the expression for $\langle \mathbf{O} \rangle$ becomes[6]

$$\langle \mathbf{O} \rangle = \frac{\sum\limits_{i=1}^{n} \mathbf{O}(X_i) P^{-1}(X_i) f(\mathcal{H}(X_i))}{\sum\limits_{i=1}^{n} P^{-1}(X_i) f(\mathcal{H}(X_i))} \qquad (6.150)$$

If we choose $P(X) = Z^{-1} f(\mathcal{H}(X))$, i.e. we assume that the probability of the system being in the configuration X is just equal to the equilibrium distribution, we get the very simple and easily computable expression

$$\langle \mathbf{O} \rangle = \frac{1}{n} \sum_{i=1}^{n} \mathbf{O}(X_i) \qquad (6.151)$$

The task of solving a statistical mechanical problem has thus been amply simplified in the MC method. The only problem that must be addressed now is: how do we generate phase-space states with the probability distributed as

$$P(X) = Z^{-1} f(\mathcal{H}(X)) \qquad (6.152)$$

This specific choice for $P(X)$ means that we must sample the states of the system in thermodynamic equilibrium. The problem is that their distribution is not known beforehand. The difficulty was resolved by Metropolis *et al.* who brought in the idea of using a Markov chain. The use of Markov chain ensures that starting from an initial state (X_0) a chain of states are generated that are ultimately distributed according to $P(X)$ of Eq. 6.152. It turns out that the Markov chain is the probabilistic analog of the trajectory that the MD simulation generates from the equation of motion.

6.7.5 Transition Probabilities and Metropolis Method

In order that states are appropriately distributed, i.e. they are indeed the states in thermodynamic equilibrium, we must first assign transition probabilities $(W(X, X'))$ from one state (X) to another (X') of the system. These probabilities must satisfy the following set of constraints:[2,6]

(i) For all $X, X', W(X, X') \geq 0$.

(ii) For all $X, \sum\limits_{X'} W(X, X') = 1$. $\qquad (6.153)$

(iii) For all $X, \sum\limits_{X'} P(X') W(X, X') = P(X)$.

The first constraint concerns positivity of the transition probabilities while the second expresses the conservation of the total probability that the system will transit to a state X from all other states. The first constraint ensures that the limiting distribution will be the equilibrium distribution. In addition there is the ergodicity restriction on $W(X, X')$, which demands the following: for all complementary pairs of sets of phase points S, S' there exists states $X \in S$ and $X' \in S'$, such that $W(X, X') \geq 0$.

Let us assume now, that we have generated the states $X_0, X_1, X_2, \cdots$ with the specified transition probabilities. The evolution equation for the probability $P(X_i)$ according to which the states are distributed may be described by the differential equation (the master equation)[6]

$$\frac{dP(X, t)}{dt} = -\sum_{X'} P(X, t) W(X, X') + \sum_{X'} P(X', t) W(X, X') \qquad (6.154)$$

At a stationary point $\frac{dP(X,t)}{dt} = 0$ and the stationary solution of the master equation satisfy the condition

$$\sum_{X'} P(X) W(X, X') = \sum_{X'} P(X') W(X', X) \qquad (6.155)$$

By condition (ii) $\sum W(X,X') = 1$, which when used in Eq. 6.155 yields

$$\sum_{X'} P(X')W(X,X') = P(X)$$

That means constraint (iii) is satisfied by the stationary solutions. It is possible to impose a stronger restriction on $W(X, X')$ by invoking principle of detailed balance, which demands

$$P(X)W(X,X') = P(X')W(X',X) \tag{6.156}$$

Now, let us turn our attention to the construction of Metropolis method for an NVT Monte Carlo simulation. We know that at equilibrium the states are distributed as

$$P(X) = Z^{-1} e^{-\frac{\mathcal{H}(X)}{kT}},$$

which when used in the detailed balancing Eq. 6.156 leads to the condition

$$\frac{W(X,X')}{W(X',X)} = \frac{P(X')}{P(X)} \tag{6.157}$$

$$= e^{-\Delta \frac{\mathcal{H}(X)}{kT}}$$

That means, the ratio of the transition probabilities depends only on the change in energy $\Delta\mathcal{H} = \mathcal{H}(X') - \mathcal{H}(X)$ involved in the passage of the system from state X to X'. This idea forms the basis of assigning transition probabilities in the Metropolis MC method. Below we describe the steps involved in a canonical MC method, exploiting Metropolis sampling:[6]

1. Define an initial configuration (X), temperature (T).
2. Create a new configuration (X') randomly.
3. Calculate the energy change $\Delta\mathcal{H}$ involving $X \to X'$ transition.
4. If $\Delta\mathcal{H} \leq 0$, accept the new configuration and return to step 2 and set $X \leftarrow X'$.
5. If $\Delta\mathcal{H} \not< 0$, compute $p = e^{-\Delta\frac{\mathcal{H}}{kT}}$.
6. Generate a random number $r \in [0,1]$.
7. If $r < p$, accept the new configuration and return to step 2: set $X \leftarrow X'$.
8. If not, keep the old configuration and return to step 2.

It is evident that by executing steps 1-8, the system is being nudged to move towards the state of minimum energy that is consistent with values of N, V and T set. That is why any configurational change that lowers the energy is accepted (see step 4) while any reconfiguring move that raises the energy is accepted with the equilibrium (Boltzmann) probability. Both the MD and MC simulation techniques are still evolving and are widely used in the modeling of materials. We refer the reader to more specialized resources for details.

REFERENCES

1. A. M. Vasilyev, *An Introduction to Statistical Physics*, (Translated from the 1980 Russian Edition by G. Leib), MIR Publishers, Moscow (1983). An excellent exposition of the Gibbs method.
2. S.-K. Ma, *Statistical Mechanics* (Translated by M. K. Fung), World Scientific, Singapore (1985).
3. K. Huang, *Statistical Mechanics*, John Wiley, New York (1963).
4. D. W. Heermann, *Computer Simulation Methods in Theoretical Physics*, Second Edition, Springer-Verlag, Berlin (1990).
5. (a) L. Verlet, 'Computer "experiments" on classical fluids. I. Thermodynamical properties of Lennard-Jones molecules', *Phys. Rev.*, 159, 98 (1967). (b) A. Rahman, 'Correlations in the motion of atoms in liquid argon', *Phys. Rev.*, 136, A405 (1964).

6. M. E. J. Newman and G. T. Barkema, *Monte Carlo Methods in Statistical Physics*, Clarendon Press, Oxford (1999).

7. C. W. Gear, *Numerical Initial Value Problems in Ordinary Differential Equation*, Prentice Hall, Englewood Cliffs, New Jersey (1976).

8. D. Beeman, 'Some multistep methods for use in molecular dynamics calculations', *J. Comput. Phys.*, 20, 130 (1976).

9. S. Toxvaerd, 'A new algorithm for molecular dynamics calculations', *J. Comput. Phys.*, 47, 444 (1982).

7

Quantum Statistical Mechanics

7.1 Introduction

The statistical mechanical theory of Gibbs described in the previous chapter is capable of handling N-particle systems assuming that the individual particles can be described adequately by classical mechanics. Central to Gibbs' formulation is the idea of a $6N$-dimensional phase space – the Γ-space and the Gibbs distribution, also called the canonical distribution, which takes the form

$$dW = e^{\frac{F-\varepsilon}{kT}} d\Gamma$$

Let us repeat what the formula conveys. It tells us that the probability of the system being in a state with energy ε represented by a point in the phase space belonging to the phase-space volume element $d\Gamma$, is dW. A point in the $6N$-dimensional Γ-space represents a state characterized by the N-particle coordinates $(\vec{r_1}, \vec{r_2}, \cdots, \vec{r_N})$ and simultaneously the corresponding momentum coordinates $(\vec{p_1}, \vec{p_2}, \cdots, \vec{p_N})$. The energy ε, just like the position $(\vec{r_i})$ and momentum $(\vec{p_i})$ coordinates is assumed to be a continuous random variables.

While trying to extend the canonical distribution of Gibbs from the classical to the quantum regime we are confronted with several difficulties. For microscopic particles we can not precisely fix the positional $(\vec{r_i} = x_i, y_i, z_i)$ and momentum coordinates (p_{xi}, p_{yi}, p_{zi}) all at the same time. Our ability to do so is limited by the operation of Heisenberg's uncertainty principle (HUP) (see Chapter 2). Consequently, the concept of Γ-space or phase space introduced in the previous chapter appears to be inapplicable to quantum systems and must be modified to accomodate uncertainty restriction. Secondly, identical microscopic particles have been shown to obey the principle of indistinguishability. Hence one can not label the identical microscopic particles like electrons as particle 1, 2, $\cdots$, etc. and treat them as distinguishable objects.[1-4]

Thirdly, the particles of the microworld are naturally divided into two groups, which are fundamentally different from each other. The difference arises from their intrinsic angular moment, called spin angular momentum. It would have been too simplistic to assume that this intrinsic angular momentum arises from the rotation of the particles about its own axis. Indeed it was established experimentally that for some of these particles, the projection of the intrinsic angular momentum on an arbitrarily chosen axes can only have half-integral values in units of $\hbar$, while for others only integral values are allowed. Electrons, for example, can assume only two values of spin angular momentum projection namely $\frac{\hbar}{2}$ and $-\frac{\hbar}{2}$ while for photons these values are $\hbar$ and $-\hbar$. Particles like electron with half-integer spin are called Fermions (Fermi particles) and particles like photons with an integral spin are labeled Bosons (Bose particles). A fundamental difference exists between the two sets of particles succinctly enunciated in the form of Pauli exclusion principle, which states that only a single Fermion can occupy a state with definite values of the relevant quantum numbers and spin. In an atom for example, not more than two electrons can be in a state with a given set of angular momentum and principal quantum numbers. That means in a state with a given (angular) momentum, the two electrons must differ from each other in one having a spin angular momentum of $\frac{\hbar}{2}$ and the other $-\frac{\hbar}{2}$. Bosons do not obey this principle and the number of Bosons that can be in any given state is unlimited. Many-fermion wavefunctions, we have seen earlier in Chapter 4 are antisymmetric under permutation of the coordinates (space + spin) of any two electrons while many Boson wavefunctions are symmetric under binary permutation of space-spin coordinates. The occupation restriction of single particle states for Fermions and the absence such restrictions for Bosons are fundamentally linked to these different permutational symmetries of many Fermion and many Boson

DOI: 10.1201/9781003244882-7

wavefunctions. It turns out that the exclusion principle is a natural law and in a subsequent section we will pursue how it affects the distribution laws relevant to the two types of particles, and their physical properties.

Finally, the quantum states have discrete energies, which also has to be factored in while we try to develop quantum statistical mechanics or simply quantum statistics of many particle systems.

7.2 The Canonical Gibbs Distribution in Quantum Statistics

Let us try to extend the classical canonical Gibbs distribution to a quantum system keeping in mind the special attributes of quantum particles that the classical particles do not possess. In classical statistical mechanics, the canonical distribution takes the form (Chapter 6)

$$dW = e^{\frac{F-\varepsilon}{kT}} d\Gamma$$

dW being the probability of the system being in one of states that a phase-point in the volume element $d\Gamma$ of the phase space represents. We recall that such points in a phase-space volume element have $3N$ positional coordinates and $3N$ momentum projections along those coordinates and represent states with different energies ε, which can be treated as a continuous random variable. A quantum system may similarly be in very many different states with different energies. The first point of departure from the classical mechanical description arises because the states encountered in quantum systems are discrete. That means, there may be a state with energy ε_1, another with energy ε_2, a third with energy ε_3, and so on. The discreteness of these states lies in the fact that no state with energy in between ε_1 and ε_2 or ε_2 and ε_3, etc. exist.

Armed with this idea of discrete quantum states, let us now consider an ideal quantum gas – a collection of non-interacting quantum particles, each individual particle supporting discrete states with energies ϵ_1, ϵ_2, ϵ_3, etc. Let n_1, n_2, n_3, $\cdots$ be the occupation numbers of these states. That means, there are n_1 number of particles in the state with energy ϵ_1, n_2 in the state ϵ_2 and so on. The total energy of the system as a whole is therefore given by

$$\varepsilon_v = n_1\epsilon_1 + n_2\epsilon_2 + n_3\epsilon_3 + \cdots \tag{7.1}$$

The subscript 'v' present in the symbol ε for total energy is the symbol for the N-particle quantum states characterized by the occupation numbers n_1, n_2, $\cdots$, n_k. That is

$$v \equiv \left\{n_1, n_2, n_3, \cdots, n_k, \cdots\right\} \tag{7.2}$$

where $\sum_{i=1} n_i = N$ = total number of particles in the system. Another N-particle quantum state v' can likewise be labeled by v' where

$$v' = \left\{n_1', n_2', n_3', \cdots, n_k', \cdots\right\} \tag{7.3}$$

We are now in a position to extend the canonical Gibbs distribution to quantum mechanical systems by claiming that the probability (W_v) of realizing a particular quantum state of energy E_v at temperature T is[1-3]

$$W_v = e^{\frac{F'-E_v}{kT}} \tag{7.4}$$

where F' is the free energy of the quantum system, W_vs must satisfy the normalization condition

$$\sum_v W_v = 1 \tag{7.5}$$

The summation runs over all the possible states and Eq. 7.5 expresses the fact that the sum of the probabilities of finding the system in any one from among all the possible states must be unity. Using Eq. 7.4 and Eq. 7.5 we find that

$$\sum_\nu e^{\frac{F'-E_\nu}{kT}} = 1$$

$$\text{or,} \quad e^{\frac{F'}{kT}}\left(\sum_\nu e^{\frac{-E_\nu}{kT}}\right) = 1 \tag{7.6}$$

$\sum_\nu e^{\frac{-E_\nu}{kT}}$ may be defined as the partition function Z' of the quantum system so that we have

$$Z' = \sum_\nu e^{\frac{-E_\nu}{kT}} \tag{7.7}$$

which when used in Eq. 7.6 leads to a simple expression for the free energy F' of the same system at temperature T which reads,

$$F' = -kTlnZ' \tag{7.8}$$

Let us recall that the partition function 'Z', free energy F of a classical system and the probability dW of the system being in one of the states with energy E in the phase-space volume element $d\Gamma$ are, respectively given by

$$Z = \int e^{\frac{-E}{kT}} d\Gamma \tag{7.9}$$

$$F = -kTlnZ \tag{7.10}$$

$$dW = e^{\frac{F-E}{kT}} d\Gamma \tag{7.11}$$

It is instructive to compare the classical and quantal expression for free energy and partition functions that Gibbs' distribution throws up for classical and quantum systems. They do appear to be similar but there are important differences concealed in the similar looking expressions:

1. The states of a system in classical mechanics form a *continuous spectrum* and the energy of a state behaves as a *continuous random variable*. In quantum systems, state energies are *discrete random variables*.
2. The quantum mechanical systems must incorporate the indistinguishability principle.
3. Quantum particles carry integral or half-integral spin angular momentum. Particles with half-integral spin (Fermions) must strictly adhere to the requirements of Pauli exclusion principle (PEP) while Bosons are not required to conform to PEP.

The features (1-3) must be grafted into the Gibbs formulation while extending it to describe quantum systems.

7.2.1 Canonical Gibbs' Distribution For Discrete States

Let us consider a quantum system with *discrete states* ignoring for the moment the presence of spin, PEP and the principle of indistinguishability. Each state of the system is represented by a point in the Γ or the phase space. Since the states are discrete, the phase points representing them are at a certain distance from one another. Let us assume that there are g_Γ number of discrete points per unit volume of the Γ-space. The number of states in the phase-space volume $d\Gamma_i$ is $g_\Gamma d\Gamma_i$. We now have to carry out the summation in Eq. 7.7, which leads to

$$Z' = \sum_i e^{-\frac{\epsilon_i}{kT}} g_\Gamma d\Gamma_i$$

$$= \int e^{-\frac{\epsilon}{kT}} g_\Gamma d\Gamma \tag{7.12}$$

and

$$Z' = g_\Gamma Z \tag{7.13}$$

where $Z = \int e^{-\frac{\epsilon}{kT}} d\Gamma$ is just the classical partition function defined in the previous chapter and g_Γ is a constant quantity called the density of states in the Γ-space. It can be shown that $g_\Gamma = \left[\frac{1}{(2\pi\hbar)^3}\right]^N$ where N is the number of particles in the system.[2]

Once Z' is defined (Eq. 7.13), we can immediately find that the free energy of the quantum (discrete) system is

$$F' = -kTlnZ'$$
$$= F - kTlng_\Gamma \tag{7.14}$$

Thus, the free energy term (F') in canonical Gibbs' distribution for a quantum system with discrete energy states is different from the corresponding classical system. How much does this difference affect the experimentally measurable quantities or thermodynamic relations? Let us briefly examine the issue in what follows:

1. For an isothermal process the change in free energy of the quantum system is (Eq. 7.14)

$$(\Delta F')_T = (\Delta F)_T = -\Delta A$$

as the second term on the right hand side of Eq. 7.14 does not change in an isothermal process. So the classical and quantal estimates of change in free energy coincide in an isothermal process.

2. Differentiating Eq. 7.14 with respect to volume, at constant T, we get

$$\frac{\partial F'}{\partial V} = \frac{\partial F}{\partial V} = -P \tag{7.15}$$

so the equation of state is determined by the same expression as in the classical case.

3. For the internal energy U, the classical Gibbs statistics yields

$$U = F - \frac{\partial F}{\partial T}$$
$$= \left(F' + kTlng_\Gamma\right) - T \times \frac{\partial}{\partial T}\left(F' + kTlng_\Gamma\right) \tag{7.16}$$
$$= F' - T\frac{\partial F'}{\partial T}$$

So, the form of Gibbs-Helmholtz equation is retained when switching over to canonical quantum Gibbs' distribution.

4. We can define a quantal expression for entropy (S') by extending the classical thermodynamic definition used in the previous chapter whence we get

$$S' = -\frac{\partial F'}{\partial T} = -\frac{\partial F}{\partial T} + klng_\Gamma \tag{7.17}$$
$$\text{i.e.,} \quad S' = S + klng_\Gamma$$

The classical and quantal estimates of entropy following from canonical Gibbs' distribution therefore differ by a constant quantity ($klng_\Gamma$). This difference needs to be explored a little further to understand the connection between quantum distribution and entropy.

7.2.2 Quantum Gibbs' Distribution and Entropy

From the Gibbs-Helmholtz equation (cf. Eq. 7.16) we have the following definition of entropy of our quantum system

$$S' = \frac{U - F'}{T} \tag{7.18}$$

The internal energy U is just the mean value of energy of the system. Using the quantum Gibbs distribution we have

$$U = <\varepsilon> = \sum_{\nu} E_{\nu} e^{\frac{F'-E_{\nu}}{kT}}$$

Therefore,

$$S' = \frac{1}{T}(U - F')$$
$$= \frac{1}{T}\sum_{\nu} E_{\nu} e^{\frac{F'-E_{\nu}}{kT}} - \frac{1}{T}\sum_{\nu} F' e^{\frac{F'-E_{\nu}}{kT}} \tag{7.19}$$

In the last step we have used the fact F' does not depend on E_{ν} and by the normalization condition $\sum_{\nu} e^{\frac{F'-E_{\nu}}{kT}} = 1$. Rearranging Eq. 7.19 we have

$$S' = -k\sum_{\nu} \frac{F'-E_{\nu}}{kT} e^{\frac{F'-E_{\nu}}{kT}}$$
$$= -k\sum_{\nu} W_{\nu} ln W_{\nu} \tag{7.20}$$

where $W_{\nu} = e^{\frac{F'-E_{\nu}}{kT}}$ is the probability that the quantum system with discrete energy states $\{E_{\nu}\}$ is in the state with energy E_{ν} at temperature T. Since W_{ν} is the probability of state ν it is a pure number; so S' in Eq. 7.20 takes the dimension of Boltzmann constant ($k = k_B$).

With classical Gibbs' distribution, entropy (S) was however, defined as

$$S = -k\int wlnwd\Gamma \tag{7.21}$$

where w is the *probability density* of states with continuous energy in the phase-space volume $d\Gamma$. How can we arrive at Eq. 7.21, which involves transition from the discrete quantum case to the continuous classical one? The correct discrete to quantum transition can be carried out by identifying the probability of the system being one of the states in $d\Gamma$ as $wd\Gamma$ and not w, which is the probability density. So, in Eq. 7.20 we make the correct replacement for W_{ν} to get

$$S' = -k\sum W_{\nu} ln W_{\nu}$$
$$= -k\sum (wd\Gamma)ln(wd\Gamma) \tag{7.22}$$
$$= -k\int (wlnw)d\Gamma - k\int ln(d\Gamma)wd\Gamma$$

The second term in Eq. 7.22 can be reduced to

$$-kln(d\Gamma)\int wd\Gamma = -kln(d\Gamma) \tag{7.23}$$

by noting that the infinitesimal volume element $d\Gamma$ in terms of which the phase space has been subdivided may be assumed to be the same and normalization condition allows us to set $\int wd\Gamma = 1$. So, the entropy (S') defined by the quantum Gibbs distribution for *discrete states* when made to undergo correct transition to classical case with *continuous states* yields[1]

$$S' = -k\int wlnwd\Gamma - kln(d\Gamma)$$
$$= -k\int w(lnw)d\Gamma + \text{constant} \tag{7.24}$$

The constant $-kln(d\Gamma)$ is arbitrary as $d\Gamma$ is essentially arbitrary. Eq. 7.24 asserts that entropy is classically determined only to within an arbitrary constant. In calculating the change of entropy ΔS; this arbitrary constant gets eliminated and the classical statistical mechanical expression for entropy encoded in Eq. 7.21 possess no problem for the calculation of ΔS, but it fails to provide absolute value of entropy itself. The quantum statistical expression for entropy given in Eq. 7.20 is, however, free from this difficulty. From an experimental perspective it appears that the quantum expression for entropy $(S' = -k\sum_\nu W_\nu lnW_\nu)$ represents the reality correctly as $S' \to 0$ as $T \to 0K$, provided that the ground state of the system is non-degenerate. At $T = 0$ K, the probability W_0 of the ground state is 1 and $W_\nu = 0$ for any other states ($\nu \neq 0$) making S' vanish. This behavior is in consonance with the observed behavior of a number of thermodynamic quantities at very low temperatures, which seems to indicate that entropy decreases on cooling the system and vanishes at absolute zero. The arbitrary constant in the classical expression for entropy must therefore be so chosen that S at $T = 0$ K (call it S_0) becomes zero. This condition has an important role in thermodynamics and has been given the status of the third law of thermodynamics. We stress again that the quantum statistical expression for entropy given in Eq. 7.20 is entirely consistent with the third law and S' vanishes at $T = 0$ K if the ground state has no degeneracy.

7.3 Entropy and the Entropy Maximal State

The entropy formula obtained from Gibbs' distribution for states with discrete energy (Eq. 7.20) coincides with information theoretical expression for entropy except for the occurrence of the Boltzmann constant 'k' in the former. We can therefore assume that S' of Eq. 7.20 provides a measure of 'disorder' or 'randomness' or 'lack of information' in a macroscopic state of the system. The question now is: which macroscopic state, out of all possible states, is realized under a given set of conditions when equilibrium is reached? It turns out that the equilibrium state is the *entropy-maximal state* that is compatible with the conditions or constraints the system has been put under. Speaking physically, an equilibrium state is the most disordered state or a state with maximum randomness. This feature of the equilibrium state is so fundamental that one can enunciate a principle of maximum entropy: 'The entropy is completely determined by the formula (Eq. 7.20) and attains a maximum value when an equilibrium state consistent with the conditions of the system, is reached'. This principle has become the corner stone of the entire enterprise of statistical physics of equilibrium states. The principle, when applied correctly leads to the Gibbs distribution, which has already been derived from probabilistic considerations.

In finding the equilibrium distribution the 'constraints' the system has been put under, play a very important role as the following example demonstrates. Let us consider an N-particle system with states of discrete energies. The entropy S' is

$$S' = -k \sum_\nu W_\nu lnW_\nu \tag{7.25}$$

To determine W_ν values at equilibrium we maximize S' with respect to W_νs under the condition of normalization of the probabilities:

$$\sum_\nu W_\nu = 1 \tag{7.26}$$

Using the standard Lagrange method of extremization of a function under a given set of constraints, we first construct an auxiliary function $L(W_\nu)$ incorporating the constraints where

$$L(W_\nu) = S'(W_\nu) - \lambda \sum_\nu W_\nu$$

$$= -k \sum_\nu W_\nu lnW_\nu - \lambda \sum_\nu W_\nu \tag{7.27}$$

λ is the undetermined Lagrange multiplier and has obviously the same dimension as 'k'. Differentiating L with respect to W_ν and setting the resultant derivative to zero we have[1,4]

$$-klnW_\nu - k - \lambda = 0,$$

$$\text{whence,} \quad W_\nu = e^{-(\frac{\lambda}{k}+1)} \quad \text{for } \nu = 0,1,2,\cdots \tag{7.28}$$

The result is absurd as it predicts equal probability for all states. With an infinite number of states and the prevailing normalization condition on probabilities it leads to, as we will see a state with infinite entropy. Thus, assuming n-available states we have

$$\sum_\nu W_\nu = 1 = nW$$

$$\text{or,} \quad W = \frac{1}{n}$$

Therefore, $W \to 0$ as $n \to \infty$ and $S' = -k\sum_\nu W_\nu lnW_\nu \to \infty$. Also, Eq. 7.28 does not coincide with the Gibbs canonical distribution function $W_\nu = e^{\frac{F'-E_\nu}{kT}}$, which clearly depends on state energy (E_ν) as well as the system temperature (T), F' being the free energy of the quantum system. Why does the entropy maximization fail to produce what we have many times referred to as the canonical Gibbs distribution? It has happened because we did not incorporate the right conditions while maximizing the entropy (S') in our search for the equilibrium state. In fact, even the temperature T of the system was not specified, and our system was, as if, completely isolated with both the number of particles and total energy remaining conserved. The 'ensemble' representing such a system is called *microcanonical ensemble*, which does not represent a real thermodynamic system at equilibrium. A real thermodynamic system at equilibrium is not perfectly or completely isolated, but, is in thermal contact with a huge 'heat bath', which exchanges energy with the system across the system – heat bath boundaries. There is no mass-transfer across these boundaries so that the number of particles in the system remains conserved. *The total energy is not conserved, but the 'average energy' of the system is.* The total energy fluctuates around the average energy with the scale fixed by the temperature. We must therefore carry out the entropy maximization under the following two constraints in our search for equilibrium probabilities, i.e. find the w_νs such that

$$\sum_\nu W_\nu = 1 \tag{7.29}$$

$$\sum_\nu W_\nu \varepsilon_\nu = U = \text{constant} \tag{7.30}$$

The first condition is the conservation of total probability while the second stems from the conservation of average energy of the system. The underlying ensemble is known as the *canonical ensemble*. The appropriate Lagrangian for the constrained entropy maximization problem is

$$L(W_\nu) = -k\sum_\nu W_\nu lnW_\nu - \lambda_1 \sum_\nu W_\nu - \lambda_2 \sum_\nu \varepsilon_\nu W_\nu \tag{7.31}$$

where λ_1 and λ_2 are the undetermined Lagrangian multipliers. Setting $(\frac{\partial L}{\partial W_\nu}) = 0$, then leads to the distribution law for probabilities in the entropy maximal state at the temperature T, which should also be the equilibrium state under the given set of conditions. It is straightforward to find that

$$W_\nu = e^{-\left(1+\frac{\lambda_1}{k}+\frac{\lambda_2}{k}\varepsilon_\nu\right)} \tag{7.32}$$

Eq. 7.32 can be brought into coincidence with the canonical Gibbs distribution if we identify or assign λ_1 and λ_2 in the following way:

$$\lambda_2 = \frac{1}{T}$$

$$\lambda_1 = (-)\left(\frac{F'}{T}+k\right) \tag{7.33}$$

With Eq. 7.33, W_νs of Eq. 7.32 exactly transforms into the canonical distribution law of Gibbs:

$$W_\nu = e^{\frac{F'-E_\nu}{kT}} \tag{7.34}$$

The free energy F' is formally brought in through the Lagrange multiplier λ_1 and the temperature T is introduced through the multiplier λ_2. Thus, the correct choice of conditions or constraints under which entropy maximization is carried leads to the correct equilibrium probability distribution, which in our case is the *canonical Gibbs distribution*. The underlying ensemble is called the *canonical ensemble*. In practice, it is more convenient to assume that the temperature T is fixed instead of assuming U (the internal energy) to be given. We can then determine F' and U as function of T by using W_νs given by Eq. 7.34 and the two constraint conditions as follows:

$$\sum_\nu e^{\frac{F'-\varepsilon_\nu}{kT}} = 1$$

$$\sum_\nu \left(e^{\frac{F'-\varepsilon_\nu}{kT}} \times E_\nu \right) = U$$

7.4 The Grand Canonical Potential

The maximum entropy principle can be used to handle systems much more complex than the ones studied in the previous section and derive distributions more general than the microcanonical or canonical Gibbs distributions. The underlying ensemble in such cases is not the canonical ensemble (CE) but the *grand canonical ensemble (GCE)*, and the distribution law that emerges is the *grand canonical Gibbs distribution*.

Let us imagine an N-particle quantum system with discrete states indexed by say, ν_N and let $W_{\nu N}$ stand for the probability of realizing the νN^{th} state of the system. The entropy of such a system is

$$S' = -k \sum_{\nu N} W_{\nu N} ln W_{\nu N} \tag{7.35}$$

The system is a more general one than the system described so far in the sense that the particle number N is not fixed or conserved in the system, although the average number of particles $<N>$ is conserved just as the average energy $<E> = U$ (but not the total energy E) is conserved. Such a system may be assumed to be in contact with a huge reservoir or bath and across the boundaries of the system and reservoir both mass (particles) and energy transfer take place, the bath being assumed to be so huge that it remains unaffected by such transfers. As for the system the energy fluctuates around the mean energy (U), which is conserved. The particle number fluctuates around a mean number of particles $<N>$ when equilibrium is established. Such systems are members of what has become known as grand canonical ensemble (GCE) and our task ahead is to find out the equilibrium grand canonical distribution law for the probabilities ($W_{\nu N}$s).

To arrive at the form of the grand canonical Gibbs distribution we invoke maximum entropy principle and maximize S' of Eq. 7.35 under the following constraints that are consistent with the nature of the ensemble :

(i) $\sum_{\nu N} W_{\nu N} = 1 =$ probability of the system being in any one of the state νN.

(ii) $\sum_{\nu N} W_{\nu N} E_{\nu N} = U =$ average energy of the system.

(iii) $\sum_{\nu N} N W_{\nu N} = <N> =$ mean number of particles in the system.

The appropriate Lagrangian for the problem is easily constructed as

$$L(W_{\nu N}) = -k \sum_{\nu N} W_{\nu N} ln W_{\nu N} - \lambda_1 \sum_{\nu N} W_{\nu N} - \lambda_2 \sum_{\nu N} W_{\nu N} E_{\nu N} - \lambda_3 \sum_{\nu N} W_{\nu N}(N) \tag{7.36}$$

The extremization of $L(W_{\nu N})$ with respect to $W_{\nu N}$ leads to the equation

$$-k \ln W_{\nu N} - k - \lambda_1 - \lambda_2 \varepsilon_{\nu N} - \lambda_3 N = 0 \qquad (7.37)$$

which has the solution

$$W_{\nu N} = e^{-\frac{(k+\lambda_1+\lambda_2 \varepsilon_{\nu N}+\lambda_3 N)}{k}} \qquad (7.38)$$

Let us now make the following identifications

$$\lambda_2 = \frac{1}{T}$$

$$-(k+\lambda_1) = \Omega^* \qquad (7.39)$$

$$\lambda_3 = -\frac{\mu}{T}$$

whence $W_{\nu N}$s of Eq. 7.38 become

$$W_{\nu N} = e^{\frac{\Omega^* + \mu N - E_{\nu N}}{kT}} \qquad (7.40)$$

The formula above (Eq. 7.40) gives us the probability of realizing the νNth state having the energy $E_{\nu N}$ in an N-particle system in equilibrium with a heat bath. This probability distribution is known as the grand canonical Gibbs distribution. Note that we have brought in three quantities viz. Ω^*, μ and T in the distribution encoded in Eq. 7.40. We will examine the thermodynamic meaning of these parameters in the next subsection. As for now, let us mention that they are formally fixed from the following conditions or constraints:

$$\sum_{\nu N} W_{\nu N} = 1 = \sum_{\nu N} e^{(\Omega^* + \mu N - E_{\nu N})/kT} \qquad (7.41a)$$

$$\sum_{\nu N} W_{\nu N} E_{\nu N} = U = \sum_{\nu N} E_{\nu N} e^{(\Omega^* + \mu N - E_{\nu N})/kT} \qquad (7.41b)$$

$$\sum_{\nu N} W_{\nu N} N = <N> = \sum_{\nu N} N e^{(\Omega^* + \mu N - E_{\nu N})/kT} \qquad (7.41c)$$

In practice, it is always more convenient to assume that μ and T are given rather than determining them for a given value of internal energy U and mean number of particles $<N>$ with the help of Eq. 7.41a–c. Once μ and T values are fixed, we can use Eq. 7.41a-c to determine Ω^*, μ and $<N>$ as functions of the parameters μ and T. Before proceeding further we must first examine what Ω^*, μ and T represent physically within thermodynamics.

7.4.1 Thermodynamic Meaning of Ω^*, μ and T

Referring to the canonical Gibbs distribution, we may recognize that Ω^* in Eq. 7.40 is akin to the free energy term F' in the canonical Gibbs distribution ($W_\nu = e^{\frac{F'-E_\nu}{kT}}$) and is a function of μ and T. Let us call Ω^* the Omega potential (or the grand canonical (GC) potential) and try to provide appropriate thermodynamic meanings to the parameters appearing in the grand canonical distribution. To this end in view let us consider two separate equilibrium subsystems characterized by the following grand canonical distributions:

$$\text{Subsystem 1:} \quad W'_{\nu' N'} = e^{\frac{(\Omega^{*'} + \mu' N' - E_{\nu' N'})}{kT'}} \qquad (7.42a)$$

$$\text{Subsystem 2:} \quad W''_{\nu'' N''} = e^{\frac{(\Omega^{*''} + \mu'' N'' - E_{\nu'' N''})}{kT''}} \qquad (7.42b)$$

If the two are treated as a composite single system, assuming that the interactions between the two individual subsystems and with their heat baths are much weaker compared to interactions operating in the subsystems themselves, the composite (1+2) system may be characterized by an appropriate grand canonical distribution

$$W_{\nu N} = e^{\frac{\Omega^* + \mu N - E_{\nu N}}{kT}} \tag{7.43}$$

The assumption that systems 1 and 2 do not interact with each other except very weakly allows us to write the following equalities:

$$N = N' + N'' \tag{7.44}$$
$$E_{\nu N} = E_{\nu' N'} + E_{\nu'' N''}$$

the first equality asserting that the total number of particles in the composite system (1 + 2) is equal to the sum of the number of particles in subsystems 1 and 2, and the second confirming that total energy of the composite system is equal to the sum of the energies of the two non-interacting subsystems. We can write a third equality connecting the probability of realizing a state (νN) of the composite system with the probabilistics of states ($\nu'N'$ and $\nu''N''$) in subsystems 1 and 2, respectively being realized. Thus,

$$W_{\nu N} = w'_{\nu' N'} \cdot w''_{\nu'' N''} \tag{7.45}$$

Using Eq. 7.42a and b together with Eq. 7.43 we get

$$e^{\frac{(\Omega^* + \mu N - E_{\nu N})}{kT}} = e^{\frac{(\Omega^{*\prime} + \mu'N' - E'_{\nu' N'})}{kT}} \times e^{\frac{(\Omega^{*\prime\prime} + \mu''N'' - E''_{\nu'' N''})}{kT}} \tag{7.46}$$

Incorporating equalities in Eq. 7.44 in Eq. 7.46 we find that

$$e^{\frac{(\Omega^* + \mu N' + \mu N'' - E'_{\nu' N'} - E''_{\nu'' N''})}{kT}} = e^{\frac{(\Omega^{*\prime} + \mu'N' - E'_{\nu' N'})}{kT}} \times e^{\frac{(\Omega^{*\prime\prime} + \mu''N'' - E''_{\nu'' N''})}{kT}} \tag{7.47}$$

Eq. 7.47 above must hold for all values of $E'_{\nu' N'}$, $E''_{\nu'' N''}$, N', N'' which demands that the following equalities must also be satisfied simultaneously

$$T = T' = T'' \tag{7.48a}$$

$$\mu = \mu' = \mu'' \tag{7.48b}$$

$$\Omega^* = \Omega^{*\prime} + \Omega^{*\prime\prime} \tag{7.48c}$$

Therefore when the two equilibrated subsystems form a composite system, which eventually also reaches an equilibrium, it is necessary that the parameters T and μ must be equal throughout. The first represents the temperature and the second the chemical potential (see later). Not only that, the omega potential of the composite system at equilibrium must be equal to the sum of the omega potentials of the subsystems, which have combined to produce the composite system.[1-3]

We must not fail to note at this point, that the parameter μ (the chemical potential) plays the same role with respect to the number of particles as T (the temperature) does with respect to energy. In all systems in equilibrium with a heat bath, the systems have the same values for T and μ as the bath. That means we must better assume that T and μ values are given or fixed and use the grand canonical Gibbs distribution to compute other quantities of the system like the omega potential, energy, etc. as functions of μ and T.

1. It is possible to establish quite a few thermodynamic connections from the normalization condition for the probabilities $W_{\nu N}$. Thus starting with

$$\sum_{\nu N} W_{\nu N} = \sum_{\nu N} e^{\frac{(\Omega^* + \mu N - E_{\nu N})}{T}} = 1 \tag{7.49}$$

and differenting with respect to μ (the chemical potential) we find that

$$<N> = -\left(\frac{\partial \Omega^*}{\partial \mu}\right)_T \tag{7.50}$$

The average number of particles in the system is thus given by the derivative of the omega potential with respect to the parameter 'μ' – the chemical potential of the system taken with negative sign.

2. Differentiation of Eq. 7.49 with respect to T (at constant μ), multiplying the resulting equation by kT^2 and making use of the Eq. 7.41a-c lead us to the following relation

$$T\frac{\partial\Omega^*}{\partial T} - \Omega^* + U - \mu < N >= 0 \qquad (7.51)$$

which can be cast in the form of a generalized Gibbs-Helmholtz equation

$$U = \Omega^* - T\frac{\partial\Omega^*}{\partial T} + \mu < N > \qquad (7.52)$$

Using Eq. 7.50 in Eq. 7.52, we are finally led to an expression for the internal energy in terms of the omega potential and its derivative

$$U = \Omega^* - T\frac{\partial\Omega^*}{\partial T} - \mu\frac{\partial\Omega^*}{\partial T} \qquad (7.53)$$

3. Similarly differentiating Eq. 7.49 with respect to volume while T and μ are held constant, and following the same line of argument that was used in Chapter 6 for deriving the equation of state for the given system. We arrive at the relation

$$P = -\left(\frac{\partial\Omega^*}{\partial V}\right)_{\mu,T} \qquad (7.54)$$

4. Finally, we have to take a look at the derivative of the omega potential with respect to temperature. We suspect from Eq. 7.53 that it could possibly be related to the entropy S' of the system. Thus starting with the expression for entropy when grand canonical equilibrium distribution holds, we have

$$S' = -k\sum_{\nu N} W_{\nu N} ln W_{\nu N}$$

$$= -k\sum_{\nu N} \frac{(\Omega^* - E_{\nu N} + \mu N)}{kT} e^{\frac{(\Omega^* - E_{\nu N} + \mu N)}{kT}}$$

whence it follows that

$$S' = \frac{1}{T}(\Omega^* - U + \mu < N >) \qquad (7.55)$$

In arriving at Eq. 7.55, we have made obvious use of Eq. 7.41a–c. Use of Eq. 7.52 in Eq. 7.55 then leads to the final identification that

$$S' = -\frac{\partial\Omega^*}{\partial T} \qquad (7.56)$$

Thus the probability of realizing a state of a quantum system with discrete energy states in equilibrium with a bath that allows both energy and mass flow across the system – bath boundaries is given by the grand canonical Gibbs distribution. The entropy S' is given by the temperature derivative of the grand canonical potential taken with a negative sign. With this, the interpretation of the grand canonical potential or the Omega potential (Ω^*) as a thermodynamic quantity is complete. If the particle number remains constant, grand canonical Gibbs distribution become canonical Gibbs distribution ($\mu = 0$, $\Omega^* = F'$). The stage is now set for working out spin-dependent statistics of quantum system with indistinguishable particles.

7.5 Quantum Statistics of Bosons and Fermions

We have so far neglected 'spin' completely and worked with discrete energy states of quantum systems without any occupation restriction on the states to arrive at canonical and grand canonical Gibbs distri-butions. It is now necessary to consider how the spin angular momentum of the constituent particles of the systems affect the probability distributions and work out particle statistics (average number of particles in a state) for Fermions (particles with half-integer spin) and Bosons (particles with integer spin). It turns out that the effect of spin-dependent occupation restriction introduces effects that are both subtle and profound. Before taking up the problem let us briefly summarize the important features of Bosons and Fermions. We have already mentioned that Bosons are microparticles with zero or integral spin angular momentum in units of $\hbar$. Some of the Bose-particles (Bosons) are photons (spin 1), phonons (spin 1), K and π mesons (spin 0), molecules like He_2 (the sum of spin angular momenta of electrons, protons and neutrons is integral), etc. Bosonic particles obey the principle of indistinguishability. Fermions, on the other hand, are particles with half-integral spin in units of $\hbar$. Electrons, positrons, protons and neutrons, some nuclei and atoms are Fermions. Fermions obey the principle of indistinguishability just like Bosons; but unlike Bosons, they obey in addition the Pauli exclusion principle. This imposes a specific restriction on the distribution of Fermions among possible and accessible quantum states. We will first consider quantum statistics of Bosons, more correctly the quantum statistics of an ideal Bose gas – a collection of many Bose particles, which are non-interacting and are indistinguishable from each other.

7.5.1 Bose-Einstein Distribution

Suppose that we have an ideal Bose gas and we are interested in understanding its properties. We have already seen in section 7.4, that the key to this understanding lies in the Omega (grand canonical) potential of the system. For this purpose, it is convenient to assume that our system of Bose particles are in equilibrium with a heat bath. Let the temperature and chemical potential of the bath be fixed at T and μ, respectively.

The calculation of Ω^* then proceeds from the normalization condition of the grand canonical probabilities $W_{\nu N}$ where

$$W_{\nu N} = e^{\frac{(\Omega^* - E_{\nu N} + \mu N)}{kT}}$$

The normalization condition demands that

$$\sum_{\nu N} W_{\nu N} = \sum_{\nu N} e^{\frac{(\Omega^* - E_{\nu N} + \mu N)}{kT}} = 1$$

$$\text{or,} \quad e^{\frac{\Omega^*}{kT}} \sum_{\nu N} e^{\frac{(\mu N - E_{\nu N})}{kT}} = 1 \tag{7.57}$$

Defining the GC partition function

$$Z^* = \sum_{\nu N} e^{\frac{(\mu N - E_{\nu N})}{kT}} \tag{7.58}$$

we arrive at (from Eq. 7.57 and Eq. 7.58)

$$\Omega^* = -kT ln(Z^*) \tag{7.59}$$

The calculation of Z^* and hence of Ω^* as well is greatly simplified, if we choose our system in a specific way. The 'speciality' of this choice lies in considering all those particles of the Bose gas, which are in a definite state, (i) i.e. all these particles have momentum components that are pre-set. We can now proceed in two steps: first calculate the omega potential (Ω_i^*) of the particular subsystem (*i*th subsystem) consisting of the 'earmarked' particles; next we sum the expression so obtained over all possible states (Ω_i^*) leading to the omega potential (Ω^*) of the entire Bose gas. This manner of handling the system implicitly assumes that we are using grand canonical distribution as the number of particles in a given state (subsystem '*i*' in our description) is not fixed. The particle numbers in our separated subsystems do vary. We must also point out, all other subsystems act as the bath with which the particular subsystem is in equilibrium.

Let us consider a definite state (*i*) of energy ϵ_i for the particles. The energy of the subsystem of our choice depends on the number of particles (n_i) in the particular state (ϵ_i) being considered. We can therefore write

$$E_{\nu N} = n_i \epsilon_i \equiv E_i \tag{7.60}$$

The different states of our subsystem merely differs in the number of particles n_i so that n_i itself can be used for labeling the states. Thus the partition function for the subsystem (i) is

$$Z_i^* = \sum_{n_i=0}^{\infty} e^{\frac{(\mu n_i - n_i \epsilon_i)}{kT}} \tag{7.61}$$

Since Bosons do not obey Pauli exclusion principle there is no restriction on the number of particles that our subsystems can accomodate, so n_i can take on any value from 0 to ∞. More importantly, in evaluating Z_i^*, we have made use of the principle of indistinguishability as we have assumed that there is only one state with energy E_i. The summation in Eq. 7.61 can be easily evaluated analytically as it reduces to a geometric progression. Thus

$$Z_i^* = \sum_{n_i=0}^{\infty} \left\{ e^{\frac{(\mu - \epsilon_i)}{kT}} \right\}^{n_i}$$

$$= \frac{1}{1 - e^{\frac{\mu - \epsilon_i}{kT}}} \tag{7.62}$$

The omega potential for the ith subsystem (Ω_i^*) is therefore given by

$$\Omega_i^* = -kT ln Z_i^*$$

$$= kT ln \left\{ 1 - e^{\frac{(\mu - \epsilon_i)}{kT}} \right\} \tag{7.63}$$

The omega potential of the entire system (Ω^*) of Bosons can then be obtained by summing over all possible states of Bosons as denoted by the index i. Carrying out the summation, we have

$$\Omega^* = \sum_i \Omega_i^* = kT \sum_i ln \left\{ 1 - e^{\frac{(\mu - \epsilon_i)}{kT}} \right\} \tag{7.64}$$

Once the omega potential is known, we can easily calculate a host of other properties of the system. Let us try to find for example, the mean number of particles (Bosons) in the ith state. Using Eq. 7.50 and Eq. 7.63 we can write

$$<n_i> = -\frac{\partial \Omega_i^*}{\partial \mu} = \frac{1}{e^{\frac{(\epsilon_i - \mu)}{kT}} - 1} \tag{7.65}$$

The formula above determines the average number of Bosons in a discrete state with energy ϵ_i where the Bosons have been assumed to be indistinguishable from each other. It is the famous Bose-Einstein distribution.

If we wish to calculate the internal energy of the system comprising of Bosons with μ and T set we can make use of Eq. 7.53 together with Eq. 7.64 and arrive at

$$U = \sum_i \frac{\epsilon_i}{\left\{ e^{\frac{(\epsilon_i - \mu)}{kT}} - 1 \right\}}$$

$$= \sum_i <n_i> \epsilon_i \tag{7.66}$$

The other thermodynamic characteristics of the system can be similarly evaluated by using results described in section 7.4. It turns out that all these quantities are ultimately determined by the mean value of the number of Bosons in a state with energy ϵ_i, which is completely determined by the Bose-Einstein distribution.

7.5.2 Fermi-Dirac Distribution

Instead of an ideal Bose gas, let us now consider the case of an ideal Fermi gas, for example, a gas consisting of electrons, which we assume do not interact with one another. The assumption seems to be a rather improbable one as electrons carry negative charge and repeal each other quite strongly. However, we may assume that each electron in the gas is moving in an 'average field' set up by all other electrons. This average field may be viewed as an equivalent external electric field. For electrons in a solid like metals or semiconductors or for electrons in a gas-discharge plasma, this equivalent field is often found to be vanishingly small because of the compensation of the negative charge on electrons by the opposite charge on positive ions, which on average, is equal to the electronic charge. Thus the concept of a non-interacting electron gas may not be an unfeasible one.

Now the system consisting of many non-interacting electrons or Fermions may be viewed as a collection of the possible states of individual electrons and distribution of the electrons among these states. To arrive at the equilibrium distribution we will proceed as we did in the case of an ideal Bose gas. We will therefore choose a single state marked by 'i' and energy 'ϵ_i' as the subsystem, which is constantly exchanging electrons with the remaining part of the system, which we take as the heat bath with μ and T set to certain values. The expression for the omega potential of the subsystem 'i' is then

$$\Omega_i^* = -kTlnZ_i^* \tag{7.67}$$

where the partition function for the many Fermion subsystem 'i' is given by

$$Z_i^* = \sum_{n_i} e^{\frac{\mu n_i - n_i \epsilon_i}{kT}} \tag{7.68}$$

where 'n_i' denotes the occupancy number of the ith state. Now the occupation number n_i for the one Fermion states is restricted by Pauli exclusion principle to assume only two values namely 0 and 1. That is, the level with energy ϵ_i is either occupied by one Fermion or is unoccupied. Thus, we can easily carry out the summation in Eq. 7.68 and write

$$Z_i^* = 1 + e^{\frac{\mu - \epsilon_i}{kT}} \tag{7.69}$$

which is then used in Eq. 7.67 to obtain the omega potential (Ω^*) of the ith subsystem of our choice. Thus,

$$\begin{aligned} \Omega_i^* &= -kTlnZ_i^* \\ &= -kTln\left\{1 + e^{\frac{\mu - \epsilon_i}{kT}}\right\} \end{aligned} \tag{7.70}$$

The grand canonical potential (Ω^*) of the whole system is readily obtained by summing over the omega potentials of all the subsystems (Ω_i^*) into which we have partitioned the system. That means,

$$\Omega^* = \sum_i \Omega_i^* = -kT \sum_i ln\left\{1 + e^{\frac{\mu - \epsilon_i}{kT}}\right\} \tag{7.71}$$

The entire thermodynamics of ideal electron gas (or any many Fermion system) can now be recovered from Eq. 7.70 and Eq. 7.71; but, let us first calculate the mean occupation number of the ith level as n_i has been assumed to fluctuate around a mean by construction. Using the general connection between $<n>$ and Ω^* (Eq. 7.50), we have, for the ith subsystem

$$<n_i> = -\frac{\partial \Omega_i^*}{\partial \mu} = \frac{1}{e^{\frac{(\epsilon_i - \mu)}{kT}} + 1} \tag{7.72}$$

The formula above determines the mean number of Fermions in a state having energy ϵ_i and is known as Fermi Dirac distribution. The internal energy of the ideal Fermion gas can be easily calculated by evaluating the sum of the products of mean occupation number of the ith state $<n_i>$ and the energy $<\epsilon_i>$ of the ith state. Thus,

$$U = \sum <n_i> \epsilon_i = \sum_i \frac{\epsilon_i}{e^{\frac{(\epsilon_i - \mu)}{kT}} + 1} \tag{7.73}$$

We could have obtained the same expression for U also by making use of the general grand canonical result that

$$U = \Omega^* - T\frac{\partial \Omega^*}{\partial T} + \mu\frac{\partial \Omega^*}{\partial \mu}$$

The total average number of Fermions in the system can be directly obtained by summing $< n_i >$ over all the subsystems. Hence[1,3]

$$< N > = \sum_i < n_i > = \sum_i \frac{1}{e^{\frac{(\epsilon_i - \mu)}{kT}} + 1} \tag{7.74}$$

The equation above can be exploited to find the chemical potential μ of the system.

7.5.3 Boson Statistics and Indistinguishability Principle

Let us recall how we used the principle of indistinguishability of particles in an ideal Bose gas. While evaluating the partition function (Eq. 7.58) we simply assumed that there is only one state with energy $E_i = n_i \epsilon_i$. It did not matter which individual particles, out of all the available ones, occupied the state ϵ_i because the particles were indistinguishable.

Had we used classical notions, the particles would have been treated as distinguishable entities and hence there could have been many states with energy $E_i = n_i \epsilon_i$. There would have been a lot of choices for the selection the set of n_i distinguishable particles out of the N particles in the system. Each of the set of n_i particles would be different at last in one individual particle. The total number of such sets is

$$\frac{N!}{n!(N-n)!}$$

The partition function (call it Z_i) then takes the form (compare Eq. 7.61 for Z_i^*)

$$Z_i = \sum_{n_i} \frac{N!}{n!(N-n)!} e^{\frac{(\mu' n_i - n_i \epsilon_i)}{kT}} \tag{7.75}$$

We have used Z_i in place of Z_i^* and μ' in place of μ to take care of the possibility that the classical partition function and chemical potential might be different from their quantum analogues (Z_i^* and μ). Since we are using the grand canonical (GC) distribution or GC-ensembles, any number of particles may enter the ith subsystem which means N must be infinitely large – a feature that must be introduced in the evaluation of Z_i of Eq. 7.75; in other words, we must, while evaluating Z_i, take the limit $N \to \infty$ and take care of the fact that $N >> n_i$ for every i. The combinatorial coefficient on the right hand side of Eq. 7.75 for Z_i, under the limiting condition ($N \to \infty$) becomes

$$\frac{N!}{n!(N-n)!} = \frac{N(N-1)(N-2)\cdots(N-n_i+1)(N-n_i)}{n_i!(N-n_i)!} \approx \frac{N^{n_i}}{n_i!} \tag{7.76}$$

With Eq. 7.76, Z_i of Eq. 7.75 becomes in the limit $N \to \infty$

$$Z_i = \sum_{n_i} \frac{N^{n_i}}{n_i!} e^{\frac{(\mu' - \epsilon_i)n_i}{kT}}$$

$$= \sum_{n_i} \frac{1}{n_i!} \left\{ e^{\frac{(\mu' + kTlnN - \epsilon_i)}{kT}} \right\}^{n_i} \tag{7.77}$$

In the limit N is infinitely large, therefore, in order that $\mu' + kTlnN$ may produce a finite expression, μ' must also be infinite (< 0) and we can under that condition write[1,3]

$$\mu = \mu' + kTlnN \tag{7.78}$$

where the finite quantity μ may be taken to represent the chemical potential of the system. Thus incorporation of Eq. 7.78 in Eq. 7.77 leads us to

$$Z_i = \sum_{n_i=0}^{\infty} \frac{1}{n_i!} \left\{ e^{\frac{(\mu - \epsilon_i)}{kT}} \right\}^{n_i}$$

$$= e^{\left\{ e^{\frac{(\mu - \epsilon_i)}{kT}} \right\}} \tag{7.79}$$

We have used the obvious fact that the sum over n_i represents the series for e^{X_i} where $X_i = e^{\frac{(\mu - \epsilon_i)}{kT}}$.

The omega potential (Ω_i) of the ith subsystem of distinguishable Bosons is then given by

$$\Omega_i = -kTlnZ_i$$
$$= -kTe^{\frac{(\mu - \epsilon_i)}{kT}} \qquad (7.80)$$

while Ω for the whole system of Bosons becomes

$$\Omega = -kT \sum_i e^{\frac{(\mu - \epsilon_i)}{kT}} \qquad (7.81)$$

Ω_i and Ω for a system of distinguishable Bosons are quite different from Ω_i^* and Ω of N-indistinguishable Bosons (Eq. 7.63 and Eq. 7.64) at temperature T. The question that becomes expedient to address is: when, if at all, does Ω_i^* coincide with Ω_i? In other words when does (if at all) the expression for Ω_i^* merge with the expression for Ω_i for distinguishable Bosons? The expression for Ω_i^* (Eq. 7.63) is

$$\Omega_i^* = kTln\{1 - e^{\frac{(\mu - \epsilon_i)}{kT}}\}$$

In the expression for Ω_i^* above, if $\mu << \epsilon_i$ so that

$$\epsilon_i - \mu >> kT$$

the exponent $\frac{(\mu - \epsilon_i)}{kT}$ of the exponential becomes a large negative number. The condition

$$\frac{(\mu - \epsilon_i)}{kT} << -1$$

holds and the exponential term in the expression for Ω_i^* itself becomes small. This allows us to write

$$\Omega_i^* = kTln\{1 - e^{\frac{(\mu - \epsilon_i)}{kT}}\}$$
$$\approx -kTe^{\frac{(\mu - \epsilon_i)}{kT}} \qquad (7.82)$$
$$\approx \Omega_i$$

Thus, for small values of the chemical potential (by small, we mean μ has a large negative value), the difference between Ω_i and Ω_i^* practically disappears. In order that the same conclusion can be made for all the states (not only the ith) or the subsystems, μ must be at least several kT smaller than the lowest energy level of the system; for then only Eq. 7.82 is satisfied for all ϵ_is of the Bosons. If the temperature is low enough the quantum behavior can not be neglected and full indistinguishability must be taken into account and Ω_i^* and Ω^* used for all calculation instead of Ω_i and Ω.

7.6 Applications of Bose Statistics to Ideal Photon and Phonon Gas

7.6.1 Photon Gas (Ideal)

We are now in a position to consider thermal radiation in a cavity of volume V. Our main task will be to calculate the omega potential of the photon gas (Ω_i^*) and use it to obtain diverse properties of the system. We will continue to use the grand canonical distribution with a modification that the chemical potential 'μ' is zero for photon gas. 'μ' appeared in our description and derivation of GC distribution to ensure, through the use of a Lagrange multiplier that the mean number of particles $<N>$ is conserved. For photon gas, the photon number is fixed by the temperature and volume of the system, making explicit imposition of the constraint $\sum_{\nu N} W_{\nu N} \cdot N = <N>$ redundant. We can therefore set the corresponding multiplier μ to be equal to zero and continue to use the GC distribution to understand properties of an ideal photon gas. The mean number of Bosons at the ith energy level is

$$< n_i > = \frac{1}{e^{\frac{(\epsilon_i - \mu)}{kT}} - 1}$$

For photons, with μ set equal to zero, and $\epsilon_i = \hbar \omega_i$ the Bose-Einstein distribution takes the form

$$< n_i > = \frac{1}{e^{\frac{\hbar \omega_i}{kT}} - 1} \tag{7.83}$$

The omega potential of the photon gas is, with μ set equal to zero is

$$\Omega^* = kT \sum_i ln\{1 - e^{\frac{-\epsilon_i}{kT}}\} \tag{7.84}$$

The summation over 'i' has to be carried out now. Before doing that we take note of a special feature of an ideal photon gas. The energy levels of the system are no doubt discrete, yet in many practically important instances, they lie so close that ϵ_i can be treated as an almost continuous quantity (ϵ) and the summation in Eq. 7.84 can be replaced by integration after introducing a function g_ϵ called the density of states in the energy interval $d\epsilon$ (g_ϵ physically measures the number of levels per unit interval of energy in the vicinity of the energy ϵ). The assumption inherent in the procedure described above is that the energy levels ϵ_i lie so close to one another that the difference between the neighboring levels is much smaller than kT. The small energy interval $d\epsilon$ can then taken to be much smaller than kT, but much larger than the interval between the adjacent energy levels. That means $d\epsilon$ contains many energy levels. The summation in Eq. 7.84 can be conveniently performed in two steps.

Step 1: We divide the full energy scale into intervals of size $d\epsilon$ and sum up the levels in $d\epsilon$. Since these levels differ from each other by an amount $<<$kT, all the terms in the summation over levels in $d\epsilon$ have approximately the same value and the sum is $g_\epsilon d\epsilon$ where g_ϵ is the number of levels in unit interval of energy – i.e. the density of states in $d\epsilon$. Once that is done, we can write (assuming $\mu = 0$ for photons)

$$d\Omega^* = kTln\{1 - e^{\frac{-\epsilon}{kT}}\}g_\epsilon d\epsilon$$

Step 2: Now we can carry out the summation over all the intervals $d\epsilon$ leading to

$$\Omega^* = kT \int_0^\infty ln\{1 - e^{-\frac{\epsilon_i}{kT}}\}g_\epsilon d\epsilon \tag{7.85}$$

For carrying out the integration, we need $g_\epsilon d\epsilon$ explicitly for photons and we will first do that.

Let us note in the present context that the energy ϵ and the momentum p of a photon we have the relation that

$$\epsilon = pc$$

c being the speed of light. The energy interval from ϵ to $\epsilon + d\epsilon$ will therefore contain all photon states with the magnitude of momentum larger than $p = \frac{\epsilon}{c}$, but smaller than $p + dp = \frac{\epsilon}{c} + \frac{d\epsilon}{c}$.

The states referred to lie in the momentum space and are depicted by points in a spherical layer of thickness dp belonging to the energy interval $d\epsilon$. The volume of such a layer is clearly $d\Omega_p = 4\pi p^2 dp$ and the number of states in $d\Omega_p$ is

$$dG = \frac{4\pi p^2 dp}{(2\pi \hbar)^3} V \tag{7.86}$$

Expressing momentum 'p' in terms of energy we have

$$dG = \frac{4\pi \epsilon^2 d\epsilon}{(2\pi \hbar c)^3} V \tag{7.87}$$

Note that by our defination $dG = g_\epsilon d\epsilon$. Therefore, we have from Eq. 7.87,

$$g_\epsilon d\epsilon = \frac{4\pi \epsilon^2}{(2\pi \hbar c)^3} V d\epsilon \tag{7.88}$$

Since we are dealing with an ideal photon gas, it is necessary to take into account the fact that there are two kinds of photons of energy ϵ, differing only in polarization and formula Eq. 7.88 holds for photons of each kind. With this additional information taken into account the omega potential of the ideal photon gas becomes

$$\Omega^* = 2kT \int_0^\infty ln\{1 - e^{-\frac{\epsilon}{kT}}\} \frac{4\pi \epsilon^2}{(2\pi \hbar c)^3} V d\epsilon$$
$$= \int_0^\infty d\Omega^*(\omega) \tag{7.89}$$

Eq. 7.89 can be expressed in terms of the frequency ω by noting that $\epsilon = \hbar\omega$ and the integration over $d\epsilon$ transformed in integration over $d\omega$ leading to

$$\Omega^* = 2kT \int_0^\infty ln\{1 - e^{-\frac{\hbar\omega}{kT}}\} \frac{4\pi \omega^2}{(2\pi c)^3} V d\omega \tag{7.90a}$$

The integral is readily evaluated by substituting $\frac{\hbar\omega}{kT} = x$ whence

$$\Omega^* = \frac{(kT)^4}{\pi^2 \hbar^3 c^3} V \int_0^\infty ln(1 - e^{-x}) x^2 dx$$
$$= \frac{(kT)^4}{\pi^2 \hbar^3 c^3} V \times (-2.15) \tag{7.90b}$$

where we have used $\int_0^\infty ln(1 - e^{-x}) x^2 dx = -2.15$. Once Ω^* has been calculated, many other properties can be calculated. However, if we are interested in the omega potential of all photons whose frequency lies in the interval $\omega \leftrightarrow \omega + d\omega$ we can write from Eq. 7.89

$$d\Omega^* = 2kTV ln\left\{1 - e^{-\frac{\hbar\omega}{kT}}\right\} \frac{4\pi \omega^2}{(2\pi c)^3} d\omega \tag{7.91}$$

which is the spectral form of the omega potential for the electromagnetic radiation in a cavity. A pertinent question that comes up at this point concerns the validity of our assumption of a continuous distribution of energy levels used in arriving at the expression for Ω^* in Eq. 7.90. The quasi-continuity approximation clearly holds true at high temperatures when kT is large, in fact much larger than the spacing between adjacent energy levels. In fact, if kT is larger than the energy of the lowest state, the quasi-continuity approximation remains quite valid, since the states of the system include those in which one, two, three or more photons are in the lowest level while all other levels are vacant or unoccupied. This is equivalent to saying that the spacing between adjacent levels must be smaller than $\hbar\omega_0$, ω_0 being the lowest possible frequency of oscillations in the system. Thus $kT >> \hbar\omega_0$ ensures that the approximation of quasi-continuity does not involve any appreciable error.

7.6.2 Phonon Gas (Ideal)

Bose statistics can be applied with equal facility to understand the properties of crystal vibrations. A gas at high temperature has little or no quantum effects modulating its properties. When the temperature is lowered sufficiently for quantum effects to become prominent, the gas turns into a solid. A solid at low temperature may reveal quantum properties that can only be properly understood in the light of quantum statistics. For that we adopt the same procedure as was developed for describing a system of ideal photon gas, i.e. radiation in a cavity.

The atoms (ions) of a solid form a crystal lattice. If an atom in the lattice is slightly displaced, it causes displacements of neighboring atoms (as they are held in equilibrium by interatomic forces) and these coordinated displacements lead to the appearence of a wave propagating along the crystal. The wave nature of the motion of atoms in a crystal lattice enables us to set up a statistical theory of crystals in much the same way as electromagnetic waves enclosed in a cavity have been treated already.

The broad contours of the statistical theory no doubt match with the statistical description of electromagnetic waves enclosed in a cavity, but there are significant, even substantial differences. The first significant point of departure comes from the discrete nature of crystal lattice, which can support different kinds of vibrations, like longitudinal and transverse ones, in the simplest case. The second point of departure arises from the anisotropy of crystal properties, which means the waves set up in the crystal may have direction-dependent features. That complicates the treatment of vibration in a crystal.

To make progress, we may replace the discrete crystal lattice with a continuous isotropic medium having the same elastic properties as the discrete lattice does. Practically, this amounts to replacing the discrete lattice with an elastic medium in which the velocity of propagation of the elastic waves (i.e. the compression, dilatation and displacement waves) are the same as in the lattice and is identical in all directions (isotropic). At the simplest level we must consider two transverse waves with two mutually perpendicular directions of polarization and one longitudinal wave. Theoretical treatment leads to the same results for vibrations of each kind except that the speeds of the transverse and longitudinal acoustic waves set up in the solid are different. Therefore, we will explicitly treat the longitudinal waves propagating with the speed c_l. The same results hold for the transverse waves with c_l replaced by c_t.

Let us now consider a crystal having the shape of a rectangular parallelepiped with the sides a, b and c. Since the dimensions are limited, only vibrations of precisely definite frequencies can be set up in the crystal with these frequencies depending on the crystal dimensions and the conditions of reflection at the free boundaries of the crystal. These frequencies together form a discrete set; but in many cases of practical importance the relevant frequencies may be considered approximately continuous. With this approximation in place, the important next step is to determine the number dG of various waves whose frequencies lie within the interval $d\omega$. If g_ω be the density of states we have

$$dG = g_\omega d\omega \tag{7.92}$$

The task now is to find g_ω. Here, analogy with electromagnetic waves can be exploited to assert that the only possible frequencies are those for which the components of the wave vector $\overrightarrow{k}$ are determined by the following relations:

$$k_x = \frac{2\pi}{a} m_x$$

$$k_y = \frac{2\pi}{b} m_y \tag{7.93}$$

$$k_z = \frac{2\pi}{c} m_z$$

where m_x, m_y and m_z are integers. The wave vector $(\overrightarrow{k})$ itself is related to the wave frequency ω and the speed of propagation c_l of the longitudinal acoustic wave in the crystal:

$$|\overrightarrow{k}| = k = \frac{2\pi}{\lambda} = \frac{\omega}{c_l} \tag{7.94}$$

The perfect analogy with electromagnetic waves ends here and now it is important to recognize a very important difference between the electromagnetic waves and the acoustic waves arising from the discrete nature of the crystal lattice. The frequencies of the electromagnetic waves are not bounded from above; so any ω from 0 to ∞ is possible. In a discrete lattice, a little reflection suggests that the smallest wavelength (λ_{min}) can not be less than $2d$ where 'd' stands for the lattice spacing. With this restriction on λ_{min} or ω_{max} ($= 2\pi c_l / \lambda_{min}$) the theory of thermal oscillations of a lattice is reduced to the theory of equilibrium acoustic (sound) waves, which in turn is very similar to the theory of equilibrium of radiation in a cavity discussed in the previous section. To fix the value of ω_{max} properly, we have to proceed from the condition that the total number of waves in a crystal is equal to the total number of degrees of freedom of the atoms that the crystal is made of. For an N-atom crystal, this number is $3N$. There are therefore N different wavelengths for each of the three types of vibrations – one longitudinal and two transverse, in simple lattices. ω_{max} can thus be fixed by condition

$$\int_0^{\omega_{max}} \frac{4\pi \omega^2}{(2\pi c_l)^3} V d\omega = N \tag{7.95}$$

whence it follows that

$$\omega_{max} = \sqrt[3]{6\pi^2}\left(\frac{N}{V}\right)^{\frac{1}{3}}c_l \qquad (7.96)$$

If we take the simplest lattice, i.e. the cubic lattice $V = Nd^3$ where 'd' is the lattice constant. This leads us to the simple expression for ω_{max}:

$$\omega_{max} = \frac{\sqrt[3]{6\pi^2}c_l}{d} \simeq \frac{2\pi c_l}{1.61d} = \frac{2\pi c_l}{\lambda_{min}}$$

That is, the minimum wavelength (λ_{min}) supported by the lattice is $1.61d$. We can now write, in analogy with the Eq. 7.90a, for thermal radiation an expression for the Omega potential of the acoustic waves of one kind (longitudinal) in the crystal:

$$\Omega^* = kT\int_0^{\omega_{max}} ln\left\{1 - e^{-\frac{\hbar\omega}{kT}}\right\}\frac{4\pi\omega^2}{(2\pi c_l)^3}Vd\omega \qquad (7.97)$$

The factor of 2 has been dropped as we have only the longitudinal waves present.

Let us now examine the expression for Ω^* of the acoustic waves in a crystal under two different conditions:

$$\begin{aligned}&(i) \quad \hbar\omega_{max} >> kT\\ &(ii) \quad \hbar\omega_{max} << kT\end{aligned} \qquad (7.98)$$

In the case (i), ω_{max}, the upper limit of integration is much larger compared to $\frac{kT}{\hbar}$. So, while integrating we will encounter frequencies $\omega >> \frac{kT}{\hbar}$, which means $e^{-\frac{\hbar\omega}{kT}} << 1$. For such frequencies

$$ln\{1 - e^{-\frac{\hbar\omega}{kT}}\} \approx -e^{-\frac{\hbar\omega}{kT}}$$

which is a rather small quantity. So, in Eq. 7.97 we can replace the upper limit of integration by ∞ without incurring large error, thereby reducing the expression for Ω^* to

$$\Omega^* = kT\int_0^{\infty} ln\left\{1 - e^{-\frac{\hbar\omega}{kT}}\right\}\frac{4\pi\omega^2}{(2\pi c_l)^3}Vd\omega \qquad (7.99)$$

which completely coincides (apart from a factor of 2) with the expression for Ω^* for electromagnetic radiation in a cavity of volume V and c (speed of light) replaced by c_l, the speed of propagation of the longitudinal acoustic waves (sound). The theory of crystal vibrations therefore becomes completely analogus to the theory of radiation in a cavity. We can now say that we are dealing with a phonon gas instead of gas of photons, the phonons being the quantum of acoustic energy (or sound) in the solid propagating with the speed c_l instead of the speed of light c. We can complete the integration in Eq. 7.99 quite easily and arrive at an expression for Ω^* (for $\omega_{max}\frac{kT}{\hbar}$) for each of the three kinds of wave set up in the crystal:

$$\Omega^* = -1.07\frac{k^4}{\pi^2\hbar^3 c_{ph}}VT^4 \qquad (7.100)$$

where c_{ph} is the speed of the corresponding wave of phonons.

Notice the similarity and difference between the expression for Ω^* of a phonon gas in Eq. 7.100 and the corresponding expression for Ω^* of a photon gas in Eq. 7.90b.

The internal energy (U) of the phonon gas can be easily obtained from the Gibbs-Helmholtz equation, which yields

$$U = \Omega^* - T\frac{\partial\Omega^*}{\partial T} = -3\Omega^*$$

$$= 3.21\frac{k^4}{\pi^2\hbar^3 c_{ph}^3}VT^4 \qquad (7.101)$$

The heat capacity of the crystal at constant volume is therefore given by

$$C_v = \frac{\partial U}{\partial T} = 12.84 \frac{k^4}{\pi^2 \hbar^3 c_{ph}^3} V T^3 \qquad (7.102)$$

From Eq. 7.102 we find that the heat capacity (C_v) of the solid rapidly falls to zero following a T^3 law as the temperature is lowered (T → 0). This is Debye's T^3 law for specific heat of solids as $T \to 0$.

(ii) It is now important to consider the other limiting case where $\hbar\omega_{max} << kT$. In this case all the frequencies (ω) that contribute to the integral in Eq. 7.97 have values that satisfy

$$\frac{\hbar\omega}{kT} << 1$$

Let us now consider, for example, the internal energy of phonons relating to the longitudinal vibrations having frequencies within the interval $d\omega$. This internal energy dU is given by

$$dU = d\Omega^* - T\frac{\partial}{\partial T}(d\Omega^*)$$

where $d\Omega^* = kT ln\{1 - e^{-\frac{\hbar\omega}{kT}}\}\frac{4\pi\omega^2}{(2\pi c_{ph})^3} d\Omega$. It is straightforward to find that

$$dU = \frac{4\pi\omega^2}{(2\pi c_{ph})^3} V \frac{\hbar\omega}{e^{\frac{\hbar\omega}{kT}} - 1} \qquad (7.103)$$

The exponential in the denominator can be approximated using the smallness of $\frac{\hbar\omega}{kT}(<< 1)$ giving

$$e^{\frac{\hbar\omega}{kT}} \approx 1 + \frac{\hbar\omega}{kT}$$

whence

$$dU = \frac{4\pi\omega^2}{(2\pi c_{ph})^3} d\omega V \cdot (kT) \qquad (7.104)$$

The first factor represents the number of states within the frequency interval $d\omega$ and the second factor reveals that the mean energy apportioned to each state is kT. So, Eq. 7.104 is in fact a restatement of the classical theorem asserting uniform distribution of energy among the degrees of freedom possessed by the system.

The internal energy U of the longitudinal photons in the limiting case, (ii) i.e. for $\frac{\hbar\omega}{kT} << kT$ is

$$U = \int_0^{\omega_{max}} \frac{4\pi\omega^2}{(2\pi c_l)^3} V k T d\omega = \frac{4}{3}\pi\omega_{max}^3 \frac{V}{(2\pi c_l)^3}(kT) \qquad (7.105)$$

Recall that we have estimated already that

$$\omega_{max} = \frac{2\pi c_l}{1.61d}$$

which when used in Eq. 7.105, yields for the energy of the longitudinal waves ($\frac{V}{N} = d^3$)

$$U = NkT$$

Since there are also two types of transverse vibration in addition to the longitudinal one, the total internal energy of the crystal in the present limiting case ($\hbar\omega_{max} << kT$) becomes

$$U = 3NkT \qquad (7.106)$$

reasserting agreement with the Boltzmann principle of equipartition of energy among the available degrees of freedom. In this case, the specific heat at constant volume is easily calculated and found to be

$$C_v = \frac{\partial U}{\partial T} = 3Nk$$

Again, the behavior is completely classical. The two limiting cases help us define a temperature θ_D – the Debye temperature for a specific crystal below which quantum response prevails and above which classical behavior takes over. Thus for

$$T << \frac{\hbar\omega_{max}}{k} = \theta_D = \frac{\hbar}{k}(6\pi^2)^{\frac{1}{3}}\frac{c_{ph}}{d} \tag{7.107a}$$

quantum behavior prevails and Bose-Einstein statistics must be invoked to describe and understand the behavior of phonons in the crystal. For

$$T >> \frac{\hbar\omega_{max}}{k} = \theta_D = \frac{\hbar}{k}(6\pi^2)^{\frac{1}{3}}\frac{c_{ph}}{d} \tag{7.107b}$$

classical description continues to hold.[1] We must note, however that the Debye temperature is different for the longitudinal and the transverse modes as the velocity of the corresponding waves are different.

7.7 Quantum Statistics for Electron Gas in a Potential Well

The electrons in a metal or semiconductor can be modeled and treated as an ideal electron gas in a potential well. The state of an electron in a potential well is characterized by the components of its momentum p_i as follows:

$$p_{x,i} = \hbar k_{xi} = \hbar\frac{2\pi}{a}m_{xi}$$

$$p_{y,i} = \hbar k_{yi} = \hbar\frac{2\pi}{b}m_{yi} \tag{7.108}$$

$$p_{z,i} = \hbar k_{zi} = \hbar\frac{2\pi}{c}m_{zi}$$

m_{xi}, m_{yi}, m_{zi} are integers, a, b, c represent the three sides of the rectangular potential well. Our main task once again boils down to calculating the omega potential (Ω^*) of the electron gas, which obeys Fermi-Dirac statistics. We are tempted therefore to make use of Eq. 7.71 and evaluate the sum over discrete states. While that is possible, it is much more convenient and practical to switch over to integration especially because the electron-states although discrete, are so closely spaced it is possible to consider the distribution of states to be continuous for all practical purposes. The passage from discrete summation to integration can be done exactly as we did in the case of Bosons by noting that the volume $\Delta\Omega$ of the momentum space that accomodates one state is

$$\Delta\Omega = \Delta p_x \Delta p_y \Delta p_z = \frac{2\pi\hbar}{a}\cdot\frac{2\pi\hbar}{b}\cdot\frac{2\pi\hbar}{c} = \frac{(2\pi\hbar)^3}{V} \tag{7.109}$$

where $V = abc$ is the volume in the conventional or the physical space. Let $d\Omega$ be the volume in the momentum space confined between $p \leftrightarrow p + dp$ in which all the states having energy in the interval from ϵ to $\epsilon + d\epsilon$ are confined. The number of such states (dG') can be obtained by computing the ratio of $d\Omega$ to $\Delta\Omega$. The actual number of states (dG) is, however twice the ratio (dG') as each state with a given momentum can be occupied by two electrons with oppositely oriented spin angular momentum. So, we find

$$dG = 2dG' = 2\frac{d\Omega}{\Delta\Omega} \tag{7.110}$$

Now $d\Omega$ is clearly $4\pi p^2 dp$ (the volume of the annular space between two concentric spheres of radii p and $p + dp$). Since $p^2 = 2m\epsilon$, it is easy to find that

$$d\Omega = 4\pi p^2 dp = 4\pi(2m)^{\frac{3}{2}}\frac{1}{2}\epsilon^{\frac{1}{2}}d\epsilon \tag{7.111}$$

Using Eqs. 7.109-7.111, we arrive at

$$dG = \frac{4\pi (2m)^{\frac{3}{2}} \epsilon^{\frac{1}{2}} d\epsilon}{(2\pi \hbar)^3} V$$

$$= g_\epsilon d\epsilon$$

(7.112)

where g_ϵ or the density of states turns out to be

$$g_\epsilon = \frac{4\pi (2m)^{\frac{3}{2}} \epsilon^{\frac{1}{2}}}{(2\pi \hbar)^3} V$$

(7.113)

The omega potential of an ideal electron gas having energy confined in the interval from ϵ to $\epsilon + d\epsilon$ can now be written using Eq. 7.112 and Eq. 7.113 in Eq. 7.71 as

$$d\Omega^* = -kTln\left\{1 + e^{\frac{(\mu-\epsilon)}{kT}}\right\} g_\epsilon d\epsilon$$

$$= -kTln\left\{1 + e^{\frac{(\mu-\epsilon)}{kT}}\right\} \frac{4\pi (2m)^{\frac{3}{2}} \epsilon^{\frac{1}{2}}}{(2\pi \hbar)^3} V d\epsilon$$

(7.114)

The total omega potential Ω^* of all the electrons can now be expressed as an integral over ϵ

$$\Omega^* = \int d\Omega^* = -kT \int_0^\infty ln\left\{1 + e^{\frac{(\mu-\epsilon)}{kT}}\right\} g_\epsilon d\epsilon$$

(7.115)

thereby completing the transition from discrete summation over ϵ_is to a continuous integration over ϵ from 0 to ∞. It is now possible to consider the properties of an ideal electron gas under two limiting situations, defined by the parameter $\xi_0 = \frac{\mu}{kT}$.

7.7.1 Non-Degenerate Electron Gas

Let $\xi_0 << -1$, i.e. $\mu << -kT$. The electron gas with its chemical potential much smaller than $-kT$ is said to be non-degenerate. The gas obeys classical statistics. To see that let us calculate the mean number of electrons with energy between ϵ and $\epsilon + d\epsilon$. From our previous discussion we can easily relate this number (dn) to the derivative of the omega potential of the electrons with energy confined in $d\epsilon$. Thus with $d\Omega^*$ defined in Eq. 7.115, we have

$$dn = -\frac{d(d\Omega^*)}{d\mu}$$

$$= \frac{1}{e^{\frac{(\epsilon-\mu)}{kT}} + 1} \times \frac{4\pi (2m)^{\frac{3}{2}} \epsilon^{\frac{1}{2}}}{(2\pi \hbar)^3} V d\epsilon$$

(7.116)

If we set $\xi = \frac{\epsilon}{kT}$ and $\xi_0 << \xi$, we have $\mu << \epsilon$. Under this condition the first term (the Fermi-Dirac distribution function) in Eq. 7.116 can be very well approximated as

$$\frac{1}{e^{\frac{(\epsilon-\mu)}{kT}} + 1} \approx e^{\frac{(\mu-\epsilon)}{kT}}$$

$$= e^{\frac{\mu}{kT}} \cdot e^{\frac{-\epsilon}{kT}}$$

(7.117)

revealing that Fermi-Dirac distribution has been reduced Maxwell-Boltzmann distribution under the given condition. The mean number of the designated electrons then becomes

$$dn = e^{\frac{\mu}{kT}} \cdot e^{\frac{-\epsilon}{kT}} \frac{4\pi (2m)^{\frac{3}{2}}}{(2\pi \hbar)^3} V \epsilon^{\frac{1}{2}} d\epsilon$$

(7.118a)

Eq. 7.118a gives the number of electrons confined to the energy interval from ϵ to $\epsilon + d\epsilon$ in the volume 'V', which is just the Maxwell-Boltzmann distribution. Eq. 7.118 can also be applied to small portion dV of total volume 'V'. Thus the mean number of electrons in the energy interval $d\epsilon$ and in the volume dV is

$$dn = e^{\frac{\mu}{kT}} \cdot e^{\frac{-\epsilon}{kT}} \frac{4\pi (2m)^{\frac{3}{2}}}{(2\pi\hbar)^3} dV \epsilon^{\frac{1}{2}} d\epsilon \qquad (7.118b)$$

The total mean number of particles in the volume 'dV' with any energy between 0 and ϵ is easily obtained by evaluating the following integral:

$$ndV = dV \int_0^\infty dn = dV \int_0^\infty e^{\frac{\mu}{kT}} \cdot e^{\frac{-\epsilon}{kT}} \frac{4\pi (2m)^{\frac{3}{2}}}{(2\pi\hbar)^3} \epsilon^{\frac{1}{2}} d\epsilon \qquad (7.119)$$

Evaluation of the integral is straightforward and leads to an expression for the chemical potential of the electron gas[1,3]

$$\mu = kT ln\left[\frac{(2\pi\hbar)^3 n}{(2\pi mkT)^{\frac{3}{2}} \times 2} \right]$$

$$= kT ln\left[\sqrt{2} n\hbar^3 \left(\frac{\pi}{mkT} \right)^{\frac{3}{2}} \right] \qquad (7.120)$$

It is pertinent to examine now the condition under which the electron gas becomes non-degenerate and obeys classical statistics. Since the gas becomes non-degenerate when $\mu << -kT$, Eq. 7.120 leads to the condition for non-degeneracy as

$$ln\left[\sqrt{2}\hbar^3 n \left(\frac{\pi}{mkT} \right)^{\frac{3}{2}} \right] << -1 \qquad (7.121a)$$

$$or, \sqrt{2}\hbar^3 n \left(\frac{\pi}{mkT} \right)^{\frac{3}{2}} << 1$$

$$n << \left(\frac{mkT}{\pi\hbar^3} \right)^{\frac{3}{2}} \frac{\sqrt{2}}{2} \qquad (7.121b)$$

Thus the condition for non-degeneracy (Eq. 7.121a) demands that the concentration of electrons must be low and the temperature sufficiently high. If the Fermions are not electrons but some other particles, then the higher the mass, the more likely is the gas to be non-degerate and behave classically at low density and high temperature.

7.7.2 Degenerate Electron Gas

Let $\xi_0 >> 1$, i.e. $\mu >> kT$. When the above mentioned condition is satisfied the electron gas is said to be degenerate and its behavior can be understood only and only by application of the appropriate quantum statistics, namely Fermi-Dirac statistics. From Eq. 7.115 and Eq. 7.112, we have

$$\Omega^* = -kT \int_0^\infty ln\left\{ 1 + e^{\frac{\mu-\epsilon}{kT}} \right\} g_\epsilon d\epsilon$$

$$= -kT \int_0^\infty ln\left\{ 1 + e^{\frac{\mu-\epsilon}{kT}} \right\} \frac{4\pi m^{\frac{3}{2}}}{(2\pi\hbar)^3} V \epsilon^{\frac{1}{2}} d\epsilon \qquad (7.122)$$

By substituting ξ for $\frac{\epsilon}{kT}$ and using ξ as the integration variable in place of ϵ we have

$$\Omega^* = -(kT)^{\frac{5}{2}} \frac{4\pi m^{\frac{3}{2}}}{(2\pi\hbar)^3} V\, erf(\xi_0) \qquad (7.123)$$

where $\zeta_0 = \frac{\mu}{kT}$ and $\zeta = \frac{\epsilon}{kT}$ and the error function erf (ζ_0) is defined as an indefinite integral

$$erf(\zeta_0) = \int_0^{\infty} ln\{1 + e^{\zeta_0 - \zeta}\}\zeta^{\frac{1}{2}}d\zeta \qquad (7.124)$$

In the present limiting case $\zeta_0 >> 1$, the dominant contribution to the integral (cf. Eq. 7.124) comes from the values of the integral at $\zeta < \zeta_0$. This happens because for $\zeta > \zeta_0$, we have

$$ln\{1 + e^{(\zeta_0 - \zeta)}\} = ln\{1 + e^{-(\zeta - \zeta_0)}\}$$
$$\approx e^{-(\zeta - \zeta_0)} << 1 \qquad (7.125)$$

Condition in Eq. 7.125 allows us to terminate the integration in Eq. 7.124 at $\zeta = \zeta_0$ and neglect unity in comparison with the exponential inside the logarithmic term in the integrand and write

$$erf(\zeta_0) = \int_0^{\zeta_0} ln\{e^{(\zeta_0 - \zeta)}\}\zeta^{\frac{1}{2}}d\zeta$$
$$= \frac{4}{15}\zeta_0^{\frac{5}{2}} \qquad (7.126)$$

Using Eq. 7.126 in Eq. 7.122, and remembering that $\zeta_0 = \frac{\mu}{kT}$, we arrive at an expression for Ω^* of the degenerate electron gas in terms of the chemical potential μ:

$$\Omega^* = -\frac{4\pi}{(2\pi\hbar)^3}(2m)^{\frac{3}{2}} \times \frac{4}{15}V\mu^{\frac{5}{2}} \qquad (7.127)$$

The equation of state for the degenerate electron gas is then found following the general prescription that

$$P = -\frac{\partial\Omega^*}{\partial V} = \frac{2}{15}\frac{(2m)^{\frac{3}{2}}}{\pi^2\hbar^3}\mu^{\frac{5}{2}} \qquad (7.128)$$

For the internal energy we can similarly proceed to obtain an expression for U by using once again the general prescription that

$$U = \Omega^* - \mu\frac{\partial\Omega^*}{\partial\mu} - T\frac{\partial\Omega^*}{\partial T} \qquad (7.129)$$

Now for the degenerate electron gas, Ω^* in Eq. 7.127 does not have any dependence on the temperature T so that the third term in Eq. 7.129 can be set equal to zero, yielding, finally U as a function of the chemical potential:

$$U = -\frac{3}{2}\Omega^* = \frac{(2m)^{\frac{3}{2}}}{5\pi^2 3}V\mu^{\frac{5}{2}} \qquad (7.130)$$

It is important now to look for explicit dependence of P and U on the temperature of the degenerate electron gas, if any. With that objective in view, let us first try to ascertain how 'μ' depends on the electron concentration (n). To do that let us recall that the mean number of electrons $< N >$ in our system is given by

$$< N >= -\frac{\partial\Omega^*}{\partial\mu} = \frac{(2m)^{\frac{3}{2}}}{5\pi^2 3}V\mu^{\frac{3}{2}}$$

The mean electron concentration is then

$$n = < N > /V = \frac{(2m)^{\frac{3}{2}}}{5\pi^2 3}\mu^{\frac{3}{2}}$$

which immediately leads us to an expression for the classical potential of the degenerate electron gas in terms of electron concentration

$$\mu = \frac{(5\pi^2\hbar^3)^{\frac{3}{2}}}{2m} \times n^{\frac{2}{3}} \qquad (7.131)$$

We may now use this expression for μ in Eq. 7.128 and Eq. 7.13 for P and U, respectively, yielding

$$P = \frac{(5\pi^2)^{\frac{5}{3}}\hbar^2}{15\pi^2 m} n^{\frac{5}{3}}$$

(7.132)

and

$$U = \frac{(5\pi^2)^{\frac{5}{3}}\hbar^2}{10\pi^2 m} V n^{\frac{5}{3}}$$

(7.133)

Thus, the pressure and the internal energy of a degenerate electron gas do not depend on the temperature (T). An immediate consequence of the temperature independence of U is that the specific heat of the degenerate electron gas is $C_v = \frac{\partial U}{\partial T} = 0$. Since the mean electron concentration is $n = <N>/V$, the pressure P (cf. Eq. 7.132) becomes equal to (c is a constant)

$$P = c \times \left(\frac{<N>}{V}\right)^{\frac{5}{3}}$$

where the constant $c = \frac{(5\pi^2)^{\frac{5}{3}}\hbar^2}{15\pi^2 m}$. The equation of state of the degenerate electron gas thus becomes

$$PV^{\frac{5}{3}} = c <N>^{\frac{5}{3}}$$

(7.134)

It is relevant to ask now what is the criterion of degeneracy of the electron gas? The condition as stated already is

$$\mu >> kT$$

which can be restated using Eq. 7.131

$$\frac{(5\pi^2\hbar^3)^{\frac{2}{3}}}{2m} n^{\frac{2}{3}} >> kT$$

(7.135)

$$\text{i.e.} \quad n >> 2^{\frac{3}{2}}\left(\frac{mkT}{\pi\hbar^2}\right)^{\frac{3}{2}} \times \frac{1}{5\sqrt{\pi}}$$

The electron gas density must satisfy the condition in Eq. 7.135 in order to be degenerate and behave as a non-classical electron gas obeying Fermi-Dirac statistics.

We will return to the quantum statistics of electron gas in the next chapter once again as we take up the problems of understanding the properties of metallic conductors and semiconductors.

7.8 Quantum Effects in Heat Capacity of Gases

We have seen in preceeding section that a degenerate electron gas has a virtually zero specific heat (C_v). A non-degenerate electron gas, however, responds classically to thermal stimulus and has specific heat $C_v = \frac{3}{2}k <N>$. This feature of the non-degenerate gas can be demonstrated easily. Thus, from Eq. 7.123,

$$\Omega^* = -(kT)^{\frac{5}{2}}\frac{4\pi(2m)^{\frac{3}{2}}}{(2\pi\hbar)^3} V \times erf(\xi_0)$$

For the non-degenerate gas ξ_0 is small ($<< -1$). Therefore, on the right hand side of the $erf(\xi_0)$

$$erf(\xi_0) = \int_0^\infty ln\left\{1 + e^{(\xi_0 - \xi)}\right\} \xi^{\frac{1}{2}} d\xi$$

We can approximate the integral as

$$ln\left\{1 + e^{(\xi_0 - \xi)}\right\} \approx e^{(\xi_0 - \xi)}$$

The exponential being a very small quantity as $\xi_0 << 1$, we can write

$$erf(\xi_0) = e^{\xi_0} \int_0^\infty e^{-\xi} \xi^{\frac{1}{2}} d\xi$$

$$= e^{\xi_0} \frac{\sqrt{\pi}}{2}$$

$$= e^{\frac{\mu}{kT}} \frac{\sqrt{\pi}}{2}$$

Ω^* therefore becomes equal to

$$\Omega^* = -(kT)^{\frac{5}{2}} e^{\frac{\mu}{kT}} \frac{\sqrt{\pi}}{2} \frac{4\pi (2m)^{\frac{3}{2}}}{(2\pi \hbar)^3} V = \Omega'(say)$$

The internal energy U, following the general formula (Eq. 7.53) and the result $-\frac{\partial \Omega^*}{\partial \mu} = <N>$ becomes

$$U = \Omega^* - T \frac{\partial \Omega^*}{\partial T} - \mu \frac{\partial \Omega^*}{\partial \mu} = \frac{3}{2} kT <N> \tag{7.136}$$

and C_v of the non-degenerate electron gas turns out to be $\frac{3}{2} k <N>$. For 1 mole of electrons, $<N> = N_0$ (Avogadro's number) and $C_v = \frac{3}{2}R$ as expected classically.

7.8.1 A Model Application of Quantum Statistics

The nature of the quantum effects on the properties of say, a gas consisting of a rigid diatomic molecules executing motion in a plane is worth exploring. The energy of the molecule has contributions from the translational kinetic energy due to its motion along the x and y axes and rotation about the z-axis. The energy ϵ of the molecule is then given by

$$\epsilon = \frac{p^2}{2m} + \frac{L_z^2}{2I_0} = \epsilon_{\text{translation}} + \epsilon_{rot} \tag{7.137}$$

where p is the linear momentum and L_z is the projection of the angular momentum on to an axis through the center of mass (CM) parallel to the z-axis, M is the mass of the molecule, I_0 is the moment of inertia relative to the axis passing through the CM at right angles to the plane of motion (xy plane). The translational motion may be assumed to be executed along the x and y axes enclosed in a rectangular box of sides a and b. Under these assumptions

$$\epsilon_{trans} = \frac{(2\pi \hbar)^2}{2M} \left\{ (\frac{n_x}{a})^2 + (\frac{n_y}{b})^2 \right\}$$

$$\epsilon_{rot} = \frac{m^2 \hbar^2}{2I_0} \tag{7.138}$$

where n_x, n_y, m are the arbitrary integers (quantum numbers). The discrete energy levels (ϵ) of the diatom are therefore given by[1]

$$\epsilon_i = \epsilon(n_x, n_y, m) = \frac{4\pi^2 \hbar^2}{2Ma^2} n_x^2 + \frac{4\pi^2 \hbar^2}{2Mb^2} n_y^2 + \frac{\hbar m^2}{2I_0} \tag{7.139}$$

Since our interest is to understand the properties of the gas consisting of the 'model' diatoms with discrete energy levels in the light of statistical mechanics for which we must compute the omega potential (Ω^*) and therefore the partition function (Z^*) of the gas.

For this purpose we will define a subsystem as a collection of all the molecules in a specific state 'i', which is characterized by a definite set of quantum numbers n_x, n_y and m. Let 'n_i' be the number of molecules in the state 'i'. Hence, we may write

$$Z^* = \sum_{n_i} n_i e^{\frac{(\mu - \epsilon_i)}{kT}} \tag{7.140}$$

If the molecules possess half-integer spin, they are Fermions and n_i can only take on values 0 and 1. For the ith subsystem, we therefore have

$$(Z_i^*)_F = 1 + e^{\frac{(\mu-\epsilon_i)}{kT}} \tag{7.141a}$$

and the corresponding subsystem omega potential

$$(\Omega_i^*)_F = -kTln(Z_i^*)_F$$
$$= -kTln\left\{1 + e^{\frac{(\mu-\epsilon_i)}{kT}}\right\} \tag{7.141b}$$

If the 'molecules' have zero or integer spin, they are Bosons and the corresponding partition function and omega potential for the ith subsystem are

$$(Z_i^*)_B = \sum_{n_i=0}^{\infty} 1 + n_i e^{\frac{(\mu-\epsilon_i)}{kT}}$$
$$= \left\{1 - e^{\frac{(\mu-\epsilon_i)}{kT}}\right\}^{-1} \tag{7.142a}$$

$$(\Omega_i^*)_B = -kTln(Z_i^*)_B$$
$$= kTln\left\{1 - e^{\frac{(\mu-\epsilon_i)}{kT}}\right\} \tag{7.142b}$$

If the 'gas' satisfies the non-degeneracy condition for Fermions (see Eq. 7.135) the expression for the subsystem omega potential in either case reduces to the same (classical) expression calculated without factoring in the indistinguishability of the molecules. This happens when the concentration of the molecules is low and the chemical potential 'μ' is negative with magnitude much larger than kT, which makes the exponential term in Eq. 7.141b and Eq. 7.142b small enough. A back of the envelope calculation shows that if the temperature is 20 K, M is the atomic mass of the hydrogen, the non-degeneracy condition is satisfied when the concentration of molecules $n < 1.7 \times 10^{28}$ m^{-3}, a concentration so high that the sample must be in a condensed state to achieve that kind of concentration.[1] At 20 K (say) therefore the 'diatomic' gas is expected to behave as a non-degenerate gas and the subsystem omega potential for Fermionic and Bosonic molecules both reduce to the same expression (manifestly classical)

$$\Omega_i^* = -kTe^{\frac{(\mu-\epsilon_i)}{kT}} \tag{7.143}$$

The total omega potential then becomes

$$\Omega^* = \sum_i \Omega_i^* = -kTe^{\frac{\mu}{kT}} \sum_i e^{\frac{-\epsilon_i}{kT}} \tag{7.144}$$

The sum over the index 'i' is for all practical purposes is reducible to a sum over the quantum numbers n_x, n_y and m (cf. Eq. 7.139) and we can write

$$\Omega^* = -kTe^{\frac{\mu}{kT}} \sum_{n_x} e^{-\frac{An_x^2}{kT}} \sum_{n_y} e^{-\frac{Bn_y^2}{kT}} \sum_m e^{-\frac{Cm^2}{kT}} \tag{7.145}$$

where A, B, C are constants specified as

$$A = \frac{4\pi^2\hbar^2}{2ma^2}, \quad B = \frac{4\pi^2\hbar^2}{2mb^2}, \quad C = \frac{\hbar^2}{2I} \tag{7.146}$$

Let the gas be at an elevated temperature (T) such that $kT >> A, B, C$. Under this condition, each of the three sums in Eq. 7.145 can be reduced to integrals of the same type. Considering the first of the three sums, we have

$$\sum_{n_x=-\infty}^{+\infty} e^{-\frac{An_x^2}{kT}} \approx \int_{-\infty}^{+\infty} e^{-\frac{An_x^2}{kT}} dn_x \tag{7.147a}$$

$$= \sqrt{\frac{kT\pi}{A}}$$

Similarly,

$$\sum_{n_y=-\infty}^{+\infty} e^{-\frac{Bn_y^2}{kT}} = \sqrt{\frac{kT\pi}{B}} \tag{7.147b}$$

and

$$\sum_m e^{-\frac{Cm^2}{kT}} = \sqrt{\frac{kT\pi}{C}} \tag{7.147c}$$

At high temperatures therefore the omega potential becomes (cf. Eq. 7.145)

$$\Omega^* = -kTe^{\frac{\mu}{kT}} \frac{(\pi kT)^{\frac{3}{2}}}{\sqrt{ABC}} \tag{7.148}$$

The mean number of molecules $<N>$ then turns out to be

$$<N> = -\frac{\partial \Omega^*}{\partial \mu} = e^{\frac{\mu}{kT}} \frac{(\pi kT)^{\frac{3}{2}}}{\sqrt{ABC}} \tag{7.149}$$

The internal energy (U) of our model diatomic gas is given by (using Eq. 7.149 in the last step)

$$U = \Omega^* - T\frac{\partial \Omega^*}{\partial T} - \mu\frac{\partial \Omega^*}{\partial \mu}$$

$$= \frac{3}{2}kTe^{\frac{\mu}{kT}} \frac{(\pi kT)^{\frac{3}{2}}}{\sqrt{ABC}} \tag{7.150}$$

$$= \frac{3}{2}kT <N>$$

If $<N> = N_0$, N_0 being the Avogadro number, the molar internal energy becomes equal to $\frac{3}{2}kTN_0 = \frac{3}{2}RT$, which is what is expected classically for a 'three degrees of freedom' system. The molar specific heat of the gas (C_v) under the condition $kT >> A, B, C$, is therefore $\frac{3}{2}R$. At sufficiently high temperature all the three degrees of freedom contribute to the internal energy and the specific heat (C_v). That takes us to the question: at what temperature, do all the degrees of freedom of the molecule reveal their presence? If we consider typical values of the constants A, B and C, it turns out that $A, B << C$. So, if $k_BT >> C$, k_BT is automatically $>> A, B$ as well, leading to the manifestation of all the three degrees of freedom in U or C_v, as we have already seen. It is possible to estimate the temperature, say T_1 above which all three degrees of freedom will be manifested from the inequality $kT >> C$, which fixes the temperature at

$$T_1 = \frac{C}{k_B} \tag{7.151}$$

Using typical value of C ($\approx 1.4 \times 10^{-21} J$) and the known value of the Boltzmann constant (k_B), we find that $T_1 \approx 100$ K, and conclude that above 100 K the model diatomic gas will have $C_v = \frac{3}{2}R$.

We must investigate now the behavior of the gas at temperature (T) much lower than T_1. If 'T' is such that $k_BT << C$, the sum over 'm', i.e. $\sum_m e^{-\frac{Cm^2}{k_BT}}$ becomes approximately equal to 1 (the only term in the sum that contributes is for $m = 0$, which becomes 1, all other terms ($|m| > 0$) becomes ≈ 0). So for $kT << C$, the omega potential takes the form

$$\Omega^* = -kTe^{\frac{\mu}{kT}} \times \frac{\pi kT}{\sqrt{AB}} \qquad (7.152)$$

Thus, at temperatures much lower than T_1, the diatomic gas behaves like one with only two degrees of translational freedom with the rotational degree being completely frozen out. Thus, its internal energy at $T << T_1$ becomes

$$U = k_B T < N > \text{ and } \quad C_v = < N > k_B$$

the expressions being obtained from the general recipe that $U = \Omega^* - T\frac{\partial \Omega^*}{\partial T} - \mu\frac{\partial \Omega^*}{\partial \mu}$ with Ω^* given by Eq. 7.152.

We can easily follow C_v as a function of temperature from one extreme condition ($T_1 >> C/k_B$ to another $T_1 << C/k_B$). The summation over m at intermediate temperatures must be evaluated by allowing m to assume values $m = 0, \pm 1, \pm 2$, instead of restricting it to $m = 0$ only, and adding them up. Only for $T >> C/k_B$, the sum can be reduced to an integral. Thus quantum statistics enables us to explore the functional dependence of heat capacity on temperature and compare it effectively with experimental observation of such dependence. We will return to this aspect once again in the next chapter while addressing issues concerning temperature dependence of specific heats of metals and semiconductors.

7.9 Bose-Einstein Condensation

We have seen that the classical ideal gas or the ideal Fermi-Dirac gas does not have a thermodynamic phase transition. The ideal Bose gas like the photon gas also does not reveal a phase transition. However, in photon gas, the photon number is not a conserved quantity. It turns out that the Bose-Einstein ideal gas in which particle number is not conserved is quite unique in that it has a thermodynamc phase transition called Bose-Einstein condensation (BEC), which is driven purely by the statistics that the particles obey and not by any interaction, which is anyway absent, the gas being ideal. At the phase transition the thermodynamic observables of the system undergo an abrupt change which in turn defines a transition temperature T_c at which the condensation sets in. In ordinary condensation, the condensed particles form a 'droplet', which gets detached from the other particles in the gas phase. In BEC, the particles forming the condensate (Bose-Einstein condensate) do not get separated from the normal particles in the physical space. Instead they are separated in the momentum space. Below the T_c, the normal gas particles and the particles in condensate co-exist in equilibrium, all the condensate particles occupying a single quantum state of zero momentum (the ground state) while the non-condensate or the normal particles are distributed among states having non-zero (but finite) momentum.[2] The condensate has been viewed by some as a new state, the fifth state of matter. BEC plays a vital role in the theory of superconductivity of metals, which will be taken up in the next chapter. Hence, we will examine the unique phenomenon of BEC at some length here.[5-6]

From the BE distribution function, the total number of particles confined in a potential well of volume 'V' at a temperature 'T' is given by

$$N = \sum_k \frac{1}{e^{\frac{(\epsilon_k - \mu)}{k_B T}} - 1} \qquad (7.153)$$

In the thermodynamic limit $V \to \infty$, the allowed k values become continuous allowing us to replace the discrete sum over k in the equation above with integration, which if valid leads to (with $\beta = \frac{1}{k_B T}$)

$$N = \frac{V}{(2\pi)^3} \int \frac{1}{e^{\beta(\epsilon_k - \mu)} - 1} d^3k \qquad (7.154)$$

Eq. 7.154 defines the number density of particles (n) as

$$n = N/V = \frac{1}{(2\pi)^3} \int \frac{1}{e^{\beta(\epsilon_k - \mu)} - 1} d^3k \qquad (7.155)$$

Introducing the idea of density of states per unit volume ($g(\epsilon)$), we have from the previous equation

$$n = \int_0^\infty \frac{1}{e^{\beta(\epsilon_k - \mu)} - 1} g(\epsilon) d\epsilon \tag{7.156}$$

The equation above defines the density 'n' as a function of T and μ. In practice, we have to work with a given density at a given temperature. We will therefore view Eq. 7.156 as defining 'μ', the chemical potential as a function of T and n and proceed to determine $\mu(n, T)$. To that end, we make use of the known expression for $g(\epsilon)$ and the substitution $Z = e^{\beta\mu}$ and $x = \beta\epsilon$ in Eq. 7.156 and thereby find that[5]

$$n = \frac{(mk_B T)^{\frac{3}{2}}}{\sqrt{2}\pi^2\hbar^3} \int_0^x \left(\frac{Ze^{-x}}{1 - Ze^{-x}}\right) x^{\frac{1}{2}} dx \tag{7.157}$$

Now, expanding the term in the integrand within parenthesis in an infinite series, we have

$$\frac{Ze^{-x}}{1 - Ze^{-x}} = \sum_{p=1}^\infty Z^p e^{-px} \tag{7.158}$$

The series is clearly covergent (check by ratio test) for $Z < 1$ and can therefore be ploughed back into Eq. 7.157 for term by term integration giving

$$n = \left(\frac{mk_B T}{2\pi\hbar^2}\right)^{\frac{3}{2}} g_{\frac{3}{2}}(Z) \tag{7.159}$$

where

$$g_{\frac{3}{2}}(Z) = \sum_{p=1}^\infty \frac{Z^p}{p^{\frac{3}{2}}} \tag{7.160}$$

In arriving at Eq. 7.159 we have used the standard result that

$$\int_0^\infty e^{-px} x^{\frac{1}{2}} dx = \frac{1}{p^{\frac{3}{2}}} \cdot \frac{\sqrt{2}}{2}$$

It turns out that $g_{\frac{3}{2}}(Z)$ is convergent for $|Z| < 1$, divergent for $|Z| > 1$ while for $Z = 1$, it is just convergent. Thus, we have for $Z = 1$,

$$g_{\frac{3}{2}}(1) = \sum_{p=1}^\infty \frac{1}{p^{\frac{3}{2}}} = \zeta(3/2) \tag{7.161}$$

where $\zeta(s)$ is the Riemann zeta function. Although $g_{\frac{3}{2}}(Z)$ is convergent at $Z = 1$ and remains finite there, $g'_{\frac{3}{2}}(Z)$, the derivative of $g_{\frac{3}{2}}(Z)$ can be easily shown to diverge at $Z = 1$, becoming infinite. This assertion can be verified by examining $g'_{\frac{3}{2}}(Z)$, which reads $g'_{\frac{3}{2}}(Z) = \sum_{p=1}^\infty \frac{Z^{p-1}}{\sqrt{p}}$. With the limiting behavior and values of $g_{\frac{3}{2}}(Z)$ thus defined, we can roughly sketch $g_{\frac{3}{2}}(Z)$ as shown in Figure 7.1. We can see that Eq. 7.159 defines the density n in terms of the function $g_{\frac{3}{2}}(Z)$. We can, however invert the relation and claim that $g_{\frac{3}{2}}(Z)$ and therefore the values of Z and hence the chemical potential μ itself (as $Z = e^{\beta\mu}$) are obtainable from the relation[4-5]

$$g_{\frac{3}{2}}(Z) = g_{\frac{3}{2}}(e^{\beta\mu}) = \left(\frac{2\pi\hbar^2}{mk_B T}\right)^{\frac{3}{2}} n \tag{7.162}$$

If the density 'n' is low or the temperature (T) is high or low, the right hand side of Eq. 7.162 is manifestly low, which permits us to replace $g_{\frac{3}{2}}(Z)$ by its small Z expansion $g_{\frac{3}{2}}(Z) \approx Z + \cdots$, thereby obtaining from Eq. 7.162,

$$e^{\beta\mu} = \left(\frac{2\pi\hbar^2}{mk_B T}\right)^{\frac{3}{2}} n$$

$$\text{or,} \quad \mu = -\frac{3}{2} k_B T ln\left(\frac{mk_B T}{2\pi\hbar^2 n^{\frac{2}{3}}}\right) \tag{7.163}$$

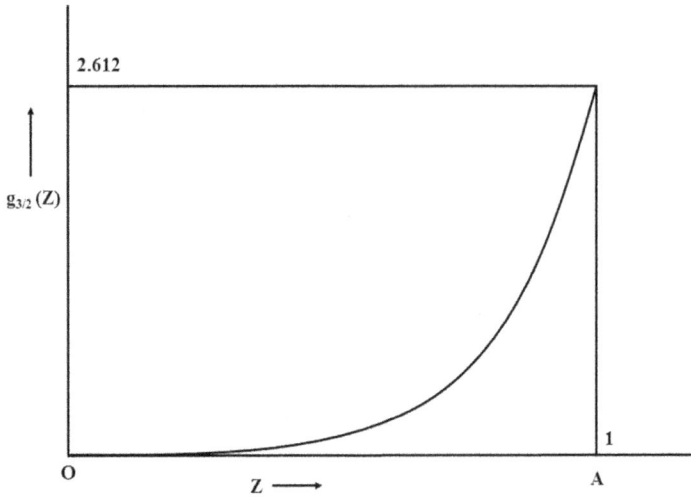

FIGURE 7.1   Graph of $g_{\frac{3}{2}}(Z)$ between $Z = 0$ and 1.

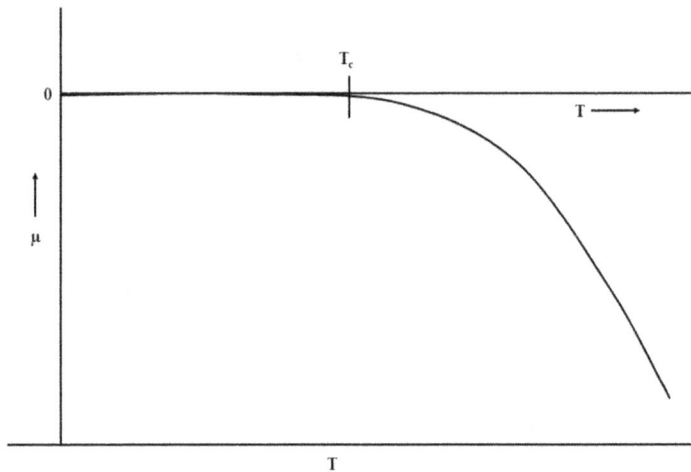

FIGURE 7.2   Chemical potential (μ) of the Bose-Einstein condensate as a function of temperature at a fixed density.

The chemical potential μ of the ideal Bose gas is therefore negative (< 0) when the temperature T is high or the density n is low, as expected classically. What would happen if the gas is cooled further to much lower temperature? As the temperature is lowered, the value of Z increases ($Z = e^{\frac{\mu}{k_B T}}$) and eventually a temperature is reached where the value of Z becomes equal to 1. That signals $\frac{\mu}{k_B T}$, i.e. μ itself must become zero at this temperature (T_c) for the fixed density (n). The behavior of μ from high to the low temperature T_c can be sketched as shown in Figure 7.2. At T_c, $Z = 1$ and $g_{\frac{3}{2}}(1) = 2.612$ as we have already seen. We can use these results in Eq. 7.162 to provide a theoretical estimate of T_c for the given density 'n':

$$T_c = \frac{2\pi \hbar^2}{k_B m} \left(\frac{n}{2.612} \right)^{\frac{2}{3}} \qquad (7.164)$$

T_c is known as the Bose-Einstein condensation temperature. What would happen if we continue to cool the ideal Bose gas below the BEC temperature T_c? Notice that the lowest quantum state corresponds to the $k = 0$ state for which the energy ϵ_0 is also '0'. Once the temperature has reached T_c and μ has become zero, the number of particles in the lowest quantum state $\epsilon_0 = 0$ will become infinite as was first pointed out by Einstein. It will be more precise to say that out of a total number of particles of the system of Bosons occupying a volume V, a macroscopic number, say N_0 will occupy the lowest energy state[6] $\epsilon_0 = 0$. A macroscopic number here signifies that N_0 will be proportional to the volume V of the system and the fraction of molecules in the ground state will be N_0/N. The volume V in the thermodynamic limit tends to infinity so that $N_0 \to \infty$ as $V \to \infty$. The BE distribution law tells us that the number of particles in the $\epsilon = 0$ state is

$$N_0 = \frac{1}{e^{\beta \mu} - 1} \qquad (\beta = \frac{1}{k_B T})$$

We can rearrange the equation above to yield ($\frac{1}{N_0} << 1$)

$$\mu = -k_B T ln\left(1 + \frac{1}{N_0}\right)$$

$$\sim -k_B T \frac{1}{N_0}$$

Hence the chemical potential 'μ' will tend to become zero as $N_0 \to \infty$ in the limit $V \to \infty$ (thermodynamic limit). That shows, as we cool the system down below the BEC temperature (T_c), μ effectively remains zero justifying the shape of the μ versus T graph in Figure 7.2.

Let us now take a fresh look at the expression for the total number of molecules as given in Eq. 7.154:

$$N = \sum_{k=0}^{\infty} \frac{1}{e^{\beta(\epsilon_k - \mu)} - 1}$$

As the temperature T drops below T_c, μ becomes zero, and the ϵ_0 state becomes hugely populated while the higher k states ($k > 0$) remain very thinly populated. It makes sense therefore to split the sum in the equation above into a term dealing only with $k = 0$, while all other terms are lumped together yielding

$$N = N_0 + \sum_{k \neq 0} \frac{1}{e^{\beta \epsilon_k} - 1} \tag{7.165}$$

$$= N_0 + N_1$$

We can, as before, introduce the density of states per unit volume $g(\epsilon)$ and replace the summation on the right hand side of Eq. 7.165 by integration and obtain the particle density n ($= N/V$) as a sum of density of particles in the condensate ($n_0 = N_0/V$) and density in the normal state ($n_1 = N_1/V$) where

$$n = n_0 + n_1$$

$$= n_0 + \frac{(mk_B T)^{\frac{3}{2}}}{\sqrt{2}\pi^2 \hbar^3} \int_0^\infty \frac{e^{-x}}{1 - e^{-x}} x^{\frac{1}{2}} dx \tag{7.166}$$

where $x = \beta \epsilon$ as before. The integration can be done easily following the method used earlier in this section to obtain for $T < T_c$ the following expression:

$$n = n_0 + 2.612\left(\frac{mk_B T}{2\pi \hbar^2}\right)^{\frac{3}{2}} \tag{7.167}$$

The fraction of particles in the BE condensate can be easily expressed in a more compact form:

$$\frac{n_0}{n} = 1 - \left(\frac{T}{T_c}\right)^{\frac{3}{2}} \tag{7.168}$$

The equation above shows that the fraction of particles (molecules/atoms) in the condensate is 1 at $T = 0$ K, signaling that all the particles are in the ground state. The fraction in the condensate falls sharply and becomes zero at $T = T_c$. The condensate disappears and $\frac{n_0}{n}$ remains zero at all temperatures $T > T_c$. It can be shown that T_c is a critical point and the BE condensation is a first order phase transition.[2,5] However, we will not address these issues here. The interested readers may consult more specialized books for these details.

The fact that below T_c, a macroscopic number of Bosons (atoms/molecules) occupy the same quantum state (ground state) is quite spectacular and unique. The condensate is a sample of matter comprising of a macroscopic number of atoms/molecules all of which are in one and the same state. It is hard to imagine that such a state, a macroscopically coherent quantum state of matter could exist. The condensate can be viewed as a new type of material – a purely quantum material with unique properties arising from its macroscopic quantum coherence. The theoretical prediction of BEC stimulated a lot of experimental work for its realization in practice. Bose-Einstein condensation was observed experimentally for the first time in 1995 in trapped ultra-cold gases made up of alkali metal atoms. The discovery came after decades of hard work by a number of groups, which led to the development of the technology of trapping and cooling atoms with magnetic and laser traps. The culmination of such work resulted in three different groups of workers[6] achieving independently the experimental realization of BEC using different isotopes of alkali metal atoms, ^{87}Rb, ^{23}Na and ^{7}Li. The T_c at which the BEC was observed depended on the specific alkali atom and the density in the trap. It lay in the range 0.5-2 μK. The discovery was greeted with the award of Noble Prize to Cornell, Ketterle and Weiman in 2001. It remains to be seen how the quantum materials, fleeting as they may be, are put to practical use.

REFERENCES

1. A. M. Vasilyev, *'An Introduction to Statistical Physics'*, (Translated from Russian by G. Leib), MIR Publishers, Moscow (1983).
2. K. Huang, *Statistical Mechanics*, John Wiley, New York (1963).
3. J. W. Gibbs, *Elementary Principles of Statistical Mechanics*, Dover Publications, Mineola, New York (1964).
4. S.-K. Ma, *Statistical Mechanics*, (Translated by M. K. Fung), World Scientific Publishing, Singapore (1985).
5. J. F. Annet, *Superconductivity, Superfluids and Condensates*, Oxford University Press (2004).
6. W. Ketterle, 'Nobel Lecture: When atoms behave as waves: Bose-Einstein condensation and the atom laser', *Rev. Mod. Phys.*, 74, 1131–1151 (2002).

8

Traditional Materials

8.1 Introduction: Atom-Based Materials

Among the traditional materials, the most prominent are the metallic materials. They are usually a combination of metals called alloys or the pure metal itself. A look at the periodic table reveals that a majority of elements are metals. They are invariably solids under the standard temperature and pressure (STP), mercury (Hg) being an exception. Metals rarely exist in the free state in nature (again, gold (Au) is an exception). They exist mostly as oxides and sulphides, sulphates, halides, carbonates, etc. They can be extracted in the pure form from the minerals by following well defined principles and processes of metallurgy.

Metals possess many common properties. Thus, they are opaque to visible light. A polished metal surface has a typical luster (metallic luster). Metals are easily deformable, yet quite strong – properties that make them suitable for structural applications. Metals are good conductors of heat and electricity. In other words they have high thermal as well as electrical conductivity, which means thermal and electrical resistivities are low for metals.

Metals possess a large number of non-localized or free electrons. These electrons are not bound to particular atoms and provide a negatively charged surrounding in which the positively charged metal ions (or ionic cores) are embedded. The resulting interactions among the oppositely charged species result in a stabilization manifested as metallic bonds. They are as strong as covalent bonds, but lack the localized and bicentric character of typical covalent bonds. In fact, many distinctive properties of metals emanate from the delocalized nearly free electrons and the metallic bonds.[1]

Metals form crystalline solids. Many of them have either the FCC- or the BCC-type of crystal structures. Copper, aluminium, silver and gold are familiar examples of metals with FCC lattice while chromium, α-iron, molybdenum, tantalum and tungsten have BCC lattice structure. Although the BCC and FCC lattices are common for metals, not all metals have cubic symmetry in the solid. The third most commonly encountered metallic crystal structure has hexagonal unit cells giving rise to the hexagonal closed packed (HCP) structure. Cadmium, magnesium, titanium and zinc are some examples of HCP structure. Metals are characterized by typically large atomic packing factor (APF) enabling maximization of the shielding effect that the free electron cloud provides.[1] The FCC metals have a coordination of 12 and APF equal to 0.74 (the maximum packing achievable for identical spheres). The BCC metals have lower coordination number (8) and APF (0.68) while the HCP metals have the same coordination number and APF as the FCC metals. Some metals may have multiple crystal structures, a phenomenon called polymorphism. Many non-metals also display polymorphism. The prevailing crystal structure is shaped by external pressure and temperature. Polymorphism when present in elemental solids is called allotropy, a familiar example of which is seen in carbon. Graphite is the stable form (polymorph) at STP whereas diamond is formed under extremely high pressure. Similarly, pure iron has a BCC crystal structure at room temperature, which gets transformed into FCC iron at 912 °C. Such polymorphic transitions are often accompanied by modification of density and other properties.

8.2 Conducting, Superconducting and Insulating Materials

On the application of an electric field to a substance (material), an electric current flows through it. The magnitude of the current density (j) is shaped by what is known as the electrical resistivity (ρ) or equivalent electrical conductivity (σ) of the material. ρ and σ are inversely related, and are defined by the following pairs of relations:

$$j = \sigma E$$
$$E = \rho j \tag{8.1}$$

The electrical resistivity (ρ) of a sample of a material is related to the electrical resistance (R) by the relation (L is the sample length and A gives the cross sectional area of the sample):

$$\rho = \frac{RA}{L}, \tag{8.2}$$

the unit of resistivity ρ being in ohm·meter ($\Omega \cdot m$) when R is expressed in ohm (Ω). If the current density is expressed in amperes per square meter (Am^{-2}) and the electric field in volts per meter (Vm^{-1}), the unit of σ turns out to be $\Omega^{-1}m^{-1}$.

On examination of electrical resistivity data of elements it turns out that pure metals have low resistivities (10^{-8} Ωm to 10^{-6} Ωm) while non-metallic elements have typically much larger electrical resistivities. Sulphur for example, has $\rho \approx 10^{15} \Omega m$. A close scrutiny of the electrical resistivity data shows that only a very few elements have $\rho >> 1$ Ωm. Such elements (materials) are called electrical insulators or simply insulators. A few elements like silicon, germanium, selenium and tellurium have resistivities intermediate between those of insulators and metals. They are classified as semiconductors. The resistivities of metals, low as they are, tend to rise with rise in temperature and fall with the fall in temperature approximately in a linear manner. For certain metals, the resistivity tends to disappear as the temperature is brought down close to 0 K. There appears to be a threshold temperature (T_c) below which such metals pass over into the zero resistivity or the superconducting state. It is curious that the coinage metals, i.e. Cu, Ag and Au, which are the best known electrical conductors in the ambient condition, do not become superconductors as the ambient temperature is brought down while metals like tin and lead do become superconductors at $T < T_c$.

8.3 Metallic Conductivity: A Rudimentary Theory

We have already developed a statistical mechanical description of an ideal electron gas in metals in Chapter 7. At 0 K, the electron gas is non-degenerate and manifests quantum features strongly. What happens when the electron gas is exposed to an external electric field? Let us recall first that in a crystalline solid like the metals there is a periodic potential (see Chapter 5), which results in the formation of energy bands. At 0 K, the electron gas is non-degenerate and obeys Fermi-Dirac statistics, which means all the energy levels (electron states) of the crystal will be filled up with electrons upto an energy level, called the Fermi level (ϵ_F) so as to account for all the electrons in the system. The surface in the reciprocal space (k-space) that separates the occupied states from the vacant ones is called the Fermi surface. The surface is spherical for free electrons. For crystalline solids, it will have a symmetry that is appropriate to the lattice. Under the action of an electric field (dc field) the only electrons that can respond are those at the Fermi surface because they can move into an adjacent empty state with a slightly different momentum ($p = \hbar k$). If the Fermi energy lies within an allowed energy band, all electrons on the Fermi surface will be able to move under the influence of the applied dc field and take part in the conduction process. The solid is then said to respond like a metal. Should the Fermi energy fall in the energy gap between two bands where the density of states is zero, the applied electric field would fail to displace or move any electron, i.e. there would be no current and the solid will behave as an insulator. If the gap is of the order of kT, there will be some electrons excited above the gap into the conduction band, where the field will act to accelerate them into vacant states. The solid is then said to respond like a semiconductor and we have a semiconducting material at hand.

Although electrons in metals residing in quantum states with energies close to the Fermi energy can make translations into unoccupied quantum states with energies just above the Fermi level, under the influence of an applied electric field, the transition is not the only process that electrons in metals undergo. There are several other processes collectively called scattering, that the electrons in metals could experience during their electric field induced organized motion. In the present context, we may define scattering as a process in which an electron in one quantum state with the wave vector $\vec{k}$ makes a transition to another quantum state with the wave vector $\vec{k'}$. For electrons in metals, the scattering process has quite a bit of randomness associated with it. It means that the initial state ($\vec{k}$) does not solely determine the final state ($\vec{k'}$) that it would be scattered into in a specific event. The scattering takes place when the quantum state occupied by an electron is no longer the right one for it to reside in. Two important scattering processes relevant for understanding the resistivity of metals are the electron-phonon and electron-impurity scattering. The former occurs when the electron in the crystal is influenced by a lattice distortion or vibration in a specific region causing it to make a transition from the state with wave vector $\vec{k}$ to a more appropriate state with the wave vector $\vec{k'}$. Since the lattice motions are quite random, $\vec{k}$ can not be completely determined by the initial wave vector ($\vec{k}$) – there is a chance element associated with the transition. Similarly, if the electron in the crystal moves into a region where it encounters an impurity atom, the charge distribution around the impurity atom will influence the electron energy forcing it to make a transition to a new state ($\vec{k'}$). These scatterings adversely influence the directed or organized motion of the electron and resistivity builds up. Had there been no scattering due to the two processes just described the electrons would continue to reside in the same state it was and the resistivity would disappear altogether – something that could be possible in a perfect crystal without any impurity or lattice motion or distortion. Let us now try to put the ideas described in the preceeding paragraph into a working model for the conductivity of metals.[2]

The application of an electric field causes an electron to aquire some acceleration as determined by Newton's law during the time interval Δt:

$$\vec{F} = \frac{\Delta \vec{p}}{\Delta t} \tag{8.3}$$

Had there been no scatterer, the acceleration would be uniform over any interval of time. Let Δt be the scattering time. The presence of scatterers will destroy the change in momentum during the interval Δt completely after each scattering event that supposedly takes place after the lapse of time equal to Δt. The change in momentum $\Delta \vec{p}$ in between two scattering events is given by

$$\Delta \vec{p} = \vec{F} \Delta t = q \vec{E} \Delta t \tag{8.4}$$

where q is the charge on the electron ($-e$). The corresponding change in the wave vector of the electron ($\Delta \vec{k}$) (using de Broglie hypothesis)

$$\Delta \vec{k} = \frac{\Delta \vec{p}}{\hbar} = \frac{q \vec{E} \Delta t}{\hbar} \tag{8.5}$$

Since 'q' is negative, $\Delta \vec{k}$ will be in a direction opposite to the direction of the applied field. The corresponding shift in velocities of electrons in the metal moving under the applied field $\vec{E}$ would be[2]

$$\Delta \vec{V} = \frac{\Delta \vec{p}}{m} = \frac{\hbar \Delta k}{m} = \frac{q \vec{E} \Delta t}{m} \tag{8.6}$$

The $\Delta \vec{V}$ in Eq. 8.6 is known as the drift velocity and is the same, on average, for all the electrons. In the absence of the applied electric field ($\vec{E} = 0$), $\vec{V} = 0$ (Eq. 8.6) and the average velocity $\vec{V_0}$ of the electrons is also zero as every direction of motion is equally probable for the electrons in the absence of a directing field. The average velocity $\vec{V}$ in the presence of the field ($\vec{E}$) is just $\vec{V}$ as can be easily verified by taking the required average. The movement of charge sets up an electric current.

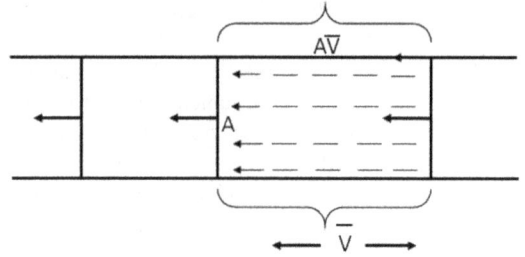

FIGURE 8.1   *A*: area; *A$\bar{V}$*: volume. Flow of electrons in a metal under an impressed voltage.

Let '$\vec{i}$' be the current flowing through the sample of cross section '*A*' and '*n*' be the number density of current carriers of charge '*q*'. Then, a reference to Figure 8.1 immediately suggests that all the carriers in the volume $A\bar{V} = A\Delta V$ will sweep across the cross section every second. The total charge flowing through the area per unit time is then $nq \times A\Delta \vec{V}$, which by defination is the current $\vec{i}$. Thus, using Eq. 8.6, we have

$$i = nqA\Delta \vec{V} = \frac{nqAq\,\vec{E}\,\Delta t}{m} = \frac{nq^2A}{m}\Delta t\,\vec{E} \tag{8.7}$$

The current density *j* is *i/A* whence we get

$$j = \frac{nq^2\Delta t}{m}\vec{E} = \sigma\,\vec{E} \tag{8.8}$$

which on comparison with Eq. 8.1 where σ and ρ were defined shows that[2]

$$\sigma = \frac{nq^2\Delta t}{m}$$
$$\rho = \frac{m}{nq^2\Delta t} \tag{8.9}$$

Eq. 8.9 have one undetermined quantity, the scattering interval, Δt. It is quite difficult to calculate Δt from scratch. We may provide a reasonable estimate of Δt based on physical considerations. Thus, assuming that interactions of the electrons with lattice vibrations or distortions causes scattering of the electrons, we may anticipate that Δt would be of the order of a typical vibrational period of the lattice, which turns out to be 10^{-13} s or so. The higher the scattering interval (Δt), the lower is the resistivity (ρ) and higher is the conductivity. Also, a higher carrier density (*n*) would mean a lower resistivity and higher conductivity. The temperature dependence of conductivity, however, poses a more complicated problem and a more complete formulation of quantum theoretical model of metallic conductivity becomes necessary. We can anticipate that electron-phonon interaction will play an important role in such a model as lattice vibration appears to be at the root of scattering of electrons.

8.4 Quantum Theory of Metallic Conductivity, Electron Phonon Interactions

Consider an ideal metallic lattice at 0 K. All the atoms are at rest and the lattice periodicity is unbroken everywhere. That means the free path of an electron in such a setting would be infinite (the scattering interval Δt of our rudimentary model $\Delta t \sim \infty$).

The electron state in the ideal solid is described by Bloch's function (Chapter 5)

$$\psi_k(\vec{r}) = U_k(\vec{r})e^{i\vec{k}\cdot\vec{r}} \tag{8.10}$$

In the absence of any perturbation, the electron will continue to remain in the state ψ_k for ever. However, at non-zero temperatures, there are thermally induced vibrations or deformation of the lattice, which could lead to scattering of the electrons. We can describe the scattering in terms of absorption and emission of lattice phonons by the electron which alters its quasi-momentum ($\hbar \vec{k}$) and gives rise to a random motion. This process generates electrical resistance to the flow of charges and is the root cause of resistivity of metals. The question now is: how to handle the problem of scattering of an electron by the acoustic or longitudinal vibrations of the lattice. We have already described a quantum statistical theory of such vibrations in the previous chapter (see photon gas, Chapter 7). We may treat the scattering problem by perturbation theory in which electrons are scattered by an effective potential ($V_{eff}(r)$), which would be a function of the quantized amplitudes of lattice vibrations.[3–5] For this purpose, let us consider an ionic crystal in which the oppositely charged ions are vibrating periodically. The displacements of the positive and negative ions are in opposite directions and are represented as[3–4]

$$\delta R_i^+ = \frac{Q}{\sqrt{N}} e^{i(k \cdot R_i^+ - \omega t)}$$

$$\delta R_i^- = -\frac{Q}{\sqrt{N}} e^{i(k \cdot R_i^- - \omega t)}$$

(8.11)

where N is number of ions in the crystal, Q represents the vibrational amplitude and k is the wave vector. The dipole moment (d) of the system of two oppositely charged ions is (neglecting the difference in the coordinates of the ions in the same unit cell)

$$d = 2qe_0 \frac{Q}{\sqrt{N}} e^{i(k \cdot R - \omega t)}$$

(8.12)

where qe_0 is the charge on an ion. If V_0 is the volume of the unit cell, the dipole moment of the system of two ions per unit volume is the polarization, P where

$$P = \frac{2qe_0}{V_0} \frac{Q}{\sqrt{N}} e^{i(k \cdot R - \omega t)}$$

(8.13)

The divergence of polarization gives the local charge density 'ρ'

$$div P = -\rho$$

(8.14)

The electrostatic potential satisfies the Poisson equation

$$\nabla^2 \Phi = -4\pi \rho = 4\pi (div P)$$

(8.15)

which yields the following expression for the electrostatic potential Φ:

$$\Phi_k = (-)\frac{8\pi e_0 q}{V_0 \sqrt{N}} \cdot \frac{ik \cdot Q}{k^2} e^{i(k \cdot R - \omega t)}$$

(8.16)

and the electron phonon interaction potential[3–4]

$$V(r) = -\sum_k e_0 \Phi_k$$

$$= \frac{8\pi e_0^2 q}{V_0 \sqrt{N}} \sum_k \frac{ik \cdot Q}{k^2} e^{i(k \cdot R - \omega t)}$$

(8.17)

$$= \sum_k D_k \frac{ik \cdot Q}{\sqrt{N}} e^{i(k \cdot R - \omega t)}$$

where $D_k = \frac{8\pi q e_0^2}{V_0 k^2}$.

For a non-ionic crystal like a metal crystal, the expression for $V(r)$ in Eq. 8.17 can be interpreted as a deformation potential for electron-phonon interaction and we can write

$$V(r) = \sum_k \frac{D_k ik \cdot Q}{\sqrt{N}} e^{i(k \cdot R - \omega t)}$$

(8.18)

If we restrict ourselves to the acoustic, i.e. the longitudinal vibrations only, k and Q are parallel so that $V(r)$ takes the form[5]

$$V(r) = \sum_k \frac{D_k ikQ}{\sqrt{N}} e^{i(k\cdot R - \omega t)} \tag{8.19}$$

The amplitude 'Q' of harmonic vibration can be expressed in terms of the phonon creation and annihilation operators (see Chapter 5) as

$$Q = \sqrt{\frac{\hbar}{2M\omega_k}}\left(a_k^+ + a_k^-\right) \tag{8.20}$$

where a_k^+ represents the creation operator for phonons with frequency ω_k, M is the mass of the vibrating atoms and a_k is the corresponding annihilation operators for phonons. Substituting the expression for Q in Eq. 8.19 the part of the electron-phonon interaction potential responsible for absorption of phonons (the a_k containing part of $V(r)$) can be expressed in the following form:

$$V^{abs}(r) = \sum_k V_K^{abs} = \sum_k D_k \sqrt{\frac{\hbar}{2NM\omega_k}} ik a_k e^{i(k\cdot R - \omega t)} \tag{8.21}$$

Let us now introduce the notation

$$V_k = ikD_k\left(\frac{\hbar}{2M\omega_k}\right)^{\frac{1}{2}} \tag{8.22}$$

We can write

$$V_k^{abs} = (V_k^{0,abs}) \cdot a_k e^{-i\omega_k t} \tag{8.23}$$

where the time-independent part of the perturbation energy $V_k^{0,abs}$ has the form

$$V_k^{0,abs} = \frac{V_k^0}{\sqrt{N}} e^{ik\cdot R} \tag{8.24}$$

We are now in a position to calculate the quantum mechanical probability of transition from the electron state 'k' to the state 'k''' due to the absorption of a phonon with frequency ω_k by invoking Fermi-golden rule formula (or by invoking time-dependent perturbation theory):

$$P_{k,k'} = \frac{2\pi n_k}{\hbar}\left|\langle k'|V_k^{0,abs}|k'\rangle\right|^2 \delta\big(\epsilon(k') - \epsilon(k) - \hbar\omega_k\big) \tag{8.25}$$

In Eq. 8.25, n_k is the number of phonons with energy $\hbar\omega_k$ and $\epsilon(k) = \frac{\hbar^2 k^2}{2m}$ is just the energy of the electron moving freely in the conduction band with momentum $\hbar k$. The calculation of the value of $P_{kk'}$ requires evaluation of the matrix element of the perturbation operator $V_k^{0,abs}$ between the unperturbed Bloch states $\psi_k(\vec{R})$ and $\psi_{k'}(\vec{R})$ of the crystal. However, the motion of electrons in the conduction band can be described by the free electron wavefunctions quite well allowing us to choose

$$\psi_k(\vec{R}) = \frac{1}{\sqrt{L^3}} e^{i\vec{k}\cdot\vec{R}}$$

With this choice for $\psi_k(\vec{R})$, the transition matrix element becomes

$$\langle k'|V_k^{0,abs}|k\rangle = \frac{V_k}{\sqrt{N}} \cdot \frac{1}{L^3} \int_{-L/2}^{+L/2} e^{i(k+K-k')\cdot R} dv \tag{8.26}$$

$$= \frac{V_k}{\sqrt{N}} \delta_{k+K,k'}$$

The phonon absorption process occurs with conservation of energy and momentum. Hence we have

$$E(k') = E(k) + \hbar\omega_k$$
$$k' = k + K \tag{8.27}$$

With these conditions, the final expression for the transition probability becomes

$$P^{abs}_{k,k+K} = \frac{2\pi}{\hbar} \frac{|V_k|^2}{N^2} n_k \delta\big(E(k+K) - E(k) - \hbar\omega_k\big) \tag{8.28}$$

In a similar manner, we can also calculate the phonon emission probability, which turns out to be

$$P^{emm}_{k,k+K} = \frac{2\pi}{\hbar} \frac{|V_k|^2}{N^2} (n_k + 1) \delta\big(E(k+K) - E(k) - \hbar\omega_k\big) \tag{8.29}$$

The total probability of $k \rightarrow k'$ transition is then obtained by adding the phonons absorption and emission probabilities after making some simplifications:

(i) Since the electron energies are much higher than the energy of a phonon, we neglect $\hbar\omega_k$ in the argument of the Dirac delta function in Eq. 8.28 and Eq. 8.29.

(ii) The number of phonons n_k is assumed to be rather large so that we can neglect 1 in comparison with n_k in Eq. 8.29.

(iii) We set $V_k = ikD_k(\frac{\hbar}{M\omega})^{\frac{1}{2}}$ as already defined in Eq. 8.22, and use $\hbar\omega_k = \hbar k v_0$, v_0 being the velocity of acoustic waves in the metal crystal.

With (i)-(iii) in place, we get an expression for the total transition probability from the state k to $k'(=k+K)$, which reads

$$p_{k,k+K} = \frac{2\pi}{\hbar} \frac{D^2 n_k}{NMv_0^2} \big(\hbar\omega_k\big) \delta\big(E(k+K) - E(k)\big) \tag{8.30}$$

Now, the average number of phonons $(<n_k>)$ of energy $\hbar\omega_k$ at temperature T is given by the Bose-Einstein distribution function

$$<n_k> = \frac{1}{e^{\frac{\hbar\omega_k}{k_B T}} - 1} \tag{8.31}$$

which can be ploughed back in Eq. 8.30 to arrive at the temperature-dependent probability of the conducting electron being scattered from a state (k) to a state $k' = k + K$, which reads

$$P_{k,k+K} = \frac{2\pi}{\hbar} \frac{D^2 \hbar\omega_k}{NMv_0^2} \frac{1}{e^{\frac{\hbar\omega_k}{k_B T}} - 1} \delta\big(E(k+K) - E(k)\big) \tag{8.32}$$

These scatterings affect the motion of the conduction electron, which tends to become randomized and this randomization of the electron momentum leads to the appearence of electrical resistance of metals. The momentum randomization rate equation reads[3-4]

$$\frac{d<\vec{k}>}{dt} = \sum_{k'} (\vec{k'} - \vec{k}) P_{k,k'} = -\frac{<k>}{\tau} \tag{8.33}$$

'τ' in Eq. 8.33 is called the relaxation time, formally introduced through the solution of momentum randomization equation

$$<k(t)> = k(0)e^{-\frac{t}{\tau}} \tag{8.34}$$

'τ' is the key quantity that modulates the electrical conductivity of metals through the relation

$$\sigma = \frac{Ne_0^2 \tau}{m} \tag{8.35}$$

where 'm' is the mass, e_0 the charge and N the number density of free electrons in metals. In comparison with σ of Eq. 8.9, we find that the scattering time Δt is now replaced by the relaxation time 'τ' in the present model of electrical resistivity/conductivity and their temperature dependence.

From Eq. 8.33, we can immediately identify that[3,5]

$$\frac{1}{\tau} = -\sum_{k'} (\vec{k'} - \vec{k}) P_{k,k'} \tag{8.36}$$

Now for the final state following the scattering we have

$$\vec{k'} = \vec{k} + \vec{K}$$

If $\vec{k}$ and $\vec{K}$ are at an angle θ, Eq. 8.36 becomes

$$\frac{1}{\tau} = -\sum_{k} P_{k,k+K} \frac{K}{k} \cos\theta \tag{8.37}$$

To determine 'τ' we must evaluate the sum on the left hand side Eq. 8.37. We will try to do that in two extreme cases:

  (i) high temperatures,
  (ii) low temperatures.

(i) In the case of high temperatures, we can set $k_B T >> \hbar\omega_k$. The Bose-Einstein distribution function in Eq. 8.32 can then be reduced to (chapter 7)

$$<n_k> = \frac{k_B T}{\hbar\omega_k} \tag{8.38}$$

Using this result and making the substitution $\omega_k = k v_0$, v_0 being the velocity of sound in the metal, the summation over k in Eq. 8.37 can be reduced to an integration over $\vec{k}$. For k, the upper limit is set at the maximum value consistent with the linear dimension of the crystal.

$$\frac{1}{\tau} = \frac{V_0}{\pi} \frac{D^2 (k_B T) m |k|}{\hbar^3 M v_0^2} \tag{8.39}$$

V_0 being the volume of the crystal. It is evident from the equation above that the relaxation time (τ) depends on the value of momentum $\hbar|k|$ of the scattered electron and hence on its energy as well. Most importantly, the resistivity 'ρ', which is equal to σ^{-1} is predicted to depend linearly on $\frac{1}{\tau}$ and therefore linearly on 'T' in the high temperature regime. This prediction is found to be entirely consistent with experimental data on the resistivity of metals as a function of temperature.

  (ii) Let us now consider the low temperature regime where we can assume $kT << \hbar\omega_k$. Now, we can no longer approximate the phonon distribution function by the expression in Eq. 8.38, but we must retain the full expression for the Bose-Einstein distribution function. Making the substitution $x = \frac{\hbar\omega_k}{k_B T} = \frac{\hbar k v_0}{k_B T}$, we can arrive at the expression[5]

$$\frac{1}{\tau} = \frac{V_0}{4\pi} \cdot \frac{D^2}{\hbar^3} \frac{m}{M v_0^2 k^3} \frac{(k_B T)^5}{(\hbar v_0)^3} \int_0^\infty \frac{x^4 dx}{e^x - 1} \tag{8.40}$$

The integral over x turns out to be

$$\int_0^\infty \frac{x^4 dx}{e^x - 1} = 24\zeta(5)$$

where $\zeta(5)$ is the Riemann zeta-function $\zeta(x)$ at $x = 5$ and has the value 1.037 approximately. Thus, in the low temperature regime, $\frac{1}{\tau}$ becomes

$$\frac{1}{\tau} = \frac{V_0}{4\pi} \cdot \frac{D^2}{\hbar^3} \frac{m}{M v_0^2 k^3} \frac{(k_B T)^5}{(\hbar v_0)^4} \times 24\zeta(5) \tag{8.41}$$

Since the electrical resistivity is proportional to $\frac{1}{\tau}$, we anticipate that ρ will disappear altogether following T^5 power law in the low temperature regime (Eq. 8.41) as $T \to 0$ while in the high temperature regime ρ will increase linearly with temperature T (Eq. 8.40). The two types of behavior is so drastically different that it would be safe to assume that the metal probably will undergo a phase transition at some low temperature $T = T_c$ where the normal metallic phase will pass over into a superconducting phase perhaps with zero resistance. General experimental results corroborate both the high and low temperature behavior of metallic resistivity predicted by the quantum theory. However, understanding the nature of the superconducting transition, the appearance of a superconducting transition temperature (T_c) and its estimate require a much deeper analysis. We will address these issues in the following section only sketchily avoiding use of much of the sophisticated mathematical apparatus that the quantum theory of superconductivity entails.

8.5 Superconductivity and Superconducting State

K. Onnes (1911)[6] observed that when mercury in solid state (under high pressure) is cooled to a very low temperature $T \approx 4.1K$, it exhibits very very low resistivity implying that a new state of the metal appears at extremely low temperatures. It appeared that in such a state, the metal would be able to carry electric current without any ohmic loss. The technological implications of this behavior was quite exciting. It was discovered two decades later that a superconductor has the unique ability to expel completely the magnetic field applied to it. The superconductor at $T < T_c$ (i.e. below the superconducting transition temperature) becomes a perfect diamagnet. This unique phenomenon came to be known as the Meissner effect. The technological possibilities of this effect can not be missed as a magnet is able to float or levitate above a superconductor kept below the critical transition temperature. The hallmarks of a superconductor are:

(i) ability to conduct electric current with no resistance or zero resistivity,

(ii) ability to eject applied magnetic field completely and display perfect diamagnetism,

(iii) existence of a superconducting transition temperature T_c at or below which the material displays the extraordinary behaviour listed in (i)-(ii).

The major theoretical developement leading to a better understanding of the phenomenon of superconductivity came in 1950 with the advancement of the phenomenological theory of Landau and Ginzburg. A microscopic theory of superconductivity however, appeared much later. The theory became known as Bardeen, Cooper and Schrieffer (BCS) theory of superconductivity and was hailed as a spectacular success of the quantum theory. The developements in the area of superconductivity are still continuing. Indeed, the BCS theory applies to the so called low temperature superconducting materials (also called type I superconductors). These materials are characterized by the superconducting transition temperature $T_c < 10$ K. Type II superconductors have a much higher T_c (≥ 40 K). We will consider them later. Some of the well-known type I superconducting materials are listed in Table 8.1, along with their transition temperatures (T_c). It turns out, quite paradoxically, that electron-phonon interaction, which was found to be responsible

TABLE 8.1

List of Type I Superconducting Metals and Their T_cs[7]

Material	T_c (K)	Material	T_c (K)
Ga	1.1	Ir	0.1
Al	1.2	Ru	0.5
In	3.4	Os	0.7
Sn	3.7	Mo	0.92
Pb	7.2	La	6.0
Rh	0	W	0.015

for the generation of electrical resistance, is also the root cause, in an entirely different way, for the sudden disappearence of electrical resistivity of metals (and other conducting materials) as they are cooled to $T \leq T_c$. The complete microscopic theory (BCS) of the transition is beyond the scope of the present work as the mathematical apparatus needed is far too complex to be introduced here. We will however, present the basic aspects of the theory in an elementary fashion.[7-9]

8.5.1 The Nature of the Superconducting State

An important concept in the theory of superconductivity is the exchange of virtual phonons between a pair of conducting electrons. It is as if, one of the two electrons causes a lattice deformation, which acts upon the other electron. The net outcome is the generation of an attraction between the electron pair, which results in the creation of a bound state of the two electrons. Formally, we may say that the emission of a phonon of energy $\hbar\omega_k$ from an electron with momentum $\hbar k$ and its absorption by another with momentum $\hbar k'$ causes an attraction between the electrons. Such pairs of electrons are called Cooper-pairs. The momentum and spins of the two electrons must be antiparallel to reach the lowest energy state. Once the pairing takes place, the nature of motion of the conducting electron undergoes a drastic change. They do not move independently any longer, but move in a correlated manner as bound electron pairs with zero spin ($s = 0$) and momenta. The minimum energy of a Cooper-pair becomes lower than the energy ϵ_F of the ordinary electron making the upper Fermi boundary of the normal conducting state unstable. The formation of Cooper-pairs then happens to be the energetically favorable process at low enough temperature.

Let us now examine the situation in a bit more detailed manner. We consider two isolated electrons in the conductor, interacting with each other, the interaction with all other electrons being disregarded. The wavefunction of the electron pair can be written as

$$\Phi(\vec{r_1}, \vec{r_2}) = \frac{1}{\Omega} e^{ik_1 \cdot (\vec{r_1} - \vec{r_2})} \tag{8.42}$$

'Ω' being the normalizing volume. In the state of the lowest energy of the pair of the following conditions hold:

(a) Total momenta is zero: $\vec{k} = \vec{k_1} + \vec{k_2} = 0$.
(b) Total spin angular momentum is zero : $\vec{s} = \vec{s_1} + \vec{s_2} = 0$.

The state function Φ represents the normal state of the two electrons with energy $E_{min} = 2E_F$ when there is no interaction between the two electrons, $E_F = \frac{\hbar^2 k_F^2}{2m}$ being the Fermi energy of the system. The question now is: what happens to the superconducting state of the two electrons when there is an interaction (form yet unspecified) between them. Note that all the levels with energy $E < E_F$ are occupied by the other electrons. Therefore, the minimum energy of the isolated pair is $E = 2E_F$ (ignoring interaction). To look for the state with the minimum energy in the presence of interaction we must replace the wavefunction $\Phi(\vec{r_1}, \vec{r_2})$ by a wavefunction constructed as a superposition of all pair states with momenta lying outside the Fermi-sphere of radius k_F. Calling the new wavefunction $\psi(\vec{r_1}, \vec{r_2})$, we have

$$\psi(\vec{r_1}, \vec{r_2}) = \sum_{|k'|>k_F} c_{k'} e^{ik' \cdot (\vec{r_1} - \vec{r_2})}$$
$$= \sum_{|k'|>k_F} c_{k'} e^{ik' \cdot \vec{R}} \tag{8.43}$$

where $\vec{R} = \vec{r_1} - \vec{r_2}$, and our target is to determine $\{c'_k\}$. If there be no interaction between the electrons in the superconducting state, the energy of the state represented by $\psi(\vec{r_1}, \vec{r_2})$ of Eq. 8.43 will clearly have an energy $E > 2\epsilon_F$. The question at this point is: can an interaction alter the situation? To answer the question, let us try to find out the superposition coefficients $\{c'_k\}$ taking into consideration an interaction V between the two electrons. If H_0 be the Hamiltonian of the system without any interaction, and V be the interaction (to be specified later) our $\psi(\vec{r_1}, \vec{r_2})$ will satisfy the energy eigenvalue equation

$$(H_0 + V)\psi = E\psi \tag{8.44}$$

H_0 is just the sum of kinetic energy operators of the two electrons. Using the expansion of $\psi(\vec{r_1}, \vec{r_2})$ as given in Eq. 8.43 in the equation above, it is straightforward to arrive at the following equation for determining the coefficients $\{c'_k\}$ of the superposition that represents $\psi(\vec{r_1}, \vec{r_2})$

$$\sum_{k' > k_F} c_{k'}(E - H_0)e^{i\vec{k'} \cdot \vec{R}} = \sum_{k' > k_F} c_{k'} e^{i\vec{k'} \cdot \vec{R}} V \tag{8.45}$$

Since

$$H_0 e^{i\vec{k'} \cdot \vec{R}} = 2\epsilon(k')e^{i\vec{k'} \cdot \vec{R}}$$

$$\text{where} \quad \epsilon(k') = \frac{\hbar^2 k'^2}{2m}$$

Eq. 8.45 takes the following form:

$$\sum_{k' > k_F} c_{k'}\left(E - 2\epsilon(k')\right)e^{i\vec{k'} \cdot \vec{R}} = \sum_{k'} c_{k'} e^{i\vec{k'} \cdot \vec{R}} \tag{8.46}$$

Multiplying both sides of Eq. 8.46 by $\Phi^* = \frac{1}{\Omega}e^{-i\vec{k} \cdot \vec{R}}$ and integrating over all space, we arrive at the equation for the determination of $c_{k'}$:

$$c_k(E - 2\epsilon(k)) = \frac{1}{\omega^2} \sum_{k' > k_F} c_{k'} \int e^{i\vec{k'} \cdot \vec{R}} V dv_1 dv_2 \tag{8.47}$$

where $\vec{K} = \vec{k'} - \vec{k}$, $\vec{R} = \vec{r_1} - \vec{r_2}$ and we have made use of the fact that

$$\int e^{i\vec{R}(k'-k)} dv_1 dv_2 = \Omega^2 \delta^2_{kk'} \tag{8.48}$$

It is not possible to arrive at a general solution of Eq. 8.47. To make progress, we have to make some assumptions. One of the assumptions may be, for example, that the interaction 'V' between the two electrons has a rather simple form, which allows us to represent the following integral in a product form

$$\frac{1}{\omega^2} \int e^{i\vec{k} \cdot \vec{R}} V dv_1 dv_2 = \gamma\, \omega_k \omega_{k'} \tag{8.49}$$

The constant γ introduced in Eq. 8.49 may take negative values ($\gamma < 0$) for representing attraction between the pair of electrons or positive values ($\gamma > 0$) for representing repulsion between them.

In either case, Eq. 8.47 is reduced to

$$c_k(E - 2\epsilon(\vec{k})) = \gamma\, \omega_k \sum_{k' > k_F} c_{k'} \omega_{k'} \tag{8.50}$$

$$= \gamma\, \omega_k a$$

where

$$a = \sum_{k' > k_F} c_{k'} \omega_{k'} \tag{8.51}$$

is a constant and independent of k. Hence, it is now possible to write

$$c_k = \frac{\gamma\, \omega_k a}{E - 2\epsilon(\vec{k})} \tag{8.52}$$

Using the above expression for c_k in Eq. 8.51, we can eliminate c_k and arrive at a condition for the existence of a non-trivial solution of Eq. 8.50, which reads[5,7]

$$\gamma \sum_{k>k_F} \frac{\omega_k^2}{\left(E - 2\epsilon(\vec{k})\right)} = 1 \tag{8.53}$$

Let us examine now the condition under which Eq. 8.53 admits of a solution with energy $E < 2\epsilon_F$. Two cases arise:

(a) The interaction potential is > 0, that is there is repulsion between the two electrons. In this case $\gamma > 0$, which immediately rules out the existence of a solution with $E < 2\epsilon_F$, for in that case, the left hand side of Eq. 8.53 would be negative violating the condition.

(b) The interaction potential and therefore $\gamma < 0$ meaning that there is attraction between the two electrons. In this case, there could be a solution with energy $E < 2\epsilon_F$ as the left hand side now remains positive as demanded by Eq. 8.53. Thus, an attractive interaction ($\gamma < 0$) between the two electrons ensures that there can be bound state of the electron pair with energy (E) below the Fermi energy of the system when there is no interaction ($2\epsilon_F$). The state is a special one the minimum energy of which is lower than the energy of the normal conducting state. Indeed, this state is a very special coherent state with pairing of momenta and spins. It is so called superconducting state. Cooper-pair formation is thus energetically feasible and advantageous, if there is an attractive interaction between the electrons mediated by phonons. Having seen that Cooper-pair formation is indeed possible, it is necessary now to estimate the typical binding energy of the Cooper-pairs and look for the mechanism of the formation of Cooper-pairs and appearance of a transition temperature T_c at or below which resistivity disappears.

8.5.2 Binding Energy of a Cooper-Pair

To estimate the binding energy 'Δ' of a Cooper-pair, let us focus our attention on Eq. 8.53 and make some conjectures on ω_k based on the energy $\epsilon(k)$ of the electrons. Specifically, we assume that $\omega_k = 0$ if $\epsilon > E_{max}$ and $\omega_k = \beta$ when $E_F \le \epsilon \le E_{max}$. Here β is a constant and $\epsilon = \frac{\hbar^2 k^2}{2m}$. Consistent with these assumptions, the summation in Eq. 8.53 has to be restricted over only a small range of k near the surface of the Fermi surface. As discussed in Chapter 7, we can now replace the summation in Eq. 8.53 by integration after introducing the density of the electron-pair states g_ϵ in the usual manner, leading to

$$\frac{\gamma \beta^2}{2} \int_{E_F}^{E_{max}} \frac{g_\epsilon d\epsilon}{(E - 2\epsilon)} = 1 \tag{8.54}$$

The factor of $\frac{1}{2}$ on the left hand side of Eq. 8.54 arises because we have chosen only the electron-pair states in which the spins are antiparallel. As mentioned already, the integration interval is small, which allows us to take g_ϵ out of the integral sign with g_ϵ set at its value at the Fermi energy E_F. Let us call it g_F. Also, we set $E_m = E_{max}$. With these, Eq. 8.54 becomes after integration

$$\frac{|\gamma|}{4} \beta^2 g_F ln \left| \frac{2E_m - 2E_F + \Delta}{\Delta} \right| = 1 \tag{8.55}$$

where $\Delta = 2E_F - E$ is the quantity, which we call the binding of the Cooper-pair. Let us introduce the notation $E_m - E_F = E_D = \hbar\omega_D$, where E_D is the Debye energy and ω_D is the Debye frequency, which we have already encountered in Chapter 7 (phonon gas) representing the maximum phonon frequency, i.e. the maximum frequency of lattice vibration.

$$\frac{|\gamma|}{4} \beta^2 g_F ln \left| \frac{2E_D + \Delta}{\Delta} \right| = 1 \tag{8.56}$$

which can be rearranged to yield following expression for Δ:

$$\Delta = \frac{2E_D}{e^{\frac{4}{|\gamma g_F \beta^2|}} - 1} \tag{8.57}$$

$$\Delta = 2E_D \cdot e^{-\frac{4}{|\gamma g_F \beta^2|}} \tag{8.58}$$

From Eq. 8.57 to Eq. 8.58 we have used the fact that the exponential term in the denominator $>> 1$, the exponent being a large positive quantity ($|\gamma \beta^2| << 1$). This shows that the binding energy of the pair is proportional to the Debye energy $E_D = \hbar\omega_D$. Thus, we arrive at the following picture. If there is attraction between the electrons moving in a crystal, they may begin to form coherent states wherein they move in pairs with their spins and momenta equal but aligned in an antiparallel orientation. These pairs have zero spin ($\uparrow\downarrow$) and they behave as Bosonic quasi-particles, all existing in an identical coherent state. A new state of matter, the Bose-Einstein condensate therefore emerges. The formation of Cooper-pairs, depends on the γG^2 term in Eq. 8.58, which is responsible for the onset of pair formation. The attractive interaction is generated by absorption and emission of longitudinal virtual phonons by electrons moving through the lattice and may be larger than the Coulomb repulsion between the electrons, but the process is manifested only over a small range of quasi-momenta ($\hbar k$) in a small neighborhood of the Fermi boundary, where the following condition holds:

$$\left| \frac{\hbar^2 k^2}{2m} - \frac{\hbar^2 k_F^2}{2m} \right| < \hbar\omega_D \tag{8.59}$$

which can be reduced to the equivalent condition

$$\left| k - k_F \right| < \frac{\omega_D}{v_F} \tag{8.60}$$

$v_F = \frac{\hbar k_F}{m}$ being the Fermi velocity.

The binding energy 'Δ' of the Cooper-pair is a small quantity and even a slight increase in temperature may destroy the pair when $kT > \hbar\omega_D$ or prevent the formation of a coherent state by inducing thermal excitation in the Bose-Einstein (BE) condensate. So far, we have considered only the ground state of the superconductor – a macroscopically coherent state in which all the Cooper-pairs are in the lowest state of energy forming a BE condensate. In the elaborate microscopic theory of low temperature or type I superconductivity, the idea of excitation is also brought in. These quasi-particles have an energy spectrum $E(k)$ where

$$E(\overrightarrow{k}) = \sqrt{\left[\frac{\hbar^2 k^2}{2m} - \frac{\hbar^2 k_F^2}{2m} \right] + \Delta^2(\overrightarrow{k})} \tag{8.61}$$

The term within the square-brackets under the square root is just the kinetic energy of the excitation measured from the Fermi surface. The quantity $\Delta(k)$ is the energy gap that depends on the quasi-momentum in a rather complex manner. The binding energy 'Δ' of the electron pair can be (as we have done) treated as a constant quantity as given in Eq. 8.57 only when the temperature 'T' has been brought down sufficiently ($T \to 0$). As 'T' rises, 'Δ' begins to decrease and at a certain critical temperature $T = T_c$ (material-dependent) becomes equal to zero. The superconducting state melts away when $T > T_c$ as the 'gap' disappears under this condition. The maximum value of Δ is related to T_c for the superconducting transition by

$$\Delta \approx k_B T_c \tag{8.62}$$

k_B being the Boltzmann constant. For low temperature or type I superconductors the values of the T_c range between 0-10 K.

The existence of an 'energy gap' in the excitation is rather critical in the theory of superconductors. The non-zero gap makes the superconducting state different from ordinary metallic state in which there are unoccupied states infinitely close to the ground state (see Chapter 5). The existence of non-zero gap makes the superconductor appear similar to the semiconductors with a forbidden energy gap of 2Δ. The gap is of magnitude 2Δ (not Δ) because the formula in Eq. 8.61 gives only the positive branch of the square root representing electron type excitation.[3–5] The negative branch of the square root, likewise represents the hole type of excitation. Thus, for electrons with quasi-momentum $\hbar \overrightarrow{k} = \hbar k_F$, the minimum value of excitation energy is 2Δ. If a current is set up in the superconductor, normal scattering processes that damp the current in ordinary conductors causing it to decay, can not affect the current set up in the superconductor as the coherent ground state is stable and can be altered only by supplying to it an energy at least equal to the forbidden gap energy 2Δ of the Cooper-pairs explaining the rather surprising permanence of the electric current set up in a superconductor.

8.5.3 Superconducting State Function

What kind of wavefunction (ψ) represents the superconducting state? We have mentioned earlier that the superconducting state is a macroscopically coherent state of a huge collection of Cooper pairs forming a BE condensate. Thus, the wavefunction ψ must represent the condensate as a collective whole and can be chosen in the following form:

$$\psi = \sqrt{\rho}e^{i\phi} \tag{8.63}$$

where $\rho(\vec{r},t)$ and $\phi(\vec{r},t)$ are real functions of coordinate and time. The wavefunction in Eq. 8.63 corresponds to the entire macroscopically coherent state formed by Cooper-pairs, each with charge 'q' equal to $2e$ and spin '0'. The probability current density for the pair in a magnetic field H = curl of A turns out to be

$$j = -\frac{i\hbar}{2m}\left\{\psi^*\nabla\psi - \psi\nabla\psi^*\right\} - \frac{qA}{m}\psi^*\psi \tag{8.64}$$

Using Eq. 8.63 in Eq. 8.64, we obtain the relation

$$j = \left\{\frac{\hbar}{m}\nabla\phi - \frac{q}{m}A\right\}\rho \tag{8.65}$$

The phase ϕ of the wavefunction (Eq. 8.63) turns out to be an observable.

8.5.4 Special Features of the Superconducting State

A very surprising phenomenon shaped by the properties of the superconducting state is the quantization of the magnetic flux passing through a superconducting ring like the one shown in Figure 8.2a.

The superconducting ring in Figure 8.2a is at a temperature (T) above the superconducting transition temperature (T_c) and placed in a magnetic field $\vec{H}$. The magnetic lines of force pass through the material that the ring is made of (Figure 8.2a). As the temperature is lowered to $T < T_c$ (close to 0 K), the magnetic field is ejected from the superconductor (Figure 8.2b) revealing what is known as the Meissner effect. Even when the magnetic field is completely switched off, a residual magnetic flux is found to be retained in the superconducting ring-hole (Figure 8.2c).

The superconducting current flowing through the ring appears to hold the magnetic flux passing through the hole unaltered. It is as if, the magnetic flux has become frozen in the superconductor. The trapped magnetic field turns out to be quantized. Using the wavefunction ψ for the superconducting state as given in Eq. 8.63 and the corresponding probability current density in the presence of a magnetic field 'H' (Eq. 8.65), it is straightforward to find that the magnetic flux trapped in the ring (ϕ) is quantized:

$$\Phi_n = \frac{2\pi\hbar c}{q}n \tag{8.66}$$

$$= n\Phi_0, \qquad n = 0,1,2,\cdots$$

where ϕ_0 is the basic quantum of magnetic flux. The quantum number n – an integer that corresponds to the complete change of the phase of the superconducting wavefunction determines the 'flux' in units of Φ_0. Φ_0 is a rather small quantity for

$$\Phi_0 = \frac{2\pi\hbar c}{q} \tag{8.67}$$

$$= \frac{2\pi\mu_B}{r_0}$$

where μ_B is the Bohr magneton and r_0 is the so called classical radius of the electron expressed as

$$r_0 = \frac{e^2}{mc^2} \tag{8.68}$$

An experiment carried out by Doll *et al.* (1961) and others[10-11] confirmed the flux quantization relation stated in Eq. 8.66 and the value of ϕ_0, which was found to be quite small (= 2.07×10^{-7} G cm^2 – only about 1% of the Earth's magnetic field). They also confirmed that the charge 'q' in Eq. 8.66 defining the quantized magnetic flux was indeed twice the electronic charge, which rather convincingly justified the concept of Cooper-pairs introduced in the microscopic theory of superconductivity (BCS theory). Yet another special attribute of superconductors is the Josephson effect displayed by them. To make the perspective clear, let us first consider a system of two metal pieces separated by an insulator of thickness δ (Figure 8.3). We can assume that the Fermi boundaries in these two metals are identical allowing the electrons to maintain a

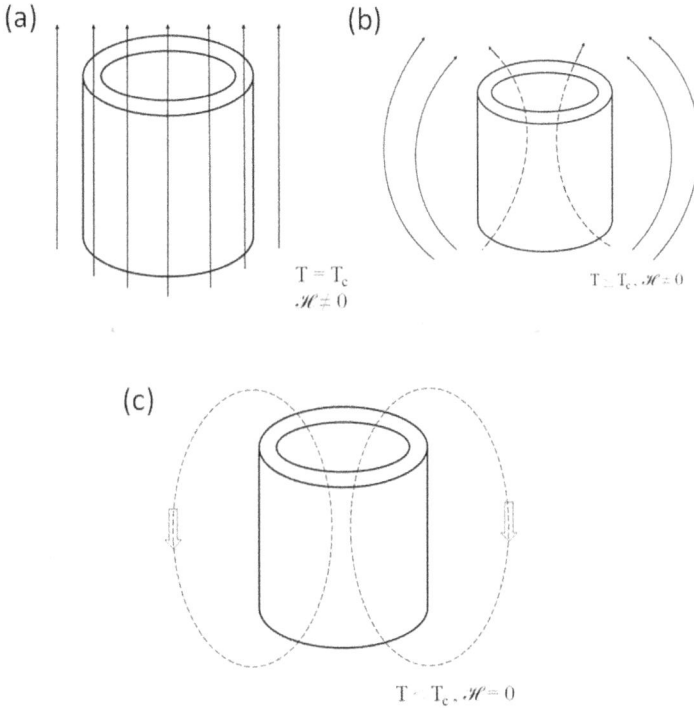

FIGURE 8.2 (a)-(c) Magnetic flux passing through a superconducting ring under different conditions.

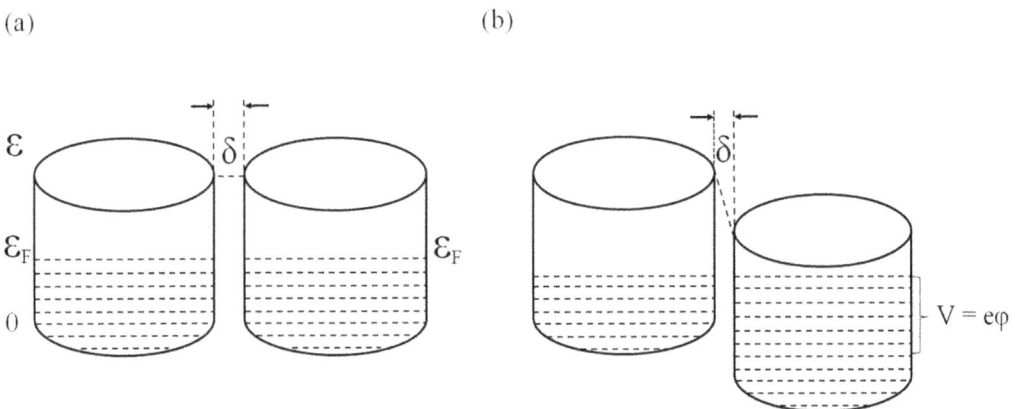

FIGURE 8.3 (a), (b) Tunneling in metallic conductors.

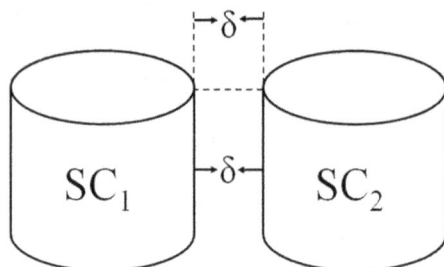

FIGURE 8.4 Tunneling between two superconductors δ is the thickness of a thin insulating film separating SC_1 and SC_2.

stable equilibrium and therefore there will be no current across the insulator. If a potential difference 'V' is applied to the metals (Figure 8.3a) a current (J) proportional to the applied potential difference ($J \propto V$) begins to flow. The application of the potential difference 'V' causes the Fermi boundaries to be displaced with respect to each other (Figure 8.3). The displacement of energy levels ΔE generates an electric field ε at the contact where $\varepsilon = \phi/\delta$, causing flow of tunneling current across δ where the tunneling current is given by

$$J = \frac{e}{\pi \hbar} \overline{D} e \Phi = \text{constant}$$

Here $\Phi = V/R$ and $R = \pi \hbar / e\overline{D}$ is the ohmic resistance, $\overline{D}$ being the transmission coefficient at Fermi-energy ϵ_F, i.e.

$$\overline{D} = e^{-\frac{2}{\hbar}\left(\int_{x_1}^{x_2} \sqrt{2m(V-\epsilon_F)}dx\right)}$$

Let us now replace the two metals by two superconductors SC_1 and SC_2 as displayed in Figure 8.4 where 'δ' is the thickness of a thin layer of an insulating material separating the superconductors. It turns out that even when there is no externally impressed potential difference across the superconductors a current $\overrightarrow{J}$ flows through 'δ' due to the passage of the correlated electron pairs (Cooper-pairs) by tunneling transitions. This has become known as the Josephson effect.[12] The flow is sustainable for an arbitrarily long period of time as long as the superconducting states are left undisturbed ($T < T_c$). The current across the Josephson junction is also called supercurrent and represents a macroscopic manifestation of quantum mechanical phenomenon.

Josephson junctions have great technological relevance in as much as they pave the way to the realization of practical use of this purely quantum phenomenon of superconductivity. Such practical uses include fabrication of superconducting tunneling diodes for the microwave and the infrared interference quantum computer based on flux qbits, SQUID magnetometers, etc. The discovery of superconductivity and its theoretical elucidation have been a great triumph of quantum theory. Superconductivity of traditional materials like the metals has also found sophisticated technological applications. However, the basic phenomenon of superconductivity of metals is manifested only at very low temperatures $T < T_c$. The highest T_c recorded for metallic superconductors is 9.2 K for niobium. Such temperatures can be realized and maintained only through the use of liquid helium and sophisticated cryogenics. It turns out to be a bit too costly in terms of energy expenditure. It is natural therefore that scientists started looking for materials like alloys, semiconductors or ceramics that could display superconductivity at temperatures much higher than the T_cs of type I (low temperature) superconductors. Thus a search was on for 'materials' that could display superconductivity at room temperature or at least at temperatures that can be created and maintained by using liquid nitrogen and associated cryogenics. The availability of such materials would have enhanced the technological viability of the phenomenon of superconductivity manyfold. The search has led to the discovery of a host of materials, collectively called type II superconducting materials or the so called high T_c-superconductors. We will take up the topic in another section after examining certain important issues relating to traditional, very important technological materials called semiconductors and insulators. As we will see later, the digression was necessary at this point.

8.6 Semiconducting Materials and Insulators

In Chapter 5, we had briefly introduced the ideas of semiconductors and insulators as crystalline solids having specific types of energy band structures. Here we will consider these materials from the point of view of statistical properties of free electrons in semiconductors and insulators when they are in the state of equilibrium with thermal vibrations of the crystal lattice. We know that the properties of systems in a state of thermodynamic equilibrium is independent of the mechanism that establishes the equilibrium. Accordingly, we do not have to worry about specific mechanism of interaction among the free electrons or holes with the lattice vibrations or the processes of thermal excitations and electron-hole recombination. Those mechanism, as we have already seen, plays a crucial role in the treatment of electrical conductivity of metals. Likewise, they play a central role in the understanding of processes involved in thermomagnetic phenomena, thermal conductivity, etc.

Let us start by recalling the behavior of an isolated electron in the periodic field of an ideal crystal. In a stationary state, the electron has a steady average velocity (non-degeneracy) and move freely throughout the crystal. It appears at the first sight that all the crystalline solids should be good electrical conductors with the number of conduction electrons being equal to the total number of electrons available in the sample. Even in metals, the actual number of electrons available for conduction is much less. In insulators, this number is practically zero (at room temperature). To grasp the situation, let us consider a monoatomic solid with 'N' unit cells each with 'n' number of atoms, each atom carrying Z number of electrons, Z being the atomic number. The sample in this case contains NnZ number of electrons. At $T = 0$ K, when the system is at its lowest energy state, the electrons in the sample will occupy the $(NnZ)/2$ energy levels or one-electron states characterized by the wave vector $\vec{k}$, in accordance with Pauli exclusion principle. Two distinct possibilities arise:

(i) The highest level occupied by the electrons merge with the upper edge of one of the allowed energy bands (Figure 8.5a).

(ii) The highest level is inside such a band (Figure 8.5 b). In the second case, on the application of even a weak electric field, the electrons occupying levels close to the band edge 'ϵ_0' would be accelerated. They would move to occupy the unoccupied states close to the band edge, establishing an electric current in the sample. In the first case, the applied electric field will fail to induce a redistribution of electrons and cause transition to higher allowed energy bands unless the field crosses a threshold of about 10^6 V/cm or so when dielectric breakdown occurs. There is thus no electric current and the sample behaves like an insulator (also called a dielectric).[13]

Each separate energy band in the crystal has $2N$ quantum states. In a simple monoatomic crystal with $n = 1$, therefore, the number of bands filled with electrons is $NZ/2N = Z/2$. Hence, if Z is odd, the number of filled bands is not an integer, which means that such odd Z elements that crystallize also in a simple lattice, is expected to be a metal. It does not automatically mean that even Z elements would turn out to be insulators. It is possible that the bands overlap enabling the emergence of metallic behavior. If the material has a more complex crystal lattice structure signifying lower symmetry and atoms of different types are present, the energy bands will have a more pronounced tendency to separate, leading to the formation of an insulator. From the point of view of energy band structure, every solid at 0 K will either be a metal or a dielectric (insulator). There is no other option. Now consider a dielectric at absolute zero with a completely

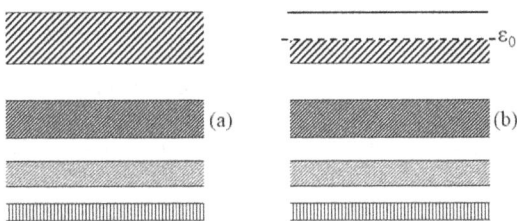

FIGURE 8.5 Band structures of two types ((a) and (b)).

filled valence band (the topmost band), which is separated from the next higher (unoccupied) band (the conduction band) by a forbidden band of width $\Delta\epsilon_g$. As the solid is heated up, the electrons start moving from the valence to the conduction band provided that the thermal stimulus is large enough to overcome the forbidden energy gap ($\Delta\epsilon_g$). If $\Delta\epsilon_g$ is ≈ 1 eV or less, the pure crystal, a dielectric at 0 K, will display appreciable electrical conductivity at room temperature or at higher temperatures. The conductivity has its origin in the motion of electrons in the conduction band and motion of holes in the valence band. Such pure materials with a rather narrow band gap ($\Delta\epsilon_g$) are called **intrinsic semiconductor**. Silicon and germanium are the most well-known and extensively used semiconductors. As we know already (Chapter 5) we will more frequently encounter what are known as impurity semiconductors in which donors act as the source of electrons for the conduction band and acceptors act as suppliers of holes for the valence band. Thus a wide variety of electrical properties can be created in a pure semiconducting material by doping with different types of impurities. Doped germanium and silicon are the most widely used impurity semiconductors in the electronic industry today.

8.6.1 Equilibrium Statistics of Electrons in Semiconductors and Metals

Certain properties of semiconductors and metals shaped primarily by the electrons and holes in a state of thermodynamic equilibrium with thermal oscillation of the crystal lattice may now be considered. We will adopt the single electron approximation or the independent particle mode in which the interaction between electrons in the crystal is taken care of at the self-consistent field level of approximation in which every electron moves independently of all other electrons. From a purely statistical mechanical point of view, the single electron approximation is equivalent to adopting an ideal electron-gas model already described in Chapter 7. In such an ideal electron gas, be it in a semiconductor or a metal, the average number of electrons (F) in a definite quantum state labeled by the quantum numbers k_1, k_2, k_3 or k_x, k_y, k_z with energy ϵ_k at a temperature T is given by the Fermi-Dirac distribution law (see Chapter 7).

$$< f(\epsilon_k) >= f(\epsilon_k) = \frac{1}{e^{\frac{(\epsilon_k - \xi)}{k_B T}} + 1} \tag{8.69}$$

The total number of electrons (N) in the volume 'V' of the crystal is obtained by summing over $f(\epsilon_k)$:

$$N = \sum_k f(\epsilon_k) = \sum_k \frac{1}{e^{\frac{(\epsilon_k - \xi)}{k_B T}} + 1} \tag{8.70}$$

We assume that the summation is performed with due care of the electron spins. ξ is the chemical potential, which is a function of the electron concentration or number density $n = N/V$ and the temperature T. The number density of the electron states in the energy interval $d\epsilon$ per unit volume is given by the following equation where due care has been taken of the electron spins:

$$n(\epsilon) d\epsilon = 2f(\epsilon)g(\epsilon)d\epsilon \tag{8.71}$$

$g(\epsilon)$ being the density of states at energy ϵ. We can represent Eq. 8.71 in k-space as

$$n(\vec{k})dk_x dk_y dk_z = 2f(\vec{k})\frac{dk_x dk_y dk_z}{(2\pi)^3} \tag{8.72}$$

k_x, k_y, k_z are the components of the wave vector $\vec{k}$. In the distribution function $f(\vec{k})$, energy $\epsilon(\vec{k})$ is expressed in terms of the wave vector ($\vec{k}$) as

$$\epsilon(\vec{k}) = \frac{\hbar^2 k^2}{2m^*} = \frac{p^2}{2m^*} \tag{8.73}$$

where m^* is the scalar effective mass of the electrons in the lattice. Using the known formula for $g(\epsilon)$ in a periodic field and the effective mass m^*, $g(\epsilon)$ takes the following form (the only change in $g(\epsilon)$ vis-a-vis what was introduced in Chapter 7 lies in the effective mass m^* as opposed to the free electron mass 'm').

$$g(\epsilon) = \frac{\sqrt{2}}{2\pi^2} \frac{(m^*)^{\frac{3}{2}}}{\hbar^3} \sqrt{\epsilon} \tag{8.74}$$

which leads to the following expression for the electron concentration n (number per unit volume)

$$n = \frac{\sqrt{2}}{\pi^2} \frac{(m^* kT)^{\frac{3}{2}}}{\hbar^3} \int_0^\infty \frac{\epsilon^{\frac{1}{2}}}{e^{\frac{(\epsilon - \zeta)}{k_B T}} + 1} d\epsilon \tag{8.75}$$

Making the satisfaction $\frac{\epsilon}{k_B T} = x$ and $\frac{\zeta}{k_B T} = y$, the expression above (Eq. 8.75) reduces to

$$n = \frac{\sqrt{2}}{\pi^2} \frac{(m^* k_B T)^{\frac{3}{2}}}{\hbar^3} \int_0^\infty \frac{x^{\frac{1}{2}}}{e^{x-y} + 1} dx \tag{8.76}$$

and defines the chemical potential ζ as a function of n, T, m^*.

The integral on the right hand side of Eq. 8.76 can not be expressed in terms of elementary function of y, although the values of integrals of the type $I_n(y) = \int_0^\infty \frac{x^n}{e^{x-y}+1} dx$ have been tabulated for a number of values of n. It is more important here to consider the relevant integral in Eq. 8.76 in specific limiting situations only. In Chapter 7 the problem was handled in one way. Here we introduce a more general approach.[13] Let us consider the example of a strongly degenerate electron gas (see Chapter 7) when $y \geq 1$, which is the case when the chemical potential $\zeta >> k_B T$. In this case $f(\epsilon) \approx 1$ for $\epsilon << \zeta$ and $f(\epsilon) \approx 0$ for $\epsilon >> \zeta$. A sharp drop in the value of $f(\epsilon)$ is noticed at $\epsilon = \zeta$ in an interval of the order of $k_B T$ (see Figure 8.6) We have displayed both $f(\epsilon)$ and $(\frac{-df}{d\epsilon})$ approaches the delta function as the temperature is lowered ($k_B T \to 0$) and is non-zero only when $\epsilon = \zeta$. Hence we can write $(\frac{-df}{d\epsilon}) = \delta(\epsilon - \zeta_0)$ where ζ_0 is the value of ζ at $T = 0$ K. Our intention now is to calculate ζ_0 (chemical potential of the electron gas in the first approximation) and see how a rise in temperature affects the chemical potential of the electron gas. To do that we have to evaluate the integral in Eq. 8.76. As a first step, let us consider the following integral:

$$\begin{aligned}
I &= \int_0^\infty \chi(\epsilon) f(\epsilon) d\epsilon \\
&= \int_0^\infty f(\epsilon) d\phi(\epsilon) \\
&= f(\epsilon)\phi(\epsilon)\Big|_0^\infty - \int_0^\infty \phi(\epsilon) \frac{df}{d\epsilon} d\epsilon \\
&= -\phi(0) + \int_0^\infty \phi(\epsilon) \left(\frac{-df}{d\epsilon}\right) d\epsilon
\end{aligned} \tag{8.77}$$

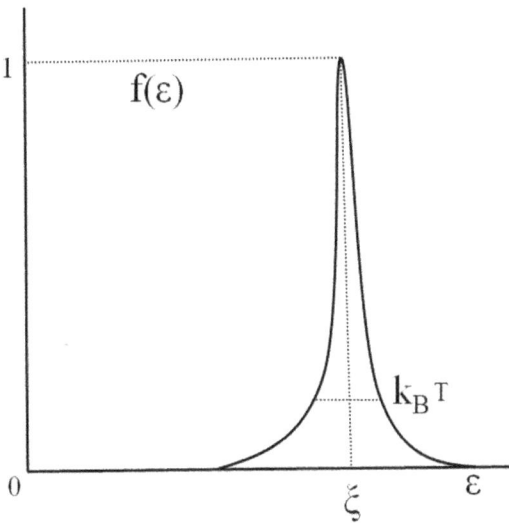

FIGURE 8.6 Fermi-Dirac distribution function $f(\epsilon)$ plotted against ϵ.

where $\chi(\epsilon)$ is an arbitrary function of ϵ. If $\phi(0) = 0$ (in most of the practical cases that is what happens), we have

$$I = \int_0^\infty \phi(\epsilon)\left(\frac{-df}{d\epsilon}\right)d\epsilon \tag{8.78}$$

We can now use this formula in Eq. 8.76 under the approximation that

$$\left(\frac{-df}{d\epsilon}\right) = \delta(\epsilon - \xi_0)$$

with the arbitrary function χ chosen as

$$\chi(\epsilon) = 2g_\epsilon$$

and complete the integration and get

$$n = \frac{\sqrt{2}}{\pi^2}\frac{(m^*)^{\frac{3}{2}}}{\hbar^3}\frac{2}{3}\int_0^\infty \delta(\epsilon - \xi_0)\epsilon^{\frac{3}{2}}d\epsilon$$
$$= \frac{(2m^*)^{\frac{3}{2}}}{3\pi^2\hbar^3}\xi_0^{\frac{3}{2}} \tag{8.79}$$

It follows therefore that the chemical potential in the first approximation is

$$\xi_0 = \frac{\hbar^2}{2m^*}\left(3\pi^2 n\right)^{\frac{2}{3}} = \frac{(2\pi\hbar)^2}{2m^*}\left(\frac{3n}{8\pi}\right)^{\frac{2}{3}} \tag{8.80}$$

which is just the maximum kinetic energy of an ideal gas at absolute zero. The chemical potential of strongly degenerate electron gas is thus independent of temperature (T) in the first approximation. In the next approximation the chemical potential turns out to be

$$\xi = \xi_0\left[1 - \frac{\pi^2}{12}\left(\frac{k_BT}{\xi_0}\right)^2\right] \tag{8.81}$$

The condition for strong degeneracy can be worked out in explicit form, using the zeroth order approximation to ξ:

$$\frac{\xi_0}{k_BT} = \frac{\hbar^2(3\pi^2 n)^{\frac{2}{3}}}{2m^*k_BT} >> 1 \tag{8.82}$$

From Figure 8.4 where the Fermi-Dirac (FD) distribution function has been displayed, it is clear that the above condition of degeneracy actually has an exponential character which means that

$$e^{\frac{\xi_0}{k_BT}} >> 1 \tag{8.83}$$

Therefore if $\frac{\xi_0}{k_BT} \approx \frac{5}{7}$, the degeneracy may be considered to be strong. The condition (Eq. 8.82) suggests that high electron concentration (n), low effective mass (m^*) and low temperature (T) conspire to facilitate degeneracy. For a typical metal with $n \approx 10^{22}/cm^3$, $m^* \approx 10^{-27}$ gm, at room temperature $\frac{\xi_0}{k_BT} \approx 10^2$ very strong degeneracy condition prevails. The free electrons in a metal will therefore be likely to be in the state of strong degeneracy possibly upto the melting point. So far, we have considered the system with positive chemical potential.

Let us now consider the opposite case of negative chemical potential satisfying the condition

$$e^{\frac{-\xi}{k_BT}} >> 1$$

The distribution function $f(\epsilon)$ in this case can be approximated as

$$f(\epsilon) = A \cdot e^{-\frac{\epsilon}{k_BT}} \tag{8.84}$$

with $A = e^{\frac{\xi}{k_BT}}$.

It is just the Maxwell-Boltzmann (MB) distribution function with 'A' as the normalization constant. The concentration of electrons in the conduction band then turns out to be

$$n = \frac{\sqrt{2}}{\pi^2} \frac{(m^*)^{\frac{3}{2}}}{\hbar^3} \int_0^\infty A \cdot e^{-\frac{\epsilon}{k_B T}} \epsilon^{\frac{1}{2}} d\epsilon$$

$$= \frac{(2\pi m^* k_B T)^{\frac{3}{2}}}{4\pi^3 \hbar^3} (A) \tag{8.85}$$

The integral in Eq. 8.85 can be evaluated easily yielding

$$A = e^{\frac{\xi}{k_B T}} = \frac{4\pi^3 \hbar^3 n}{(2\pi m^* k_B T)^{\frac{3}{2}}} \tag{8.86a}$$

$$\xi = k_B T \ln\left[\frac{4\pi^3 \hbar^3}{(2\pi m^* k_B T)^{\frac{3}{2}}}\right] \tag{8.86b}$$

The FD distribution function in this case becomes

$$f(\epsilon) = \frac{4\pi^3 \hbar^3 n}{(2\pi m^* k_B T)^{\frac{3}{2}}} e^{\frac{-\epsilon}{k_B T}} \tag{8.87}$$

Let us note that under condition assumed the chemical potential (ξ) is a fairly strongly temperature-dependent quantity (Eq. 8.86) in sharp contrast with what was found for the strongly degenerate electron gas. It is pertinent here to work when does classical statistics apply? The applicability of the classical statistics is determined by the non-degeneracy condition:

$$e^{\frac{-\xi}{k_B T}} = \frac{1}{A} = \frac{(2\pi m^* k_B T)^{\frac{3}{2}}}{4\pi^3 \hbar^3 n} >> 1 \tag{8.88}$$

Low density (n), high effective mass (m^*) and high temperature (T) facilitate the satisfaction of the non-degeneracy condition. Thus, with $n \approx 10^{17}/\text{cm}^3$, $m^* \approx 10^{-27}$g, at room temperature $\frac{1}{A} \approx 3 \times 10^2$, which means that the non-degeneracy criterion is rather easily satisfied. If we set $\frac{1}{A} \approx 1$, the maximum electron concentration for which non-degeneracy condition is satisfied is $\approx 10^{19}/\text{cm}^3$. It is therefore possible to determine easily under what conditions quantum statistics must be invoked.

8.6.2 Equilibrium Statistics of Electron Gas in Semiconductors

The equation obtained for electron gas in metals continue to hold for electron gas in semiconductors as well. The utility of the equation, however is rather limited in semiconductors. This is because the electron concentration 'n' in the conduction band is itself a temperature-dependent quantity in semiconducting materials and we do not know the nature of the temperature dependence of 'n'. The source of temperature dependence of 'n' can be traced to thermal excitation of electrons on the impurity levels and in the valence band. We have already mentioned (Chapter 5) that the excitations of electrons in the valence band and their subsequent transition to the conduction band creates positively charged holes in the valence band. These holes together with the electrons contribute to the transport phenomenon. Before proceeding further, let us examine a typical energy level diagram of an impurity semiconductor (Figure 8.7). ϵ_g is the forbidden energy gap. The donor and acceptor levels are below the bottom edge of the conduction band and are ϵ_D and ϵ_A, respectively, all the energies being measured from the zero level, which is supposed to be coincident with the bottom edge of the conduction band ('0' marked Figure 8.7). The Fermi-Dirac distribution function for the holes in the semiconductor is

$$f_+(\epsilon) = 1 - f(\epsilon)$$

$$= \frac{1}{e^{\frac{\xi-\epsilon}{k_B T}} + 1} \tag{8.89}$$

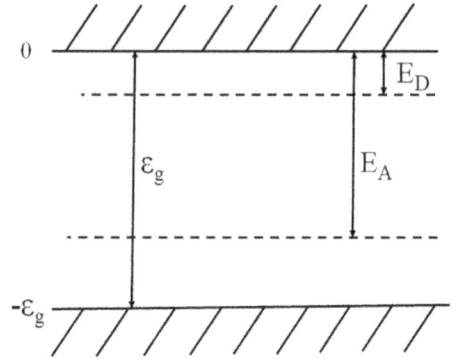

FIGURE 8.7   Schematic energy level diagram of an impurity semiconductor.

where $f(\epsilon)$, as already stated, is the FD distribution function for the electrons. If all the energies are measured from the bottom edge of the conduction band, the energy of the electrons in the conduction band is ($m_- = $ mass of the electron)

$$\epsilon = \frac{\hbar^2 k^2}{2m_-}$$

while on the donor and acceptor it is $\epsilon = -\epsilon_D$ and $\epsilon = -\epsilon_A$, respectively. In the valence band, the energy is $\epsilon = -\epsilon_g - \epsilon'$ where $\epsilon' = \frac{\hbar^2 k^2}{2m_+}$ is the kinetic energy of the hole. ϵ_g is the forbidden energy gap, m_+ being the mass of the hole. The hole distribution function can now be written as

$$f_+(\epsilon) = \frac{1}{e^{\frac{(\epsilon' - \zeta')}{k_B T}} + 1} \tag{8.90}$$

where $\zeta' = -\epsilon_g - \zeta$ is the chemical potential of the hole.

The chemical potential 'ζ' can be fixed by imposing the electroneutrality condition, which demands that the following equality holds: the number of electrons in the CB + the number of electrons in the acceptor levels = the number of holes in the VB + the number of holes in the donor levels.

Using the respective distribution functions and the density of states g_ϵ and g'_ϵ, we can translate the condition for electroneutrality mathematically into the following equality:

$$\int_{CB} \frac{2g_\epsilon d\epsilon}{e^{\frac{(\epsilon - \zeta)}{k_B T}}} + \sum_{(A)} \frac{1}{e^{\frac{(-\epsilon_A - \zeta)}{k_B T}} + 1} = \int_{VB} \frac{2g_{\epsilon'} d\epsilon'}{e^{\frac{(\zeta + \epsilon_G + \epsilon')}{k_B T}}} + \sum_{(D)} \frac{1}{e^{\frac{(\epsilon_D + \zeta)}{k_B T}} + 1} \tag{8.91}$$

Since the concentration of electrons in the CB and the concentration of holes in the VB are low, and the temperature is not low, we can assume that the laws of classical statistics hold and assert that $B = e^{\frac{\zeta}{k_B T}} \ll 1$ and $e^{\frac{\zeta + \epsilon_G}{k_B T}} \ll 1$. The integrals in the equality above (Eq. 8.91) can be evaluated as in Eq. 8.85, yielding[10]

$$\frac{(2\pi m_- k_B T)^{\frac{3}{2}}}{4\pi^3 \hbar^3} \times B + n_A \left(\frac{1}{B} e^{\frac{-\epsilon_A}{k_B T}} + 1 \right)^{-1} = \frac{(2\pi m_+ k_B T)^{\frac{3}{2}}}{4\pi^3 \hbar^3} \cdot \frac{1}{B} e^{\frac{-\epsilon_G}{k_B T}} + n_D \left(B e^{\frac{\epsilon_D}{k_B T}} + 1 \right)^{-1} \tag{8.92}$$

In Eq. 8.92, n_A and n_D are the concentrations of the acceptor and donor, respectively. It turns out from the condition Eq. 8.92 above, that the electron concentration 'n' in the CB and the concentration of holes in the VB are given by[13]

$$n_- = n = \frac{(2\pi m_- k_B T)^{\frac{3}{2}}}{4\pi^2 \hbar^3} \times e^{\frac{\zeta}{k_B T}} \tag{8.93a}$$

$$n_+ = p = \frac{(2\pi m_+ k_B T)^{\frac{3}{2}}}{4\pi^2 \hbar^3} \times e^{\frac{-(\zeta + \epsilon_G)}{k_B T}} \tag{8.93b}$$

The condition of electroneutrality represented in Eq. 8.92 determines the classical potential $\xi = k_B T ln B$, which in general leads to an algebraic equation in 'B' of degree 4. Although graphical or numerical solution is possible, we will consider only several limiting cases, where the solutions can be obtained from much simpler equations. Let us consider these cases one by one:

(a) **The case of intrinsic semiconductor**: Here $n_A = n_D = 0$ (no impurity atoms). In this case Eq. 8.92 reduces to

$$B = \left(\frac{m_+}{m_-}\right)^{\frac{3}{4}} e^{\frac{-\epsilon_G}{k_B T}}$$

$$\xi = k_B T ln B = -\frac{\epsilon_G}{2} + k_B T \times \frac{3}{4} ln\left(\frac{m_+}{m_-}\right) \tag{8.94}$$

Since the factor $\frac{3}{4} ln(\frac{m_+}{m_-}) \sim 1$, the second term in Eq. 8.93 is $\sim k_B T$ and ξ, the chemical potential of an intrinsic semiconductor coincides with the middle of the band gap to within $k_B T$. Only if $m_+ = m_-$, $\xi = -\frac{\epsilon_G}{2}$, coinciding exactly with the middle of the forbidden band and is obviously temperature-independent if the gap energy is temperature-independent. The concentration of electrons in the CB and that of holes in the VB turn out to be identical and we can write down (from Eq. 8.93)

$$n_\pm = \frac{(2\pi \sqrt{m_+ m_-} k_B T)^{\frac{3}{2}}}{4\pi^2 \hbar^3} \times e^{\frac{-\epsilon_G}{2 k_B T}}$$

$$= n_T^0 \times e^{\frac{-\epsilon_G}{2 k_B T}} \tag{8.95}$$

If we set $m_+ = m_- = 0.9 \times 10^{-27}$ gm, $T = 294$ K, n_T^0 turns out to be $\sim 2.44 \times 10^{19}$ cm^{-3}. However, n_T^0 rises to infinity as $T \to \infty$, which is not consistent with the finite number of electrons in the VB, a signature of the fact that the finite size of the VB was never taken into consideration while arriving at Eq. 8.93.

(b) **The cases of impurity semiconductors**: For a donor (n-type) semiconductor $n_D \neq 0$, $n_A = 0$. We assume further that the band gap is wide ($\epsilon_G >> \epsilon_D$). From Eq. 8.92, we then have

$$n_D = \frac{(2\pi m_- k_B T)^{\frac{3}{2}}}{4\pi^3 \hbar^3} B\left[B e^{\frac{\epsilon_D}{k_B T}} + 1\right] \tag{8.96}$$

In arriving at Eq. 8.96 we have neglected the term proportional to $e^{\frac{-\epsilon_G}{k_B T}}$ as $\epsilon_G >> \epsilon_D$. Two subcases now arise:

(i) $B e^{\frac{\epsilon_D}{k_B T}} >> 1$: This is possible when $\epsilon_D >> k_B T$ as $\beta << 1$. Neglecting unity in Eq. 8.96, we arrive at the following expression for B and the chemical potential:

$$B = \frac{(2\pi \hbar)^{\frac{3}{2}} n_D^{\frac{1}{2}}}{\sqrt{2}(2\pi m_- k_B T)^{\frac{3}{2}}} e^{\frac{-\epsilon_D}{2 k_B T}} \tag{8.97a}$$

$$\xi = -\frac{\epsilon_D}{2} + \frac{1}{2} k_B T ln\left\{\frac{(2\pi \hbar)^3 n_D}{2(2\pi m_- k_B T)^{\frac{3}{2}}}\right\} \tag{8.97b}$$

If we take $n_D = 10^{17}$/cc, $m_- = 10^{-27}$g and $T = 300$ K, the term in Eq. 8.97 turns out to be $\sim k_B T$. That means, the chemical potential is roughly midway between the bottom edge of the conduction band and the donor level. The concentration of the conduction electrons turns out to be (using Eq. 8.97 in Eq. 8.93)

$$n = \frac{(2\pi m_- k_B T)^{\frac{3}{4}} n_D^{\frac{1}{2}}}{(2\pi \hbar)^{\frac{3}{2}}} e^{\frac{-\epsilon_D}{2 k_B T}} \tag{8.98}$$

Eq. 8.98 shows that $n \propto \sqrt{n_D}$.

(ii) In the second case, $Be^{\frac{\epsilon_D}{k_B T}} << 1$: Since $B <<1$, this situation is always realized for $\epsilon_D \leq k_B T$. If we neglect the $Be^{\frac{\epsilon_D}{k_B T}}$ term within the square brackets in Eq. 8.96 we have[13]

$$B = \frac{4\pi^3 \hbar^3 n_D}{(2\pi m_- k_B T)^{\frac{3}{2}}}$$ (8.99a)

and

$$\xi = k_B T ln \left\{ \frac{4\pi^3 \hbar^3 n_D}{(2\pi m_- k_B T)^{\frac{3}{2}}} \right\}$$ (8.99b)

Now the concentration of electrons (n) turns out to be equal to n_D. Thus, if $k_B T \geq \epsilon_D$, practically all donors will be ionized. The chemical potential is seen to be negative (Eq. 8.99b) and located below the donor level.

8.6.3 Semimetals

A question of importance concerns the effect that lowering the temperature to zero Kelvin has on the concentration of the electrons in a semiconductor. From Eq. 8.93 or Eq. 8.98, it turns out that $n \to 0$ as $T \to 0$, which means that the resistivity of the semiconductor rises without limit and should become infinitely large at $T = 0$ K. In practice, the resistance of many of the semiconductors remains finite as the $T \to 0$ limit is approached. This happens because the wavefunctions of electrons localized on the impurity center begin to overlap when the concentration of the impurity atoms is substantial leading to some degree of widening of the energy levels of the electrons involved. That results in the formation of an impurity band, which in turn enables the impurity electrons to move through the crystal leading to the emergence of impurity conductivity. The mobility of electrons in the impurity band $<<$ the mobility of electrons in the conduction band and accordingly they contribute much less to the overall conductivity of the semiconductor. However, the impurity conductivity depends little on the temperature and diminishes rapidly as the impurity concentration is reduced. Nevertheless, impurity conductivity can be observed even with small concentration of impurity centres. For example, impurity conductivity has been observed in germanium (Ge) even at impurity concentration of 10^{15}/cm^3. At still higher concentrations the impurity band merges with the conduction band and is occupied by electrons at all temperatures, the concentration of electrons being independent of temperature. This gives rise to a new class of materials called semimetals.[13] The overlapping of the wavefunctions of the impurity electrons begins to take place at progressively lower concentrations, if the Bohr radius of the electron is larger. That is, greater the dielectric constant and lower the effective mass, higher is the overlapping at lower concentration. That is why materials with high dielectric constant (ϵ) and small effective mass (m^*) turn into semimetals at relatively low impurity concentrations.[13] Thus, in n-InSb, with $n_D \approx 10^{15}$/cm^3, all the impurity atoms remain ionized whatever the temperature is. A similar effect is observed in n-Ge only if the impurity concentration exceeds 5×10^{18}/cm^3. Although we have arrived at the concept of semimetals by examining the band structure and resistivity of doped semiconductors, there is scope for further generalization.

We can define semimetals as materials or solids with a very small overlap between the top of the VB and the bottom of the CB (Figure 8.8). This small overlap ensures that there is no direct band gap as in a semiconductor but there is a negligibly small (but non-zero) density of states at Fermi energy. In contrast metals have zero band gap and significant density of states at ϵ_F as the conduction band is partially filled.

A number of elemental solids fit the description of semimetals given in the preceeding paragraph. Some of these elements, which separate the typical metallic elements from the non-metallic ones in the periodic table, are better known as metalloids. The classic semimetallic elements are arsenic, antimony, bismuth, α-tin and graphite. Arsenic and antimony are also classified as metalloids. However, tin, bismuth and graphite are not regarded as typical metalloids. While metalloids are always elemental solids, semimetals can be chemical compounds like tellurides, e.g. mercury telluride (HgTe) or even polymers. Semimetals have low thermal and electrical conductivities quite expectedly as the carrier concentrations ($n_\pm$) are

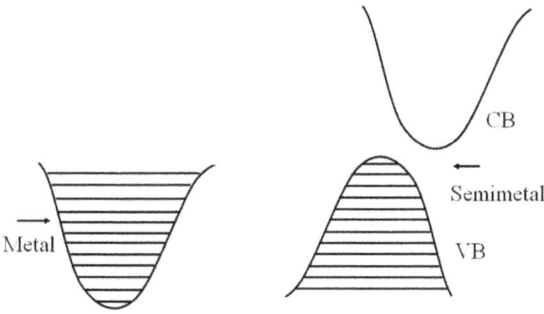

FIGURE 8.8   Model semimetallic and metallic band structures.

low. They have typically higher Seebeck coefficients than metals. They are important materials in thermoelectrics. Bubonova *et al.*[14] have measured the thermoelectric properties of various poly(3,4-ethylene dioxy thiophene) samples and found a marked rise in the Seebeck coefficient when the electrical conductivity is enhanced through molecular organization, initiating the transition of the polymer from a **Fermi glass to semimetals**. The high value of Seebeck coefficient, the metallic conductivity at room temperature and the absence of unquenched electron spins make the 'polymer semimetal' an attractive candidate for thermoelectrics and spintronics. Reed *et al.*[15] observed a transient semimetallic layer in detonating nitromethane, normally a wide band gap insulator. The transition appears to be associated with generation of charged decomposition products. Metalloids, on the other hand, find extensive use in semiconductor technology and electronics and as alloying agents. Thus, germanium and silicon constitute the backbone of solid state electronics based on transistors and have been in use since 1940. Boron finds use as a dopant (hole donor) in semiconductors and being extremely hard is often used as a bonding agent in permanent rare earth magnets. Arsenic also is used as a dopant (electron donor) in semiconductors.

8.6.4 Compound Semiconductors

So far we have considered only elemental semiconductors like Si and Ge with or without doping. They have band gaps less than 4 eV. Materials with band gaps > 4 eV are routinely classified as insulators. Although both Si and Ge are elemental semiconductors, their band gaps have a certain difference. Thus, Ge and Si are both indirect band gap materials with band gap of 1.12 and 0.67 eV, respectively. Diamond on the other hand is a material with a direct band gap of 5.6 eV. Let us recall that the band gap represents the energy difference between the bottom of the conduction band and the top of the valence band. If we examine the band energies $E(k)$ as a function of the crystal momentum $\hbar k$, it turns out that the minimum of the conduction band and the top of the valence band do not always appear at the same value of k. If they (the maximum of VB and minimum of CB) appear at the same point in the k-space, we have a direct band gap material while if they appear at different parts of the k-space, a material with an indirect band gap results. If the band gap is less than 4 eV, the material that we have in the first case is a direct band gap semiconductor while in the second case, we have an indirect band gap material. The situation can be visualized with reference to Figure 8.9a-b in which we have plotted the band energy $E(k)$ as a function of k in a one-dimensional system.

Let us mention here that Si and Ge have much more complex energy band structures. In *n*-germanium and *n*-silicon the energy spectrum is made up of a set of equivalent minima symmetrically distributed in the Brillouin zone. The number of equivalent minima in Ge is 4 and the number for silicon is 6. The corresponding $E(\vec{k})$ versus $\vec{k}$ plots are therefore much more complex than the one-dimensional plots of Figure 8.9a-b. As semiconductors for making transistors, Si and Ge are absolutely fine. Although Si has a larger band gap (1.12 eV) than Ge (0.67 eV), it has become the mainstay of the electronic industry and technology because of its abundance and easy processability. The problem with indirect low band gap materials becomes evident in the context their usefulness as material in photovoltaics, optoelectronics, LASER, etc. In all such applications, direct band gap compound semiconductors are

(a)   (b)

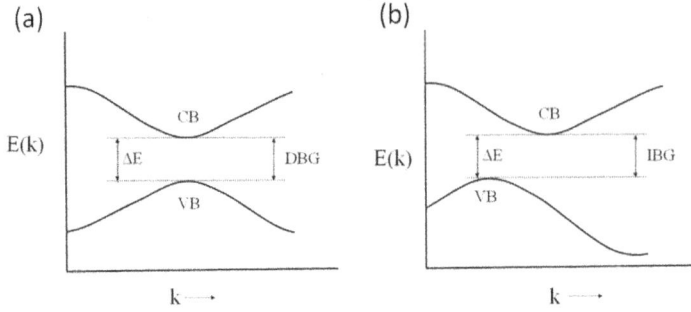

FIGURE 8.9 $E(k)$-k profiles for direct (a) and indirect (b) band gap materials.

the automatic choice. The reasons are not far to seek. They provide access to a wide choice of materials with tunable band gap. If a photon excites an electron into the conduction band of a direct band
gap semiconductor, creating an electron-hole pair, there is no major requirement for further adjustment
of the electron's momentum as the top of the valence band and the bottom edge of the conduction
band occur at the same value of the crystal momentum ($\hbar k$). Therefore, the process of photon mediated electron-hole pair formation or that of electron-hole recombination producing photon (light) tend
to occur fast. In an indirect band gap material the photon induced electron-hole pair formation or the
reverse process of electron-hole recombination with the emission of a photon tends to become much
slower as the electron momentum must be further adjusted after excitation into the conduction band
where the crystal momentum is different from its value at the top of the valence band. The adjustment
requires the electron to interact with the phonon bath leading either to absorption or emission of one
or more phonons, which slows down the electron-hole pair formation or electron-hole recombination
processes. In fact an assortment of compound direct band gap materials are now available for different
applications.

Silicon and germanium are both members of group IV of the periodic table and just like diamond are
covalently bonded. The valence bands in Si and Ge correspond to sp^3-hybridized energy levels of the
isolated Si and Ge atoms. These hybridized valence bands are fully occupied at 0 K or low temperatures.
The situation is the same for carbon in diamond except that the sp^3-hybridized bonds are perfectly tetrahedral and stronger than those in Si/Ge so that the band gap is much higher (5.6 eV) compared to Si or Ge.
If compound formation takes place between elements of groups IIIA and VA, say between gallium and
arsenic forming gallium arsenide (GaAs), or between indium and antimony producing indium antimonide
(InSb), these group III-V compounds are found to have semiconducting properties with direct band gaps,
which are lower than that in either Si or Ge. The bonds formed are still covalent, but have significant ionic
character, causing increase in band gap. A similar scenario unfolds when we consider compounds formed
by combining group IIB and group VIA elements. For example cadmium sulphide (CdS) or zinc telluride
(ZnTe). These molecules also give rise to semiconductors with band gaps larger than that of Si or Ge.
Table 8.2 lists a number of such direct band gap compound semiconductors, which can be at the forefront
of '**beyond silicon electronics**'. Silicon has so far been at the root of the digital revolution. We are witnessing and is likely to dominate the space of digital technology for a while. However, silicon-based transistors
are fast approaching its limit of miniaturization. Already the Si-transistors are getting too small to be oper-
ated with high efficiency and stability. A drawback of silicon is its relatively poor carrier mobility, which
limits the speed at which silicon transistors can work. In addition it does not work so well at higher temperatures. Neither does it respond to photons that well. The response to light is important as 'photonics'
will almost certainly play an important role in computer electronics for enhancing computing power in
the near future. All these make the search for new compound semiconducting materials so important and
exciting.

TABLE 8.2

Some Compound Semiconductors and Their Band Gaps (NSM Archive - Physical Properties of Semiconductors)

		Band gap (eV)	Type
Group IIIA-VA	GaP	2.25	indirect
	GaAs	1.42	direct
	InSb	0.17	direct
	GaN	3.44	direct
	InAs	0.36	direct
Group IIB-VIA	CdS	2.40	direct
	ZnTe	2.26	direct
	ZnO	3.37	direct
	CdTe	1.49	direct

8.7 Insulators

Electrical insulators, also called dielectrics are materials that appear as poor conductors of electricity at ordinary temperatures when the applied electric field is weak or small. The electric current flowing through a sample of dielectric materials is very very small, but not zero, even at low fields. The primary action of the applied field (if we neglect the very small current set up by the field) is to polarize the dielectric material.[1] The polarization field opposes the external field thereby weakening it. The extent of this weakening is measured by what is known as the dielectric constant ϵ also called relative dielectric permittivity of the material. If the polarization is simply proportional to the applied field strength, the material is called a **linear dielectric material**. The electronic energy band structure of such materials is of the same type as seen in semiconductors, namely a filled valence band separated from an empty conduction band by a band gap (ΔE_g). This band gap is much larger in insulators ($> 4\,\text{eV}$) than in semiconductors ($< 4\,\text{eV}$). At normal temperatures only very few electrons can be thermally excited into the conduction band ($k_B T << \Delta E$). Hence, the electrical conductivity of such large band gap materials is expected to be low and the relative dielectric permittivity to be high. Generally, ionic materials, plastics (covalently bonded polymers) have large band gaps and are good insulators. Table 8.3 reports dielectric constants (ϵ) of a number of solid insulating materials including some non-metallic solids. In general, they have high ϵ and low conductivity (σ) or high electrical resistivity (ρ). For many technological applications materials with high electrical resistivity is required. Hence, good insulators are in high demand. We note, however, these insulating materials tend to develop higher conductivity as the temperature is raised and eventually may display electrical conductivity higher than a semiconductor at high temperatures. The behavior can be understood easily as the probability p of exciting an electron into the conduction band is given by $p \propto e^{\frac{-\Delta E}{k_B T}}$. Hence, the insulators usually display increased electrical conductivity at higher temperatures, which may be of the same order of magnitude as seen in typical semiconductors. So, for insulating materials operating temperatures must be carefully determined and specified. A simple, order of magnitude estimate of the dielectric constant of a solid (insulator) can be obtained from the theory of dielectric polarization of a non-polar gas for which orientation polarization is absent and $\epsilon_g = 1 + \frac{n_g \alpha}{\epsilon_0}$ where n_g is the number density of the gas and α is the polarizability of atoms/molecules of the gas. In a solid we can similarly write $\epsilon_{solid} \approx 1 + \frac{n_{solid}\alpha}{\epsilon_0}$, by assuming the solid to be a highly dense gas. Typically $n_{solid} >> n_{gas}$ so that a material in the solid state will have much higher dielectric constant than the corresponding liquid has.[1]

If the applied electric field strength is gradually increased, eventually a stage is reached when the electric current begins to rise quite spectacularly, finally destroying the insulating material. The phenomenon is known as **dielectric breakdown** and the critical field strength at which the breakdown sets in is known as the **dielectric strength**. Typical values of dielectric strengths of some common insulating materials are reported in Table 8.4. For specific technological application, the dielectric strength of a material is

TABLE 8.3

Typical Values of Dielectric Constant of Insulators

Materials	Dielectric constant (ϵ)
Non-Metals	
Silicon (Si)	11.9
Germanium (Ge)	16.0
Sulphur (S)	3.5
Phosphorus (P)	4.1
Ceramics/Glass	
Alumina (Al_2O_3)	8.5
Quartz (SiO_2)	4.5
Lead glass (PbO, SiO_2)	7.0
Barium titanate ($BaTiO_3$)	3.8
Strontium titanate ($SrTiO_3$)	200
Strontium zirconate ($SrZrO_3$)	38
Plastics	
Polyethylene	2.3
Polytetrafluoroethylene	2.1

TABLE 8.4

Dielectric Strength of Some Insulators

Material	Dielectric strength (Vm^{-1})
Alumina (Al_2O_3)	$10\text{-}35 \times 10^6$
Quartz (SiO_2)	$25\text{-}40 \times 10^6$
Beryllia (BeO)	$10\text{-}14 \times 10^6$

an important parameter that must be known beforehand. The detailed mechanism of the dielectric breakdown is complex and will not be addressed here. However, we may understand the process through a simple analysis. Let us note that the density of charge carriers in an insulator is low at equilibrium if both the temperature and the applied field are low. As the field strength increases these carriers (electrons) are accelerated. The highly accelerated carriers collide with carriers in its neighborhood and thereby creates a region which is locally at a higher temperature than the equilibrium temperature. It is likely therefore that such locally hotter regions will generate additional carriers and eventually set off an 'avalanche' of carriers leading to a steep rise in the current and an eventual electrical breakdown in the insulator when the field strength exceeds a threshold value. Typically, the breakdown field is of the order of 10^7 Vm^{-1}, which is substantially lower than the effective electric field experienced by electrons in the solid material. The presence of cracks and impurities in the sample provides additional mechanisms for the onset of breakdown. A look at Table 8.3 reveals that the dielectric constants of the common insulators are in the range 1-16 while there are some mixed oxide materials for example (SrO, TiO_2) also expressed as $SrTiO_3$(strontium titanate) and (SrO, ZrO_2), i.e. strontium zirconate ($SrZrO_3$), which have dielectric constant that are approximately 2-10 times higher than the dielectric constants of ordinary insulators. If we try to estimate ϵ by using the approximate relation $\epsilon_{solid} = 1 + \frac{n_{solid}\alpha}{\epsilon_0}$, we find that the causative factor for the anomalously large value of ϵ must be found in a higher value of atomic or molecular polarizability α since the number density n_{solid} can not become so high (by a factor of 10-20) in specific solids. The large value of molecular polarizability signifies that the possibility of a rather large electrical polarization occuring in such solids by even a small applied electric field is quite high. That signals the possibility of even a spontaneous electric polarization taking place in some such ionic solids with highly polarized ionic components. Such solids are called ferroelectric materials – a special type of ionic (oxide) solid or insulator, which reveals electric polarization even in the absence of an external electric field. The possibility is realized in practice in what are known as ferroelectric materials.

8.7.1 Ferroelectric Materials

Ferroelectric materials constitute a class of dielectric materials called paraelectrics, which display spontaneous electric polarization – that is polarization even in the absence of an external electric field. Like any other paraelectric material, their field induced polarization (P) is a non-linear function of the applied field. In addition their spontaneous polarization can be reversed by applying a sufficiently strong electric field in the opposite direction. They are the dielectric analogs of ferromagnetic materials, which exhibit spontaneous magnetization (see Chapter 10), i.e. permanent magnetic behavior without the application of external magnetic field. The ferroelectric behavior demands that there must exist permanent electric dipoles in such materials which are not mutually completely compensated by their random orientations. How is that possible? We will try to explain the existence of a permanent electric dipole in ferroelectric materials with reference to barium titanate ($BaTiO_3$) the oldest ferroelectric substance known to us.[1] The spontaneous electric polarization in barium titanate has its origin in the crystal structure of $BaTiO_3$, which is an ionic solid comprising of Ba^{+2}, Ti^{+4}, O^{-2} ions. The location of these ions within the tetragonal unit cell are such that each unit cell acquires an electric dipole moment of its own. Figure 8.10a displays a unit cell of tetragonal barium titanate along with the locations of the three different ions in the unit cell. It turns out that the Ba^{+2} ions are located at the corners of the unit cell bearing tetragonal symmetry. The O^{-2} ions are located close to, but a little above the centers of each of the six faces while the Ti^{+4} ion is slightly displaced upwards from the center of the unit cell. The relative displacements *pf* the Ti^{+4}, O^{-2} ions from their ideal symmetric positions give rise to a permanent dipole moment associated with each unit cell. However, when the sample of $BaTiO_3$ is heated to 120 ° C, the ferroelectric behavior abruptly disappears. The unit cell becomes cubic and all the ions occupy appropriate symmetric positions within the cubic unit cell generating or producing what is known as perovskite crystal structure (Figure 8.10b). Ba^{+2} ions are located at each of the eight corners of the cubic unit cell, O^{-2} ions occupy the center of each of the six faces while the Ti^{+4} ion is situated at the cubic center. The structure is centrosymmetric. There is thus no longer a non-zero dipole moment associated with the unit cell of barium titanate and it ceases to be ferroelectric. The temperature T_c at which this transition takes place (both structural, tetragonal to cubic and electric, ferroelectric to paraelectric) is a critical temperature called the ferroelectric Curie temperature.

The spontaneous electric polarization of barium titanate in the tetragonal phase can now be understood as a consequence of interaction between adjacent electric dipoles, which align themselves in the same direction in order to minimize the energy, creating ferroelectric ordering (see Chapter 10, for ferromagnetic ordering). This ordering means that the relative displacements of Ti^{+4} and O^{-2} ions are in the same direction for all unit cells within a certain volume of the sample of the material. Many other materials have been discovered to display ferroelectric behavior, for example, potassium dihydrogen phosphate

(a)   (b)

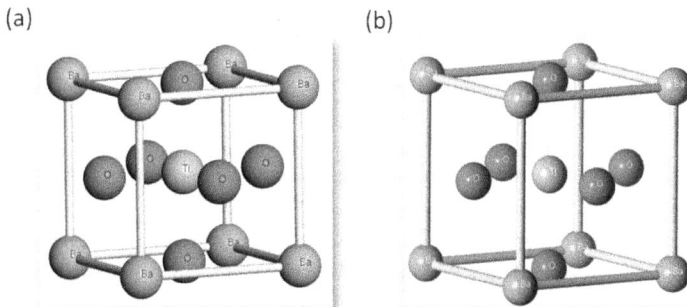

FIGURE 8.10   (a) Tetragonal structure of $BaTiO_3$. (b) Cubic perovskite structure of $BaTiO_3$.

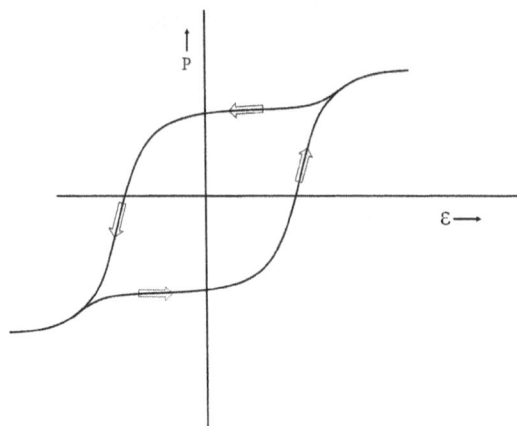

FIGURE 8.11   Hysteresis loop in a ferroelectric material.

(KH_2PO_4), strontium titanate ($SrTiO_3$), potassium niobate ($KNbO_3$) and lead zirconate-titanate ($Pb[ZrO_3, TiO_3]$) and some others. Ferroelectric materials have very high dielectric constants (ϵ) in dc as well as low frequency applied electric fields. Thus, at ambient temperature ϵ for barium titanate can be as high as 5000. ϵ of some other ferroelectric materials has been reported in Table 8.3. The high dielectric constants of these materials have been profitably utilized in making high capacitance capacitors much smaller in size than equivalent capacitors made from other materials. The spontaneous polarization in ferroelectric materials point to the existence of a hysteresis effect (see Figure 8.11), which can be exploited as a memory function to make ferroelectric random access memory systems for computers. Symmetry considerations suggest that a ferroelectric material should be piezoelectric and thermoelectric as well. The combination of memory function, piezo and thermoelectricity render the ferroelectric capacitors extremely useful in sensor applications. Is it possible to have materials that are both ferroelectric and ferromagnet at the same time in the same phase? If so, can we control ferroelectricity magnetically or ferromagnetism electronically? We will return to these question in Chapter 10 where ferromagnetism will be the central issue to address. It perhaps suffices to mention here that multiferroics, i.e. materials with coexisting ferromagnetism and ferroelectricity have been the target of many researchers and developments are still continuing. Multiferroics belong to a class of futuristic materials.

8.8 High Temperature or Type II Superconductors

Let us start by reviewing[1] what we have already learnt about materials, mostly metals displaying normal (slow temperature) superconductivity. Many high purity metals, when cooled down to temperatures close to 0 K, have their resistivities lowered to some very small but non-zero value as $T \rightarrow$ O K (Cu, Au). This low but non-zero resistivity is metal specific. Some metals or alloys of metals exist (Al, Sn, Pb, Hg, for example) for which the resistivity abruptly drops to a virtually zero value at some low but non-zero critical temperature T_c (see Figure 8.12) with T_c ranging between 1 and 20 K. These superconducting materials (type I) are diamagnetic in nature ejecting an applied magnetic field (H) from the body of the material as long as they exist in the superconducting state (**Meissner effect**). On increasing the magnetic field strength, the diamagnetism is retained until a critical magnetic field strength (H_c) is reached. At the critical field strength, normal electrical conduction is restored and complete penetration by the magnetic flux density takes place. This critical field strength (H_c) depends on the temperature ($T < T_c$) and material. It further turns out that a critical current density (J_c) exists above which the superconducting state is destroyed.

The magnetic response of superconducting materials may be conveniently used as a diagnostic tool to to classify them into type I and type II superconducting materials. Type II materials are perfectly diamagnetic at low field strength and ejection of the magnetic field is total. On increasing the strength of the applied magnetic field, the transition to the normal conducting state is not abrupt as in type I materials, but takes

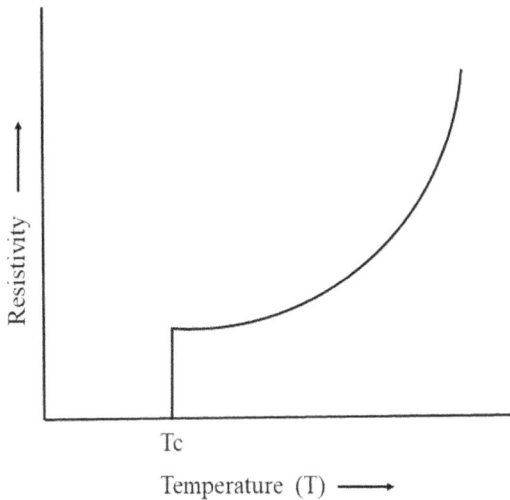

FIGURE 8.12 Typical temperature dependence of resistivity of normally superconducting materials.

place gradually between a lower ($H_<^c$) and an upper critical field strength ($H_>^c$). The magnetic flux density starts penetrating into the body of the sample at $H_<^c$ and the penetration is complete at $H_>^c$. The type II superconducting materials are better suited to practical applications on two counts: (i) they have much higher critical temperatures; and (ii) the critical magnetic field strength they can withstand are higher compared to type I superconducting materials. Several alloys of niobium reveal type II superconductivity. Among the most frequently used superconducting materials of the type II category were alloys of niobium and zirconium, niobium and titanium and the niobium-tin compound (Nb_3Sn).

The hunt for high T_c superconductors received a huge boost when Bsdnorz and Muller discovered in 1986 that some ceramic materials, which would have normally been expected to be electrical insulators, are actually superconductors with spectacularly high critical temperatures (see Section 8.8.2). The discovery was honoured with the award of Nobel Prize to the discoverers in 1987. These materials are recognized as the most important class of high T_c superconductors ($T_c > 90$ K). We will take up the phenomenon of high T_c superconductivity of these materials for analysis and comprehension later in this chapter after getting ourselves acquainted with the structural features of ceramic materials of different kinds. It is important as structure plays a key role in shaping high T_c superconducting ceramics.

8.8.1 Ceramics and Their Structures

Ceramics are compounds formed by the combination of metallic and non-metallic elements so that the interatomic bonds are either purely ionic or ionic with partial covalent character. Thus they have large band gaps. The traditional ceramic materials were made from clay as the primary material and include porcelain, glasses, china, etc. It is during the past six decades or so that some advancement has been made towards understanding the basic nature of the ceramics and their structural principles and how they shape the emergence of their unique properties. Since ceramics involve at least two kinds of atoms, they are quite expectedly more complex from the structural point of view, than pure metals, the interatomic bonding type ranging from purely ionic to covalent with variable degrees of ionic character. What actually happens is decided by the electronegativity difference between the interacting atoms, their relative sizes, etc. A ceramic material has a stable crystal structure if all the anions surrounding a cation are in contact with the cation, which in turn is possible provided the cation-anion radius ratio has a critical or minimum value. Depending on the radius ratio, the coordination number and nearest neighbor geometries vary. Below we list some of the most common structural types of ceramic materials.[16]

Rock-salt structure: Ceramic materials with equal proportion of cations and anions belong to the *AX*-type. Most common *AX*-type crystal structure is of the rock-salt-(or sodium chloride) type, which can be viewed as a combination of two interpenetrating FCC lattices, one made up of the cations and the other

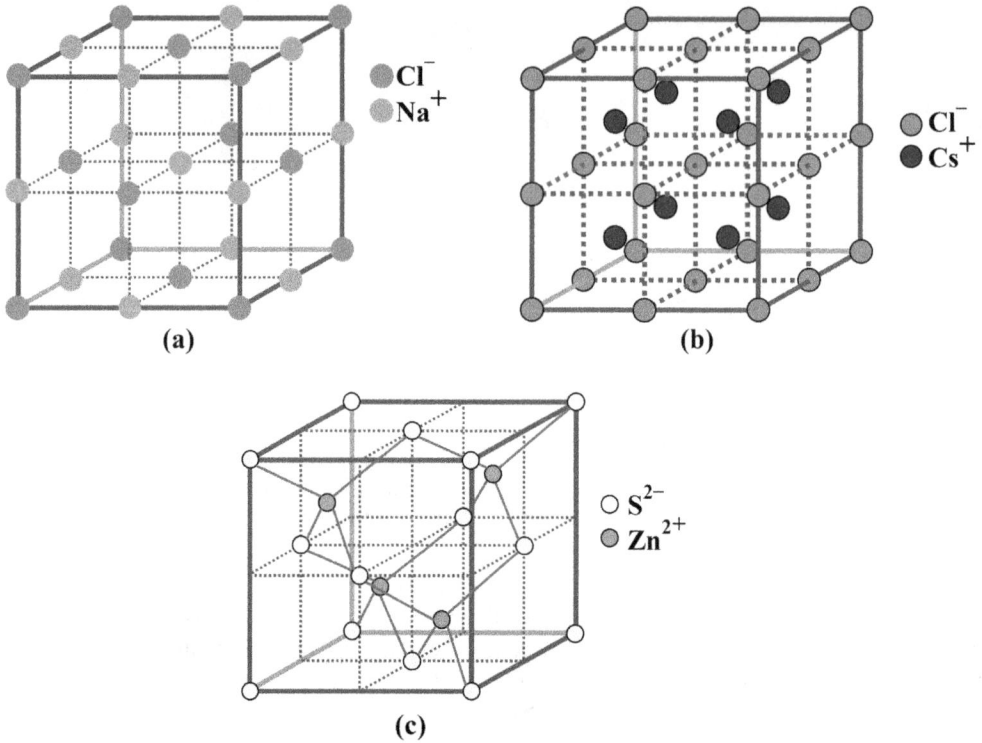

FIGURE 8.13 (a) Rock-salt-(NaCl) type of crystal structure. (b) CsCl structure. (c) Zinc blende structure.

composed of the anions (Figure 8.13). This form of structure appears in ceramic materials like FeO, MgO, MnS, LiF, etc.

Cesium chloride structure:
Figure 8.13 displays the CsCl structure. Notice that both the cations and anions have the same coordination number equal to 8, the anions being located at each of the corners of a cube, with the cation being at the cube center. Exchange of anions with cations and vice-versa will produce the same structure.

Zinc blende structure:
It is the third structural possibility for *AX* ceramics (Figure 8.13). ZnS, ZnTe, SiC, etc. are important ceramics belonging to this structural type. In Figure 8.13 all the corner and face positions of the cubic cell are occupied by S atoms while the Zn atoms take up interior tetrahedral positions. If Zn and S positions are interchanged, an equivalent structure results. ZnS, SiC are ceramics with zinc blende-type of crystal structure.

$A_m X_n$-type ceramics:
An example of this common crystal structure is found in fluorite (CaF_2, $m = 1$, $n = 2$). The Ca^{+2} ions are located at the centers of the cubes with F^- ions positioned at the corner. The chemical formula and charge neutrality suggest that only half of the cube centers are taken up by the cations with one unit cell consisting of eight such cubes. Ceramics with fluorite structure include ThO_2, UO_2, PuO_2, etc.

$A_m B_n X_p$ ceramics and crystal structures:
A ceramic compound may have more than one type of cation. If the ceramic has two cations *A* and *B*, *X* with $m = n = 1$ and $p = 3$, we get ABX_3-type of ceramic materials. An important example is provided by

the well-known ferroelectric material called barium titanate ($BaTiO_3$), which we have already discussed (section 8.7). The material has a perovskite crystal structure. Other ceramics bearing perovskite structure are strontium zirconate ($SrZrO_3$), strontium stanate ($SrSnO_3$), etc.

If $m = 1$, $n = 2$ and $p = 4$, we get AB_2X_4-type of ceramics. With $X = O$, we have magnesium aluminate ($MgAl_2O_4$) or spinel and ferrous aluminate ($FeAl_2O_4$) bearing the spinel structure in which Fe^{+2} or Mg^{+2} have the coordination number of 6 while Al^{+3} have the coordination number 4 with O^{-2} having coordination number of 4. The spinel structure is typically found for magnesium aluminate (spinel). In spinels, O^{-2} ions form an FCC lattice with Mg^{+2} ions filling up the tetrahedral sites and Al^{+3} residing in octahedral positions. Magnetic ceramics – the ferrites have crystal structure that is a variation of the spinel structure.

Silicate ceramics:

Instead of classifying these materials in terms of unit cells it is more convenient to classify them in terms of various arrangements of SiO_4^{4-} tetrahedron in which each atom of silicon is bonded to four oxygen atoms which occupy the corners of the tetrahedron with the silicon atom, taking up the tetrahedron center. Various silicate structures then arise depending on how the SiO_4^{4-} units are combined to form one-, two-, and three-dimensional arrangements. Si-O bonds are regarded as predominantly covalent, so silicate ceramics are not usually considered as ionic ceramic materials.

The various complex silicate minerals are formed by the sharing of one, two or three corner oxygen atoms of the SiO_4^{4-} tetrahedra. They can be represented by formula SiO_4^{4-}, $Si_2O_7^{6-}$, $Si_3O_9^{6-}$, $Si_6O_{18}^{12-}$, $\cdots$, $(SiO_3)_n^{2n-}$, etc. The structurally most simple silicates involve isolated tetrahedra (Figure 8.14) like in forsterite (Mg_2SiO_4), which has equivalent of two Mg^{+2} ions associated with each SiO_4 tetrahedron in a way that ensures every Mg^{+2} ion to have six oxygen atoms as the nearest neighbors. The $Si_2O_7^{6-}$ ion is formed when two SiO_4 tetrahedra share a common oxygen atom (Figure 8.14). This is illustrated by akermanite ($Ca_2MgSi_2O_7$) – a mineral having the equivalent of two Ca^{+2} and one Mg^{+2} ion bonded to each $Si_2O_7^{6-}$ unit thereby maintaining charge neutrality.

Layered silicates:

With $(Si_2O_5)^{2-}$ as the repeating unit, a two-dimensional sheet or a layered structure can be generated by the sharing of three oxygen ions (O^{2-}) in each of the tetrahedra as shown in Figure 8.15. The net negative charge is supposed to be carried by the nonbonded oxygen atoms dangling out of the plane of the paper. This negative charge is neutralized by the excess of cations on the second planar sheet structure, with the cations linking to the unbonded oxygen atom from the Si_2O_5 sheet. This is how the layered or the sheet silicate materials are formed as found in clays and other minerals. As an example, we may take the common clay mineral called kaolinite [$Al_2(Si_2O_5)(OH)_4$], in which the silica (tetrahedral) layer represented by ($Si_2 O_5)^{2-}$ units is charge neutralized by an adjacent $Al_2(OH)_4^{2+}$ layer. Silicates are among the most complex of all the inorganic materials, although their structural principles are well understood.[1]

Silica and silica glasses:

Silica (SiO_2) is structurally a three-dimensional network generated by the sharing of every corner oxygen atom in each tetrahedron by the adjacent tetrahedra. The resulting structure is electroneutral. More importantly, it ensures that every atom is in a stable electronic configuration. If these tetrahedra are arranged in a regular, ordered fashion, a crystalline structure emerges, the primary crystalline polymorphic forms

(a)

(b)

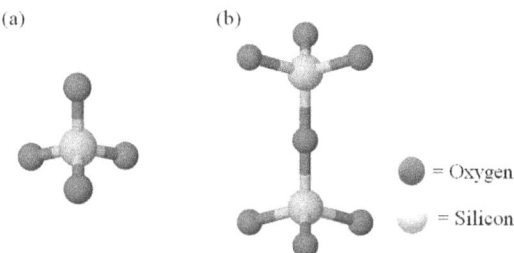

= Oxygen

= Silicon

FIGURE 8.14 (a) SiO_4^{4-}, (b) $Si_2O_7^{6-}$ structures.

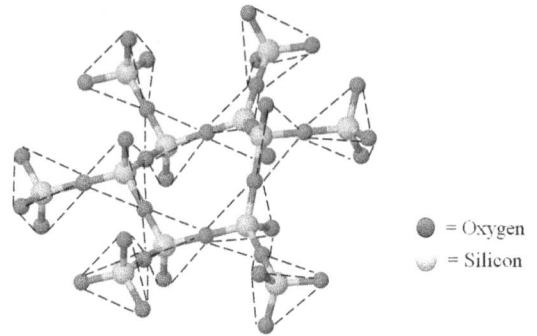

FIGURE 8.15   Two-dimensional silicate sheet structure with a repeat unit $(Si_2O_5)^{2-}$.

of silica being quartz, cristobalite and tridymite. These are rather open but complex structures. The Si-O bonds are very strong so that these crystalline polymorphs are low density solids of relatively high melting points.

Silica can also exist in the form of a non-crystalline solid or glass wherein a liquid like atomic randomness prevails. Such a material is known as fused or vitreous silica.[1] As observed with crystalline silica, the basic unit here too, is the SiO_4^{4-} tetrahedron. Beyond this basic structural unit, disorder exists. Polyhedral networks are also formed by boric oxide (B_2O_3) and GeO_2 and others. Like SiO_2 they are called network builders. By adding oxides like CaO, Na_2O to molten SiO_2, one gets the commonly used silicate glasses for containers, window-pane, etc. CaO or Na_2O do not form polyhedral networks like SiO_2 or Ba_2O_3; but their cations (Ca^{2+}, Na^+) act as modifiers of the SiO_4^{4-} networks by getting incorporated within. Al_2O_3 or TiO_2, on the other hand, can substitute for silicon, stabilize the network and also become a part of it. They are called intermediates. The addition of network modifiers and intermediates lowers the melting point and viscosity of the glassy material, which makes it easier to work with them.

The ceramics are all expected to be of electrically insulating type from a purely electronic band structural point of view. In electronic band structure theory, as we have seen already electrons are assumed to move independently in the average field provided by the nuclei and all other electrons. The instantaneous Coulomb interactions between electrons are disregarded in the band description. In this sense, the band structure is an approximate description of the reality. It is not surprising that the band theory of solids work quite well in many cases. It is not surprising either that it fails even spectacularly in some cases. One such situation arises in the band theory-based classification of solids as metals, semiconductors and insulators. Thus, a solid with a band gap $\Delta E < 4$ eV is predicted to be a semiconductor while $\Delta E > 4$ eV is taken as the signal that the material in question will electrically respond as an insulator. Zero band gap signals metallic conduction. It turns out that the prediction or expectation based on band theory results may not always come true. It is found that some transition metal oxides, which are expected to respond like metals based on band structure theory actually behave like an insulator. Nickel oxide, for example, illustrates this somewhat perplexing behavior. It was pointed out by Mott and Peierls[17] that Coulomb repulsion between $3d$ electrons on nickel atoms of neighbouring unit cells was responsible for the unexpected behavior of NiO. The electron-repulsion causes localization of the electrons, preventing electrical conduction. Such a material is called a Mott insulator. Mott and Peierls visualized the conduction process as transfer of an electron from one $Ni^{2+} O^{2-}$ unit to a neighboring $Ni^{2+} O^{2-}$ as follows : $(Ni^{2+} O^{2-})^2 \rightarrow Ni^{3+} O^{2-} + Ni^{1+}$ O^{2-}. The repulsion between two $3d$ electrons on neighboring Ni atoms is much smaller than that between two electrons on the same nickel atom as the lattice spacing 'a' $>> a_B$ where a_B is the Bohr radius of a nickel atom. That means the electron hopping (delocalization) between neighboring metal atoms will not take place. The electrons will remain localized on its own site and an energy gap (band gap) will be produced. This process has become known as Mott transition – a 'metal to insulator transition' mediated by electron repulsion. This 'transition' can be formally captured in the Hubbard model.[18] The Hubbard Hamiltonian (H_h) for a 1D chain of atoms each containing one electron in an S-state reads

$$H_h = (-)t \sum_{k,s} \left(\hat{C}_{k,s}^+ \hat{C}_{k+1,s} + H \cdot C \right) + U \sum_k \hat{n}_k^+ \hat{n}_k^- \qquad (8.100)$$

where $\hat{C}_{k,s}^{+}$ is the creation operator for an electron of spin s in an orbital on the kth site of the chain while $\hat{C}_{k+1,s}$ is the destruction operator for an electron with spin s in an orbital on the site $k+1$ adjacent to the site k. 't' is the transfer or hopping integral the magnitude of which determines the extent of delocalization of an electron onto the neighboring sites. 'U' is the interelectronic repulsion integral. $\hat{n}_{k,s} = \hat{C}_{k,s}^{+}\hat{C}_{k,s}$ is the spin density operator for spin s (up or down) on the site k. If we set $U = 0$, H_h becomes the tight-binding Hamiltonian. Then there is only hopping and a half filled S band is produced. The 'chain' becomes metallic. If U/t is increased, say by reducing the interatomic spacing 'a' (lattice spacing), eventually a stage is reached where 'hopping' becomes negligible and a 'metal-to-insulator' transition takes place, explaining the formation of a Mott insulator. This shows that interelectronic repulsions can dramatically change the predictions of a band theoretical calculation on the electrical behavior of a material like nickel oxide.

8.8.2 High T_c Superconducting Materials

Among the pure metal BCS superconductors niobium has the highest superconducting transition temperature ($T_c = 9$ K). By late 1970, several alloys and intermetallic compounds of niobium were discovered to have T_c much higher than the T_cs of pure metal superconductors. Thus, NbTi, Nb_3Sn, Nb_3Ge were shown to have higher T_cs (≈ 20 K). Bednorz and Muller discovered in 1986 that the oxide ceramic systems, $La_{2-x}Ba_xCuO_4$ (lanthanum barium cuprate) becomes a superconductor at a much higher transition temperature.[19] For $x = 1$, one get the cuprate $LaBaCuO_4$, which has its resistivity dropping to zero at 30 K. This cuprate is an example of a high T_c superconductor although the T_c is not high enough to enable maintenance of the superconducting state by cooling with liquid nitrogen (77 K). However, the discovery of a ceramic cuprate superconductor was a breakthrough that spawned vigorous research activity focused on discovering new materials with higher superconducting transition temperatures.

Schilling *et al.*[20] discovered in 1993 that the cuprate of mercury, barium and calcium has, at normal atmospheric pressure, the superconducting transition temperature $T_c = 134$ K, the highest T_c known at that time. The crystal structure is tetragonal with three CuO_2 planes in unit cell and the chemical formula is $HgBa_2Ca_2Cu_2O_8$. Two other superconducting cuprates, namely the bismuth based ones and the thallium based cuprates have been extensively investigated. High T_c superconductivity in the Bi-Ca-Sr-Cu-O systems was investigated by Hazen *et al.* (1988)[21], which led to the discovery of three superconducting phases with $T_c = 20$, 85 and 110 K, respectively. Sheng and Hermann (1988)[22] discovered several thalium based cuprates having T_cs above liquid nitrogen temperature. The highest T_c (125 K) was found in the bismuth based system with the formula $Bi_2Ba_2Ca_2Cu_3O_{18}$.

The first superconducting cuprate with T_c above liquid nitrogen temperature to be discovered was the cuprate of barium and yttrium having the composition $YBa_2Cu_3O_{7-x}$. The maximum T_c (92K) was observed for $x \approx 0.15$ while the superconductivity disappeared at $x = 0.6$ along with a structural transformation from orthorhombic to tetragonal. The monopoly of cuprates in the arena of high T_c superconductor was finally broken with the discovery of certain families of iron-based high T_c superconductors. Notable in these families are the $LnFeAsO_{1-x}$ (Ln = lanthanide) systems with T_c values upto a maximum of 56 K.[23] A fluoride variant (O replaced by F) has also been found with similar high T_c-features.

There is a second iron-based family $(BaK)Fe_2As_2$ and similar materials with T_c values upto a maximum of 38 K, which have drawn some attention away from the cuprates.[24] Similar iron arsenides with K replaced by Cs and Ba by Sr were studied by Sasmal *et al.*[26] They recorded T_c upto 37 K in this series of iron arsenides. The highest T_c in the family of iron-based superconductors has been found in thin films of FeSe ($T_c \approx 100$ K).[27]

Several other high T_c superconducting materials not related to the cuprates or the iron-based systems have been discovered. Magnesium diboride, for example, has a $T_c \approx 39K$ although it has been more generally regarded as a normal or the BCS, type superconductor in which two separate bands are present at the Fermi-level, which accounts for the rather unexpectedly high T_c for a BCS-type superconductor. A fulleride compounds Cs_3C_{60} has been found to have $T_C = 38K$.[28] Fullerides with other alkali metals too have been found to superconductor although at lower temperatures.

Recently, it has been found that many metallic hydrides become superconducting at temperatures (T_c) that are quite high for a BCS or a type I superconductor. With their discovery, attention has once again been focused on the search for traditional superconductors with high T_cs. Drozdov *et al.* (2019)[25] on the basis of theory guided experiment discovered a superconducting lanthanum hydride (LaH_{10}) with $T_c = 250$ K under high pressure of 170 GPa. In fact, DFT calculations together with the BCS and Migdal-Eliashberg theories of traditional superconductivity first predicted the possible existence of a new family of superconducting hydrides with clathrate like structures in which the cage is formed by hydrogen atoms with the metal atom at the center. For YH_{10} and LaH_{10}, the onset of superconductivity was predicted to occur at T_cs of 240 and 320 K, respectively under megabar pressure. These are exotic materials, not likely to enter the advanced technology space in the near future.

8.8.3 Understanding High T_c Superconductivity

Traditional superconducting materials with low or high transition temperatures (like the metal hydrides or fullerides, etc.) can be satisfactorily understood on the basis of the BCS theory and its strong coupling generalization. The development of understanding has reached a stage where it has become possible to exploit DFT-based electronic structure calculations together with the BCS and Migdal-Eliashberg theories to predict the T_c of a traditional superconducting candidate material. The situation is completely different when it comes to understanding why do the cuprate superconducting materials behave the way the do? What is the origin or mechanism of superconductivity in these ceramic materials? Before we briefly address the theoretical issues that we are confronted with while trying to understand the high T_c superconductivity in cuprates, let us have a look at the structural features of the cuprates or similar ceramics. Structural studies have confirmed that these materials have complex perovskite structures (see Figure 8.16). They are usually classified as distorted multilayered oxygen deficient perovskites. One of the distinguishing structural features of cuprates or other oxide ceramics is the presence of alternating multiple-layers of CuO_2-planes with superconductivity taking place between them. The higher the number of CuO_2 planes the higher is the superconducting transition temperature (T_c). The electrical conductivity in these materials, be it the normal conductivity or the superconductivity, is highly anisotropic with much higher conduction in directions parallel to the CuO_2-planes and much lower conduction in directions perpendicular to the CuO_2-planes. It also appears that electric current is carried by holes at the oxygen centers of the CuO_2-planes. In general, the T_c depends on chemical composition, especially cation substitution and oxygen content. Structurally,

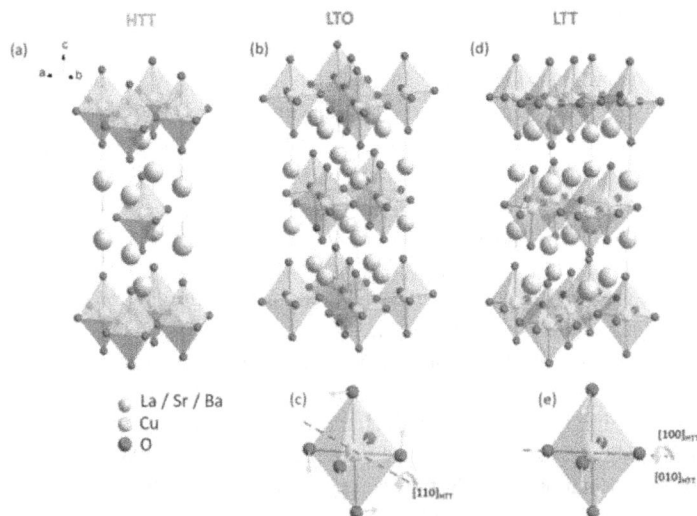

FIGURE 8.16   Complex perovskite structures of cuprate superconductors.

these materials are recognized as stripes. The theoretical understanding of the ceramic superconductors (high T_c superconductivity) is still evolving and complete clarity is yet to emerge.

P. W. Anderson (1987)[29] suggested that a connection exists between the resonating valence bond and the superconducting states of lanthanum cuprate (La_2CuO_4) – a ceramic superconductor. From the observed special features of the cuprate superconductors Anderson argued that a common mechanism must be operative. The cuprate ceramics are Mott insulators at ordinary temperatures and the superconductivity tends to occur or actually occurs in the vicinity of the metal to insulator transition point where the metallic cuprate turns over into an odd electron insulator with special magnetic properties. The insulating phase was sought to be identified as the resonating valence bond state representing a '**magnetic spin liquid**'. According to Anderson, this magnetic phase of the Mott insulator appears to be favored by low spin, low dimensionality and magnetic frustration. As to the formation of charge carriers in the superconducting state, Anderson suggested that the pre-existing magnetic singlet pairs of the insulating phase turn into the superconducting 'charge pairs' when the insulator has been strongly doped. The mechanism of superconductivity appears to be dominantly mediated by electronic and magnetic interactions. A weak phonon-coupling may also be operative, perhaps playing a rather minor role unlike in the metallic superconductors.

The view that superconductivity in ceramic cuprates is linked to their Mott insulator nature, and the superconducting transition taking place in the neighborhood of metal-insulator transition has gained traction from a recent work by Phillips *et al.* (2020)[30] who have, for the first time exactly solved a toy model of a doped Mott insulator. The model exhibits both Mottness and superconductivity and has been known as the Hatsugai-Kohmoto model.[31] It represents a prototypical Mott insulator. The exact calculations done by the authors reveal that the system turns into a non-Fermi liquid metal with a superconducting instability brought in either by doping or by reducing the interaction strength. It turns out that in the presence of a weak pairing interaction, the instability produces a thermal transition to a new superconducting phase that is quite distinct from the BCS superconducting state of a metal as is clearly demonstrated by the gap to transition temperature ratio which exceeds the universal BCS limit. The authors further note that elementary excitations of this prototypical cuprate superconductors are not the familiar Bogoliubov quasi-particles, but are composite excitations of a different kind (superposition of doublons and holons, to be precise). It appears that the superconducting ground state of a doped Mott insulator inherits the non-Fermi liquid character of the normal state.[32] Since the authors[30] have been able to write down an exact wavefunction describing the superconducting state of the cuprate superconductors modeled by the HK Hamiltonian it raises hope that the extension to the problem of superconductivity of the real cuprates will be possible soon. Before closing the discussion on high T_c superconductors we would like to present the results of a recent experiment (2019), which could have significant ramifications on the still raging debate on the mechanism of superconductivity of cuprate superconductors, in general.

In a large number of members of the family of cuprate high T_c superconductors charge density waves (CDWs) have been observed. CDWs are real space oscillation of the crystal electron density accompanied by lattice distortion. In their simplest form, CDWs can be represented as BCS condensate of electron-hole pairs that breaks the translational symmetry. Do the CDWs help or hinder the emergence of high T_c superconductivity? In principle, any atomic scale disorder should suppress both high T_c superconductivity and CDWs. M. Leroux *et al.*[33] has recently contradicted this assertion on the basis of their experimental findings. These authors used proton irradiation to induce disorder in crystals of $La_{2-x}Ba_xCuO_4$ ($x = 1/8$) and observed a 50% rise in the T_c accompanied by suppression of the CDW. This observation is in sharp contrast with what is expected of a d-wave superconductor for which both magnetic and non-magnetic defects would be expected to bring down T_c. The authors have claimed that irradiation induced disorder may even be used as a significant tuning parameter for many families of superconductors with coexisting density waves (CDWs or the more elusive pair density wave (PDW)). This possibility has been practically realized and demonstrated in dichalcogenide superconductors and an alloy, $Lu_5Ir_4Si_{10}$. The cuprates are layered materials. An intriguing feature of layered high T_c superconductors has been the quite dramatic variations in the maximum superconducting transition temperature (T_c) as one moves from one family to another. Thus we have T_c ranging from 26 K in single CuO_2-layered oxychlorides to 135 K in 3-CuO_2 layered mercury compounds. The range of variation becomes even larger if materials under high pressure are also included. Constructing a coherent explanation for the observed variations has been a challenge to the theoreticians because the low energy electronic structure of these materials has been known to be due to a single band of carriers, which evolve from the hybridization of $3d_{x^2-y^2}$ orbital of copper with the $2p_x$ and $2p_y$ orbitals on four coplanar coordinated oxygen atoms. This feature is shared by all the cuprates. The variation of the crystal structure between the CuO_2 layers is significantly different for different members of the cuprate family. Could this difference be responsible for the dramatic effect seen in the T_c? The question assumes significance because this variation (between the layers) has only an indirect effect on the electronic structure in the CuO_2 plane.

Let us note that the stereochemistry around the copper atom can be planar or pyramidal or octahedral. The non-planar oxygen atoms that coordinate to the Cu atom are known as apical oxygen atoms. Pavarini *et al.*[34] in a land mark paper used band theoretical results to argue that the observed T_c- structure correlation originates from the indirect effect of the apical oxygen atoms and other atoms not lying on the CuO_2-plane. The authors noted that the interlayer coupling arises primarily from the hybridization of the $4s$ orbital of copper atom with, for example, the $2p_z$ orbitals on the apical oxygen atoms. The hybridized orbital thus formed affects the various hopping integrals operating in the CuO_2-plane and ultimately influences electronic structure in the CuO_2-plane. To obtain further insight Slezak *et al.*[35] had undertaken STM studies of the cuprate $Bi_2Sr_2CaCu_2O_8$ under zero magnetic field. The cuprate exhibits a mismatch between BiO and CuO_2-planes. This mismatch gives rise to a rather complex supermodulation – a modulation of the superstructure that is incommensurate with the underlying CuO_2 lattice. Structural studies had indicated earlier that this supermodulation significantly affects the apical oxygen atoms but does not have noticeable influence on the atoms on the CuO_2-planes. The question therefore arises if the supermodulation and superconductivity of the cuprate are intertwined. Slezak *et al.*[35] measured the gap energy (Δ) at zero magnetic field as a function of the position, Fourier transformed it and found that the gap energy displayed a modulation with the same period that the supermodulation had. The significance of the observation is that the Cooper-pairs in this cuprate exhibits a pair-density wave, a feature that was earlier predicted by Chen *et al.*[36]

Very recently, Victor Lakhano[37] constructed a new translation invariant bipolaron theory of high T_c superconductivity, which avoids the high mass of small radius bipolaron that affects the T_c in ordinary bipolaron theory. The final word about the origin or mechanism of high T_c superconductivity has not yet been spoken. The developments during the last few years have been quite exciting and the curtain may be expected to be rung down the topic in the near future.

8.9 Metal Alloys

Initially, material science was solely devoted to the study of metals, alloys, intermetallic compounds, etc. Even a sample of very high purity metal will invariably contain impurity to the extent of $10^{22} - 10^{23}$ atoms per cubic meter. The occurence of these impurity atoms is natural. However, if the impurity atoms have been introduced deliberately with a bid to create specific properties in the resulting material, the metal-impurity combination is recognized as an alloy. In general, alloying is used to improve the mechanical properties and corrosion resistance of the metal.[1] Sterling silver is an example. It is an alloy of 92.5% silver and 7.5% copper. Pure silver under the normal (ambient) conditions is highly corrosion resistant but rather soft. Alloying with copper makes the resulting material (sterling silver) mechanically much stronger without any appreciable degradation of its corrosion resistance. The addition of impurity atoms of one kind to the host metal results in the formation of a solid solution and/or emergence of a new phase. What actually happens is dictated by the nature of the impurity atoms, their concentrations and the temperature. The solvent or host in a solid solution is the metal in large excess, while solutes are the entities present in minor concentrations. Just as it happens in a liquid solution, the solid solutions are compositionally homogeneous with the solute atoms randomly and uniformly dispersed within the body of the solid. The solid solutions are of two types namely substitutional and interstitial. In the substitutional solid solution solute or the impurity atoms substitute for the host atom in specific locations. The extent to which a solute will dissolve in the solvent (host) is determined by a number of factors. Some of them are listed below:

TABLE 8.5

Atomic Valence and Structural Parameters of Nickel and Copper

	Nickel	**Copper**
Atomic radius	1.28 Å	1.25 Å
Electronegativity	1.9 eV	1.8 eV
Valence	+2	+1(+2)
Crystal type	FCC	FCC

1. An appreciable amount of a solute may be found in the solid solution if the size differential (measured by the difference between the atomic radii) is less than about ± 15%.

2. The solute will appreciably dissolve in the host solid, only if the crystal structures of the metals involved are the same.

3. If the electronegativity difference between the solute and solvent atoms is high, intermetallic compound formation is preferred to the formation of substitutional solid solution.

4. If all other factors are held constant, a metal will be more prone to dissolve in another metal (solvent) of higher valency. The rules stated above are empirical thumb rules based on which a quick decision can be made while choosing a solute or a solvent components of a solid solution. As an example of substitutional solid solution, an alloy of nickel and copper serves to illustrate the thumb rules 1–4 of solid solubility (Table 8.5). The two metals are soluble in each other in all proportions (Table 8.5).

Interstitial solid solutions are formed when the impurity atoms (solute) occupy the voids among the host atoms. If the host metal has high packing factor, the sizes of these voids are small. Hence, the solute atoms must be relatively much smaller in size and the maximum solute concentration can only be small. Carbon forms an interstitial solid solution with iron. The carbon content is restricted to a maximum of 2% only.[1]

8.9.1 Ferrous Alloys

Alloys with iron as the host element are called ferrous alloys. These alloys display a rather wide range of mechanical and physical properties as functions of the nature and concentration of the solute or the impurity atoms and are very important materials for engineering and construction work. They are however, rather prone to corrosion. We take a look into some of them.

8.9.2 Steels

They are primarily iron-carbon alloys that may also contain significant amounts of other alloying elements. Plain carbon steels have only a residual amount of other impurities together with a little manganese. Alloy steels contain intentionally added alloying elements in significant amounts in addition to carbon. The low carbon steels contain less than 0.25% by wt of carbon and are not responsive to heat treatment. Microstructures in low C steel consist of ferrite and pearlite primarily and are soft and weak, but tough and ductile at the same time, are machinable, wettable and have low production cost. They are the steels produced in the largest quantities.

Medium carbon steels have carbon content lying between 0.25 and 0.50 wt%. Microstructures of tempered martensite appear following tempering. Addition of chromium, nickel and molybdenum improves the response to heat treatment. The heat treated medium carbon steels are stronger than low carbon steels but have lower ductility. High carbon steels have carbon content ranging from 0.60 to 1.4 wt% and are the hardest and strongest, yet least ductile of the carbon steels. High carbon alloy steels contain chromium, vanadium, tungsten and molybdenum. The alloying elements combine with carbon to produce very hard and wear resistant carbides like $Cr_{23}C_6$, V_4C_3 and WC.

Stainless steels:
They contain chromium as the dominant alloying element at least to the extent of 11 wt%. They are resistant to corrosion and the corrosion-resistance can be enhanced further by the addition of nickel and molybdenum. These steels are divided into three classes depending the predominant phase of the microstructure, namely martensitic, ferritic or austenitic. Both martensitic and ferritic stainless steels are magnetic, while the austenitic stainless steels are non-magnetic but highly corrosion resistant.

Cast irons:
These belong to a class of ferrous alloys with carbon content above 2.14 wt%. The actual carbon content in cast irons used in industry ranges between 3.0-4.5 wt%. Other alloying elements are present in much lower amounts. Alloys within the mentioned composition range pass over into the liquid phase at temperatures roughly between 1150 and 1300 °C, much lower than steels. Cast irons are thus easily melted and suitable for casting. Fe-C system has a very rich phase diagram, which we do not discuss here.

8.9.3 Non-Ferrous Alloys

Ferrous alloys including steel are consumed in huge quantities all over the world as they offer a rather wide range of mechanical properties. They can be fabricated with ease and are also cost-effective. Their high density, low electrical conductivity and great susceptibility to be wasted by corrosion put them in a somewhat disadvantageous position. That is why technology has looked for alternatives in the form of non-ferrous alloys be it cast alloys or wrought alloys.

Copper and its alloys:
The most common copper alloys are the brasses, which are alloys of copper and zinc, Zn being a substitutional impurity. In the copper-zinc phase diagram (not shown here) the α phase remains stable upto 35 wt% of zinc and has an FCC crystal structure. The α-brasses are comparatively soft, ductile and can be cold-worked. On increasing the Zn content further α and β' phases coexist at room temperature. The β' phase has a BCC crystal structure. It is both stronger and harder than the α phase. So, $\alpha + \beta'$ brasses are usually hot-worked. Another family of copper alloy of importance are the bronzes, which are alloys of copper, tin and elements like aluminium, silicon and nickel. Somewhat harder than the brasses, they are endowed with a good degree of corrosion resistance. Beryllium coppers are alloys of copper with beryllium (beryllium content ranges between 1-2.5 wt%), which are heat treatable and have high tensile strength, excellent electrical properties and corrosion resistance. They are used extensively in the aviation industry.

Aluminium and its alloys:
Aluminium alloys are marked by their low density, high electrical and thermal conductivity and a good degree of corrosion resistance in the ambient atmosphere. The main alloying elements are Cu, Mg, Si, Zn and Mn. Alloying increases the mechanical strength, but reduces corrosion resistance. The alloys of aluminium with the other low density metals like Mg and Ti are receiving increasing attention in the automobile and aviation industry for reducing fuel consumption. Low melting temperature of Al, however, is a drawback.

Magnesium and its alloys:
The density of magnesium (1.7 gm/cm^3) is the lowest among all structural metals. So, Mg alloys are used wherever light-weight structure is a necessity. Like aluminium, Mg has a low melting temperature. Magnesium alloys are relatively unstable and susceptible to corrosion.

Titanium and its alloys:
Ti and its alloys are comparatively new materials with extraordinary combination of properties. Ti has a low density (4.5 gm/cm^3), high melting point (1668°C), high elastic modulus. Ti-alloys are extraordinarily strong. However, it becomes chemically active compared to other materials at higher temperatures. At

ordinary temperatures, they however, offer extreme resistance to corrosion. The alloying elements are Al, tin, molybdenum and vanadium.[1]

8.9.4 Special Materials

The refractory metals: The refractory group of metals include niobium, molybdenum, tungsten (W) and tantalum (Ta), all having melting points in the range between 2468°C (Nb) - 3410°C (W). The interatomic bond strengths in these metals are very high, being responsible for the high melting temperatures, large elastic moduli and high hardness and strength. Tantalum is virtually immune to chemical attack in any environment below 150°C. The applications of the refractory metals are extremely varied.

The noble metals:
In the noble group, there are eight metals, all precious, soft, ductile and resistant to oxidation. They are silver, gold, platinum, palladium, rhodium, ruthenium, iridium and osmium. Ag, Au, and Pt are very commonly used in jewelry – the first two are strengthened by alloying with copper when a solid solution is formed. Electrical connection in some integrated circuits are of gold.

The super alloys:
Technological demand has continuously driven the search for materials with an extraordinary combination of properties. For example, materials for components of turbines of aircraft must be able to resist oxidation in severely oxidizing environments, tolerate high temperatures for a reasonable period of time, and retain mechanical integrity. For turbine applications, the density is also a key factor as low density materials can reduce the centrifugal stress. The super alloys are designed with cobalt, nickel or iron as the host and the refractory metals, chromium and titanium as alloying elements. We will address the design issues for materials in Chapter 13.

REFERENCES

1. W. D. Callister, Chapters 1–3, In *Materials Science and Engineering – An Introduction*, John Wiley, New York (2003).
2. M de Podesta, Chapters 6–7, In *Understanding the Properties of Matter*, Second Edition, Taylor and Francis, London and New York (2002).
3. A. S. Davydov, *Theory of Solids*, Nauka, Moscow (1976).
4. J. Ziman, *Principles of the Theory of Solids*, Second Edition, Cambridge University Press (1972).
5. A. A. Sokolov, L. M. Ternov and V. Ch. Zhukovskii, *Quantum Mechanics*, (Translated from Russian by Ram. S. Wadhwa), MIR Publishers, Moscow (1984).
6. H. K. Onnes, 'The resistance of pure mercury at helium temperatures', *Comm. Phys. Lab. Univ. Leiden*, 120 (1911).
7. J. F. Annett, *Superconductivity, Superfluids and Condensates*, Oxford University Press (2004).
8. P. W. Anderson, *Basic Notions of Condensed Matter Physics*, Benjamin/Cummings, Melno Park (1985).
9. A. S. Alexandrov, *Theory of Superconductivity from Weak to Strong Coupling*, Institute of Physics Publishing, Bristol (2003).
10. R. Doll and M. Nabuer, 'Experimental proof of magnetic flux quantization in a superconducting ring', *Phys. Rev. Lett.*, 7, 51 (1961).
11. B. S. Deaver, Jr. and W. M. Fairbank, 'Experimental evidence for quantized flux in superconducting cylinders', *Phys. Rev. Lett.*, 7, 43 (1961).
12. B. D. Josephson, 'Possible new effects in superconductive tunneling', *Phys. Rev. Lett.*, 1, 251–253 (1962).
13. A. Anslem, *Introduction to Semiconductor Theory*, (Translated from Russian by M. M. Samokhvalov), MIR Publishers, Moscow (1981).
14. O. Bubonva, U. Khanzia and H. Wang, 'Semimetallic polymers', *Nat. Mater.*, 13, 190–194 (2014).
15. E. J. Reed *et al.*, 'A transient semimetallic layer in detonating nitromethane', *Nat. Phys.*, 4, 72–76 (2007).

16. W. Hauth, 'Crystal Chemistry in Ceramics: 8 Papers', Iowa Engineering Experiment Station, Iowa State College (1951).

17. N. F. Mott and R. Peierls, 'Discussion of the paper by de Boer and Verwey', *Proc. Phys. Soc.*, 49, 72 (1937).

18. J. Hubbard, 'Electron correlation in narrow energy bands', *Proc. Roy. Soc.* (London) 276, 238–257 (1963).

19. J. G. Bednorz and K. A. Muller; 'Possible high T_c superconductivity in the Ba-La-Cu-O system', *Z. Physik B* 64, 189–193 (1986).

20. A. Schilling, M. Cantoni and J. D. Guo, 'Superconductivity above 130 K in the Hg-Ba-Ca-Cu-O system', *Nature*, 363, 56–58 (1993).

21. R. Hazen *et al.*, 'Superconductivity in the high T_c Bi-Ca-Sr-Cu-O system: Phase identification', *Phys. Rev. Lett.*, 60, 1174–1177 (1988).

22. Z. Z. Sheng and A. M. Hermann, 'Superconductivity in the rare-earth-free Tl-Ba-Cu-O system above liquid-nitrogen temperature', *Nature*, 332, 55–58 (1988).

23. Zhi-An Ren *et al.*, 'Superconductivity and phase diagram in iron-based arsenic-oxides ReFeAsO$_{1-\delta}$ (Re = rare-earth metal) without Fluorine doping', *EPL* (Europhysics Letters), 83, 17002 (2008).

24. (a) G. Wu *et al.*, 'Superconductivity at 56 K in samarium-doped SrFeAsF', *J. Phys.: Condensed Matter*, 21, 142203 (2009). (b) M. Rotter, M. Tegel and D. Johrendt, 'Superconductivity at 38 K in the iron arsenide $(Ba1-_xK_x)Fe_2As_2$', *Phys. Rev. Lett.*, 01, 107006 (2008).

25. A. P. Drozdov *et al.*, 'Superconductivity at 250 K in lanthanum hydride under high pressures', *Nature*, 569, 528–531 (2019).

26. K. Sasmal, B. Lv, B. Lorenz, A. M. Guloy, F. Chen, Y. Y. Xue and C. W. Chu, 'Superconducting Fe based compounds $(A_{1-x}Sr_x)Fe_2As_2$ with A = K and Cs with transition temperatures up to 37 K.', *Phys. Rev. Lett.*, 101, 107007 (2008).

27. J. F. Ge, Z. L. Liu, C. Liu, C. L. Gao, D. Qian, Q. K. Xue, Y. Liu and J. F. Jia, 'Superconductivity above 100 K in single-layer FeSe films on doped SrTiO$_3$', *Nat. Mater.*, 14, 285–289 (2015).

28. A. Y. Ganin, Y. Takabayashi, Y. Z. Khimyak, S. Margadonna, A. Tamai, M. J. Rosseinsky and K. Prassides, 'Bulk superconductivity at 38 K in a molecular system', *Nat. Mater.*, 7, 367–371 (2008).

29. P. W. Anderson, G. Baskaran, Z. Zou and T. Hsu, 'Resonating-valence-bond theory of phase transitions and superconductivity in La$_2$CuO$_4$-based compounds', *Phys. Rev. Lett.*, 58, 2790 (1987).

30. P. W Phillips *et al.*, 'Exact theory for superconductivity in a doped Mott insulator', *Nat. Phys.*, 16, 1175–1180 (2020).

31. Y. Hatsugai and M. Kohmoto, 'Exactly solvable model of correlated lattice electrons in any dimensions', *J. Phys. Soc. Japan*, 61, 2056–2069 (1992).

32. G. Baskaran, 'An exactly solvable fermion model: Spinons, holons and a non-fermi liquid phase', *Modern Phys. Lett. B*, 5, 643–649 (1991).

33. M. Leroux *et al.*, 'Disorder raises the critical temperature of a cuprate superconductor', *Proc. Natl. Acad. Sci.* (USA), 116, 10691–10697 (2019).

34. E. Pavarini, I. Dasgupta, T. Saha-Dasgupta, O. Jepsen and O. K. Andersen, 'Band-structure trend in hole-doped cuprates and correlation with T_{cmax}', *Phys. Rev. Lett.*, 87, 047003 (2001).

35. J. A. Slezak, J. Lee, M. Wang, K. McElroy, K. Fujita, B. M. Andersen, P. J. Hirschfeld, H. Eisaki, S. Uchida and J. C. Davis, 'Imaging the impact on cuprate superconductivity of varying the interatomic distances within individual crystal unit cells', *Proc. Natl. Acad. Sci.*, (USA), 105, 3203–3208 (2008).

36. H. D. Chen, J. P. Hu, S. Capponi, E. Arrigoni and S. C. Zhang, 'Antiferromagnetism and hole pair checkerboard in the vortex state of high T_c superconductors', *Phys. Rev. Lett.*, 89, 137004 (2002).

37. V. Lakhano, 'Superconducting properties of a non-ideal polaron gas', *Physica C: Superconductivity and its Applications*, 561, 1–8 (2019).

9

The Advent of Smart Materials

9.1 Introduction

The advancement of technology requires better and smarter materials for implementing the advanced technologies. Similarly, availability of advanced or smart materials fuels development of technologies that can fully exploit the relatively more advanced or smart features or properties of the materials. Thus, there is a synergy between advanced technologies and advanced materials. Smart materials belong to a special class of advanced materials that can apparently sense changes in the environment around it and respond by altering one or more of its properties. Such materials are indeed very useful for the development of sensors of various kinds.

The smart materials fall under a number of categories depending on the type of stimulus that brings in a specific type of response. Thus, there are electrochromic materials that change color on the application of an external electric field. Such materials are a boon to smart display technology. There are photochromic materials that change color in response to light. Such materials are important in modern building technology. Photochromic glasses are nowadays extensively used for controlling the level of illumination inside a building and keeping it at a constant level. There are pH sensing materials, pressure sensing materials, temperature sensing materials, etc. Smarter variants of these materials are the so called shape memory materials that appear to remember the shape before deformation and return to the original shape on heating. In what follows we will briefly discuss some of these 'smart materials' and try to understand the origin of their smart functionalities.

9.2 Electrochromic (EC) Materials

These materials undergo a change of color (or absorption or reflection) in response to an applied electric field in a persistent but reversible manner due to electrochemically induced redox processes. The phenomenon has become known as electrochromism and materials displaying electrochromism are called electrochromic materials. As the system switches between the redox states, electronic absorption bands in different visible regions of the spectrum is generated, producing a change in color. The color change commonly takes place from a transparent or a 'bleached' state to a colored state, or between two colored states. It is possible that a particular material has a number of colored redox states that are electrochemically accessible. In that case, the electrochromic (EC) material may display different colors giving rise to the phenomenon called polyelectrochromism. Such materials are termed polyelectrochromic materials.[1−3]

Not too many electrochromic materials exist. The oxides of transition metals, like WO_3, MoO_3, InO_2, NiO and V_2O_5 are the most well-known and commonly used electrochromic materials. Structurally, crystals of all the well-known EC oxide materials are made up of MO_6 octahedra (M = transition metal) in all kinds of corner or edge sharing arrangements. When these materials are fabricated in the form of their films, both columnar and cluster-type microstructures involving MO_6 units are seen to have been formed. The octahedral coordination environment of the metal ions is essential for electrochromism at least in WO_3 and is consistent with the electronic band structure (schematic) shown in Figure 9.1.[1−2] At least for defect perovskite and rutile lattices, the band structure can explain the presence or absence of cathodic or anodic electrochromism. WO_3 for example, can be viewed as being made up of W^{6+} ions and O^{2-} ions. The bonding is not completely ionic, there being significant covalent character. The band structure

DOI: 10.1201/9781003244882-9

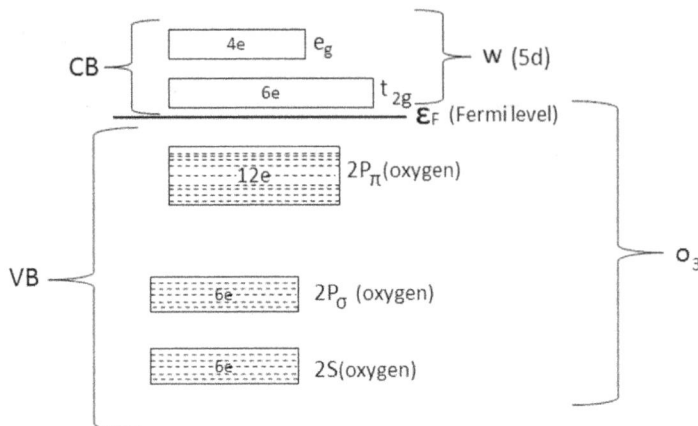

FIGURE 9.1   Schematic band structure of transition metal oxide (WO_3). $5d$ (W) band is split in octahedral field.

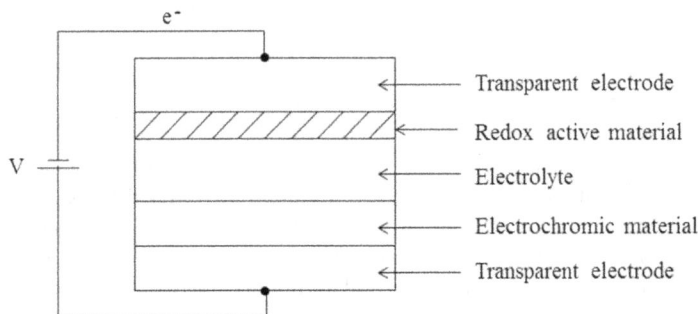

FIGURE 9.2   A simple electrochromic set up.

reveals that the filled oxygen $2p$ bands (part of the valence band) are well separated from the metallic $5d$ band (conduction band), which is split into t_{2g} and e_g bands in the octahedral environment. In WO_3 the O_{2p} bands are filled, while the metallic $5d$ band is empty. The Fermi level is positioned in the middle of a wide gap between the filled O_{2p} and the empty W $5d$ bands. The gap is large enough to make pure WO_3 transparent. Application of a low voltage to transport ions and electrons between the two electrochromic layers (Figure 9.2) makes both colored while transport in the opposite directions makes both appear bleached. The coloration appears to be due to the insertion of charge balancing electrons in the empty $5d$ band. By combining both anodic and cathodic oxide films, a neutral visual appearence can also be created. The most widely used EC-oxides are based on tungsten (W) and nickel (Ni), which display cathodic and anodic electrochromism, respectively. Schematically, the electrochemical reactions can be represented as follows.[1-3]

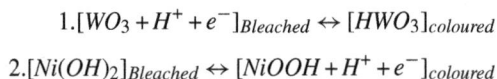

$$1. [WO_3 + H^+ + e^-]_{Bleached} \leftrightarrow [HWO_3]_{coloured}$$

$$2. [Ni(OH)_2]_{Bleached} \leftrightarrow [NiOOH + H^+ + e^-]_{coloured}$$

As noted already, pure WO_3 has a filled O_{2p} band (VB) and an empty d band (CB) with a sizable band gap. Injecting small ions (H^+, for example) and charge balancing electrons appears to lead to partial filling of the d band along with optical absorption, the detailed mechanism of which in the electrochromic oxides in general, being not completely understood as yet. For WO_3 however, charge transfer and polaron absorption appear to be responsible for all the important spectral features.

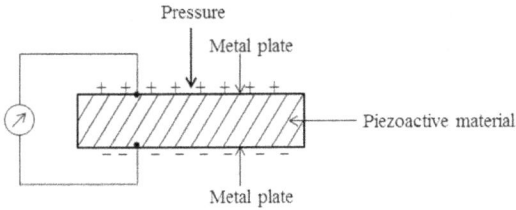

FIGURE 9.3   Piezoelectric response of a solid.

9.3 Piezoelectric Materials

Certain crystalline materials, ceramics and biological materials are known to produce electric charges in response to an applied mechanical stress (Figure 9.3). Conversely, an electric field applied across such materials is known to produce structural deformation or strain. These two phenomena are collectively called **piezoelectric effect** (**PE**) and the materials displaying the effects are called **piezoelectric materials**. It appears that the linear electrochemical coupling between mechanical and electrical states of the solid is at the root of PE. The symmetry of the crystal class is an important determinant for PE to be realized. Out of the 32 classes of crystals, 21 are devoid of inversion symmetry. Twenty (among the 21) of the non-centrosymmetric crystal classes are known to display piezoelectric effects. Only ten of these crystal classes belong to the polar crystal class, which are endowed with spontaneous electrical polarization without any applied mechanical stress. The unit cells of such crystals are associated with non-zero dipole moment and are known as ferroelectric materials, which have been already discussed in Chapter 8. The other members of the non-polar piezoelectric class have electric polarization $P \neq 0$ only when a mechanical stress is applied.[4–5]

Gallium orthophosphate ($GaPO_4$), a quartz like crystal, lithium niobate ($LiNbO_3$), lithium tantalate ($LiTaO_3$), quartz, Rochelle salt, lead titanate ($PbTiO_3$) are some of the crystalline solids that are known to generate piezoelectricity. Among the ceramics lead zirconate titanate $[Pb(Zr_xTi_{1-x})]O_3$, $0 \leq x \leq 1$, also called PZT, potassium niobate, sodium tungstate ($NaWO_3$), single crystalline ZnO in its wurtzite structure are both piezo and pyroelectric. Lead free piezoelectric materials have been in demand to avoid toxic contamination of the environment with lead in PZT.[6–7] As a substitute for PZT potassium sodium niobate $(KNa)NbO_3$ has attracted a great deal of attention as it is lead-free. It has piezoelectric effect close to what is displayed by PZT in addition to possessing mechanical properties that are superior to those of PZT. It is proving itself to be an ideal material for high power resonance applications such as piezoelectric transformer. Other such materials are barium titanate, bismuth ferrite, etc.

Group II-V and group II-VI semiconductors with wurtzite structures have been found to display strong piezoelectric response. We have already mentioned ZnO. The others in this category are gallium and indium nitrides (GaN, InN) along with aluminium nitride. A number of polar semiconductors have been found to possess strongly non-linear piezoelectric response.

So far, piezoelectric materials in use have been mainly inorganic in nature. However, polymers like poly vinylidene fluoride (PVDF) exhibit piezoelectric response much stronger than that of quartz or quartz like systems. PVF in addition has good thermal stability enabling its use in piezo-shock wave sensors, and pressure sensors. High electromechanical coupling strength, its semicrystalline nature along with mechanical flexibility have made PVDF the most popular choice for fabrication of sensors that are robust as well as sensitive to pressure variations. In addition, PVDF is non-toxic, inert, tolerant to water absorption, bio-compatible. These properties make PVDF ideal for use in hydrophones and implantable biomedical devices.[8]

Hybrid organic-inorganic piezomaterials have attracted contemporary attention. Jacques Curie and Pierre Curie discovered piezoelectric effect while working with quartz. The converse effect, however was theoretically predicted by Lippmann 30 years later on the basis of thermodynamic argument. The requirement for the piezomaterials to have non-centrosymmetric structure has restricted the design options for these materials. We would like to mention here that for personal low power electronic devices and body-worn sensors, piezoelectric form of harvesting of energy from mechanical movement appears to be one of the most dependable alternative to the rechargeable batteries.

9.4 Shape Memory Materials (SMM)

Certain solid materials have the unique ability to return to their original (pre-deformation) shape from a rather significant and apparently plastic deformation under the action of an external stimulus (e.g. mechanical force) after the stimulus is withdrawn and another stimulus (e.g. thermal/magnetic, etc.) is applied. It appears, as if, the material retains the information about its original shape in 'memory' and recovers it when the appropriate stimulus is applied. This phenomenon is known as shape memory effect (SME) and the materials displaying SME are called shape memory materials (SMMs).[9]

SME was discovered in 1932 by Arne Olander in an alloy of gold and cadmium (Au Cd). The first shape memory alloy (SMA) was not much noticed. The interest in shape memory alloys rapidly grew after the discovery of 'nitinol', an alloy of Ni and Ti (Ni Ti) and a wide variety of SMAs and SMMs were developed. With the development of new materials the use of SMAs became widespread in biomedical devices (stents), acrospace engineering (deployable structures, morphing wings), robotics (flat robots). Only three alloy systems, e.g. NiTi, CuAlNi/CuZnAl, and a number of iron-based systems have been commercially developed so far.

The ability of shape memory alloys to undergo large reversible deformation under loading and thermal cycles and their ability to generate large thermomechanical force is quite spectacular but somewhat counterintuitive. It turns out that the SME has its roots in the ability of the relevant alloys to undergo reversible changes in their crystal structures depending on the temperature and the state of stress.[10] These structural changes have been interpreted as reversible martensitic transformations between a crystallographically more ordered parent phase of the alloy called the austenite phase (A-phase) and a crystallographically less ordered product phase called the martensitic phase or the M-phase. The M-phase turns out to be the thermodynamically favored phase at lower temperatures $(T_<)$ and typically has monoclinic or orthorhombic crystalline structure (the so called B'_{19} or B_{19}). These M-phase structures do not have enough slip systems. So, when stressed, they tend to deform by detwining, twining being already present in the M-phase. To understand, let us note that the austenite phase is favoured at higher temperatures $(T_>)$ and is typically cubic (B_2) in crystallographic disposition. The two types of structures do not only have different symmetrics, but also have different lattice sizes. So, when the 'austenite' is cooled into 'martensite', the size and symmetry mismatch between two phases leads to the development of large strain-energy in the martensitic phase. The system tries to relieve the strain by 'twining'. The shape memory alloys are manufactured typically at higher temperatures, so that they are already in a highly twined state.[10] On applying force to the system in the martensite phase, which is already existing in a twined state, there is an onset of 'detuning', which provides a facile path to the deformed state. The atoms at every site retain their positions, no new bonds are formed or old bonds are broken in the deformation process. On raising the temperature post deformation, the austenite – the thermodynamically favored phase at higher temperatures, appears, the atoms on every site quickly reordering into the B_2 (austenite) structure. The reorganization takes place very fast as no diffusion of atoms is involved, recovering the pre-deformation macroscopic geometry.

The transition from the M-phase to the A-phase depends only on the stress and temperature. It may be noted that the martensite or austenite structures also exist in steel, and the transition from the A-phase to M-phase does take place in steel. But, steel does not have any shape memory functionality. This is because $A \rightarrow M$ transition that takes place in steel is not reversible. It is instructive to examine the heating and cooling induced transition diagram in a typical shape memory alloy. The transition diagram is obtained by plotting the fraction $f_M(T)$ of the material in the martensite state as a function of temperature. A typical diagram is shown in Figure 9.4. It can be seen clearly from the figure that there is hysteresis. It is an indication that there is a loss of some amount of mechanical energy in the process.

There are certain alloys of nickel, manganese and gallium (Ni-Mn-Ga), which display magnetic shape memory effect and are called magnetic shape memory alloys (MSMA). They can produce deformation

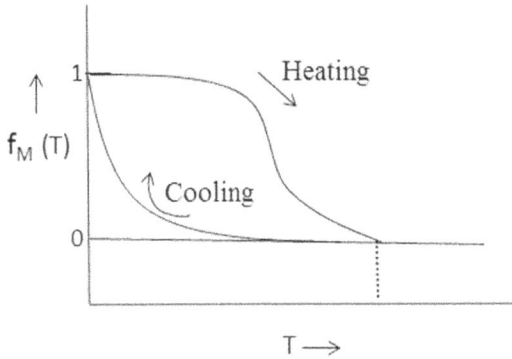

FIGURE 9.4 Hysteresis in shape memory alloys.

and forces under moderate magnetic field.[11] In single crystalline nickel-manganese-gallium shape memory alloy the extent of deformation produced can be as high as 6%. The rather large magnetic field induced strain produced and the quick response time make these materials very attractive for designing and fabricating novel actuators. The magnetic field is not the only stimulus to which the MSMA responds. These materials also respond to the thermal stimulus. The shape memory effect in these materials occurs in the martensitic phase with tetragonal elementary cells. On heating beyond the M↔A transition temperature, the alloy gets into the austenite phase with cubic elementary cells and the shape memory effect is lost. The MSM alloys can alter their magnetic properties depending on the deformation.[12] This feature has been exploited to make speed sensors and mechanical energy harvesters out of MSMAs. The mechanism of producing the large strain can be traced to the magnetically induced reorientation of the magnetization vector of individual cells along the applied field. Somewhat counterintuitively, this is achieved by geometric reorientation of the cells themselves rather than by rotation of the magnetization vector within a cell.

Since ferromagnetic shape memory alloys (FSMAs) respond to applied magnetic field, the magnetic field could be used as an instrument of motion generation in systems involving the FSMAs. However, in practical applications, the thermoresponsive SMAs are still the favored option. Thus, thin films of nitinol (Ni Ti alloys) made by sputtering deposition techniques are extensively used for motion generation in micro or nano electromechanical systems. Some shape memory alloys display temperature memory effect (TME) in addition to the shape memory effect. Thus, the highest temperature (within the traditional range) reached in the previous heating cycle can be accurately recorded and subsequently revealed in the next heating cycle.

Of late, shape memory polymers (SMPs)[13,14] have been recognized as an important class of smart materials that possess the ability to recover the original shape after deformation, on the application of an appropriate stimulus. The SMPs are fundamentally different from the SMAs if the origin of the shape memory effects is considered for classification. The SME in SMPs arise because of a glass or melting transition of the polymer from a hard to a soft phase while in SMAs, the shape memory effect has its roots in the martensitic-austenitic phase transformation. The advantage of SMPs lies in the fact that they are capable of undergoing large elastic deformation, to the extent of 200% in a majority of the cases, and still recover the original shape when the appropriate stimulus is applied. Their low cost and density, broad range of operating temperatures, good biocompatibility make them attractive candidates for many applications, notably, biomedical applications. Polyurethanes, block copolymers of polyethylene terephthalate and polyethylene oxides and many others are currently under assessment.[15]

9.5 Photochromic Materials (PM)

Photochromic materials (PM) display photochromism, which loosely means that the material changes color when illuminated with light (photons) and regains the original color when the source of illumination is cut off. The change is reversible and occurs quickly. These materials appear transparent initially and become colored when illuminated. They can be optical glasses or plastic materials to which a photochromic substance has been added. In a molecular material, the photochromism may be traced to the molecules that constitute the basic molecular units of the material. Referred to the molecular units, photochromism represents a reversible transformation of a molecule induced by light, between the two or more isomeric forms. The isomers have distinctly different absorption spectra and produce distinct visual response (colors). The reversible photoinduced transformation is usually accompanied by changes in some observable physical properties like absorption and fluorescence emission spectrum, electronic conductivity, refractive index, magnetic properties, dielectric constant, structural disposition, etc. Such properties are called phototunable properties.

The possibility of phototuning properties of such materials has evoked keen interest in photochromic molecules or compounds or materials, in the context of designing photochemical molecular switches, which can act as switching units in various optoelectronic devices and functional materials. The switches are activated or stimulated with light causing switching between two specific states. Photochromic molecules and materials have significant possibilities in constructing molecular logic gates, data recording and storage. Photoswitchable materials also found important use in sensing, self-assembly and photocontrolling biological systems. Three important classes of molecular chromophores have been used as building blocks of organic photochromic materials:

1. Dithienylethene.
2. Spiropyran.
3. Azobenzene.

Several possibilities have been and are being explored with molecular materials built from these molecules.

1. Photochromic switches as multi addressable materials.
2. Photochromic switches capable of sensing ions.
3. Self assembly based on photochromic materials. A recent review by Zheng *et al.*[16] provides an excellent resource for assessment of possibilities that exist in this hot field of research, the existing challenges that are to be overcome to develop the photoswitchable materials into practical applications.

The most extensive use of PCM or photochromism has been in glass.[18] A significant amount of photochromism effect can be produced in glass by adding special ingredients to the melt and proper melting and heat treatment. However, it has not yet been possible to prepare glasses that are fatigue-free (color bleach cycles are unlimited). Strong photochromic effects can be produced by using silver halides or copper halides as light absorbing sources. Silver halide-based technology of PC glasses is the most mature one. Among the most commonly encountered applications of PCMs are those in ophthalmic (lenses for glasses), actinometry, cosmetics, optical memory units, photo-optical switches, filters, etc. Their use in ophthalmic glasses exploits the large change in opacity of PC glass depending on the UV-content of radiation to which the glasses are exposed. Thus lenses darken when exposed to bright sunlight and becomes transparent or colorless in low light. The photochromism here is switched by the UV part of light.

MoO_3 is the most extensively studied material for photochromic applications. When exposed to UV radiation ($h\nu$ > band gap) electrons and holes are formed. Photogenerated electrons (e^-) are injected into the conduction band (CB) of MoO_3 while the holes (h^+) react with adsorbed water producing hydrogen ions (H^+), which diffuses into the CB of MoO_3 producing what is known as hydrogen molybdenum bronze accompanied by a change of color – from transparent to blue. The reactions leading to the observed photochromic response in MoO_3 may be summarized as follows:

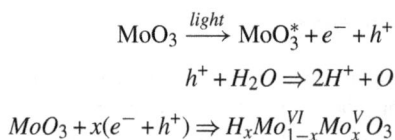

$$MoO_3 \xrightarrow{light} MoO_3^* + e^- + h^+$$

$$h^+ + H_2O \Rightarrow 2H^+ + O$$

$$MoO_3 + x(e^- + h^+) \Rightarrow H_x Mo_{1-x}^{VI} Mo_x^V O_3$$

MoO_3-based PCM have found good use in smart window glass, erasable optical storage devices, display units, etc.

WO_3 is very useful material for photochromic applications. When amorphous WO_3 is exposed to UV radiation, the color changes from transparent to intense blue or brown. Much of the mechanistic details is similar to what has been described for MoO_3, the end product being $H_xW^{VI}_{1-x}W^V_xO_3$ (hydrogen tungsten bronze). The photosensitivity of WO_3 is limited by its band gap of 3.17 eV, which corresponds to the near UV range of the spectrum. The sensitivity can be tuned and extended into the visible range by doping a small amount of K^+ ions in WO_3 crystals and Li^+ ions in MoO_3 through cathodic polarization. The introduction of potassium ions leads to the formation of K_xWO_3 accompanied by small structural deformation. That results in the appearence of intermediate metastable trap states in the band gap region. These states are accessible to the visible light.

TiO_2 is yet another photochromic oxide material with a band gap of 3.2 eV. It has long term stability against light, is resistant to chemical corrosion and is biologically inert. TiO_2 films doped with vanadium (V-doped TiO_2) when exposed to UV-radiation undergoes a color change from beige to brownish violet. It takes place only in vanadium-doped TiO_2. In the absence of V-doping no color change takes place. It turns out that V-doping creates energy levels below the conduction band edge of TiO_2, which act as trap centers for photogenerated electrons when V-doped TiO_2 films or nanoparticles are irradiated with UV light. As before, UV-light generates electron-hole pairs. The excited electrons get trapped in V^{5+} ion trap centres whereby V^{5+} is reduced to V^{4+} state with accompanying change of color.

9.6 Quantum Tunneling Composites (QTC)

These composites can be regarded as a kind of smart pressure sensors. The composite behaves as a perfect insulator in the normal state. When compressed it turns into a more or less normal electrical conductor (electrical resistance drops dramatically) allowing electric current to pass through it. In the normal state, the conducting elements of the composite are too far apart and no electric current can pass through the material. When pressed, the conducting elements come closer and soon a point is reached where they are near enough for electrons to tunnel through what was an insulator in the normal, uncompressed state establishing a tunneling current. The effect is more dramatic than could be expected only on the basis of classical theory or classical effect as classical electrical resistance varies linearly with the length (l) of the conducting element. On the other hand, quantum tunneling (and therefore tunneling current) is capable of growing exponentially as the tunneling distance or barrier width decreases due to compresssion (chapter 2). In fact, the current can register even a thousand-fold increase or more.

QTC, first produced in 1996 by D. Lussay is a composite material made from micron-sized conducting particles of metals like nickel (Ni) as fillers combined with a non-conducting elastomeric binder, like silicone rubber, for example. The composite behaves as a complex flexible polymer endowed with a unique electrical response comapred to other materials that are electrically conducting. A typical QTC sheet has one layer of QTC material, one layer of a conducting material (Al, for example) and a third layer of plastic insulator (Figure 9.5). The set up can be used as an extremely sensitive pressure sensor, specially in robotics, for recognizing and responding to even feather-light touch. High production cost has so far restricted its general purpose use. It is indeed a quantum material as it uses quantum tunneling for its functionality.

9.7 Quantum Materials (QMs)

QMs are a broad class of materials that are endowed with exotic properties having roots in purely quantum mechanical features like tunneling, strong electron correlation, coherence, entanglement, etc. We have already referred to some of these materials in Chapter 8, e.g. superconducting materials, cuprates (high temperature superconductors), giant magneto resistance (GMR)/colossal magneto resistance (CMR) materials in Chapter 10, and quantum tunneling composites in the present chapter (Section 9.6). Some of these

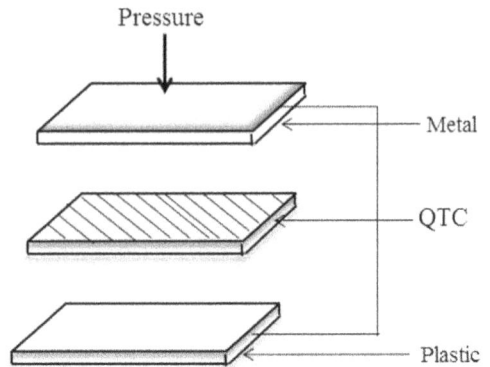

FIGURE 9.5   Schematic diagram of a pressure sensor based on quantum tunneling composites.

materials are already in use in critical areas. For example, MRI machines are a common, critical diagnostic equipment used in every hospital. Today's computers in every home make use of hard disc drives that employ sensors based on the so called GMR materials.

In a way all materials are quantum materials as the microscopic particles of matter in any form interact and form states that are governed by the laws of quantum mechanics. However, in most cases, normal materials and their functionalities can be understood and explained by invoking elementary quantum mechanics. Quantum materials are very different in this respect. We need to deploy deep and exotic features of quantum mechanics to make sense of the true quantum materials and their properties.[21] In general, the states of quantum systems exist as coherent superpositions of eigenstates that the system (material) supports. In most materials, these superpositions of many states do not last for a long enough time or extend over a large area or volume of space to be able to affect properties of the system. Any interaction with the environment destroys the coherence and the superposition collapses. Quantum materials behave in an entirely different manner. There, the coherent superpositions can survive or persist for milliseconds and over an extended region of space even in the presence of interaction with the environment. The Bose-Einstein condensates for example are purely quantum materials where macroscopic coherences exist for a long enough time and macroscopic tunneling can be observed.

A more exotic form of quantum material reveals itself in the form of topological insulators.[22] In topological materials, the electronic wavefunctions are wrapped up forming something akin to complicated knots. These wavefunctions can be deformed in many ways keeping certain topological features invariant. A sample of topological insulator can be cut in many ways, keeping the topology of the electronic bands invariant. The consequence is that the surface will conduct electricity irrespective of the way the sample is sliced. The direction of movement of the electrons on the surface depends on its spin direction (up or down), which enables one to design new devices exploiting the spin-dependent surface conduction of electricity. Technologies based on the topological properties of quantum materials can be expected to be more tolerant to defects and disorder as the technologies will almost certainly be developed around the invariant features of the topology.

We have seen in Chapter 8 that semi metals like metals do not have band gaps, but have fewer charge carriers (e^-, h^+) per unit volume compared to metals. Dirac semimetals have an hour glass like disposition of electronic states because of symmetry, imparting high mobility to the electrons. Some of the Dirac semimetals are graphene, cadmium arsenide (Cd_3As_2), zirconium silicon sulphide, etc. If the semimetal is doped with magnetic ions or if an external magnetic field is applied, the Dirac points can split into mirror image pairs forming what has been called a Weyl semimetal in which spins of the carrier electrons depend on their direction of travel (because of spin-linear momentum or magnetoelectric coupling). Tantalum arsenide is a well-known and thoroughly studied example of a Weyl semimetal, which reveals a bulk signature of Weyl Fermions in the form of negative magnetoresistance of the material. Such materials can be very useful in switching, in memory formation as well as sensing.[23]

Spin is a unique quantum attribute. We will see in Chapter 10 how magnetism is inextricably linked to spin angular momentum. The spin angular momentum can assume only two orientations – spin up ($\uparrow$)

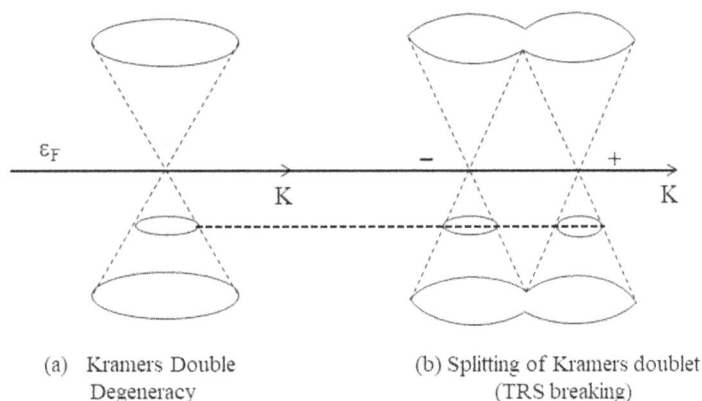

(a) Kramers Double
Degeneracy

(b) Splitting of Kramers doublet
(TRS breaking)

FIGURE 9.6 (a) Dirac semimetal (both time reversal and space inversion symmetry are present). (b) Weyl semimetal (breaking of time reversal symmetry splits Kramers doublet).

and spin down ($\downarrow$) and interact following certain rules. A number of quantum materials like spin liquids, spin glasses, spin ices have been conjured up. They display rather exotic electronic properties because the geometry of structures of such materials does not allow spins to align themselves following the standard rules of alignment causing what is known as 'frustration'. This 'frustration' can be leveraged as a design principle to produce materials in which many electron states with the same energy can exist. Spins are 'entangled' in such states (a local spin orientation fixes the spin orientation of a distant electron). Materials of this type can sustain coherence over an extended region of space and can be exploited in data-storage, in memory systems of computers, specially in the context of quantum computing.

Spin can be used in a different fashion called spin tuning in designing soft magnetic materials. Thus spin crossover is a rarely feasible event in a normal hard solid because change in spin states of the atoms/ions in the solid is usually accompanied by quite significant change of atomic/ionic sizes, which can not be accommodated in a normal 'hard' solid. Materials with organic components are soft and flexible allowing spin crossover to take place without hindrance resulting in changed electrical properties. Such soft magnetic materials can be useful in forming magnetic memory systems, as sensors and in quantum computing.[24] Quantum 2D materials led by graphene are also receiving keen attention in this context (see Chapter 11).

9.8 Organic Superconductors

By conducting materials, we generally refer to solids formed by metals and alloys of metals and non-metals. They display what is known as metallic conductivity and may become superconducting at low enough temperature (see Chapter 8). Organic compounds have been traditionally classified as insulators or at best semiconductors with room temperature conductivity around 10^{-8} mho cm^{-1}. Melby and her group[25] were the first to discover that 7, 7, 8, 8 tetra cyano quinodimethane (TCNQ) has an electrical conductivity of about 100 mho cm^{-1}. The conductivity is admittedly much lower than that of a typical metal like copper, which has an electrical conductivity of $\approx 5 \times 10^8$ mho cm^{-1}. Yet the discovery aroused a lot of interest. Little[26] suggested that linear chains of organic molecules with polarizable side chains around them could lead not only to metallic conductivity, but even to superconductivity. Although superconductivity was not established, interest in organic conductors grew rapidly and TCNQ and TCNQ-like molecules became objects of keen attention. Ferraris *et al.*[27] were able to synthesize tetrathiafulvalene-TCNQ, which has an electrical conductivity of about 100 mho cm^{-1} at room temperature, which increased rapidly to $\approx 10^4$ mho cm^{-1} as the temperature was lowered. The conductivity temperature profile turns out to be remarkably metal-like. HMTSF-TCNQ (hexamethylene tetra selano fulvalenium -TCNQ) was the first organic compound discovered to exhibit metallic conductivity at all temperatures. Electrical properties of HMTSF-TCNQ is quite similar to those of the quasi-one-dimensional inorganic polymer $(SN)_x$, which shows a superconducting transition at low temperature.[28] This similarity inspired researchers to continue their search of organic superconductors. The search continues even today.

TCNQ turns out to be a strong electron acceptor and undergoes facile charge transfer complex formation with electron donors like TTF. The CT-complex thus formed is salt-like and can be grown into crystals. The crystals of the salt-like TTF-TCNQ complex has parallel conducting strands of TTF(+) and TCNQ(-) ions organized along the *b*-axis of the crystal. The highest occupied orbitals (one-electron wavefunctions) are π-orbitals, which are oriented perpendicularly with respect to the plane of the molecule. The intermolecular spacing along each stack in the direction of the *b*-axis is such that the π-MOs are able to overlap in that direction allowing electrons to move freely along a stack. That means, electron conduction can take place along the parallel conducting chains of TTF and TCNQ units in the crystal mainly along the *b*-axis. A much weaker interaction between chains of unlike molecules prevails along the *a*-axis determining the location of the Fermi level, which in turn determines if the highest occupied energy band is partially filled or not. If it is partially filled, the right condition for displaying metallic conductivity prevails. However, the salt has high crystalline anisotropy and the metallic conductivity is predominantly along one direction (*b*-axis), leading to the designation of TTF-TCNQ and likes of them being one-dimensional metals.

The average conductivity of TTF-TCNQ at room temperature is about 500 mho cm^{-1}, which slowly increases at lower temperatures, attaining a maximum at 59 K. Thereafter, conductivity decreases rapidly with breaks at 53 K and 38 K. These breaks are associated with phase transitions mediated by the so called Peierls instability expected of a one-dimensional metal with partially filled band. The instability triggers lattice distortion and opening of a band gap around the Fermi energy, and a metal to insulator transition takes place, the salt becoming a small band gap semiconductor.[28–30]

Tetramethyl tetrathiafulvalene (TMTTF) salts like $(TMTTF)AsF_6/ClO_4/PF_6$ etc. are known as Fabre salts while those of tetramethyl tetraselano fulvalene (TMTSF) are called Bechgaard salts. In the crystalline state they behave as one-dimensional metals. Only one of the salts investigated so far namely, $(TMTTF)_2/ClO_4$ has been found to be superconducting at 1.4 K under ambient pressure. In addition to the Fabre and Bechgaard salts, quasi-two-dimensional $(BEDT-TTF)_2X$ salts with a variety of anions are being investigated for locating superconducting transition temperatures (BEDT $\equiv$ bis-ethylene dithio tetra thiafulvalene also abbreviated as ET). We have not included conducting polymers in our review as it is a specialized topic in itself.

REFERENCES

1. C. G. Granqvist, 'Electrochromic materials: Microstructure, electronic bands, and optical properties', *Appl. Phys A*, 57, 3–12 (1993).
2. G. A. Niklasson and C. G. Granqvist, 'Electrochromics for smart windows: Thin films of tungsten oxide and nickel oxide, and devices based on these', *J. Mater. Chem.*, 17, 127–156 (2007).
3. C. G. Granqvist, *Electrochromic Materials and Devices*, R. J. Mortimer, David R Rosseinsky and Paul M. S. Monk, Eds., John Wiley VCH, Weinheim, Germany (2015).
4. J. Curie and P. Curie, 'Contractions and expansions produced by voltages in hemihedral crystals in inclined faces', *Competes Rendus* (in French) 91, 383–386 (1881).
5. G. Lippmann, 'Principle of conservation of electricity', *Annales de Chimie et. de. Physique*, 24, 145 (1881).
6. B. Jaffe, W. R. Cook and H. Jaffe, *Piezoelectric Ceramics*, Academic, New York (1971).
7. W. Jigang, 'Perovskite Lead free piezoelectric ceramics', J. Appl. Phys., 127, 19 (2020).
8. K. Sappati *et al.* 'Piezoelectric polymers and paper substrate: A review', *Sensors*, 18, 3605 (2018).
9. H. Funakubo, *Shape Memory Alloys*, Gordon and Breach Science Publishers, New York (1987).
10. K. Otsuka and C. M. Wayman, Eds., *Shape Memory Materials*, Cambridge University Press (1999).
11. M. Wuttig, L. Liu, K. Tsuchiya and R. D. James, 'Occurrence of ferromagnetic shape memory alloys', *J. Appl. Phys.*, 87, 4707 (2000).
12. C. Carciunescu *et al.* 'Martensitic transformation in Co_2NiGa ferromagnetic shape memory alloys', *Scripta Mater.*, 47, 285–288 (2002).

13. A. Lendlein and S. Kelch, 'Shape memory polymers', *Angew. Chem. Int. Ed.*, 41, 2034–37 (2002).
14. A. Lendlein *et al.*, 'Light induced shape memory polymers', *Nature*, 434, 879–882 (2005).
15. W. Voit *et al.* 'High strain shape memory polymers', *Adv. Funct. Mater.*, 20, 162–171 (2010).
16. J. Zhang, Q. Zou and H. Tian, 'Photochromic materials: More than meets the eye', *Adv. Mater.*, 25, 378–399 (2013).
17. M. Irie, 'Photochromism: Memories and switches-introduction', *Chem. Rev.*, 100, 1683–1684 (2000).
18. G. P. Smith, 'Photochromic glasses: Properties and applications', *J. Mater. Sci.*, 2, 139–152 (1967).
19. D. Bloor, A. Graham, E. J. Williams, P. J. Laughlin and D. Lussay, 'Metal-polymer composite with nanostructured filler particles and amplified physical properties', *Appl. Phys. Lett.*, 88, 102103 (2006).
20. L. Wang, 'Piezoresistive sensor based on conductive polymer composite with transverse electrodes', *IEEE Transactions on Electron Devices*, 62, 1299–1305 (2015).
21. B. Keimer and J. E. Moore, 'The physics of quantum materials', *Nat. Phys.*, 13, 1045–1055 (2017).
22. J. E. Moore, 'The birth of topological insulators', *Nature*, 464, 194–198 (2010).
23. F. Arnold, *et al.*, 'Chiral Weyl pockets and Fermi surface topology of the Weyl semimetal TaAs', *Phys. Rev. Lett.*, 117, 146401 (2016).
24. J. M. Silveyra, E. Ferrara, D. L. Huber and T. C. Monson, 'Soft magnetic materials for a sustainable electrified world', *Science*, 362, 6413, eaao0195 (2018).
25. L. R. Melby, R. J. Harder, W. R. Hertler and R. E. Benson and W. E. Mochel, 'Substituted quinodimethans. II. Anion-radical derivatives and complexes of 7, 7, 8, 8-tetracyanoquinodimethan', *J. Am. Chem. Soc.*, 84, 374–3384 (1962).
26. A. Little, 'Possibility of synthesizing an organic superconductor', *Phys. Rev. A*, 134, 1416–24 (1964).
27. J. Ferraris, D. O. Cowan, V. Waltka and J. H. Perlstein, 'Electron-transfer in a new highly conducting donor-acceptor complex', *J. Am. Chem. Soc.*, 95, 948–949 (1973).
28. T. O. Poehler, 'Organic-conductors', *APL Technical Digest*, 15, 13–21 (1976).
29. A. N. Bloch, T. F. Carruthers, T. O. Poehler and D. O. Cowan, *Chemistry and Physics of One Dimensional Metals* (pp. 47–85), H. J. Keller, Ed., Plenum Press, New York (1977).
30. A. G. Lebed, Ed., *The Physics of Organic Superconductors and Conductors*, Springer Series in Materials Science, 110 (2008).

10

Magnetic Materials

10.1 Introduction: Magnetic Materials

Materials that have the characteristics of a magnet or can be converted into a magnet (magnetized) or are attracted by a magnet are called magnetic materials. Iron (Fe), cobalt (Co) and nickel (Ni) are the most prominent pure (elemental) magnetic substances. They are not natural magnets, but can be easily processed and turned into magnets. Many alloys of Fe, Co and Ni are also magnetic and are used to fabricate magnets, large or small. The Fe-Co-Ni triad are called ferrromagnetic metals. The only naturally occuring material that acts like a magnet and attracts pieces of iron is ferroso-ferric oxide (Fe_3O_4) – the magnetic oxide of iron. In fact 'lode stone' – the first natural magnet to have been discovered thousands of years ago, is a piece of 'magnetite' – a mineral containing the magnetic oxide of iron. Apart from Fe, Co, Ni, ruthenium (Ru) appears to be the fourth single (pure) element that acts like ferromagnetic material at room temperature. A tetragonal phase of ruthenium has been claimed to have been created by adopting ultra-thin filament growth method by Quaterman *et al.*[1] In fact, the search for newer ferromagnetic materials is ever increasing as magnetic materials find extensive use in the fabrication of hard disk drives, random access memory systems, magnetic field sensing devices and in many other sundry appliances like magnetic doors of refrigerators, television sets, motors, generators, etc. In view of their technological relevance, it is important to understand why ferromagnetism and ferromagnetic materials are relatively so rare and what is at the root of the emergence of ferromagnetism. Ferromagnetism is not the only form of magnetism that exists in nature. There are other types of magnetism and magnetic materials as well. Before we start looking into different forms of magnetism and magnetic materials, let us first get acquainted with several important properties of magnetic materials in terms of which they can be classified into several types from a purely operational point of view.[2] Once the classification has been done we will address the more fundamental issues involving connection of electronic structure with magnetic properties.

10.2 Important Magnetic Vectors

A magnet carries a magnetic moment ($\vec{m}$). If the magnet is small enough, $\vec{m}$ represents what is commonly called the magnetic dipole moment – that is the component of the magnetic moment that originates from an equivalent magnetic dipole. The magnetic dipole moment is measured by the torque ($\vec{\tau}$) that the magnetic entity experiences when placed in a magnetic field $\vec{H}$:

$$\vec{\tau} = \vec{m} \times \vec{H} \tag{10.1}$$

The magnetic moment measures the inherent magnetic strength of magnetic substance and is therefore an important attribute of a magnetic material. In SI units it is expressed in $A.m^2$ (Ampere meter squared) units on equivalently in Joules per Tesla unit. An important magnetic vector is the intensity of magnetization or simply the magnetization ($\vec{M}$), which is defined as the magnetic moment per unit volume:

$$\vec{M} = \vec{m}/V \quad \text{(unit: Ampere/meter or A/m)} \tag{10.2}$$

DOI: 10.1201/9781003244882-10

The magnetic field ($\vec{H}$) is another magnetic vector measured in the same (A/m) unit. Thus if we consider a loop carrying a current $\vec{i}$, the magnetic field produced at the center of the loop is

$$\vec{H} = \vec{i}/2\tau \quad \text{A/m} \tag{10.3}$$

The magnetic moment ($\vec{m}$) associated with the current carrying loop is given by

$$\vec{m} = \vec{i} \times \text{area of the loop} \quad (\text{A} \cdot \text{m}^2) \tag{10.4}$$

and hence $\vec{m}/V = \vec{M}$ has the same unit as $\vec{H}$ (A/m). Similarly, we can introduce another quantity ($\vec{\sigma}$), which is defined as the magnetic moment per unit mass:

$$\vec{\sigma} = \vec{m}/\text{mass} \quad (\text{A} \cdot \text{m}^2/\text{Kg}) \tag{10.5}$$

The ratio of magnetization ($\vec{M}$) and the magnetic field $\vec{H}$ produces a dimensionless quantity k called the volume susceptibility.

$$k = M/H \quad \text{(dimensionless)} \tag{10.6}$$

A similar quantity is mass susceptibility or simply susceptibility (χ) represented by the ratio of σ and H.

$$\chi = \sigma/H \quad (\text{m}^3/\text{kg}) \tag{10.7}$$

The susceptibility is a measure of the magnetizability of a substance (material) and can be used to classify magnetic materials. A third magnetic vector of magnetic substance is the magnetic induction ($\vec{B}$) where $\vec{B}$ is related to the magnetic field and the magnetization ($\vec{M}$) of the material:

$$\vec{B} = \mu_0(\vec{H} + \vec{M}) \tag{10.8}$$

μ_0 being the permeability of free space ($4\pi \times 10^{-7}$ Henry/meter).

10.3 Types of Magnetism and Magnetic Materials

Commonly, we associate magnetism with ferromagnetic (Fe-, Co-, Ni-based) materials. Ferromagnetism is but only one type of magnetism displayed by magnetic materials. In a broader sense, all materials are magnetic although the nature and intensity of the response of materials exposed to a magnetic field vary widely. The most common type of magnetism displayed by materials is called diamagnetism and such materials are called diamagnetic. Diamagnetism is a fundamental property of matter, although it is very weak. Diamagnetic materials are weakly repelled in an external magnetic field. The external magnetic field induces a magnetic field in the diamagnetic material that opposes the external field. Many elements in the periodic table turn out to be diamagnetic in the free state, e.g. hydrogen (molecule), helium, boron, carbon, nitrogen, Ne and all other inert gases, phosphorus, antimony, arsenic, bismuth, silicon, germanium, lead, all the members of the halogen family, sulphur, selenium, telurium, polonium, galium, indium, thalium, metals like copper, silver and gold, zinc, cadmium and mercury and many of their compounds are diamagnetic in character. Experimental data indicate that magnetization (M) in diamagnetics is proportional to the external magnetic field (H) and the proportionality constant – the diamagnetic susceptibility (χ_a) is small and negative. It is a temperature-independent quantity, which does not depend on the external magnetic field also. For example, χ_d, for quartz is -0.62 and that for water is -0.90 in units of $10^{-8}\text{m}^2/\text{Kg}$. The magnetization ($M$) of diamagnetic substances when plotted against the applied field reveals a linear graph ($M = \chi H, \chi < 0$) with negative slope (Figure 10.1a).

Clearly, the magnetization opposes the external field and disappears ($M = 0$) as the field becomes zero. $\chi_d = M/H$ is clearly negative, albeit very small.

Some elements in the periodic table display another form of magnetism called paramagnetism. Paramagnetic susceptibility also (χ_p) is small, but positive being at least ten times larger than the typical diamagnetic susceptibility.[2-4] Paramagnetic materials are weakly attracted in a magnetic field. Oxygen,

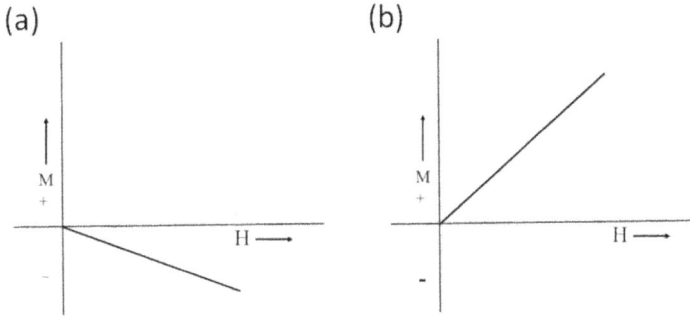

FIGURE 10.1 (a) *M-H* plot for a typical diamagnetic substance. (b) *M-H* plot for a typical paramagnetic substance.

unlike the other group VIB elements is paramagnetic in the free state (molecular ground state). Alkali metals and rare earth elements too, are paramagnetic. Among the alkali earth elements (group IIB), all the elements excepting beryllium are diamagnetic. Among the first transition series elements Sc, Ti, V and Mn are paramagnetic. In the second transition series, all the elements from yttrium Y ($Z = 39$) to Pd ($Z = 46$) are paramagnetic. The trend continues in the third transition series of elements where too all the elements starting with La ($Z = 57$) and the elements from hafnium (Hf, $Z = 72$) to platinum (Pt, $Z = 78$) turn out to be paramagnetic. A typical plot of magnetization (M) against H for paramagnetic substances is shown in Figure 10.1b. The graph is linear with small but positive slope. The paramagnetic susceptibility (χ_p) varies typically in the range of 10^{-15}-10^{-3} m³/kg. The plot also indicates that magnetization disappears as the field (H) is withdrawn. In sharp contrast with diamagnetic substances, the paramagnetic susceptibility (χ_p) turns out to be a temperature-dependent quantity (see later) and tends to saturate at higher field strength. The field and temperature dependence of magnetization (M) of paramagnetic materials is governed by Curie's law according to which

$$\overrightarrow{M} = \frac{C}{T}\overrightarrow{B} \tag{10.9}$$

where 'C' is a material-dependent constant known as Curie's constant. On empirical evidence Curie's law appears to hold strictly at temperatures that are high and for magnetic fields that are weak. An immediate consequence of Eq. 10.9 is the inverse temperature dependence of paramagnetic susceptibility

$$\chi_p = \frac{C}{T} \tag{10.10}$$

A typical χ_p-T plot is displayed in Figure 10.2. In a later section, we will address several issues concerning rationalization of Curie's law from a theoretical point of view.

Ferromagnetic (FM) materials as a class can have non-zero magnetization (M) even when the external magnetic field is zero ($\overrightarrow{H} = 0$). They are said to display spontaneous magnetization. A typical example is the mineral called magnetite (Fe_3O_4). In all such materials a net magnetization exists in a uniformly magnetized tiny volume even at zero field. The magnetization increases rapidly with the increase in the strength of the applied field, saturating, however at much lower field strengths compared to the saturation fields observed in paramagnetic materials (see Figure 10.3). Beyond the field stregth $H = H^*$, no further increase in magnetization takes place even at $H >> H^*$. At a constant magnetic field, the magnetization decreases as the temperature rises and above a critical temperature, called the Curie temperature (T_C) magnetization drops to zero (Figure 10.4) – signaling complete loss of magnetization. The behavior of ferromagnetic susceptibility (χ_F) is fairly accurately modelled by Curie-Weiss's law according to which

$$\chi_F = \frac{C}{T - \theta} \tag{10.11}$$

where the constant C (the Curie constant) is material dependent and T is the temperature in Kelvin. θ is called the Weiss constant. The constant θ in Eq. 10.11 can assume positive or negative values[2-4] or can

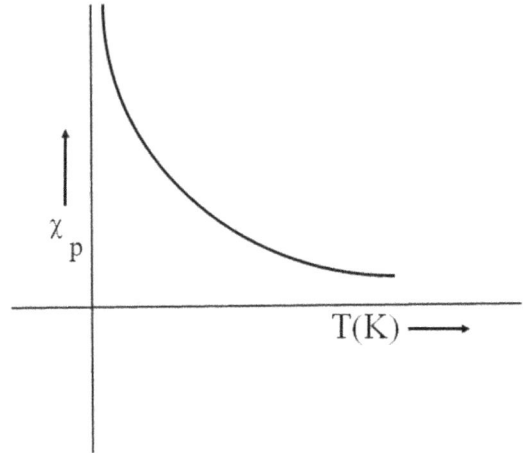

FIGURE 10.2   Typical χ_p-T plot for paramagnetic substances.

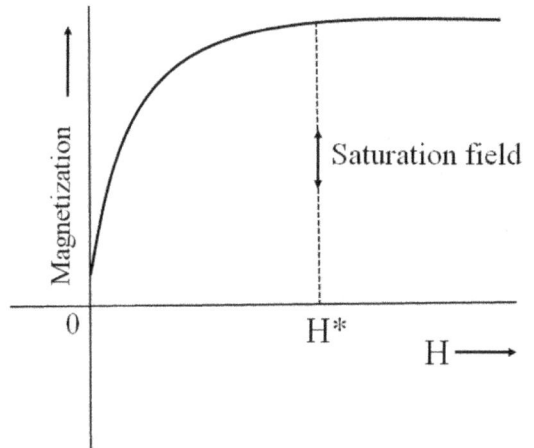

FIGURE 10.3   Typical magnetization – field plot for ferromagnetic materials.

be zero. If $\theta = 0$, we recover Curie's law. If $\theta \neq 0$, it can either be positive ($\theta > 0$) or negative ($\theta < 0$). In the first case, the material behaves ferromagnetically below a certain transition temperature $T = T_C$, i.e. the Curie temperature and the ferromagnetic susceptibility is accurately modeled by the following form of Curie-Weiss law (with $\theta = T_C$)

$$\chi_F = \frac{C}{T - T_C} \tag{10.12}$$

The form of Eq. 10.12 indicates that at T_C a critical phenomenon is perhaps taking place. For magnetite, the Curie temperature has been experimentally found to be 848 K. T_C is a fairly distinctive parameter that bears the signature of a material and can be used as an identifier for a mineral although it is possible that another mineral with the same Curie temperature may also exist. Above $T = T_C$, the material responds paramagnetically and Curie's law takes over. A plot of χ^{-1} against the temperature T is called Curie-Weiss plot. The plot should be linear if Curie-Weiss's law is valid. The inverse of the slope of the plot yields Curie constant and the intercept gives the Weiss constant (Figure 10.5).

If $\theta < 0$ (negative θ), the material is found to respond antiferromagnetically below a transition temperature $T = T_N$, called the Neel temperature while above T_N ($T > T_N$) it behaves paramagnetically. In contrast with ferromagnetic materials, 'θ' (<0) appears to be unconnected with the Neel temperature (T_N). Transition metals like chromium (Cr) and manganese (Mn) and many of the oxides and sulphides of some transition metals (MnO, NiO, Cr_2O_3, MnS, $MnSe$, etc., for example) display antiferromagnetism.[2–4]

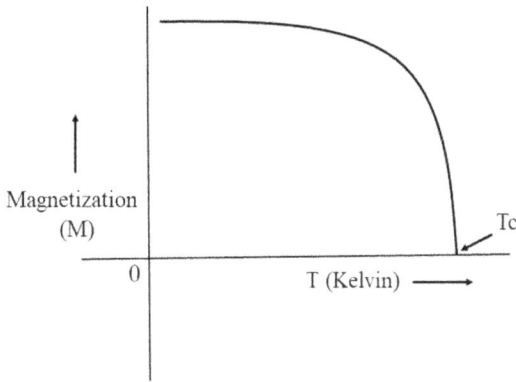

FIGURE 10.4 Magnetization – temperature plot for ferromagnetic material.

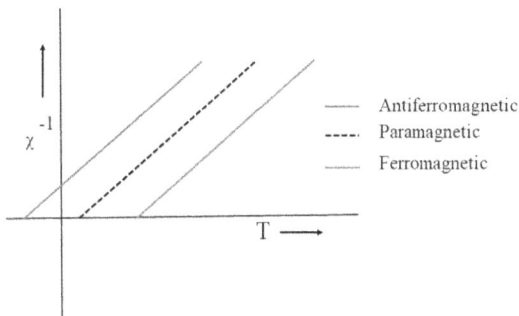

Antiferromagnetic
Paramagnetic
Ferromagnetic

FIGURE 10.5 Curie-Weiss plots for three types of magnetic materials.

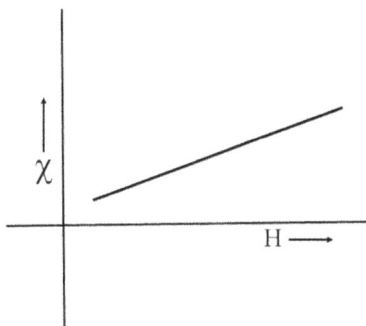

FIGURE 10.6 Antiferromagnetic susceptibility – field plot.

Antiferromagnetic susceptibility is positive, but small. The χ-H plot is therefore linear with small positive slope (Figure 10.6).

10.4 Types of Magnetism: Theoretical

So far, we have, from empirical evidence considered different types of responses from materials when they are placed in an external magnetic field. Two key quantities, e.g. magnetic susceptibility and magnetization and their behavior as functions of the applied field strength and temperature have led to the classification of magnetic materials into classes like diamagnetic, paramagnetic, ferromagnetic and antiferromagnetic materials. The question to be addressed now concerns the origin of a particular type of magnetism or magnetic response. In simple terms we ask: why does a magnetic material behave the way it does?

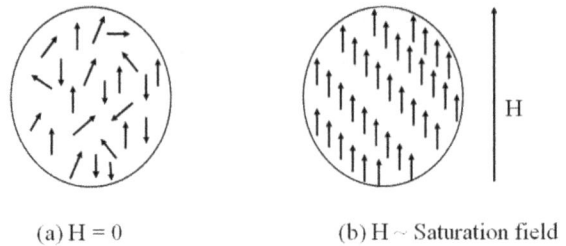

FIGURE 10.7 (a), (b) The orienting influence of the external magnetic field on the assembly of tiny magnetic dipoles.

(a) H = 0   (b) H ~ Saturation field

The answer seems to be simple. We can in fact trace that the root cause of the magnetic behavior lies in the presence or absence of a non-zero magnetic (dipole) moments ($\vec{m_i}$) of the constituent atoms or ions of a given material and their interactions. If $\vec{m_i}$ is zero for each of the constituent particle, the material as a whole can not have a non-zero net magnetic moment. Such materials are what have traditionally been known as non-magnetic materials or diamagnetic materials. In a strong external magnetic field, such materials can be magnetically polarized. The polarization creates an induced magnetic field in the material, which opposes the inducing (external) field. These materials appear to be weakly repelled in an external field – the hallmark of diamagnetism as we have stressed in the preceeding section. The diamagnetic susceptibility is thus small and negative. Even if the individual constituent particles (atoms or ions) have non-zero magnetic moment ($\vec{\mu_i} \neq 0$), the material as a whole may still have zero net magnetic moment, if the magnetic moments do not interact with neighboring moments and are randomly oriented throughout the sample. In that case, the average over all the individual magnetic moments would be zero. So, there would be no net magnetic moment or magnetization in the absence of an external magnetic field (Figure 10.7a). The tiny, individual randomly oriented magnetic dipoles can, however get oriented and aligned in the direction of an applied external magnetic field (Figure 10.7b). Once that happens, the average over all the individual magnetic dipoles can produce a non-zero net magnetic moment and magnetization unless the temperature is high enough to disrupt the external field induced magnetic ordering completely. Such materials have positive magnetic susceptibility but the magnetization disappears when the field is withdrawn completely. At ambient temperature, more and more of the individual magnetic dipoles begin to get aligned along the field direction as the strength of the external field is increased leading to an increase in magnetization, which eventually saturates. The saturation takes place when the field is strong enough to align all the magnetic dipoles completely. Thus, there appears the phenomenon of saturation magnetization. The reader will recall that paramagnetic materials behave in the manner described hare (see Section 10.3). Paramagnetic materials may be viewed theoretically to be made up of atoms/ions/molecules with non-zero but non-interacting magnetic moments randomly oriented, which can be magnetically ordered in an external magnetic field at a given temperature leading to the emergence of saturation magnetization and saturation field.[2-4]

A ferromagnetic material like magnetite has non-zero magnetization even in the absence of any external magnetic field ($H = 0$). It indicates, that the individual magnetic moments in the solid are arranged on a lattice and interact with neighbors strongly enough so as to produce a spontaneous alignment of the magnetic dipoles leading to the emergence of zero field magnetization (Figure 10.8a). If this be the case, the alignment can perhaps be completely destroyed by heating the substance or the material to a temperature where the thermal energy becomes high enough to overcome the intrinsic magnetic field induced ordering seen at room or lower temperature. The temperature – a critical temperature beyond which a ferromagnetic solid losses its magnetization completely is known as the Curie temperature.

The fourth category of magnetic materials are the antiferromagnetic materials. If the solid is such that there are two kinds of points on the lattices on which atoms or ions are arranged and they have magnetic moments that are exactly equal in magnitude but oppositely directed, the net magnetization at low temperature would be zero as the moments are all mutually quenched and the total magnetic moment is zero. Unlike paramagnetic materials there is no random orientations of the atomic or ionic moments – but there is a specific type of magnetic ordering called antiferromagnetic ordering (Figure 10.8b) because of interactions among neighboring atomic moments associated with the two types of lattice points, called sublattice *A* and *B* (see later). The signature of antiferromagnetic coupling is revealed at a critical temperature, called

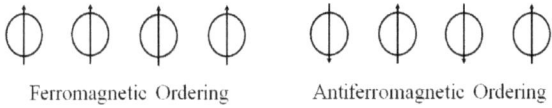

Ferromagnetic Ordering   Antiferromagnetic Ordering

FIGURE 10.8 (a) Ferro and (b) antiferromagnetic ordering.

Neel temperature (T_N) above which the susceptibility (χ) varies with temperature in a manner consistent with Curie-Weiss law for paramagnetic materials. The intercept of χ^{-1} versus T plot for AF materials turns out to be negative (on the T-axis) in stark contrast with the behavior of similar plots for paramagnetic materials. The types of interactions between neighboring magnetic moments must somehow be very different in the ferro and antiferromagnetic materials (see later). The fifth category of magnetic materials are known as ferrimagnetic materials. Here too, we assume that there are two kinds of lattice points and atomic magnetic moments belonging to them are not exactly equal although they are aligned in antiparallel manner. A complete quenching of the magnetic moments does not take place and the ferrimagnetic materials display weak zero field magnetization. Thus we have the rudiments of a theoretical model based on magnetic moments of the atomic or ionic constituents of a solid substance and interactions among them for understanding the behavior of the five types of magnetic materials that are normally found. We will now turn our attention to the basic question: what is the source of the atomic/ionic magnetic moments and their interactions? It turns out that the source is angular momenta of the electrons.

10.5 Exchange Interaction, Heisenberg's Exchange Hamiltonian and Magnetic Hamiltonian

Atoms or ions have electrons, which are quantum mechanical entities. As we discussed in Chapter 2, electrons in atoms/ions carry orbital angular momentum $(\hat{L})$ as well as spin angular momentum $(\hat{S})$. The former has a classical analog (the angular momentum vector $\vec{L}$ due to orbital motion), but the latter is a purely quantum mechanical quantity having no classical counterpart. Let us briefly recall that the eigenfunctions of $\hat{L}^2$ are spherical harmonics $(Y_{l,m}(\theta,\phi))$, which satisfy the eigenvalue equation

$$\hat{L}^2 Y_{l,m}(\theta,\phi) = l(l+1)\hbar^2$$

'l' being the angular momentum quantum number: an integer taking on values 0, 1, 2, $\cdots$. The quantum number 'm' takes all integer values between $-l$ and $+l$, $m\hbar$ being the magnitude of the angular momentum projected on the Z-axis (L_Z) or some axis parallel to the Z-axis. For $\hat{L}_Z$, we have the eigenvalue equation

$$\hat{L}_Z Y_{l,m}(\theta,\phi) = m\hbar Y_{l,m}(\theta,\phi)$$

One can then see that the length of the angular momentum vector is

$$|\vec{L}| = \sqrt{l(l+1)}\hbar$$

The associated magnetic moment is

$$|\vec{\mu_l}| = \gamma \sqrt{l(l+1)}\hbar$$

where γ is the gyromagnetic ratio:

$$\gamma = \frac{e}{2m} \quad \text{(in SI units)}.$$

The corresponding relation for the spin angular momentum of an electron are also similar

$$S^2 \chi_{\text{spin}} = s(s+1)\hbar^2 \chi_{\text{spin}}$$

$$S_Z \chi_{\text{spin}} = m_s \hbar \chi_{\text{spin}}$$

's' being the number quantizing the spin angular momentum of an electron. It takes on the value $s = \frac{1}{2}$. The Z-projection of the spin angular momentum takes on values $m_s \hbar$, m_s being either $\frac{1}{2}$ or $-\frac{1}{2}$. Let the magnetic moment associated with the spin angular momentum be $\overrightarrow{\mu_s}$ where

$$|\overrightarrow{\mu_s}| = \gamma |\overrightarrow{s}| = \gamma \hbar \sqrt{s(s+1)} = \mu_B \sqrt{s(s+1)}$$

where μ_B – the Bohr magneton – is equal to $\gamma \hbar = \frac{e\hbar}{2m}$.

The total angular momentum $\overrightarrow{J}$ is the vector sum of the spin and orbital angular momentum

$$\overrightarrow{J} = \overrightarrow{L} + \overrightarrow{S}$$

'J' is a quantum mechanical entity, the total angular momentum, the operator for which satisfies the eigenvalue equation

$$J^2 \chi_j = j(j+1)\hbar^2 \chi_j$$

where χ_j is the eigenfunction of $\hat{J}$, j is the quantum number for the total angular momentum. The length of the total angular momentum vector is given by

$$|\overrightarrow{j}| = \sqrt{j(j+1)}\hbar$$

The magnetic moment μ associated with the total angular momentum is

$$|\mu| = \mu_B \sqrt{j(j+1)}$$

If $L = 0$, then $|\mu| = |\mu_s|$, where μ_s is the spin-only magnetic moment. We have thus been able to establish that the individual atoms/ions in a material can have magnetic moments ($|\mu_l|$ or $|\mu_s|$ or $|\mu|$) and can therefore respond to an external magnetic field. When there is no external magnetic field, the individual electrons on different atoms can interact with each other electrostatically. It turns out that the interaction energy has a spin-dependent energy component, which gives rise to a phenomenological exchange interaction Hamiltonian as described below.

Suppose that we have two electrons occupying two orthogonal orbitals ϕ_a and ϕ_b on the atoms 'A' and 'B', respectively. The two-electron system is assumed to be described by the Hamiltonian

$$H(1,2) = H_A(1) + H_B(1) + \frac{e^2}{|\overrightarrow{r_1} - \overrightarrow{r_2}|}$$

It is easy to find (see Chapters 3 and 4) that the two-electron two orbital system gives rise to one singlet and one triplet state (three-fold degenerate) (Figure 10.9), the energies of which are

$$E_{singlet} = E_a + E_b + K_{12} + J_{12} = E_0 + K_{12} + J_{12}$$
$$E_{triplet} = E_a + E_b + K_{12} - J_{12} = E_0 + K_{12} - J_{12}$$

(10.13)

where $E_0 = E_a + E_b$ represents the unperturbed energies of the two electrons on atoms A and B, K_{12} is the **Coulomb integral** and J_{12} is the **exchange integral** (note that we are using the valence bond notation), defined as follows

$$K_{12} = \iint \phi_a(1)\phi_b(2)\frac{e^2}{r_{12}}\phi_a(1)\phi_b(2)dv_1 dv_2$$

(10.14)

$$J_{12} = \iint \phi_a(1)\phi_b(2)\frac{e^2}{r_{12}}\phi_a(2)\phi_b(1)dv_1 dv_2$$

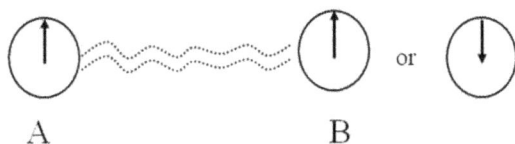

FIGURE 10.9 Interaction between two localized electrons on sites *A* and *B*.

Notice that the electron interaction operator is spin independent but the energy expressions for the singlet and triplet still carry a signature of the coupling of spins in the total wavefunction $\psi_{singlet}(1,2)$ and $\psi_{triplet}(1,2)$ (see Chapter 4). The two expressions for the energies of the singlet and triplet states can be converted into a single expression with a spin-dependent term – i.e. a term that depends on the coupling of the two spins in the total wavefunction. Thus, we can write

$$E_{S \text{ or } T} = E_0 + K_{12} - \frac{1}{2}J_{12} - 2J_{12}\vec{s_1} \cdot \vec{s_2} \tag{10.15}$$

where

$$\vec{s_1} \cdot \vec{s_2} = \frac{1}{2}(S^2 - s_1^2 - s_2^2) \tag{10.16}$$

For the singlet coupling of the two spins (s_1, s_2), $S = 0$ and $s_1^2 = s_2^2 = \frac{1}{2}(\frac{1}{2}+1)\hbar^2 = \frac{3}{4}\hbar^2$, so that $-2J_{12}\vec{s_1} \cdot \vec{s_2} = +\frac{3}{2}J_{12}$, where we have set $\hbar = 1$. Thus, we recover the singlet energy expression

$$E_S(S=0) = E_0 + K_{12} - \frac{1}{2}J_{12} + \frac{3}{2}J_{12}$$

$$= E_0 + K_{12} + J_{12}$$

For the triplet state, $S = 1$, $s_1^2 = s_2^2 = \frac{3}{4}\hbar^2$ and $S^2 = 1(1+1)\hbar^2 = 2\hbar^2$. Once again, we set $\hbar = 1$ and get

$$-2J_{12}s_1 \cdot s_2 = -J_{12}$$

so that we recover the triplet energy expression as well

$$E_T(S=1) = E_0 + K_{12} - \frac{1}{2}J_{12} - \frac{1}{2}J_{12}$$

$$= E_0 + K_{12} - J_{12}$$

So, Eq. 10.15 does indeed represent the singlet as well as the triplet state energies arising out of the same orbital configuration $\phi_a(1)\phi_b(2)$.

The spin-dependent part of the energy expression for a two-electron system given in Eq. 10.15 can now be generalized to cover any number of electron pairs. The generalized expression of the spin-dependent part of the energy also called exchange interaction energy takes the following form:

$$H_{exch} = -2\sum_{i>j} J_{ij}\vec{s_i} \cdot \vec{s_j} \tag{10.17}$$

Eq. 10.17 constitutes the formal basis for the Heisenberg-Dirac-Van Vleck vector model for magnetic susceptibilities.[5–7] When such a system is placed in a magnetic field ($\vec{B}$), the magnetic moments of individual atoms interact with the external magnetic field. For a two-electron system the total Hamiltonian now takes the form:

$$\hat{H} = \hat{H}_a + \hat{H}_b + H_{exch} + H_{1,field} + H_{2,field} \tag{10.18}$$

$$\hat{H} = \hat{H}_a + \hat{H}_b - 2J_{12}\vec{s_1} \cdot \vec{s_2} + \vec{B_1} \cdot \vec{\mu_1} + \vec{B_2} \cdot \vec{\mu_2} \tag{10.19}$$

In the simplest case, we can write

$$\vec{\mu_1} = \gamma \vec{s_1}, \quad \vec{\mu_2} = \gamma \vec{s_2}$$

The magnetic part of the total Hamiltonian H due to spin only magnetic moment then takes the form

$$\hat{H}_{magnetic} = -2J_{12}\vec{s_1} \cdot \vec{s_2} + \frac{\mu_B}{\hbar}\vec{B} \cdot \vec{s_1} + \frac{\mu_B}{\hbar}\vec{B} \cdot \vec{s_2} \tag{10.20}$$

We can now think of a more general system consisting a large collection of individual atomic magnetic moments ($\vec{\mu_i}$) arising from the spin of atomic electrons. The magnetic Hamiltonian of the system becomes

$$H_{magnetic} = -2\sum_{ij(i<j)} J_{ij}\vec{s_i} \cdot \vec{s_j} + \frac{\mu_B}{\hbar}\sum_i B\hat{s}_i$$

which can be reorganized to read

$$\hat{H}_{\text{magnetic}} = \sum_i \left\{ -\sum_j 2J_{ij}\,\vec{s_j} + \frac{\mu_B}{\hbar}\,\vec{B_{ext}} \right\} \cdot \vec{s_i} \tag{10.21}$$

The expression for the magnetic Hamiltonian in Eq. 10.21 is simple, but it will be perhaps simpler to convert it into a more compact form. Thus, instead of visualizing many magnetic dipoles within the sample of a material undergoing pair-wise exchange interactions (Eq. 10.21), it could be more convenient to think of a mean magnetic field arising from a collection of neighboring magnetic dipoles, which then acts on the individual magnetic dipoles. Thus, we can express the first term in Eq. 10.21 within the curly brackets as

$$-\sum_j 2J_{ij}\hat{s}_j = \frac{\mu_B}{\hbar}\,\vec{B_{exch}} \tag{10.22}$$

$\vec{B}_{exch}$ may be interpreted as molecular (atomic) exchange field. With the introduction of the molecular exchange field ($\vec{B}_{exch}$) the H_{magnetic} takes a more compact form:

$$H_{\text{magnetic}} = \frac{\mu_B}{\hbar} \left[\sum_i (\vec{B}_{exch} + \vec{B_{ext}}) \cdot \vec{s_i} \right] \tag{10.23}$$

Now we are in a position to formalize the theoretical classification of magnetic materials described in section 10.4 in terms of the magnetic Hamiltonian H_{magnetic} of Eq. 10.23.

10.5.1 Diamagnetic Materials

If there is no atom in the material carrying an electron with uncompensated spin, all the $\vec{s_i}$ in Eq. 10.23 can be set equal to zero. The magnetic Hamiltonian H_{magnetic} then becomes zero. It essentially means that the individual constituent particles (atoms or molecules) of the material have zero magnetic moment $\vec{\mu_i}$. Therefore the net magnetic moment of the sample is also zero. Such a material is called diamagnetic. We note here that the electrons carry orbital angular momentum ($l\hbar$) and therefore if $l \neq 0$ a non-zero orbital magnetic moment ($\vec{\mu_l}$) will be present even when $s_i = 0$. The orbital moment gives rise to the diamagnetic susceptibility χ_d (see later).

10.5.2 Paramagnetic Material

Suppose that $\vec{s_i} \neq 0$, but $\vec{B}_{exch} = 0$ due to random orientation of spins. So in the absence of an externally applied magnetic field ($\vec{B}_{ext} = 0$), the individual magnetic dipoles $\vec{\mu_i}$ compensate each other, and there is no spontaneous magnetization in the material. In other words, no spontaneous magnetic ordering takes place. On the application of an external magnetic field ($\vec{B}_{ext} \neq 0$), the individual atomic or molecular magnetic dipoles start getting aligned in the direction of the applied field. Of course this ordering tendency is opposed by the thermal randomizing effects on the individual magnetic moments. Such materials are said to be paramagnetic. However at a given temperature, an external field can always be designed to be strong enough to overcome the thermal resistance and induce complete magnetic ordering. The material is then said to display saturation magnetization at that temperature and the corresponding external field is known as the saturation field. These materials display temperature-dependent magnetic susceptibilty $\chi_p(T) > 0$.

10.5.3 Ferromagnetic, Antiferromagnetic and Ferrimagnetic Materials

The question now is to ascertain on the basis of H_{magnetic}, under what conditions spontaneous magnetic ordering can take place and the extent and types of ordering that can emerge.

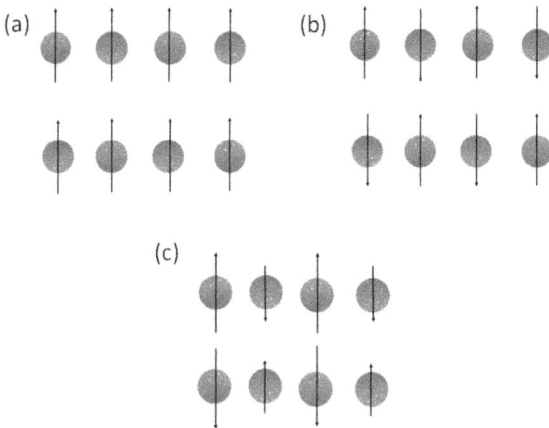

FIGURE 10.10 Different types of magnetic ordering: (a) $J(J_{ij}) > 0$, spontaneous ferromagnetic ordering below $T = T_c$, (b) $J(J_{ij}) < 0$, spontaneous antiferromagnetic ordering, (c) ferrimagnetic ordering.

TABLE 10.1

Curie Temperature (T_C) of a Number of Ferromagnetic Materials

Material	Curie temperature (K)
Ferric oxide (Fe_2O_3)	895
Iron (Fe)	1043
Cobalt (Co)	1388
Nickel (Ni)	627
Gadolinium (Gd)	293

10.5.3.1 Ferromagnetic Ordering

Let us assume that at $\overrightarrow{B_{ext}} = 0$, but the molecular exchange field $\overrightarrow{B}_{exch} \neq 0$. Thus

$$H_{magnetic} = \sum_i \overrightarrow{B}_{exch} \overrightarrow{s_i} = -\sum_{i<j} 2J_{ij} \overrightarrow{s_i} \cdot \overrightarrow{s_j} \tag{10.24}$$

If all the J_{ij}s > 0, and are independent of i, j, the molecular exchange interaction energy will be minimum if $\overrightarrow{s_i}$ and $\overrightarrow{s_j}$ are parallel. In isolated molecules, the exchange integral J_{ij} for electrons residing in orbital ϕ_i and ϕ_j can be shown to be positive.[6] So, minimization of magnetic interaction energy leads to the emergence of spontaneous and complete magnetic ordering (see Figure 10.10) giving rise to what are known as ferromagnetism.

The ordering, however, is accompanied by entropic loss (lowering of entropy) and can take place only if the sample temperature T is below a threshold temperature T_C, called the Curie temperature. Above T_C, $k_B T > \Delta E_{magnetic}$ and the spontaneous magnetic ordering abruptly disappears and the ferromagnetic susceptibility $\chi_F(T)$ diverges. At temperatures $T > T_C$, the material responds just like a paramagnetic material. We have listed the observed Curie temperatures of a number of ferromagnetic substances materials (Table 10.1).

10.5.3.2 Antiferromagnetic Ordering

As before let $\overrightarrow{B_{ext}} = 0$ and $\overrightarrow{B}_{exch} \neq 0$. If the exchange interaction terms $J_{ij} = J < 0$ for all i, j, the magnetic interaction energy $\Delta E_{magnetic} = -\sum_{i<j} 2J_{ij} \overrightarrow{s_i} \cdot \overrightarrow{s_j}$ is minimized only if $\overrightarrow{s_i}$ and $\overrightarrow{s_j}$ are antiparallel, i.e. $\overrightarrow{s_i} \cdot \overrightarrow{s_j} = -1$ for each pair of 'i' and 'j'. In this case, a complete long range magnetic ordering called antiferromagnetic ordering takes place (Figure 10.10b).

The total magnetization $M = 0$ in the absence of an external magnetic field. Here too, the ordering lowers entropy and can take place only till the sample temperature T is less than a critical temperature T_N, called the Neel temperature. Once $T > T_N$, the thermal energy $k_B T > \Delta E_{magnetic}$ and the ordering disappears completely as the magnetic dipoles (spins) on individual atoms get randomly oriented. Above T_N, the material responds like a paramagnetic substance. The magnetic susceptibility of an antiferromagnetic (AFM) material reaches a maximum at the Neel temperature (T_N) without displaying any divergence. Transition metal oxides, e.g. nickel, manganese and iron oxides display antiferromagnetism. Metallic chromium, iron-manganese alloy like FeMn also exhibit antiferromagnetic ordering below Neel temperature. Antiferromagnetic coupling has even been observed in some organic radicals such as 5-dehydro m-xylene. The Neel temperature of some of the antiferromagnetic materials are reported in Table 10.2. Antiferromagnetism plays an important role in the emergence of giant magnetoresistance (see later).

10.5.3.3 Ferrimagnetic Ordering

In the absence of an external magnetic field ($\overrightarrow{B_{ext}} = 0$) another type of long range spontaneous magnetic ordering takes place in some materials. In this case also, $\overrightarrow{B}_{exch} \neq 0$ and $J_{ij} = J < 0$, so that magnetic interaction energy minimization can take place only if the neighboring magnetic moments or spins (S_is) are antiparallel. The difference with AFM materials is that the magnetic moments on neighboring sites are not exactly equal (there are two types of lattice points or sublattices). Hence their antiparallel coupling does not result in zero magnetic moment for the pair and the net magnetic moment or magnetization (M) is small but not zero. Such materials are called ferrimagnetic materials. Ferrimagnetic ordering leads to a reduction of entropy and therefore remains viable only below a threshold temperature (T_C or T_N). Above the threshold temperature, thermal effect wins over the exchange coupling. Common examples are the ferrites. The Curie temperature (T_C) for some of the well-known ferrites are listed in Table 10.3.

The transition temperatures in ferrites are generally lower than in metals, the exchange interaction in ferrites being weaker. Figure 10.10c displays a typical ferrimagnetic ordering.

TABLE 10.2

Neel Temperature of a Number of Antiferromagnetic Materials

Material	Neel temperature (K)
Cr	308
NiO	525
Cr_2O_3	307
NiI_2	75
FeO	198
MnO	116

TABLE 10.3

Curie Temperature of a Number of Ferrites

Material	Transition temperature (T_C or T_N, K)
$NiFe_2O_4$	858
$MnFe_2O_4$	573
$Y_3Fe_5O_{12}$	553
$Y_3Fe_4AlO_{12}$	430
$Gd_3Fe_5O_{12}$	564

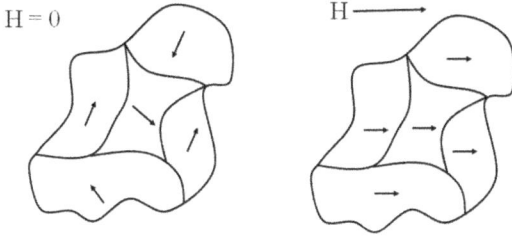

FIGURE 10.11 The orienting influence of the external magnetic field acting on super paramagnetic materials.

10.5.4 Superparamagnetism

A small sample (nanosized) of a ferromagnetic or a ferrimagnetic material can display a special form of magnetism, called superparamagnetism. If the sample size is small enough (1 μm or less) the magnetization can flip direction due to thermal agitation. The time gap between two consecutive flips is known as the Neel relaxation time (τ_{Neel}). If $\overrightarrow{B_{ext}} = 0$ and the measurement is carried out over a time 't_m' $>> \tau_{Neel}$, the average value of magnetization of the sample turns out to be zero. The nano-particle is then said to be in the 'superparamagnetic state'. When an external magnetic field is applied ($\overrightarrow{B_{ext}} \neq 0$) to the sample, the field can magnetize the sample in much the same way as a paramagnetic sample is magnetized (Figure 10.11). The magnetic susceptibility of a superparamagnetic material is much higher than the susceptibility of a typical paramagnetic substance.

Superparamagnetism is thus a time-scale-dependent small particle magnetism dominated by two key features:[8] (i) there is a size-limit beyond which the specimen can no longer gain in energy by breaking itself up into domains, hence it exists with one domain only; (ii) thermal energy for small enough size of the specimen can decouple the magnetization from the particle itself. The magnetic moment, which can be quite large as it is cooperatively built up from the coupling of individual spins, is rigidly bound to the particle at temperatures below the Curie temperature. At or above the Curie temperature thermal energy ($k_B T$) disrupts the bonding of the magnetic moment to the particle. The moment then becomes free to rotate and respond to an applied magnetic field independent of the particle. The critical size (D_s) for this to happen has been estimated from a dimensional analysis and has been found to be

$$D_S = \frac{\gamma}{M_S^2} \qquad (10.25)$$

where M_S is the saturation magnetization and γ is the domain-wall energy.

Superparamagnetic nanosized particles can be dispersed into an aqueous solution and held stable by coating them with an appropriate layer, forming what has been known as a ferrofluid. Ferrofluids have found important applications in biology both *in vivo* and *in vitro*.[8]

10.6 Paramagnetic Susceptibility of Gases and Conduction Electrons of Metals

Classical Langevin model of an ideal paramagnetic gas: The molecules of such a gas can be viewed as a collection of freely rotating molecules each of which has a permanent magnetic moment μ. If the gas is placed in a magnetic field $\overrightarrow{B}$, the molecular magnetic moments will tend to get aligned in the direction of the field. The thermal motion, chaotic as it is, will however, try to destroy the alignment. Viewed classically, the molecular magnetic moments (μ) can make any angle (θ) with the direction of the field and the interaction energy is given by ϵ where

$$\epsilon = \mu \cdot B = \mu B \cos\theta$$

The probability that μ makes an angle 'θ' with the applied field is

$$d\omega = A \cdot e^{\frac{\mu_B \cos\theta}{k_B T}} \sin\theta d\theta \qquad (10.26)$$

where $2\pi \sin\theta d\theta = d\Omega$ is the solid angle that the interval $d\theta$ corresponds to and A is a constant to be fixed by the normalization condition on the probability

$$\int d\omega = 1 \qquad (10.27)$$

This leads to condition

$$A \int_0^{\pi} e^{\frac{\mu_B \cos\theta}{k_B T}} \sin\theta d\theta = 1 \qquad (10.28)$$

Introducing the substitution

$$x = \cos\theta \quad \text{and} \quad \alpha = \frac{\mu B}{k_B T} \qquad (10.29)$$

Eq. 10.28 reduces to

$$A \int_{-1}^{+1} e^{\alpha x} dx = 1$$

$$\text{or,} \quad A = \frac{1}{\int_{-1}^{+1} e^{\alpha x} dx} \qquad (10.30)$$

The average value of the projection of the magnetic moment μ on the direction of the field ($\vec{B}$) is

$$<\mu> = \int \mu \cos\theta d\theta d\omega$$

$$= \mu \frac{\int_{-1}^{+1} x e^{\alpha x} dx}{\int_{-1}^{+1} e^{\alpha x} dx} \qquad (10.31)$$

$$= \mu L(\alpha)$$

where

$$L(\alpha) = \frac{e^{\alpha} + e^{-\alpha}}{e^{\alpha} - e^{-\alpha}} - \frac{1}{\alpha} = \coth\alpha - \frac{1}{\alpha} \qquad (10.32)$$

is the Langevin function and α as already introduced is equal to $\frac{\mu B}{k_B T}$.

If the field B is weak and T is not too low, $\mu B << k_B T$, i.e. $\alpha << 1$ condition is easily satisfied and in that sense, we can expand $\coth\alpha$ in a series in α to have[8-9]

$$L(\alpha) \approx \frac{\alpha}{3} \qquad (10.33)$$

Let us now assume that the concentration of the magnetic molecules in the paramagnetic gas to be 'n'. The magnetization M then becomes

$$M = n <\mu> = n \cdot \mu \cdot \frac{\alpha}{3} = \frac{n\mu^2 B}{3k_B T} \qquad (10.34)$$

The paramagnetic susceptibility χ_p of the gas is therefore given by[9]

$$\chi_p = M/B = \frac{\mu^2 n}{3k_B T} = \frac{C}{T} \qquad (10.35)$$

where $C = \frac{\mu^2 n}{3k_B}$ is a constant independent of the temperature. The paramagnetic susceptibility of an interaction-free gas of magnetic molecules is thus predicted to be inversely proportional to the absolute temperature of the gas, a behavior entirely consistent with Curie's law ($\chi \propto \frac{1}{T}$).

10.6.1 Quantum Model for Paramagnetic Susceptibility

A quantum version of the theory just described can be easily worked out once we graft into the theory certain important quantum features. They are:

    (a) the space quantization of the angular momentum of the electron;
    (b) the spin angular momentum of the electron.

Let us recall that an electron in any one of the stationary states of an atom has definite angular momentum projections $L_Z = m\hbar$ ($m = -l, \cdots, 0, \cdots, l$) and has a magnetic moment $\mu_Z = \mu_B m$, where μ_B the so called Bohr magneton is equal to $\frac{e\hbar}{2m_0 c} = 0.927 \times 10^{-20}$ erg/°C (m_0 = electron mass). That means,

$$\frac{\mu_Z}{L_Z} = \frac{e}{2m_0 c}$$

    It was confirmed by experiments and rationalized by theory that a free electron carries an magnetic moment equal to μ_B and an extra angular momentum S_Z, the projection of this extra angular momentum on a specific direction (Z-axis, say) being equal to $\pm \frac{\hbar}{2}$. The ratio μ_B and $< S_Z >$ then gets the value $\frac{e}{m_0 c}$ which is twice the μ_Z and L_Z ratio (and hence called anomalous). An immediate consequence of the anomalous μ_B and $< S_Z >$ ratio is that the directions of the vector sum of orbital and spin angular momenta and the total magnetic moment of the atom will not be coincident. This results in the precession of the magnetic moment about the resultant angular momentum $\overrightarrow{J}$ (in the language of the vector model) thereby creating and effective magnetic moment in the direction $\overrightarrow{J}$. Its magnitude is given by

$$\mu_j = g\mu_B j \tag{10.36}$$

where j is the quantum number for the total angular momentum $\overrightarrow{J}$ and g is the so called Lande splitting factor, where

$$|\overrightarrow{J}| = \sqrt{j(j+1)}\hbar \tag{10.37a}$$

and

$$g = 1 + \frac{j(j+1) + s(s+1) - l(l+1)}{2j(j+1)} \tag{10.37b}$$

s is the quantum number for spin angular momentum (assumes values $\pm\frac{1}{2}$), l is the quantum number for the orbital angular momentum $\overrightarrow{L}$ where

$$|\overrightarrow{L}| = \sqrt{l(l+1)}\hbar$$

The total or resultant angular momentum vector of the electron ($\overrightarrow{J}$) can take on only $2j + 1$ discrete orientations making angles θ_{jB} with the magnetic field with

$$\cos\theta_{jB} = \frac{m}{j} \tag{10.38}$$

where 'm' is the magnetic quantum number that assumes all integer values between $-j$ and $+j$ including zero. The energy (ϵ_μ) of the magnetic moment μ_j in the magnetic field is given by

$$\epsilon_\mu = -\mu_j B \cos\theta_{jB} = -\mu_B g m B \tag{10.39}$$

The average value of the magnetic moment $< \mu >$ can be calculated by evaluating the following ratio of discrete sums:[9]

$$< \mu > = \frac{\sum \mu_j \cos(\theta_{jB}) e^{\frac{-\epsilon_\mu}{k_B T}}}{\sum e^{\frac{-\epsilon_\mu}{k_B T}}} \tag{10.40}$$

where the summations are over all the possible discrete orientation of μ_j.

Now we may use the expression for ϵ_μ given in Eq. 10.39 in the right hand side of Eq. 10.40 together with the expression for μ_J in Eq. 10.36 to arrive at

$$< \mu > = g\mu_B \frac{\sum\limits_{-j}^{+j} m e^{\alpha m}}{\sum\limits_{-j}^{+j} e^{\alpha m}} \qquad (10.41)$$

where

$$\alpha = \frac{\mu_B g B}{k_B T} \qquad (10.42)$$

In weak magnetic fields we may assume that $\alpha << 1$ so that $e^{\alpha m} \approx 1 + \alpha m$, which when used in Eq. 10.41 leads to

$$< \mu > = \frac{\mu_B^2 g^2 j(j+1)}{3 k_B T} B \qquad (10.43)$$

The magnetic susceptibility χ_p can now be estimated by writing

$$\chi_p = n \cdot \frac{< \mu >}{B} = \frac{\mu_{eff}^2 n}{3 k_B T} \qquad (10.44)$$

where 'n' is the concentration of the gas molecules and $\mu_{eff} = \mu_B g \sqrt{j(j+1)}$ is the effective magnetic moment. The interesting point to note here is that the quantum discreteness does not destroy the inverse 'T' dependence of χ_p, i.e. Curie's law is recovered.

10.6.2 Paramagnetism of a Free-Electron Gas

The spin magnetic moments of the electrons in a free electron gas can cause a paramagnetic response characterized by a paramagnetic susceptibility. The model described in the previous subsection still remains applicable with the change that now the orbital angular momentum is zero ($l = 0$) and therefore $j = s = \frac{1}{2}$. The Lande g-factor then becomes equal to 2 ($g = 2$) as can be seen from Eq. 10.37b. The paramagnetic susceptibility of such an electron gas then would be given by Eq. 10.43 with $j = \frac{1}{2}$, $g = 2$ and we have

$$\chi_p \text{ (free electron gas)} = \frac{\mu_B^2 n}{k_B T} \qquad (10.45)$$

Once again Curie's law is apparently seen to hold. Eq. 10.45 seems to indicate that the free electron gas has a non-negligible and strongly temperature-dependent paramagnetic susceptibility. However, experiments seem to contradict this claim. In fact, experimental results somewhat counter intuitively reveal that the paramagnetic susceptibility of the electron gas in metals is rather small and practically temperature independent. W. Pauli attributed this apparent contradiction between theoretical predictions and experimental findings to the failure to incorporate in the theoretical models the correct quantum probability distribution function. It was his view that the conduction electrons in metals form a strongly degenerate Fermi gas (see Chapter 7) and therefore, Fermi-Dirac (not Boltzmann) distribution function must be employed for evaluating different average quantities. Pauli's suggestion laid the very foundation of the quantum theory of metals.[9–10]

10.6.3 Paramagnetism of Conduction Electrons

In a magnetic field ($\vec{B}$) the spin magnetic moment of a conduction electron in a free electron gas can either point to the direction of the field (parallel orientation) or against the field (antiparallel orientation). In the parallel orientation the electron energy is $\epsilon - \mu_B B$ while in the antiparallel orientation the energy is $\epsilon + \mu_B B$ where ϵ is the energy of the electron in the field free state. Let us now calculate the total magnetic moment of all the conduction electrons with magnetic moments pointing in the field direction (M_+). It is easy to find that (Chapter 7)

$$M_+ = \mu_B \int f(\epsilon - \mu_B B)g(\epsilon)d\epsilon \tag{10.46}$$

where $g(\epsilon)$ is the density of states of all electrons with parallel spins, $f(\epsilon)$ is the Fermi-Dirac distribution function. Similarly we can write an expression for M_-:

$$M_- = \mu_B \int f(\epsilon + \mu_B B)g(\epsilon)d\epsilon \tag{10.47}$$

where M_- is the total magnetic moment of all conduction electrons with antiparallel spins. The resultant magnetic moment (M) is given by[9]

$$
\begin{aligned}
M &= M_+ - M_- \\
&= \mu_B \int \left\{ f(\epsilon - \mu_B B) - f(\epsilon + \mu_B B) \right\} g(\epsilon)d\epsilon
\end{aligned}
\tag{10.48}
$$

where

$$f(\epsilon - \mu_B B) = f(\epsilon_-) = \frac{1}{e^{\frac{\epsilon_- - \zeta}{k_B T}} + 1} \tag{10.49a}$$

and

$$f(\epsilon + \mu_B B) = f(\epsilon_+) = \frac{1}{e^{\frac{\epsilon_+ - \zeta}{k_B T}} + 1} \tag{10.49b}$$

ζ being the chemical potential per conduction electron of the metals.

To evaluate M, $f(\epsilon \pm \mu_B B)$ are expanded in a power series of $\mu_B B$, assuming that the applied magnetic field is weak, and only terms linear in $\mu_B B$ are retained, yielding

$$M = \mu_B^2 B \times 2 \int \left(-\frac{\partial f}{\partial \epsilon} \right) g(\epsilon)d\epsilon \tag{10.50}$$

Since

$$\frac{\partial f}{\partial \epsilon} = -\frac{\partial f}{\partial \zeta} \tag{10.51}$$

Eq. 10.50 becomes

$$M = \mu_B^2 B \times 2 \int \left(\frac{\partial f}{\partial \zeta} \right) g(\epsilon)d\epsilon \tag{10.52}$$

The susceptibility χ then turns out to be

$$
\begin{aligned}
\chi &= \frac{M}{B} = \mu_B^2 \times \frac{\partial}{\partial \zeta} \int 2 \times fg d\epsilon \\
&= \mu_B^2 \frac{\partial n}{\partial \zeta}
\end{aligned}
\tag{10.53}
$$

where 'n' is the concentration of free charges. Now returning to the case of strongly degenerate conduction electrons of metals, the integral in Eq. 10.50 can be evaluated with some effort. We will, however, give only the final result[9]

$$\chi = \frac{M}{B} = \frac{3^{\frac{1}{3}}}{\pi^{\frac{4}{3}}} \frac{\mu_B^2 m^* n^{\frac{1}{3}}}{\hbar^2} \left[1 - \frac{\pi^2}{12} \left(\frac{k_B T}{\zeta_0} \right)^2 \right] \tag{10.54}$$

where $\zeta_0 = \frac{\hbar^2}{2m^*}(3\pi^2 n)^{\frac{2}{3}}$.

The main result is that the paramagnetic susceptibility χ of the strongly degenerate conduction electrons has its principal part independent of temperature (first term on the right hand side of Eq. 10.54). This result is in complete agreement with the experimental findings. There is a small temperature dependence of χ coming from the second term in Eq. 10.54, i.e. from $(\frac{k_B T}{\varsigma_0})^2$, which turns out to be $\approx 10^{-4}$ at room temperature. Comparing Eq. 10.54 with Eq. 10.45, it turns out that the thermal energy $k_B T$ of molecules in the Langevin equation gets replaced by the quantized energy $\frac{\hbar^2}{m^* d^2}$, where $d = n^{-\frac{1}{3}}$ is the average spacing between particles, m^* is the effective mass, when we are dealing with a strongly degenerate Fermi-gas.

For the non-degenerate electron gas in a semiconductor, we have

$$\frac{\partial n}{\partial \zeta} = \frac{\partial}{\partial \zeta}\left(e^{\frac{\zeta - \epsilon}{k_B T}} \right) = \frac{n}{k_B T}$$

so that Eq. 10.53 leads to $\chi = \frac{\mu_B^2 n}{k_B T}$ recovering the result expressed in Eq. 10.45.

10.6.4 Paramagnetic Resonance

The energy level of an atomic electron is split by a magnetic field into equidistant sublevels with a spacing of $g \mu_B B$ where 'g' is the Lande factor. The splitting of an electron level in a semiconductor or of a conduction electron in a crystal is expected to be equal to $g \mu_B B$ where 'g' now stands for the spectroscopic splitting factor and accounts for both the orbital angular momentum of the electron and its interaction with the lattice. Hence 'g' for a conduction electron in a crystal is generally different from '2' and may even be anisotropic, assuming different values for different orientations of the magnetic field with respect to the crystal.

The energy levels of a conduction electron (free electron) splits up into two sublevels because of the spin angular momentum of the electron ($s = \pm\frac{1}{2}\hbar$). If there is a multiplet splitting of the electrons energy' level due to the magnetic field, selection rule permits only transitions between adjacent subenergy levels corresponding to $\Delta m = \pm 1$. The resonance absorption of high frequency radiowaves takes place at the frequency ω where

$$\hbar\omega = g \mu_B B \tag{10.55}$$

The determination of the positions of the resonance peaks therefore leads directly to the 'g' value of the electron and helps one understand the possible electron states in a crystal. The width and shape of the resonance peaks can also shed light on the interaction of the impurity electron with the magnetic moments of atoms occupying neighboring sites and with the lattice vibrations. That is why there has been considerable interest in studying paramagnetic resonance of conduction electrons and electrons at an impurity center in a crystal. In view of the importance of these resonances in solid materials, we will briefly describe the semiphenomenological theory of paramagnetic resonance proposed by Bloch.

Let $\vec{L}$ be the total angular momentum of the electrons in unit volume of the material and $\vec{M}$ be the magnetization vector (M is the total magnetic moment per unit volume of the material). Then the evolution of $\vec{L}$ in a magnetic field $\vec{B}$ is described by

$$\frac{d\vec{L}}{dt} = \vec{M} \times \vec{B} \tag{10.56}$$

From quantum mechanics, it has been generally established that

$$\vec{M} = \gamma \vec{L}$$
$$\text{where} \quad \gamma = g \cdot \frac{e}{2mc} \tag{10.57}$$

The evolution equation for $\vec{M}$ takes the following form:

$$\frac{d\vec{M}}{dt} = \gamma \vec{M} \times \vec{B} \tag{10.58}$$

Now to solve the equation above, we have to specify B. Let us suppose that a strong static magnetic field has been applied in the direction of the Z-axis, so that $B = B_Z = B_0$. If simultaneously, a weak alternating high frequency magnetic field B_x is set up along the x-axis where $B_x = B_1 e^{i\omega t}$, with $B_1 << B_0$, we can set out to find the solution to Eq. 10.58 in the following form:

$$M_x = M_{1x} e^{i\omega t}$$

$$M_y = M_{1y} e^{i\omega t} \qquad (10.59)$$

$$M_z = M_{0z} + M_{1z} e^{i\omega t}$$

where M_{1x}, M_{1y} and M_{1z} are of the order of B_1 and M_{0z} is of the order of B_0. Using Eq. 10.59 in Eq. 10.58 we arrive at the following equations for M_x, M_y and M_z:

$$i\omega M_x = \gamma M_y B_z$$

$$i\omega M_y = \gamma (M_z B_x - M_x B_z) \approx \gamma (M_{0z} B_x - M_x B_z) \qquad (10.60)$$

$$\frac{dM_z}{dt} = i\omega M_{1z} e^{i\omega t} = -\gamma M_y B_x \approx 0$$

In arriving at Eq. 10.60 from Eq. 10.59, we have assumed that terms second order in B_1 can be neglected.[9] Eliminating M_y from the first two equations in Eq. 10.59 we get

$$\frac{M_x}{B_x} = \chi_x = \frac{\chi_0}{1 - (\frac{\omega}{\omega_0})^2} \qquad (10.61)$$

where $\chi_0 = \frac{M_{0z}}{B_z}$ and $\omega_0 = \gamma B_0$.

χ_x is the so called 'alternating susceptibility', which tends to become infinite as the condition of resonance, i.e. $\omega \approx \omega_o$ is reached. In practice, χ_x does not become infinite as the interaction with the lattice brings 'frictional forces' into play and the precessional motion of magnetic moment gets damped.

10.7 Diamagnetism of Atoms and Conduction Electrons

In the beginning of this chapter, we mentioned that there are substances for which the susceptibility $\chi < 0$ meaning that, the induced magnetic moment is in opposition to the applied magnetic field. Such substances are called diamagnetic. It turns out, as we will see in what follows, that all quantum systems of moving charges have an inherent diamagnetism, which in some cases, is overwhelmed by a much stronger paramagnetic response. To understand the nature of the diamagnetic response, let us consider an isolated atom with nuclear charge +Ze placed in an external magnetic field ($\vec{B}$). The atom has Z number of electrons by virtue of its electroneutrality. The nucleus of the atom is assumed to coincide with the origin of the Cartesian coordinate system being used while the z-axis lies in the direction of the magnetic field $\vec{B}$. Now Larmor's theorem tells us that at first order the effect of the magnetic field on the electrons is to cause a uniform rotation of the system of electrons as a whole about the z-axis with an angular velocity ω_L where

$$\omega_L = \frac{eB}{2mc} \qquad (10.62)$$

ω_L is the well-known Larmor frequency. On looking in the positive direction along the z-axis, the electrons are seen to revolve clockwise and therefore the magnetic moment μ_0 associated with the corresponding current will be directed against the magnetic field. The universal relation between μ_0 and the angular momentum l is

$$\mu_0 = \frac{e}{2mc} \cdot l \qquad (10.63)$$

The angular momentum in this case is

$$l = m\omega_L \sum_{i=1}^{z} (x_i^2 + y_i^2)_V \qquad (10.64)$$

where $(x_i^2 + y_i^2)_V$ is the sum of the squares of the coordinates of the ith electron averaged over the atomic volume (V). Using Eq. 10.62 and Eq. 10.64 in Eq. 10.63, it follows that

$$\mu_0 = \frac{e^2 B}{4mc^2} \sum_{i=1}^{z} \left(x_i^2 + y_i^2\right)_V \tag{10.65}$$

Hence the diamagnetic susceptibility (χ_{dia}) of a gas made up of such atoms (atomic number Z) is

$$\chi_{dia} = \frac{n\mu_0}{B} = \frac{e^2 n}{4mc^2} \sum_{i=1}^{z} \left(x_i^2 + y_i^2\right)_V \tag{10.66}$$

where n is the number of electrons per unit volume. When calculated with the help of quantum mechanics, $\sum_{i=1}^{z}(x_i^2 + y_i^2)_V$ turns out to be of the order za^2 in magnitude where 'a' is the linear dimension of the atom. As we have already seen the paramagnetic susceptibility of gas of atom each carrying a magnetic moment equal to μ_B (at a temperature T) is

$$\chi_{para} = \frac{\mu_B^2 n}{3k_B T} \tag{10.67}$$

The ratio of the diamagnetic and paramagnetic susceptibilities is

$$\frac{\chi_{dia}}{\chi_{para}} \approx \frac{ne^2 Z a^2}{mc^2} \Big/ \frac{n\mu_B^2}{k_B T} = Zk_B T \Big/ \frac{e^2}{a} \tag{10.68}$$

where we have equated 'a' to the Bohr radius $\frac{\hbar^2}{me^2}$ and put $\mu_B = \frac{e\hbar}{2mc}$. Now $\frac{e^2}{a}$ is an energy quantity, which is of the order of atomic energy. The ratio in Eq. 10.68 turns out to be of the order of 10^{-2} at room temperature assuming that the atomic number 'Z' is not too high. The value of the ratio makes it clear why diamagnetic susceptibility is detectable only for atoms, which do not carry unquenched spin, i.e. atoms devoid of any constant magnetic moment and hence without paramagnetic response.

Let us now turn to the diamagnetism of free electrons or conduction electrons in metals.[9] As a first approximation, we may try to calculate χ_{dia} by the method adopted for atoms. If a free electron moving with a velocity v is acted upon by an external magnetic field B, the electron is constrained to move in a circular orbit of radius r in a plane perpendicular to the magnetic field where

$$r = mcv^{\perp} \big/ eB = \frac{v^{\perp}}{\omega_c} \tag{10.69}$$

where $v^{\perp}$ is the component of the velocity of the electron in the plane perpendicular to the magnetic field, and ω_c is the cyclotron frequency given by

$$\omega_c = \frac{eB}{mc} = 2\omega_L \tag{10.70}$$

The magnetic moment associated with the circular motion of the electron is

$$\mu = \frac{erv^{\perp}}{2c} = \frac{m(v^{\perp})^2}{B} \tag{10.71}$$

where we have used $r = \frac{mcv^{\perp}}{eB}$. If the free electron gas is described by classical statistics, we can use $\frac{1}{2}m(v^{\perp})_{av}^2 = k_B T$ so that $\mu = \frac{k_B T}{B}$ and hence

$$\chi = \frac{\mu}{B} n = n\frac{k_B T}{B^2} \tag{10.72}$$

where as before n is the number of electrons per unit volume. The above expression for χ suffers from two serious absurdities:

1. $\chi \propto \frac{1}{B^2}$;
2. χ is independent of the electronic charge.

The source of this absurdity lies in the neglect of the surface of the body inside, which the electrons move. Bohr suggested that close to the surface electrons can not move in complete circles resulting in the appearence of a surface current in the sample. This surface current completely quenches the magnetic moments of electrons moving in complete circular orbits inside the specimen and χ becomes zero.

Let us consider the problem of calculating the magnetic susceptibility of an ideal electron gas by invoking classical statistical mechanics (chapter 6). Let the magnetic field be $\overrightarrow{B}$ where

$$\overrightarrow{B} = curl\,\overrightarrow{A} \tag{10.73}$$

where $\overrightarrow{A}(x, y, z)$ is the vector potential. From Eq. 10.73, we have

$$B_x = \frac{\partial A_z}{\partial y} - \frac{\partial A_y}{\partial z}$$

$$B_y = \frac{\partial A_x}{\partial z} - \frac{\partial A_z}{\partial x} \tag{10.74}$$

$$B_z = \frac{\partial A_y}{\partial x} - \frac{\partial A_x}{\partial y}$$

If $\overrightarrow{B}$ is directed along the z-axis, $B_z = B$ and B_x, $B_y = 0$. In that case A can be chosen in the form $A_x = 0$, $A_y = 0$, $A_z = 0$.

The Hamiltonian for the electron with charge $-e$ in a magnetic field $\overrightarrow{B}$ is then given by

$$\hat{H} = \frac{1}{2m}(\overrightarrow{p} + \frac{e}{c}\overrightarrow{A})^2 + u(x,y,z) \tag{10.75}$$

where $u(x,y,z)$ is the potential energy function for the electron. The right hand side can be easily expanded to give

$$H = \frac{1}{2m}(p_x + \frac{e}{c}A_x)^2 + \frac{1}{2m}(p_y + \frac{e}{c}A_y)^2 + \frac{1}{2m}(p_z + \frac{e}{c}A_z)^2 + u(x,y,z) \tag{10.76}$$

The free energy 'F' of the ideal electron gas, when calculated by classical statistical mechanics (Chapter 6) then becomes

$$F = -k_B T ln Z \tag{10.77}$$

where the partition function (also called the statistical integral) is given as

$$Z = \left[\int\int\int\limits_{-\infty}^{+\infty} dp_x dp_y dp_z \int\int\int dxdydze^{-\frac{H}{k_B T}}\right]^N \tag{10.78}$$

Introducing three new variables ω_x, ω_y, ω_z where

$$\omega_x = p_x + \frac{e}{c}A_x$$

$$\omega_y = p_y + \frac{e}{c}A_y \tag{10.79}$$

$$\omega_z = p_z + \frac{e}{c}A_z$$

We find that $dp_x dp_y dp_z = d\omega_x d\omega_y d\omega_z$, the range of integration remaining the same, for the new variables. Since the definite integral on the right hand side of Eq. 10.78 does not depend on the integration variables, Z in Eq. 10.78 and automatically F of Eq. 10.77 can not explicitly depend on the magnetic field B. Hence the magnetic susceptibility χ_{dia} where

$$\chi_{dia} = -\left(\frac{\partial^2 F}{\partial B^2}\right)_{T,B=0} \tag{10.80}$$

must be zero for the ideal electron gas. Proceeding in the same manner, it can be proved that χ_{dia} of the ideal electron gas described by Fermi-Dirac statistics also turns out to be zero. In summary, the ideal electron gas does not respond diamagnetically in a quasi-classical description. Landau (1930) showed that quantum mechanical description of the motion of free electrons in a magnetic field changes the result dramatically as the motion of the electrons get quantized. It turns out that the diamagnetic susceptibility of an electron gas in a magnetic field now becomes $\frac{1}{3}$ of its paramagnetic susceptibility. A quantum statistical treatment shows that the ratio of χ_{dia} and χ_{para} of the free charge carriers is the same, both for a strongly degenerate electron gas and for non-degenerate electron gas in semiconductors.

10.8 Ferromagnetic Susceptibility

As we have already seen magnetic oxide of iron (magnetite) displays spontaneous magnetic ordering of atomic magnetic moments below the Curie temperature (T_c). At $T > T_c$, the spontaneous magnetization is completely lost. T_c therefore represents a phase transition temperature – a temperature above which the spontaneous magnetic order is replaced by magnetic disorder. The spontaneous magnetization was shown to be due to exchange interaction of magnetic moments residing at different lattice points. Here, in this section we will exploit the same basic model to calculate the ferromagnetic susceptibility as a function of temperature after introducing additional approximation to keep the calculation simple. Our objective is to probe whether contact with the Curie-Weiss law can be established. For the analysis we will make use of canonical partition function and a mean-field approximation based on the well-known Ising model.

Let each site (i) in the magnetic material be occupied by one spin (s_i) where s_i is assigned the value (+1) if the spin is up (↑) at the ith site and a value of (-1) if the spin at the ith site points down (↓). The energy $E\{s\}$ of the N-spin system can be easily calculated in the presence of a magnetic field $\vec{B}$ (directed along the positive z-axis) as

$$E\{s\} = -\frac{1}{2}\sum_{i \neq j}^{N} J_{ij}s_i s_j - B\sum_{i=1}^{N} s_i \qquad (10.81)$$

J_{ij}s represent the strength of exchange interaction between the spins at the sites 'i' and 'j'. A simpler version of the model keeps only the nearest neighbor interactions, which are assumed to have the same interaction strength [$J = J_{ij}$]. This approximate model yields the much simpler energy expression

$$E\{s_i\} = -\frac{1}{2}J\sum_{<ij>}^{N} s_i s_j - B\sum_{i=1}^{N} s_i \qquad (10.82)$$

In Eq. 10.82, $<ij>$ sums over the nearest neighbors. The macroscopic magnetic moment (μ) of the entire system of N Ising spins is

$$M\{s_i\} = \sum_{i=1}^{N} s_i \qquad (10.83)$$

The canonical partition function Z for the system is

$$Z = \sum_{s_1,s_2,\cdots,s_N} e^{-\beta E\{s_i\}} \qquad (10.84)$$

where the summation is over all possible spin microstates. The average value of energy of the system $< E >$ is then given by ($\beta = \frac{1}{k_B T}$)

$$< E > = \frac{1}{Z} \sum_{s_1,s_2,\cdots,s_N} E\{s_i\}e^{-\beta E\{s_i\}}$$

$$= -\frac{\partial}{\partial \beta} lnZ \qquad (10.85)$$

Average magnetization $< M >$ can be similarly calculated from the canonical partition function

$$< M > = \frac{1}{Z} \sum_{s_1, s_2, \cdots, s_N} M\{s_i\} e^{-\beta E\{s_i\}}$$

$$= \frac{1}{\beta} \frac{\partial lnZ}{\partial B} \tag{10.86}$$

Since the Helmholtz free energy $F = -k_B T ln Z$, it is straightforward to find that

$$< M > = -\frac{\partial F}{\partial B} \tag{10.87}$$

The magnetic susceptibility χ of the N spin system is

$$\chi = \frac{1}{\beta} \frac{\partial^2 lnZ}{\partial B^2} \tag{10.88}$$

where

$$Z = \sum_{s_1, s_2, \cdots, s_N} e^{\left(\beta \frac{J}{2} \sum_{i \neq j} s_i s_j + \beta B \sum_i s_i \right)} \tag{10.89}$$

Even with all the approximations already made, evaluation of Z is still quite difficult. We can make further progress by invoking the mean field approximation.[10]

At the mean field level, we rewrite the energy expression in Eq. 10.82 as

$$E\{s_i\} = -\frac{1}{2} \sum_i s_i \sum_{<j>i} s_j - B \sum_i s_i \tag{10.90}$$

In the expression above, summation over all the spin pairs on neighboring sites has been split into a sum over spins at all sites 'i' and a sum over all the spins at the nearest neighbors of the ith site, these spins being represented as $<j>_i$. The sum $\sum_{<j>i} s_j$ may be viewed as a local magnetic field arising from all the neighboring spins (all the nearest neighbors of site 'i') that acts on the spin s_i at the ith site. A further assumption is now introduced by setting[10]

$$\sum_{<j>i} s_j = \gamma m \tag{10.91}$$

m being the average magnetization per spin and γ being the number of nearest neighbors of the ith spin (i.e. the ith site). With these additional approximations, the mean-field energy[10] of the system becomes

$$E\{s_i\} = -\frac{\gamma m J}{2} \sum_{i=1}^{N} s_i - B \sum_{i=1}^{N} s_i$$

$$= -\left(\frac{\gamma m J}{2} + B \right) \sum_{i=1}^{N} s_i \tag{10.92}$$

The partition function can now be calculated on the basis of the energy expression in Eq. 10.92 ($\beta = \frac{1}{k_B T}$):

$$Z = \sum_{s_1, s_2, \cdots, s_N} exp\left\{ \beta \left(\frac{\gamma m J}{2} + B \right) \sum_{i=1}^{N} s_i \right\}$$

$$= \sum_{s_1, s_2, \cdots, s_N} \Pi_{i=1}^{N} exp\left\{ \beta \left(\frac{\gamma m J}{2} + B \right) s_i \right\}$$

Noticing that s_i can take on only the two values, +1 or –1, we can rearrange the above expression for Z into

$$Z = \Pi_{i=1}^{N} \sum_{s_i=-1}^{+1} exp\left\{\beta\left(\frac{\gamma\,mJ}{2} + B\right)s_i\right\}$$

$$= \left[2\cosh\{\beta\left(\frac{\gamma\,mJ}{2} + B\right)\}\right]^{N} \tag{10.93}$$

Therefore, we have

$$lnZ = ln2 + Nln\left\{\cosh\beta\left(\frac{\gamma\,mJ}{2} + B\right)\right\}$$

With lnZ thus defined it is now easy to calculate

$$m = <M>/N = -\frac{1}{N\beta}\frac{\partial lnZ}{\partial B}$$

$$= \tanh\left[\beta\left(\frac{\gamma\,mJ}{2} + B\right)\right] \tag{10.94}$$

Eq. 10.94 above is a transcendental equation. Let us see whether the solutions of Eq. 10.94 suggest the possibility of spontaneous magnetization. With that objective in mind, we set $B = 0$. A graphical solution indeed suggests that for certain values of exchange coupling strength, J, temperature ($\beta = \frac{1}{k_BT}$) and coordination number (γ) non-zero values of 'm' appear as solutions of Eq. 10.94, indicating that spontaneous magnetization is a reality in the mean-field Ising model of a ferromagnetic system. Let us examine, if we can arrive at an approximate analytical expression for 'm'. Thus, for $B = 0$, and small values of 'm', we can expand the right hand side of Eq. 10.94 in powers of 'm' and write (neglecting terms beyond m^3)

$$m \approx \frac{\beta Jm\gamma}{2} - \frac{(\frac{\beta mJ\gamma}{2})^3}{3}$$

$$= m\left(\frac{T_c}{T}\right) - \frac{m^3}{3}\left(\frac{T_c}{T}\right)^3 \tag{10.95}$$

where $T_c = \frac{J\gamma}{2k_B}$.

One solution of Eq. 10.95 is obviously $m = 0$, which corresponds to zero spontaneous magnetization. The other two solutions are

$$m = \pm\sqrt{3}\left(\frac{T}{T_c}\right)\left(1 - \frac{T}{T_c}\right)^{\frac{1}{2}} \tag{10.96}$$

which point to the possible appearance of spontaneous magnetization in the system at $T < T_c$. T_c can therefore be accepted as a critical temperature. Close to $T = T_c$, $m = \pm\sqrt{3}(1 - \frac{T}{T_c})^{\frac{1}{2}}$, which tells us that at $T = T_c$, $m = 0$, that is a magnetic phase transition takes place with the abrupt disappearence of spontaneous magnetization at $T = T_c$. If the system is cooled below T_c ($T < T_c$), spontaneous magnetization begins to grow from $m = 0$ as $(1 - \frac{T}{T_c})^{\frac{1}{2}}$. The mean-field Ising model therefore predicts that the order parameter (m) for ferromagnetic phase transition in the vicinity of the T_c varies as $(1 - \frac{T}{T_c})^{\frac{1}{2}}$. The critical exponent is thus predicted to be 0.5. Experimental data for Ni, Co and Fe indicates that the value is close to 0.33. Therefore, in spite of drastic approximations in the mean-field model described here, the basic physics of ferromagnetism has not been lost.

We can now have a look at the ferromagnetic susceptibility χ_F that the mean-field Ising model predicts. Proceeding from Eq. 10.94, with $B \neq 0$ we have,

$$m = \tanh\left(\frac{\beta Jm\gamma}{2} + B\right)$$

$$= \tanh\left(\frac{mT_c}{T} + \frac{B}{k_BT}\right)$$

with $T_c = \frac{\gamma J}{2k_B}$, $\beta = \frac{1}{k_B T}$.

The ferromagnetic susceptibility χ_F can now be obtained by evaluating

$$\chi = \left(\frac{\partial M}{\partial B}\right)_{B=0}$$

which yields (for $T > T_c$, $m = 0$ if B is set 0)

$$\chi = \left[\chi\left(\frac{T_c}{T}\right) + \frac{1}{k_B T}\right] \times \frac{1}{\cosh^2(\frac{mT_c}{T})}$$

$$= \frac{1}{k_B} \cdot \frac{1}{T - T_c} = \frac{C}{T - T_c}$$

(10.97)

which is just the Curie-Weiss law.

For $T < T_c$, we can not write down an expression for 'm' in a closed or analytical form, hence χ can not be given an analytical formula for $T < T_c$.

10.9 Giant Magneto Resistance (GMR)

Grunberg and Fert[11–12] observed that the magnetization of adjacent ferromagnetic films separated by a thin non-magnetic metallic layer spontaneously align parallel or antiparallel depending on the thickness of the non-magnetic metallic layer. They found that the electrical resistance of the system is strongly affected by the orientation of the magnetization in the ferromagnetic layers. If the magnetization vectors are parallelly oriented, the state that emerges has low electrical resistance while antiparallel orientation of the magnetization vectors in the ferromagnetic layers creates a state of much higher electrical resistance (Figure 10.12). This phenomenon has become known as GMR. Formally stated the effect represents the change of electrical conductivity in a system of metallic layers when an external magnetic field changes the directions of magnetization of the ferromagnetic layers relative to each other. A parallel alignment ($\uparrow\uparrow$) has a lower electrical resistance ($R_{\uparrow\uparrow}$) than an antiparallel orientation ($\downarrow\uparrow$), which has a higher electrical resistance ($R_{\uparrow\downarrow} > R_{\uparrow\uparrow}$). The magnitude of the effect is calculated as follows:

$$\frac{\Delta R}{R} = \frac{R_{\uparrow\downarrow} - R_{\uparrow\uparrow}}{R_{\uparrow\uparrow}}$$

Magnetic metallic multilayers like Fe/Cr or Co/Cu in which the ferromagnetic layers are separated by non-magnetic spacer layers that are only a few nm thick, display GMR that can be even 100% at low temperatures.

Peter Grinberg and Albert Fert were awarded Nobel Prize in physics (2007) for the discovery of GMR effect. The effect observed was much larger than the other magnetoresistive effects that were already known, hence the effect was called giant magneto resistance. The effect has found use in magnetic recording, data storage and sensor applications. The origin of the effect lies in spin-dependent transport of

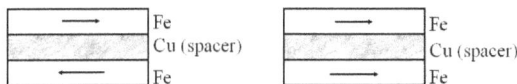

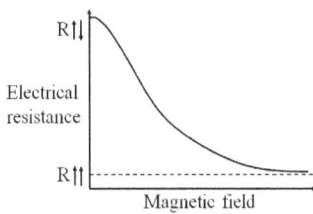

FIGURE 10.12 Electrical resistance as a function of the applied magnetic field in a typical GMR structure.

electrons. We can try to understand the phenomenon qualitatively in terms of a simple model developed in another context by Mott.[13] According to this model, electrical conductivity in metals can be described in terms two fairly independent channel of conduction – one, by the up spin (↑) and the other, by the down spin (↓) electrons. The probability of spin non-conserving scatterings is much lower than the probability of scatterings that conserve spin. Thus, the (↑) and (↓) spins do not mix up over fairly long distances of charge transport along a conductor.

In ferromagnetic metals, the rates of scattering of the up (↑) and down (↓) spin electrons differ quite significantly, independent of the nature of the scatterer. The difference in scattering rates has its root in the sizable difference in the density of states for the up and down spin electrons at or around the Fermi energy (due to exchange splitting of the energy bands in ferromagnetic metals) and the fact that scattering rates are proportional to the density of states. The upshot is that the scattering rates and hence resistivities of ferromagnetic metals are different for electrons with different spins. The ideas described above can be utilized to come up with an explanation of the GMR effects. If we assume that the scattering is strong for electrons with spin antiparallel to the axis of magnetization and rather weak for electrons with spin aligned parallel to the magnetization axis, the assumption is consistent with the observed asymmetry in the DOS at Fermi energy and the GMR effect can be rationalized qualitatively. Thus for a parallel alignment of the magnetization of the magnetic layers, the up (↑) spin electrons can pass through the magnetic layers without being hindered by scattering while the down spin (↓) electrons get strongly scattered. As stated already, the conduction process occurs in two independent spin channels, hence the overall conductivity of the structure is mainly shaped by the highly conducting up spin (↑) electrons, the resistivity becoming low for the structure with parallel alignment of magnetization. For the antiparallel alignment of magnetization vectors in the ferromagnetic layers, both up (↑) and down spin (↓) electrons are scattered extensively within one of the ferromagnetic layers or the other. The resistivity of the corresponding structure is therefore higher compared to the other structure in which the magnetization vectors are parallel. In most of the applications of GMR, specially in the read head of hard discs, magnetic multilayers of the 'spin valve' type are used. The details can be found in the excellent review articles by Parkin (2002)[14] and by Chappert (2007).[15]

Colossal magneto resistance (CMR) is another variant of magneto resistive effect in a certain class of solid materials – the so called CMR manganites. G. H. Jonker and J. H. Van Santen discovered in 1950 that certain ferromagnetic compounds of manganese with perovskite structures displayed high magnetoresistive effects.[16] It was after R. Von Helmolt *et al.* (1993)[17] and S. Jin *et al.* (1994)[18] discovered large magneto resistive effects in La-Ba-Mn-O and La-Ca-Mn-O systems, that keen interest in colossal magneto resistance began to grow rapidly. Nowadays, it is generally believed that the mechanism underlying the generation of colossal magneto resistivity is closely related to the very nature of CMR manganites, which are strongly correlated many electron systems in which the lattice, spin, charge and orbital degrees of freedom are intricately linked. A host of processes like the double exchange interaction, Jahn-Teller distortion, charge ordering, electronic phase transition, etc. appear to contribute to shape the colossal magneto resistivity in manganite perovskites. The CMR effect is measured by the quantity $\frac{\rho(H)-\rho(O)}{\rho(H)}$ where $\rho(H)$ is the resistivity of the manganite in a magnetic field H and $\rho(O)$ is the field free resistivity of the same material. In La-Ca-Mn-O thin films at 77 K the effect was found to be negative, $\rho(H)$ being lower than $\rho(O)$ and as large as 100% or more. The structure of the $RE_{1-x}AE_xMnO_3$ (RE = rare earth metal, AE = alkaline earth metal) manganite is very close to the cubic ABO_3 perovskites in which the A-sites are occupied by the bigger RE^{+3} and AE^{+2} ions with 12-fold oxygen coordination. Mn^{+2} ions occupy the B-sites at the centre of oxygen octahedron and experience a six-fold oxygen coordination. The Mn $3d$ orbitals are therefore split into t_{2g} and e_g orbitals in the octahedral field and are further split by JT distortion of the MnO_6 octahedron. In the $RE_{1-x}AE_xMnO_3$ manganites the valence state of the Mn ions is a mixture of Mn^{+3} ($t_{2g}^3 e_g^1$) and Mn^{+4} (t_{2g}^3) configurations or states in x:1-x ratio and the transfer of e_g electron from Mn^{+3} to Mn^{+4} is supposed to be the basic electrical conduction mechanism operative in the rare earth alkaline earth manganites. The transfer is mediated by the double exchange (DE) mechanism, which involves simultaneous hopping of the e_g electron of Mn^{+3} ion to the oxygen $2p$ orbital and simultaneous transfer of an electron with the same spin from oxygen $2p$ orbital to the empty e_g orbital of the Mn^{+4} ion producing ferromagnetic ordering. In some of the manganites a strong DE interaction causes the e_g electron to delocalize extensively at low temperature leading to the emergence of metal like behavior at appropriate levels of doping. The DE interaction can be further enhanced by an applied magnetic field, which works to align the spins of adjacent Mn-O-Mn ions and increase the electrical conductivity. Thus $\rho(H) < \rho(O)$ and the CMR effect will have a negative sign as observed. The DE mechanism however, fails to explain the emergence of insulator like resistivity above the transition temperature.[19] The CMR manganites are strong candidates for spintronics applications and interest in them persists.[20] A complete theoretical and quantitative description is still not available.

10.10 Materials with Ferromagnetic plus Ferroelectric Order

Multiferroics are materials that exhibit more than one of the so called primary ferroic properties in the same phase, namely ferromagnetism and ferroelectricity. The former is switchable by an external magnetic field while the latter refers to an order that is turned on by an externally applied electric field. The two requirements can lead to a contradiction, which must be circumvented to design multiferroic materials. A material is classified as ferroelectric if it has a spontaneous electric polarization, which can be switched by an externally applied electric field. As we have already seen (in Chapter 8), such electric polarization is usually brought in by structural distortions of the parent centrosymmetric phase that breaks its inversion symmetry. In the prototypical ferroelectric material $BaTiO_3$, the parent phase is an ABO_3 perovskite in which the B-sites are occupied by Ti^{+4} ions octahedrally coordinated by neighboring oxygen atoms. There is thus, inversion symmetry and no electric polarization appears. In the ferroelectric phase, the Ti^{+4} ions are shifted away from the center of the oxygen octahedron breaking the inversion symmetry. That gives rise to electric polarization. A rule of thumb asserts that such distortions are more likely to occur when the B site cation has an empty d-shell (i.e. d^0 configuration), which favors formation of covalent bonds between the B-site cations and its oxygen neighbors. That brings in its wake distortions destroying inversion symmetry. The requirement for the B-site cation to have d^0 configuration to facilitate inversion symmetry breaking and generate ferroelectric order, can prove to be an impediment to the emergence of ferromagnetic order in the same phase. It is because partially filled d-subshells of transition metals are known to be responsible for their magnetism. That means, the ferroelectric order in multiferroics must be created by some other route. We briefly describe different mechanisms of realizing coexisting ferroelectric and ferromagnetic order in the same phase.[26]

(i) **The Lone-Pair Mediated Multiferroics (MF)**: These MFs are perovskite oxides in which the A-site cation has a lone pair of electrons. The B-site cations have partially occupied d-subshells. Bismuth ferrite and lead vanadate are good examples. The Bi^{+3} or Pb^{+2} ions have $6s^2$ lone pairs, which hybridize with the empty $6p$ orbitals. These hybrids then form covalent bonds with the $2p$ orbitals of neighboring oxygen atoms facilitating the loss of inversion symmetry.

(ii) **Geometric Multiferroics**: The structural phase transition leading to the emergence of the ferro-electric state in these materials is linked to rotational distortion of the polyhedra. Such rotational distortion is common in transition metal oxide perovskites when the A-site cations are small. Because of 3D connectivity among these polyhedra, such rotational distortion does not create any electrical polarization as the rotation of one is compensated by rotation of another connected poly-hedra in the opposite sense.[26] In layered materials, however, such rotational distortion can cause electric polarization without switching ferromagnetic order. Well-known examples are materi-als like $BaMF_4$ with M standing for transition metals like Mn, Fe, Co, Ni with partially filled d-subshell. At 1000 K, there is a ferroelectric phase transition (loss of inversion center) and lower down at 50 K there is a transition to an antiferromagnetically ordered state, thus produc-ing a multiferroic material. A second example is the hexagonal family of rare-earth manganites ($REMnO_3$) with RE $\equiv$ Ho, Y, Lu. A structural phase transition at 1300 K followed by a transition to a triangular, antiferromagnetically ordered state at 100 K generates a multiferroic material.

(iii) **Multiferroics Through Charge Ordering**: In compounds of a metal in which the metal ions exist in mixed valance states, charge ordering phenomenon can take place on cooling. Electrons, which are delocalized at high temperature, begin to get localized on different cationic sites at low temperature in an ordered fashion producing an insulating phase. If the pattern of localization of electrons is polar, a ferroelectric state emerges. The metal ions in such cases are usually magnetic, so a material with coexisting ferroelectric and ferromagnetic order is produced. $LuFe_2O_4$, for example, has a charge-ordered state at 300 K comprising of Fe^{+2} and Fe^{+3} ions. The ferromagnetic phase appears at 240 K.

In some materials which are non-centrosymmetric, long range magnetic order, can induce macroscopic electric polarization. Thus, $TbMnO_3$ below 280 K forms a non-centrosymmetric spiral magnetic state with a small electric polarization caused by weak spin-orbit coupling between the NCS spin structure and the crystal lattice. Much larger electric polarization takes place in orthorhombic $HoMnO_3$, because of strong magnetoelectric coupling.

Multiferroics have so far evoked keen but mainly academic interest. It is only recently that these materials are being pursued seriously for fabricating switches, actuators, magnetic field sensors and novel electronic memory devices. On the academic front, multiferroic quantum criticality associated with the merger of two distinct quantum critical points has strong implications for fundamental physics and low temperature applications.[26]

10.11 Molecular Magnets

Some metal-organic or organo-metallic compounds display superparamagnetic behavior at the molecular scale below a critical temperature. These metal-organic molecules, called single molecule magnets (SMM), often display typical magnetic hysteresis curves, which have purely molecular origin. Unlike the conventional magnetic materials in which a collective long range magnetic order produces magnetism (ferromagnetic or antiferromagnetic), SMMs do not require any long range magnetic order. They are magnetic at the molecular scale. They display magnetic bi-stability, i.e. the magnetic moment has only two antiparallel orientations separated by a barrier. The barrier height determines the relaxation rate.

The first SMM is believed to be the so called 'Mn_{12}' magnet, a molecule with the composition $Mn_{12}(OAc)_{16}(H_2O)_4$, which was discovered in 1991 in which a central $Mn^{IV}O_4$ cube is surrounded by a ring of 8 Mn units connected through bridging oxo-ligands (OAc stands for the acetate group).[22] As long as the temperature is low enough, e.g. around 4 K, the rate of magnetic relaxation is quite slow. So any practical application based on the Mn_{12}-SMM would require the system to be maintained at least at the aliquid helium temperature. There is nothing special about manganese. SMMs based on iron-clusters also exist, e.g. $[Fe_8O_2(OH)_{12} (tacn)_6]^{8+}$, reported by Gateschi (2000).[22] Efforts are on to design single molecule magnets with transition temperatures as high as 77 K (liquid nitrogen temperature). The typically large bistable spin anisotropy in the SMMs, make them futuristic candidates for fabricating the smallest units of magnetic memory systems, which can become the building blocks for the core of 'quantum computers'. Levenberger and Loss[23] have theoretically shown that molecular magnets like 'Mn_{12}' or 'Fe_8' can be used to build dense and efficient memory devices based on Grover's algorithm. A single crystal can serve as the storage unit of a dynamic RAM device. With the help of fast electron spin resonance pulses, it would be possible to decode and read out the stored numbers with access time as low as 10^{-10}s.

10.12 Soft Magnetic Materials

For industrial or technological applications, it is often necessary to have materials that can be easily magnetized or demagnetized. Easily magnetizable or demagnetizable materials are called soft magnetic materials. High relative permeability (μ_r) and low coercivity are typical features of the soft magnetic materials. Depending on application, the electrical conductivity and saturation magnetism of the material may be the other important attributes of soft magnetic materials. Soft magnetic materials are used for AC as well as DC applications. In DC application, the material is magnetized before application and demagnetized after the application. In AC applications the cycle of magnetization and demagnetization continue over the entire period of application. In either case high permeability is required property. When the material is used for generating a magnetic field saturation magnetization is an important property of the material that would be helpful. For AC application, hysteresis loss, loss due to eddy current and anomalous loss (due to movement of domain walls) are important factors that require to be reduced. The anomalous loss can be minimized by using highly homogeneous materials, hysteresis loss can be reduced by reducing the coercivity while eddy loss can be minimized by reducing the electrical conductivity and by laminating the material.[24]

Thus iron-silicon alloys are used for transformer cores – the alloying increases the electrical resistivity (approximately by a factor of 4 at 3 wt % of silicon). The addition of Si also reduces the magnetostriction and magnetocrystalline anisotropy.

Alloys of Fe, Co, Ni with the addition of one or more of the elements like boron, carbon, phosphorus and silicon have very low coercivity and therefore have lower hysteresis loss. These alloys, specially in their nanocrystalline forms are in great demand. Nickel iron alloys, also known as 'permalloy' are versatile, the nickel-content being optimizable for specific applications. Nickel-Fe alloys with small amounts of copper or chromium have zero magnetostriction, zero magnetic anisotropy, have high relative permeability and extremely low coercivity (example, mu-metal).

For high frequency applications metallic soft magnetic materials are unsuitable due to high eddy current loss. For such applications soft ferrites are the materials of choice. These are ferrimagnetic materials with cubic crystal structure, and have the composition $MoFe_2O_3$ where M is a transition metal. Mn-Zn ferrite, for example, is a suitable for upto 10 MHz applications and is used in signal receivers and transmitters. Microwave ferrites (yttrium-Fe-garnet) are used in microwave devices, in constructing waveguides where the applicable frequency range is from 100 MHz to 500 MHz. Such materials are in demand and are in a state of continuous evolution.[25]

REFERENCES

1. P. Quaterman *et al.*, 'Demonstration of Ru as the 4*th* ferromagnetic element at room temperature', *Nature Commun.*, 9, 2058 (2018).
2. D. Jiles, *Introduction to Magnetism and Magnetic Materials*, Nelson Thornes, Cheltenham, UK (2001).
3. R. S. Teeble and D. J. Craik, *Magnetic Materials*, Wiley-Interscience, New York, (Reprinted by Books on Demand, Arin Arbor, MI) (1969).
4. W. D. Callister, Chapter 20, In *Materials Science and Engineering An Introduction*, John Wiley, New York (2003).
5. H. Van Vleck, Chapter 12, In *'Theory of Electric and Magnetic Susceptibilities'*, Oxford University Press, New York (1932).
6. J. C. Slater, *Quantum Theory of Atomic Structure*, Volume I, App. 19, McGraw-Hill, New York (1960).
7. M. Tinkham, *Group Theory and Quantum Mechanics*, Tata McGraw-Hill, New Delhi (1974).
8. C. M. Sorensen, 'Magnetism', In *Nanoscale Materials in Chemistry* (pp. 169–221l), K. J. Klabunde, Ed., Wiley Interscience, New York (2001).
9. A. Anselm, Chapter 6, In *Introduction to Semiconductor Theory*, (Translated from Russian by M. M. Samokhvalov), MIR Publishers, Moscow (1981).
10. (a) A. S. Davydov, *Quantum Mechanics*, Second Edition, Pergamon Press, Oxford (1976). (b) S.-K. Ma, *Statistical Mechanics*, (Translated by M. K. Fumg), World Scientific, Singapore (1984). (c) P. M. Chaikin and T. C. Lubensky, *Principles of Statistical Mechanics*, Cambridge University Press, New York (1995).
11. A. Fert, P. M. Levy and S. Zhang, 'Electrical conductivity of magnetic multilayered structures', *Phys. Rev. Lett.*, 65, 1643–1646 (1990).
12. G. Binasch *et al.*, 'Enhanced magnetoresistance in layered magnetic structures with antiferromagnetic interlayer exchange', *Phys. Rev. B.*, 39, 4828 (1989).
13. N. F. Mott, 'The electrical conductivity of transition metals', *Proc. Royal Soc.* (London) 153, 699 (1936).
14. S. S. P. Parkin, *Applications of Magnetic Nanostructures in Spin Dependent Transport in Magnetic Nanostructures*, S. Meekawa and T. Shinja, Eds., Taylor and Francis, New York (2002).

15. C. Chappert, A. Fert and F. Nguyen Van Dau, 'The emergence of spin electronics in data storage', *Nat. Mater.*, 6, 813 (2007).

16. G. H. Jonker and J. H. Van Santen, 'Ferromagnetic compounds of Mn with perovskite structures', *Physica*, 16, 337 (1950).

17. R. Von Helmolt et al., 'Giant negative magnetoresistance in perovskite like $La_{23}Ba_{13}$-MnO_x, 71, 2331–2333 (1993).

18. S. Jin *et al.*, 'Thousand fold change in resistivity in magnetoresistive in La-Ca-Mn-O films', *Science*, 264, 413–415 (1994).

19. J. N. Lalena and D. A. Cleary, *Principles of Inorganic Material Design*, Second Edition, John Wiley, New York (2010).

20. A. Fert, 'Nobel lecture on origin, development and future of spintronics', *Rev. Mod. Phys.* 80, 1517–1530 (2008).

21. A. Caneschi *et al.* 'Alternating current susceptibility, high field magnetization, and millimeter band EPR evidence for a ground S = 10 state in $[Mn_{12}O_{12} (CH3COO)_{16}(H2O)_4]$. 2CH3COOH. 4H$_2$O', *J. Am. Chem. Soc.*, 113, 5873 (1991).

22. D. Gateschi *et al.*, 'Single-molecule magnets based on Iron(III) Oxo clusters dedicated to the memory of Professor Olivier Kahn', *Chem. Comm.*, 9, 725–732 (2000).

23. M. N. Levenberger and D. Loss, 'Quantum computing in molecular magnets', *Nature*, 410, 789–793 (2001).

24. J. M. Silveyra, E. Ferrara, D. L. Huber and T. C. Monson, 'Soft magnetic materials for a sustainable and electrical world', *Science*, 362, 6413 (2018).

25. K. H. J. Buschow and F. R. De Boer, 'Soft-magnetic Materials', In *Physics of Magnetism and Magnetic Materials*, Volume 7, Springer, Boston (2003).

26. S. Dong *et al.*, 'Magnetoelectricity in multiferroics: A theoretical perspective', *Nat. Sci. Rev.*, 6, 629–641 (2019).

11

Low-Dimensional Materials

11.1 Introduction: The New Age Materials

The present 'Silicon Age' uses silicon transistors in most of the microelectronics, which have really improved the standard of living by reducing both the size and weight of electronics goods. Over the past few decades, the properties of silicon devices have been improved to an astonishing extent, enabling the shifting of clunky old desktop computers into sleek smartphones. But this silicon revolution must reach its limit as imposed by the fundamental physical limitations, which restrict silicon to scale beyond 10 nm technology node without severely compromising a device's performance. This simply means that the steady march toward faster, smaller, lighter products with more and more functionality can not continue within the existing framework. Besides the issue of lighter and smaller products, there is another important concern regarding electronic goods; use of microelectronics is expanding so rapidly that, in the near future, a large portion of the world's energy would be consumed by information technologies. Noticeably, this consumption is not sustainable. So, to maintain and improve our global standard of living, we must step beyond the Silicon Age, and to do this we need new age materials to keep up with the pace of progress. Now, what are these new age materials? These are low-dimensional materials; particularly single-layer two-dimensional (2D) nanomaterials, such as graphene, phosphorene, transition metal dichalcogenides, are extremely promising in this context. They offer unique electrical, optical, mechanical and chemical properties. In addition, they feature excellent electrostatic integrity and inherent scalability, which makes them attractive from a technological standpoint.

Now what makes them so special? To understand this we have to look at the properties of those low-dimensional materials, collectively called nanomaterials. The dimension of these materials lie in the intermediate zone between atoms and bulk materials. To answer the question why nanomaterials are so special, one has to understand electronic energy levels of these materials as a function of their size, composition and shape.

In this era, graphene has been considered as a revolutionary 2D material with distinct electronic and structural properties that surpass conventional semiconducting materials as well as metals. The superlative qualities of graphene make it the reference material for post-silicon and complementary metal oxide semiconductor (CMOS) technology. Furthermore, graphene also shows quite unusual electronic behavior; its charge carriers behave like Dirac particles, i.e. massless and relativistic in nature. Such exotic electronic properties have compelled the theorists to revisit the conceptual basis for the theory of metals. In fact, graphene seems to be unveiling a new era in science and technology where materials with one atom thickness but 200 times stiffer than steel are available. Thus, since the discovery of graphene by Andre Geim and Kostya Novoselov at The University of Manchester in 2004, it has attracted tremendous technological attention in semiconductor industries. Starting the journey with graphene it has now evolved a large number of 2D materials like hexagonal boron nitride (commonly known as white graphene), silicine, phosphorene, their intercalated 2D materials, transition metal dichalcogenides (TMDs) with the latest modifications including metal and covalent organic frameworks. The major advantage of these materials is that their electronic properties are easily tunable by means of defect formation, doping foreign elements and most importantly, functionalizing the materials with different functional groups. In the forthcoming sections, we will review the systematic development of such material properties from the first-principles theory as well as experimental techniques.

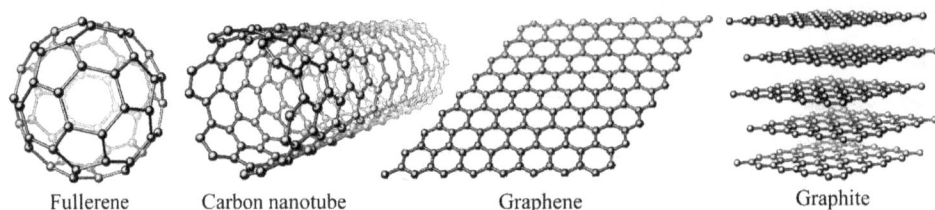

FIGURE 11.1 Schematic representations of different graphitic materials.

11.2 Graphene

Graphene is an atomically thin planar sheet of sp^2-hybridized carbon atoms, constituting a 2D honeycomb lattice. Structurally, it is the monolayer of graphite where the monolayers are stacked by means of van der Waals interaction, keeping a fixed interlayer distance. It is also the basic building block for all the other graphitic materials (Figure 11.1). Graphene can be rolled into one-dimensional (1D) nanotubes or completely wrapped up into zero-dimensional (0D) fullerenes. Thus, graphene is termed as 'mother of all graphitic form'. Unfortunately it was discovered after fullerene and carbon nanotube (CNT)!

Discovery of graphene is in fact very strange; for a long time, it had been strongly believed that 2D crystals were thermodynamically unstable and could not exist. People used to think that a divergent contribution of thermal fluctuations in low-dimensional crystal lattice at any finite temperature should lead to a huge atomic displacement, in the range of interatomic distance, which would completely break down the crystal lattice. However, unilayer carbon (graphene) had been theoretically studied a long time ago to describe the properties of various carbon-based materials. But, it was presumed to be an 'academic' material, not to exist in the free state and considered to be unstable with respect to the formation of curved structures such as fullerenes and nanotubes. Fortunately, such a vintage model ultimately turned into reality when graphene was experimentally realized by Jeim and Novoselov in 2004. The sp^2-hybridized carbon atoms in the graphene sheet utilize their three hybrid orbitals to form strong C–C σ-bonds in a hexagonal framework. On the other hand, the remaining unhybridized p-orbitals on each carbon, staying perpendicular to the sheet, create an extended π-conjugation. Such a unique crystal structure and long range delocalized π-conjugation result in some exciting properties like high thermal conductivity, unusual quantum Hall effect, high electron mobility, which immediately triggered the graphene gold rush. Presently, our daily life is highly influenced by the applications of graphene; for example automobile and airplane components, solar cells, rechargeable batteries, supercapacitors, flexible display and touch panels, high speed transistors, sensors, LED lights, conductive inks, etc. Such huge applications come from its excellent electronic properties, such as the high specific surface area of 2600 m^2 g^{-1}, the extremely high room temperature electron mobility of 2×10^5 cm^2 (V.s)$^{-1}$, the outstanding heat conductivity of 3000 W (m.K)$^{-1}$ and very high in-plane stiffness with Young's modulus estimated around 1000 GPa.

11.2.1 Geometry and Crystal Structure

Crystal structure of graphene is described by a hexagonal lattice, which has also been identified by transmission electron microscopy (TEM) suspended between bars of a metallic grid. However, some of the TEM images show a 'rippling' of the flat sheet, with amplitude of about one nanometer. These ripples might be intrinsic to the material as a result of the instability of 2D crystals or may originate from the ubiquitous dirt seen in all TEM images of graphene. Photoresist residue, which must be removed in order to obtain atomic-resolution images, may be the 'adsorbates' observed in TEM images and these can also explain the observed rippling. The 2D honeycomb lattice is not the Bravais lattice, however, the underlying rhombohedral cell could be considered as the unit cell with two-point basis. As shown in Figure 11.2, the 2D honeycomb lattice can also be viewed as a combination of two triangular sublattices each having one carbon atom. The primitive lattice vectors can be represented as

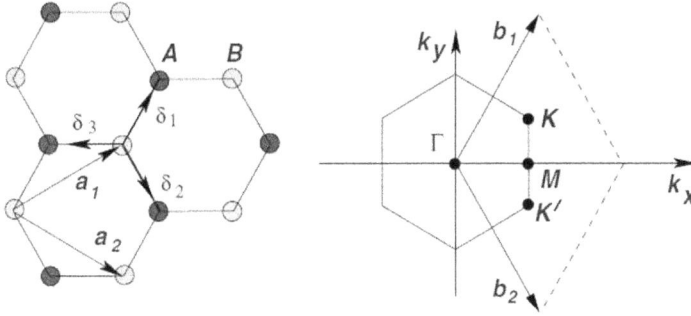

FIGURE 11.2 Crystal structure of graphene, showing direct and reciprocal lattice vectors and the first Brillouin zone.

$$\vec{a_1} = \frac{a}{2}(3, \sqrt{3}), \qquad \vec{a_2} = \frac{a}{2}(3, -\sqrt{3}) \tag{11.1}$$

where 'a' is the nearest-neighbor interatomic distance. The reciprocal-lattice vectors are then given by (see Chapter 5)

$$\vec{b_1} = \frac{2\pi}{3a}(1, \sqrt{3}), \qquad \vec{b_1} = \frac{2\pi}{3a}(1, -\sqrt{3}) \tag{11.2}$$

There are two points K and K' at the corners of the Brillouin zone (BZ) of graphene, named as Dirac points, which are of particular importance for the unique properties of graphene. In momentum space, K and K' are described by

$$\mathbf{K} = \left(\frac{2\pi}{3a}, \frac{2\pi}{3\sqrt{3}a} \right), \qquad \mathbf{K'} = \left(\frac{2\pi}{3a}, -\frac{2\pi}{3\sqrt{3}a} \right) \tag{11.3}$$

The three nearest-neighbor vectors in real space are given by

$$\delta_1 = \frac{a}{2}(1, \sqrt{3}) \qquad \delta_2 = \frac{a}{2}(1, -\sqrt{3}) \qquad \delta_3 = -a(1, 0) \tag{11.4}$$

while the six second-nearest neighbors are located at $\delta_1' = \pm a_1$, $\delta_2' = \pm a_2$, $\delta_3' = \pm(a_1 - a_2)$.

11.2.2 Electronic Structure of Graphene

As already mentioned, in the honeycomb lattice of graphene, each C atom is trigonally connected with other three C atoms through strong σ-bonding. Thus, one may expect that the resultant σ and σ^* bands in graphene are widely separated in energy. The π and π^* bands, which are formed by the lateral interactions among the p-orbitals of individual carbon atoms, are solely populated around the Fermi level. Thus, the electronic behavior of graphene is governed by these π and π^* bands. Let us understand the problem from the stand point of a simpler basis. Consider the unit cell with two atom basis, the contributing two p-electrons give rise to two π bands, one bonding and the other antibonding in nature. The dispersion relation for these two bands can be derived by using tight-binding approximation, which can be expressed upto second-nearest neighbor hopping as follows:

$$E_{\pm}(\vec{k}) = \pm t\sqrt{3 + f(\vec{k})} - t' f(\vec{k}) \tag{11.5}$$

where

$$f(\vec{k}) = 2\cos(\sqrt{3}k_y a) + 4\cos(\frac{\sqrt{3}}{2}k_y a)\cos(\frac{3}{2}k_x a) \tag{11.6}$$

and t, t' are the nearest-neighbor and next-nearest-neighbor hopping amplitudes. The positive (+) and negative (−) signs indicate the upper (π^*) and lower (π) bands, respectively. The dispersion relationships for these two bands with finite values of t and t' are shown in Figure 11.3(a), where the bands meet at K and

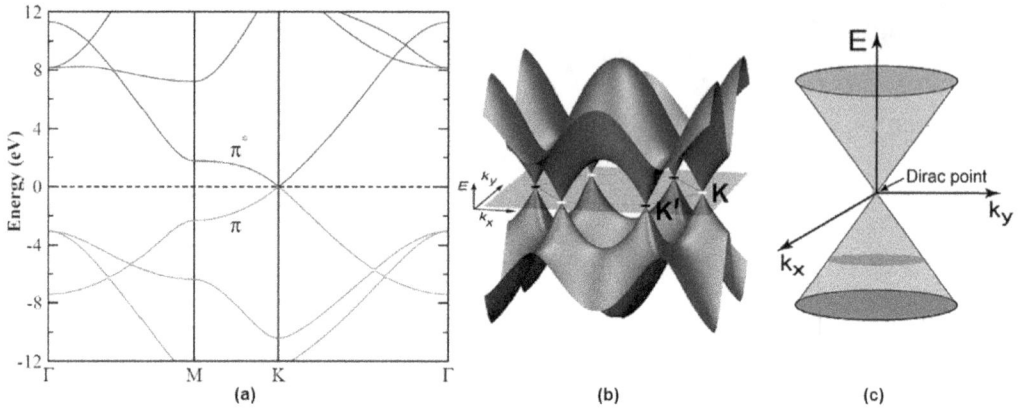

FIGURE 11.3 Band structure of graphene within the first Brillouin zone in 2D (a) and 3D (b) representations and the corresponding Dirac cone (c).

K' points. The dispersion relation in the close vicinity of the point K is given by $E_{\pm} = \pm \hbar k v_F$, where $v_F = \frac{3ta}{2\hbar}$ is known as Fermi velocity. The significance of this equation is that it represents a linear dispersion of E just around the points K and K' unlike parabolic dispersion in case of other 2D systems. Linear dispersion around the K points gives rise to two cones at each and every K point in three-dimensional reciprocal space (Figure 11.3(b)), which touch at the Fermi level. Due to such behavior, electrons and holes just around the K points behave like relativistic particles described by the Dirac equation for spin 1/2 particles. That is why charge carriers of graphene are called Dirac Fermions, the cones are called Dirac cones and K points are labeled as Dirac points. This implies that electrons in graphene have zero rest-mass and the corresponding Hamiltonian can be written in the form of a relativistic Dirac Hamiltonian

$$H = v_F \sigma . \kappa \hbar \qquad (11.7)$$

where σ is a spinor-like wave function, v_F is the Fermi velocity of graphene, and κ is the wave vector of the electron. However, the spinor character of the graphene wavefunction arises not from spin but from the pseudospin corresponding to the hopping of the electron between the two sublattices of graphene unit cell. Thus, electrons in graphene can be described in the same way as relativistic particles with energy

$$E = \sqrt{m^2 c^4 + p^2 c^2} \qquad (11.8)$$

where m is the rest mass of the particle, p is its momentum and c its velocity. Dirac Fermions behave in an unusual way as compared to ordinary electrons when subjected to magnetic field, leading to interesting properties such as anomalous integer quantum Hall effect, which is also observed experimentally. The most interesting fact is that the integer quantum Hall effect in graphene can be observed at room temperature because of the large cyclotron energies for relativistic electrons. Another important point is that in general, Fermi velocity is expressed by $v_F = \sqrt{2E/m}$, but as obtained from tight-binding method, here, Fermi velocity describing linear bands around K is given by $v_F = \frac{3ta}{2\hbar}$, which is estimated to be $\sim 10^8$ cm/s or $\frac{1}{300}$ th of speed of light. The high Fermi velocity of graphene gives rise to many unusual properties due to quantum electrodynamic effects. Due to the presence of Dirac Fermions, graphene shows several unusual transport phenomena, such as anomalous integer quantum Hall effect, high Fermi velocity, etc. The most surprising fact is that, though the carrier density is zero at the Dirac point, 2D conductivity, σ_{xx} as well as resistivity, ρ_{xx} have some non-zero values. The reported dubbed 'minimum' conductivity for bi-layer graphene is 6-9 kΩ. Actually, the minimum conductivity can never fall below e^2/h even when the carrier density is zero, which may arise from linear excitation spectrum of Dirac Fermions or may be due to the effect of impurity or due to rippling of the graphene sheet.

Figure 11.3(a) shows the simulated electronic band structure of graphene within the first Brillouin zone. The π and π^* states form the valence and conduction bands, respectively. The figure reveals that the valence and conduction bands are widely separated at Γ point, but the gap between these two bands rapidly decreases toward *M* point. After *M* point the bands smoothly approach each other upto the *K* point where they just touch. So, graphene is considered to be a zero gap semiconductor or semimetal or gapless semiconductor. Linear dispersion just around the *K* point, which results in Dirac crossing at the *K* point at the Fermi level is distinct from the band structure, as already discussed. Furthermore, as the band structure is symmetric about the Dirac point, electrons and holes in pure, free-standing graphene should have the same properties.

11.3 Graphene Nanoribbons

Now, if the motions of the electrons are further confined by cutting the 2D sheet of graphene into 1D stripes, it leads to graphene nanoribbons (GNRs). Due to the particular honeycomb structure of graphene sheet, there are two different possibilities of choosing the cutting direction, leading to armchair and zig-zag edged GNRs. However, mixed edged GNRs could also be formed as shown in Figure 11.4. As is obvious, for a particular edge symmetry, there are possibilities of different GNRs with varying ribbon width. The width of an armchair edged GNR (aGNR) is defined by number of dimer lines across the width, while number of zigzag chains across the width of a zigzag edged GNR (zGNR) defines its width. The edges of GNRs may be bare or could be passivated by other atoms or groups; however, the basic properties of a particular type of GNR depend on the edge symmetry.

11.3.1 Electronic Structure of aGNRs

Numerous experimental and theoretical studies have been undertaken to establish the electronic structure of GNRs with proper edge symmetry. It turns out that armchair graphene nanoribbons are semiconducting in nature and the band gap decreases with increasing ribbon width. In general, the band gap of group IV semiconducting materials gradually decreases with increase in size. However, this decreasing tendency, in the case of aGNR, is little bit different; it follows three different series, $3n$, $(3n + 1)$ and $(3n + 2)$ for H-passivated aGNR, where 'n' represents the width of the ribbons (number of C–C dimer lines). In fact, the order of decrease in band gap follows the hierarchy $(3n + 1) > 3n > (3n + 2)$. Quantum confinement

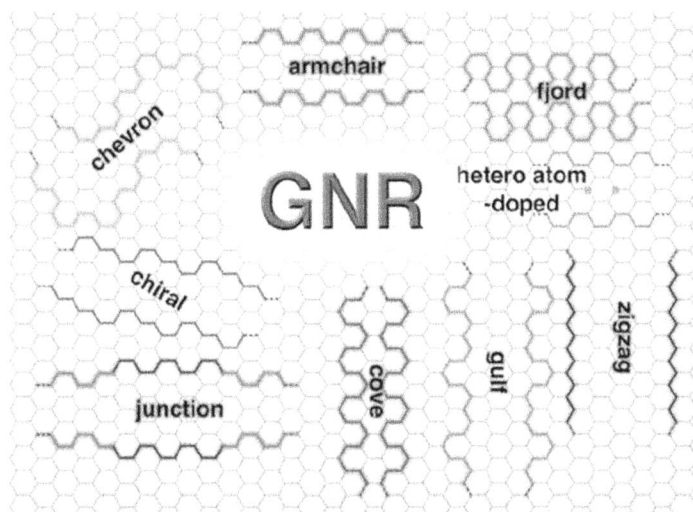

FIGURE 11.4 Schematic representations of different types of GNRs. (Reprinted with permission from Y. Yano *et al.*, J. Org. Chem., 2020, 85, 4–33. @ 2019, American Chemical Society).

plays a crucial role in determining semiconducting behavior of aGNRs and band gap hierarchy. The size dependence of band gap (BG) is well characterized by an inverse relation, $B_g \propto 1/(w + w_0)$, under a hard-wall boundary condition, where w is width of the ribbon and w_0 is a constant (2.4 Å). It is noteworthy that the band gap of aGNR varies with the edge passivating group as well as with applied electric field. It has been reported that there exists an optimum electric field, beyond which the band gap begins to decrease with increasing field strength. However, the band gap decreases at a faster rate for $(3n + 1)$ series as compared to that of $3n$ series and thus $(3n + 1)$ series has a greater band gap tunability. This may be due to delocalized wave functions in $(3n + 1)$ series ribbons and hence any perturbation affects the band structure more than $3n$ series ribbons. One can reduce the band gap of aGNRs upto a few meV by applying an appropriate electric field. What happens is that the valence band and conduction band are shifted in the upward and downward directions, respectively with increase in the field strength and as a consequence the band gap decreases.

11.3.2 Electronic Structure of zGNRs

Electronic properties of zGNRs are a bit more complicated because they are associated with edge magnetization. It has been revealed that hydrogen-passivated zGNRs possess finite local magnetic moment on each terminal C atom. The local magnetic moments on each edge are ferromagnetically (FM) coupled, however, they are coupled antiferromagnetically (AFM) across the edges, resulting an overall net zero magnetic moment in the ground state. It is notable that the excited spin polarized state lies only a few meV higher in energy than the ground state. The energy difference between FM and AFM states decreases as ribbon width increases and ultimately becomes insignificant if width is remarkably large. zGNRs are direct band gap semiconductors having degenerate bands for up and down spin states. Similar to aGNR, here also the band gap decreases with increasing ribbon width. The variation of band gap can be fitted to a functional form of $1/(w + \delta)$, where δ is a constant having the dimension of length.

The small band gap opening, in the case of zGNR, may be inferred to be coming from the interaction of local magnetic moments confined to the edges. Furthermore, these magnetic edge states contribute to the valence band top (VBT) and conduction band minimum (CBM) and the contributions get centered on edge carbon atoms as one moves towards the zone boundary. That is why the energy gap at the zone boundary is insensitive to the width of the zGNRs. The degeneracy of up and down spin states can be lifted under the application of external transverse electric field (E_{ext}). This is so because with the influence of external electric field one spin state (aligned with the E_{ext}) gets stabilized, whereas the other spin state is destabilized. Thus, under an appropriate electric field one can create a situation where one spin state is metallic and the other is semiconducting and such a system is called half-metal. Half-metallic materials are a very important class of materials as electrical current is carried by any one of the spin states, i.e. it gives rise to spin polarized current and hence they are very important in designing spintronic devices. Obviously, the direction of applied field determines the nature of conducting spin channels, i.e. whether it is up or down spin channel. Note that the optimum electric field required for achieving half-metallicity of zGNRs is dependent on ribbon width; it is almost inversely proportional to the ribbon width (w). This is because of the fact that the electrostatic potential difference between two edges is proportional to the width. As the energy shift of edge states depend on potential difference between two edges, the variation of energy gap as a function of E_{ext} shows a universal behavior. The first-principles calculation reveals that the required optimum field is 3.0 V/w(Å).[1]

Not only an external electric field, the electronic properties of zGNRs can also be modulated by introducing charge polarization such as by doping foreign elements or by adsorbing polar organic moieties onto the graphene surface. Doping can be made isoelectronic by replacing two carbon atoms by one boron and one nitrogen atom. It has been observed that position of doping and dopant concentration both greatly influence the magnetic as well as electronic properties. Following the same strategy one can ultimately lead to boron nitride nanoribbon (BNNR) where all carbon chains are replaced with B-N chains. Surprisingly, although zGNRs and zBNNRs are isoelectronic, their electronic properties are completely different; the details of which would be discussed in subsequent sections. Only N-doped defective zGNRs possess pyridine-like substructures. Theoretical investigations have revealed that nitrogen-doping at the edge is energetically favorable, while that in the interior sites is not. It has also been observed that edge nitrogen-doping removes the degeneracy of the spin states, retaining its semiconducting property. However, the CBM of the up spin channel and VBT of the down spin channel touch each other at the Fermi level. Thus, as a whole, the system behaves like a spin gapless semiconductor (SGS), i.e. no energy is necessary to excite electrons from valence band to conduction band, and the excited electrons are 100% spin polarized. Most interestingly, the spin gapless semiconducting properties of these doped ribbons are independent of the ribbon width. On the other hand, symmetrical doping by two nitrogen atoms at the opposite edges of the ribbon, keeping the σ_v plane of symmetry, makes zGNRs nonmagnetic and semimetallic in nature.

11.3.3 Transport Properties of GNRs

Electronic structure of GNRs or similar graphene contacts could be verified from the measurement of their transport properties. In a nanocontact, as the contact length is smaller than the mean free path of electron, it is expected that it would experience ballistic transport properties where the transport channels are a direct reflection of electronic states. As the electronic properties of GNRs are dependent on edge symmetry, i.e. whether it is zigzag or armchair, their transport properties also strongly depend on edge configuration. Now, from the practical measuremental point of view, the synthesized nanostructures have to be attached with suitable leads and they must be eventually integrated into a circuit. That bears the key problem because of the difficulty in contact formation. However, recent experimental techniques have resolved this issue. As for example, the use of 'molecular alligator clips' (which connect the molecular wires to the electrode surface) and scanning tunneling microscope (STM) setup enable one for transport measurement. Note that, depending upon the nature of nanocontact, it may be an electron or a hole conduction. However, the charge carrying rate across a nanowire could be measured by using spectroscopic techniques such as angle-resolved photoemission spectroscopy and gated four-tip STM technique, which provide current-voltage $(I - V)$ characteristics.

Here, it should be pointed out that nanocontacts belong to the group of mesoscopic systems. So, the theory describing electron transport properties of these materials should be different from that of the macroscopic systems or bulk materials. As mentioned earlier, the conductivity of such mesoscopic systems depend upon the length of the conducting material due to quantum effects and the transport is characterized as ballistic. Nevertheless, the nanocontacts experience contact resistance at the interface of material and the lead. This is simply because the current is carried in the contact lead by an infinite number of transverse modes but inside the nanoconductor only a few of transverse modes participate, therefore requiring a redistribution of current among the current carrying modes at the interface. That means, the charge carriers are scattered at the interface. Furthermore, there involves interference among the electronic waves, which is also responsible for contact resistance.

It requires exclusive knowledge of the electronic structure of the nanocontacts to determine their transport properties. However, the major problem is that most of the electronic structure theories deal either with the finite or infinite periodic model systems where electronic motions are treated as if they are in equilibrium. In contrast, the model of a nanocontact, consisting semi-infinite electrodes and finite nanostructure, must be infinite but non-periodic in nature. Secondly, upon the application of finite voltage, which drives the electronic motion through the subsystem must be treated as non-equilibrium in character. Therefore, the appropriate model for treating electron transport phenomena through nanocontact within density functional theory (DFT) is coupled with nonequilibrium Green's function (NEGF) formalism.

Initially, the electronic structure is calculated under the condition that the Hamiltonian interactions are strictly zero beyond some distance, which makes it easier to partition the total system, namely left electrode (L), scattering region (C) and right electrode (R) as shown in Figure 11.5. Due to this partitioning, all the matrix elements of the Hamiltonian or overlap integrals involving the atomic orbitals situated at L and R parts become zero. The Hamiltonian and density matrices are expected to converge to the bulk values of L and R parts but they only differ from the bulk values in the regions close to C, specifically designated as C-L, and C-R as shown in Figure 11.5. Thus, for describing the electron transport phenomena through the nanocontact, it requires to consider the discrete part, L-C-R of the infinite system as demonstrated in the Figure 11.5(b). The electronic distribution within this finite region can be described in terms of Green's function (G) matrix, which is obtained by inverting the finite Hamiltonian matrix

(a)

(b)

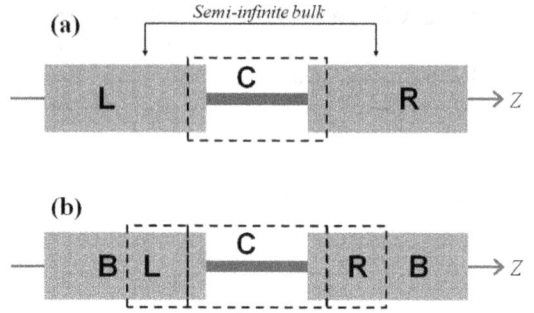

FIGURE 11.5   Two-probe model for calculating the transport property of a nanojunction.

$$\begin{bmatrix} H_L + \Sigma_L & V_L & 0 \\ V_L^\dagger & H_C & V_R \\ 0 & V_R^\dagger & H_R + \Sigma_R \end{bmatrix}$$

where H_L, H_R and H_C are the Hamiltonian matrices for the L, R and C regions, respectively; V_L and V_R are the interaction terms of the contact region with L and R parts, respectively and Σ_L (Σ_R), the self-energy term represents the coupling of L (R) to the remaining parts of the semi-infinite electrode. Now, Green's functions, which depend on the electronic energy (E) as well as applied bias (V_b), can be calculated in terms of advanced (G_A) and retarded (G_R) ones for the right and left parts of the scattering region, respectively. Obviously, they are interrelated by $G_R = G_A^\dagger$.

Having Green's functions, the electronic transmission function ($T(E, V_b)$) can then be expressed by the Landauer formula,[2]

$$T(E, V_b) = Tr[\Gamma_L G_R \Gamma_R G_A] \tag{11.9}$$

where $\Gamma_{L(R)}$ is the coupling matrix of the left (right) electrode and the total current is expressed by the Landauer-Buttiker formula,[2]

$$I(V_b) = \frac{2e}{h} \int_{\mu_L}^{\mu_R} T(E, V_b)[f(E, \mu_L) - f(E, \mu_R)]dE \tag{11.10}$$

where $f(E, \mu_{L(R)})$ is the Fermi-Dirac distribution function for left (right) electrode, and $\mu_{L(R)}$ is the electrochemical potential of the left (right) electrode such that $eV_b = \mu_L - \mu_R$. The fact is that the current passing through the nanocontact is proportional to the transmission probability of electrons through the nanocontact. At zero-bias, the conductance can be expressed in terms of conductance quantum, G_0 ($=2e^2/h$) as $G(E, V_b = 0) = T(E, V_b = 0)G_0$. Now, considering the spin degrees of freedom as an additional parameter, one can calculate spin-dependent transport properties as well.[3]

Transport properties of differently edge-modified GNRs have been studied by using graphene, carbon nanotubes or metal surfaces as contact leads. In fact, GNRs are reported to be excellent candidates for making different kinds of nanocontacts. This has become possible after the discovery of chemical vapor deposition (CVD) and lithographic techniques, which enable development of the designed nanocontacts *in situ*. Furthermore, the electronic properties of the nanocontacts could easily be tuned through edge modification and/or doping. As already mentioned, doping of foreign elements like N and B converts semiconducting aGNRs into metallic ones, which severely affects their charge transport properties. It has been demonstrated that it yields resonant backscattering for the doped GNRs and the phenomenon strongly depends on the position of the dopant element as well as the symmetry of the ribbons. Theoretical calculations[4,5] have demonstrated that B-doping in GNR influences spin-anisotropic scattering process. aGNR nanojunctions are reported to show negative differential resistance (NDR) behavior. Simultaneous doping of B and N to aGNR, which induce *p*- and *n*-type of effects, respectively, could lead to a *p-n* junction diode.

Graphene based *p-n* junction was first realized by Marcus and his group,[6] showing gate-induced carrier type and density, which opens a revenue in the field of bipolar nanoelectronics. Doped-GNR *p-n* junctions had been modeled to show current rectification property and such kinds of devices have recently been fabricated and used as a photodetector with high detectivity at room temperature. Graphene *p-n* junction has also been implemented in electron optics, which directly measures angle-dependent transmission coefficient. Continuous efforts are being made for utilizing graphene-based nanocontacts for optoelectronic devices. Apart from aGNRs, different kinds of zGNR nanojunctions also show interesting electron transport behavior.

Now, transport properties of different GNRs could be correlated with their corresponding electronic energy gap at Fermi level. For aGNRs, the transport gap is almost equal to the corresponding band gap. Thus, the region of zero conductance is reduced with increasing ribbon width. Contrarily, transport properties of zGNRs are controlled by edge states, which are basically the frontier eigenstates. As mentioned earlier, the edge states of hydrogen-passivated zGNRs possess finite local magnetic moments and thus they can be implemented in spintronics devices. Note that, zGNRs can not provide efficient spinfiltering effect due to the symmetric edge structures. However, symmetry breaking by functionalization with organic moiety, introducing defect or doping may turn zGNRs into promising spin-filtering materials.

Defect formation in graphene severely affects the charge transport properties to a great extent as the transport properties are very sensitive to the underlying symmetry of the defective graphene and the disorder potential generated due to defect formation. Introducing point defects in graphene creates localized states, resulting in backscattering at the specified resonant eigenchannels. Subsequently, resonance transmission drops and even in some cases, there may be complete suppression of transmission eigenchannels. It has been demonstrated that randomly distributed bulk vacancies in GNRs strongly suppress the ballistic nature and conductivity of electrons; however, for the edge vacancies, the quasi-ballistic nature of electron conductivity of GNRs is preserved. Some doped defective GNRs are reported to provide very high spin-filtering effects, originating from the appearance of non-degenerate edge states close to the Fermi level. A precise control over simultaneous defect formation as well as doping is really challenging; nevertheless, the interesting features of the electronic structure and magnetic ordering due to simultaneous doping and defect formation may provide a new and advanced technological window.

11.4 Carbon Nanotubes (CNTs)

We know that wrapping of graphene sheet produces CNT. The first discovered nanotube in 1919 was made of several concentric cylindrical shells, separated from each other by a distance of 3.4 Å. So, it was basically a multi-walled CNT. Soon afterwards, single-walled CNTs were synthesized in laboratory using arc discharge method. Similar to GNRs, electronic motions in CNTs are confined in two directions and thus they are 1D systems. The diameter of single-walled CNT may vary from 1 nm to few hundreds of nm and they can have a length-to-diameter ratio greater than 10,00,000.

11.4.1 Geometric Features of CNTs

The geometries of CNTs can be described in terms of unit vectors of graphene. A CNT is defined by its circumferential vector, sometimes called chiral vector ($\vec{C}_h$), which is defined in terms of primitive vectors of graphene by the relation $\vec{C}_h = n\vec{a}_1 + m\vec{a}_2$ as shown in Figure 11.6. The pair of integers (n, m) thus, represents the geometry of a CNT. The circumference of (n, 0) CNTs looks zigzag in nature so these are called zigzag CNTs, while (n, n) CNTs are defined as armchair CNTs as armchair pattern is observed along their circumference. All other configurations of CNTs are designated as chiral. Thus, $\vec{C}_h$ basically defines the chirality of CNTs. The diameter of a CNT can be calculated as $d_t = \frac{|\vec{C}_h|}{\pi} = \frac{a}{\pi}\sqrt{n^2 + nm + m^2}$, where a ($= \sqrt{3}l_{C-C}$, l_{C-C} being C-C bond length) is the lattice constant of graphene. The chiral angle (θ) between chiral vector and primitive vector is given by $\cos\theta = \frac{\vec{C}_h \cdot \vec{a}_1}{|\vec{C}_h||\vec{a}_1|} = \frac{2n+m}{2\sqrt{n^2+nm+m^2}}$. The chiral angle, θ ranges from 0 to 30° due to hexagonal symmetry of graphene sheet. Noteworthy, $\theta = 0$ and 30° respectively give rise to two different types of nanotube, namely zigzag and armchair. The smallest lattice vector along the periodic direction of nanotube, also known as the translational vector ($\vec{T}$), which is perpendicular to chiral vector and can be expressed in terms of primitive vectors of graphene as $\vec{T} = t_1\vec{a}_1 + t_2\vec{a}_2$, by using the relation

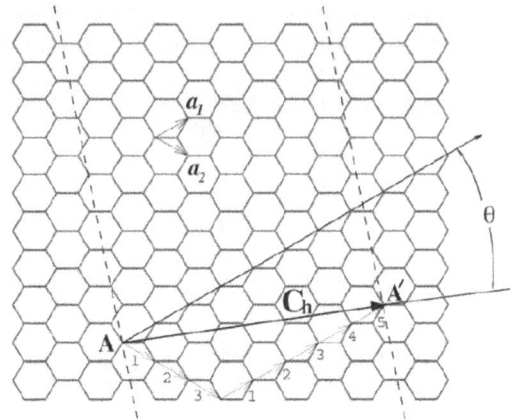

FIGURE 11.6 Formation of CNT from graphene surface and its chiral axis.

$\vec{C}_h \cdot \vec{T} = 0$ and $t_1 = (2m + n)/N_R$, $t_2 = -(2n + m)/N_R$, where N_R is the greatest common divisor of $(2m + n)$ and $(2n + m)$. The length of the translational vector can be obtained as $|\vec{T}| = \sqrt{3}a\sqrt{n^2 + nm + m^2}/N_R$ and finally the number of carbon atoms in a unit cell is given by $N_C = 4(n^2 + nm + m^2)/N_R$. Thus, once n and m are fixed, everything is defined precisely.

11.4.2 Electronic Structure of CNTs

It has been revealed that electronic structures of CNTs are very much sensitive to their geometries and chirality.[7] Depending upon the chirality, CNTs may be metal or semiconductor. Let us have a look at the electronic structures of CNTs on the basis of tight-binding approximation. As mentioned in the previous section, the main advantage of nearest-neighbor approximation in the tight-binding scheme is that it gives a very simple expression for energies of π and π^* states of graphene. Thus, in the case of CNTs, it extends the idea to have the electronic properties with the help of zone folding approximation, without considering the curvature effect. Due to confinement along the circumferential direction of the CNTs, the allowed wave vectors in this direction are quantized; they can have only a set of discrete values. On the other hand, the wave vector along the periodic direction is continuous in nature. The basic idea behind zone folding approximation is to obtain electronic band structure of CNTs by superposition of energy bands of graphene along the allowed k-lines. Within this approximation, the electronic band structure of CNT is the π-π^* bands of graphene along the allowed k-lines folded onto Γ-X direction of one-dimensional Brillouin zone of CNT, where X is the zone boundary.

The zone folding approximation provides the metallic and semiconducting properties of CNTs depending upon their chirality (n,m). It is revealed that a CNT is metallic if $n - m = 3p$ and it is semiconducting if $n - m = (3p \pm 1)$, where p is an integer. Thus, 2/3 of CNTs are predicted to be semiconducting and only a small fraction (1/3) are metallic. So armchair CNTs (n,n) are in general metallic and a subset of zigzag CNTs $(n, 0)$ with n, multiple of 3 is also metallic. For these metallic CNTs, the dispersion relation near the Fermi energy is $E^{\pm}(\delta k) \approx \pm(\sqrt{3}a/2)\gamma_0|\delta k|$, where γ_0 is the nearest neighbors' overlap integral. Obviously, the equation represents a linear energy-momentum relation. Actually, whenever the allowed k-vectors include the K point, it provides metallic character with a non-zero density of states at the Fermi level and when K point is excluded, the systems become semiconducting with a very small band gap. The second condition, i.e. when $n - m = 3p \pm 1$, it leads to band gap opening at the Fermi level with an energy gap $\Delta E_g^1 = \frac{2a\gamma_0}{d_t}$. Clearly, the band gap decreases with increase in tube diameter. Thus, a nanotube with a large tube diameter is expected to be a zero gap semiconductor.

Electronic structures of CNTs have been verified experimentally with the help of scanning tunneling microscope (STM). It has been revealed that the band gap of semiconducting nanotubes is almost inversely proportional to d_t, which agree well with the tight-binding-based theoretical result, ($\Delta E_g^1 = \frac{2a\gamma_0}{d_t}$).

Furthermore, the theoretical prediction that about $\frac{1}{3}$ of nanotubes are metallic and remaining $\frac{2}{3}$ are semi-conducting, have also been verified. Resonances in density of states is observed experimentally on both metallic and semiconducting nanotubes, whose diameter and chiral angles were determined using STM. Other experimental techniques such as resonant Raman scattering, optical absorption and emission spectroscopy have been used for determining the electronic properties of carbon nanotubes; the experimental results nicely corroborate theoretical predictions.

11.4.3 Effect of Curvature on Electronic Structure of CNTs

So far, we have discussed the electronic structures of CNTs within the zone folding model and without considering the effect of curvature on them. However, the carbon atoms of CNT are placed onto a cylindrical wall, which induces a number of changes.

(i) The C-C bonds parallel and perpendicular to the cylindrical axis experience slight difference in bond length;

(ii) unlike graphene, CNTs lack the planar symmetry and hence interaction between π and σ states occur;

(iii) the involved carbon atoms experience partial sp^2 and sp^3 characters and most importantly, (iv) change in bond lengths leads to differences in three hopping terms between a carbon atoms and its three neighbors. Obviously, such effects are not considered in the zone folding approach as π states can not mix with σ states because of their parity mismatch. In the subsequent section, we will discuss the effect of curvature on electronic structure of CNT.

Due to curvature in CNTs, the C-C bonds experience a difference in their lengths and thus the hopping parameters vary. This leads to shifting of Fermi vector (k_F) from K point of the Brillouin zone of graphene. In the case of armchair CNTs, k_F shifts along an allowed line of the Brillouin zone of graphene thereby retaining their metallic character. But in he case of zigzag CNTs, k_F moves away from the K point perpendicularly to the allowed k-lines such that the allowed 1D sub-band no longer passes through k_F. This results in an opening up of a very small band gap at the Fermi level. For small gap semiconducting nanotubes, a secondary band gap arises due to chiral angle and the diameter of the nanotube, which varies as a function of $1/d_t^2$. For quasi-metallic zigzag nanotubes, the secondary band gap can be expressed as $\Delta E_g^2 = \frac{3a^2\gamma_0}{4d_t^2}$, which is so small that thermal excitation at room temperature is enough to promote electrons from valence band to conduction band. Thus, all $n - m = 3l$ nanotubes can be regarded as metallic at room temperature. Curvature effect for small diameter nanotube is so prominent that the C atoms experience intermediate hybridized states in between sp^2 and sp^3. This leads to mixing of σ and π states. Consequently, the band gap of small nanotubes are decreased by more than 50%.

11.5 Graphene Quantum Dots (GQDs)

Electronic confinement in graphene gives rise to an energy gap at the Fermi level as observed in the case of GNRs. This idea is further implemented in graphene flakes or graphene quantum dots (GQDs). These are obtained by cutting graphene sheet into a particular shape and size and passivating all the edge C atoms by hydrogen or any other functional groups. GQDs may have different geometric shapes such as triangular, rectangular, hexagonal, etc.; also there may be different types of edge symmetries like zigzag, armchair and combination of both. Due to hexagonal symmetry of graphene, triangular and hexagonal GQDs can have either zigzag or armchair edges, whereas rectangular GQDs simultaneously have both types of edge symmetries. Figure 11.7 represents different types of GQDs with various shapes and edge symmetries. As shown in the figure, dimension of a GQD is defined by number of hexagonal motifs (N) along edge direction. They have also been characterized as hexagonal (H), trigonal (T), armchair (A), zigzag (Z), etc. A combination of shape (H/T) and edge symmetry (A/Z) specifically defines a GQD

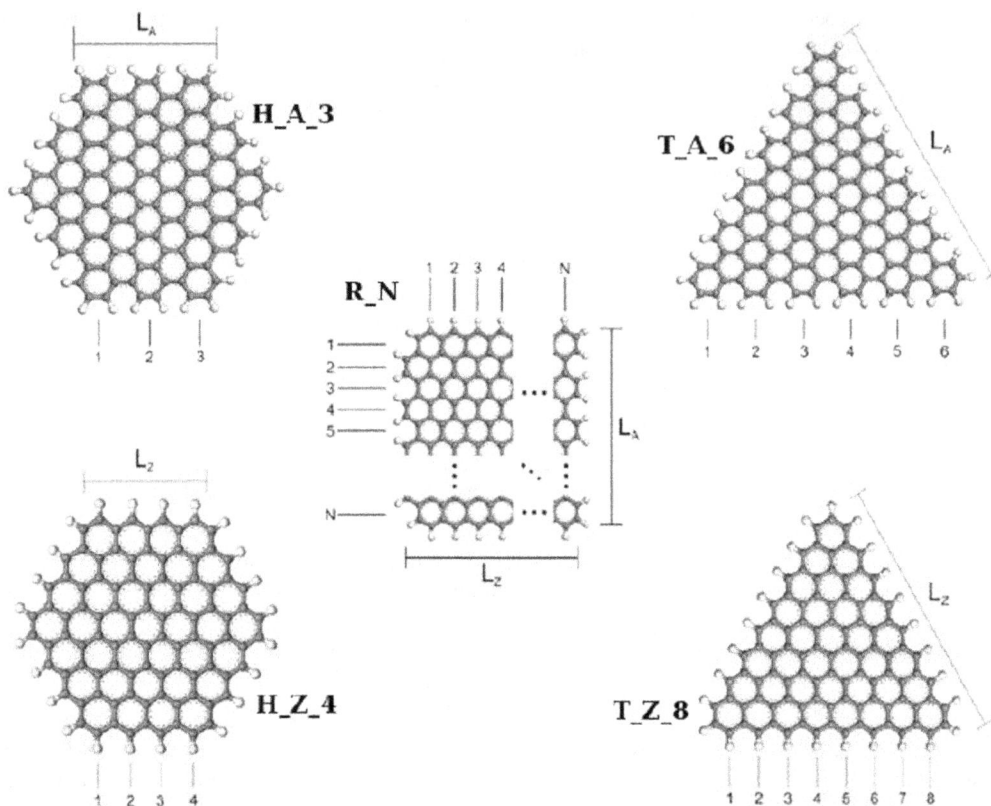

FIGURE 11.7   Structures of different types of GQDs.

of a particular size (N). As for example, H_A implies a hexagonal GQD with armchair edge. Thus, as a whole, H_A_6 represents a hexagonal GQD having $N = 6$ hexagonal motifs along armchair edge. An abbreviation R implies a rectangular quantum dot. It can be shown that the total number of C and H atoms in hydrogen-passivated GQDs are second and first degrees of polynomial of N, respectively. Further, the number of carbon atoms with a given value of N, is maximum for H_A quantum dot, followed by H_Z, T_A, R and T_Z GQDs in decreasing order, while the number of hydrogen atoms follow the order, $H\_A > H\_Z = T\_A > R > T\_Z$. It should be mentioned here that the point symmetry groups of trigonal and hexagonal GQDs are D_{3h} and D_{6h}, respectively, while rectangular GQDs belong to the point symmetry group of C_{2v} and C_{2h} for odd/even values of N, respectively.

11.5.1 Electronic Structure of GQDs

First-principles calculations reveal that most of the GQDs are semiconducting in nature and the HOMO-LUMO gap varies as a function of size and edge configuration. It has been reported that with an increasing number of carbon atoms for almost all the types of GQDs, the HOMO and LUMO energy levels are shifted up and down, respectively, resulting in a decrease in band gap. The largest variation is observed for the T_A type of GQDs, while the T_Z type of GQDs show the least variation. For the latter ones, net spin polarization might have some effect, which prohibits the band gap variation with size. In fact, all the GQDs except T_Z are having non-magnetic ground states, while T_Z_N GQDs show a ferromagnetic ground state with magnetic moment equivalent to exactly N number of unpaired electrons. It turns out that symmetry plays a crucial role in determining total energy, electronic structure and HOMO-LUMO gap of T_Z GQDs.

Researchers have studied energy spectra of triangular and hexagonal GQDs with the help of tight-binding model and Dirac equation approach. For the continuum model, the Dirac-Weyl equations are solved imposing three different boundary conditions, armchair, zigzag and infinite-mass boundary conditions. It has been reported that the energy spacing between successive energy levels decreases as the size of GQDs increases and ultimately it tends to zero, reflecting the energy spectrum of graphene sheet. External magnetic field affects the energy spacing between successive energy levels in the energy spectrum of different GQDs. It has been revealed that the energy gap of hexagonal quantum dots decreases rapidly as magnetic flux increases, whereas the gap of triangular dots reduces smoothly with increasing magnetic flux. Thus, energy gaps of hexagonal GQDs can be more easily tuned by an external magnetic field, as compared to trigonal GQDs. The tight-binding Hamiltonian with nearest-neighbor interaction reveals that for hexagonal GQDs with armchair edges, the HOMO-LUMO energy gap varies inversely with the square root of the number of carbon atoms ($E_{gap} \propto \frac{1}{\sqrt{n_C}}$).[8-9] This is quite expected for confined Dirac Fermions with photon-like linear energy dispersion. However, for similar GQDs with zigzag edge symmetry, the energy gap rapidly decreases as n_C increases. This may be attributed to the localized states at the zigzag edges of the quantum dots. It has also been reported that the energy gap of GQD increases on changing the geometrical shape from hexagonal to triangular, keeping the edge symmetry fixed. The triangular zigzag quantum dots (T_Z GQDs) are observed to possess a shell of degenerate states at the Fermi level, although the energy gap is also inversely proportional to the square root of n_C. The presence of these partially occupied degenerate states in between a well defined energy gap provides a unique way to simultaneously control magnetic as well as optical properties of T_Z GQDs. This in turn gives rise to possibilities of using them for optoelectronics, spintronics and intermediate-band photovoltaic devices. The size-dependent energy gap for different quantum dots have been measured by using scanning tunneling spectroscopy. The experimental data excluding metallic quantum dots can be fitted in a power series in the least-squares sense giving $E_{gap} = 1.57 \pm 0.21 \ eV \ nm/L^{1.19 \pm 0.15}$. This is in close agreement with the predicted scaling trend, $E_{gap} = 1.68 \ eV \ nm/L$ resulting from quantum confinement and linear dispersion of monolayer graphene.[10]

11.6 New 2D Carbon Allotropes: Defected Graphenes and Pentagraphene

As already mentioned, the cross-linking property of carbon leads to a large number of allotropes, especially in the 2D regime, only a few of them have been discovered to date. Graphene is the most stable 2D carbon allotrope, however, experimental studies have revealed that defect formation is inherent in graphene. Now, if the defect is formed periodically then it leads to a new allotrope with new crystal structure and physico-chemical properties.

Fullerene, although not a 2D system, is a famous example of allotropic carbon consisting hexagons and pentagons and the pentagons are periodically distributed over the surface of fullerene. Over the last few decades a large number of 2D carbon allotropes has been proposed based on theoretical calculations and surprisingly, some of them have also been isolated experimentally. Pentaheptite has been proposed with a planar structure of carbon composed of pentagons and heptagons, the stability of which is comparable to that of C_{60} fullerene. Theoretical calculations reveal that pentaheptite is metallic in nature. T-graphenes[11] are also one of such a kind of predicted 2D structures, consisting of tetragon and octagon. First-principles calculations demonstrate that although T-graphenes are energetically in a metastable position with respect to graphene, these are dynamically stable. Between the two possible structures, buckled T-graphene possesses Dirac-like fermions with semimetallic character and high Fermi velocity, whereas planar T-graphene is metallic in nature. Although calculated Young's modulus of T-graphenes is less than that of graphene, their mechanical strength is appreciable. It is reported that semimetallic T-graphene can be converted to a semiconductor with a certain band gap through substitutional doping of boron-nitrogen pairs. Calculations also reveal that T-graphenes can be employed for hydrogen storage material via Ti-adsorption.[12]

A new family of planar carbon sheets consisting of five, six and seven membered rings has been proposed and named haeckelites. They are supposed to be energetically more stable than C_{60} but in a metastable position with respect to graphite. In this family, the most symmetric structure consisting of pentagons surrounded by hexagons and heptagons is predicted to be the most stable one. Haeckelite sheets can also be rolled into nanotubes of different diameters and chirality. Tight-binding theory predicts that all the haeckelite sheets and their corresponding nanotubes with any diameter and chirality are metallic in nature. Calculated mechanical properties of the haeckelites are similar to that of graphene.

2D carbon sheets comprising of octagons and pentagons as building blocks is known as OP graphenes (OPGs).[13–14] Two members in this community are known, which are designated as OPG-L and OPG-Z. Both the structures are energetically and kinetically viable as predicted from their zero temperature energy, finite temperature free energy and phonon structure, on the basis of first-principles calculations. It has been demonstrated that OPG-L is metallic in nature, while OPG-Z is a gapless semimetal. The latter one is remarkably anisotropic with a pair of anisotropic Dirac points very close to the Fermi level. Apart from these, another metallic 2D carbon allotrope, namely W sheet, consisting of squares, hexagons and octagons is also known.[15] The corresponding nanoribbons are predicted to be semiconducting in nature.

Periodic defect formation in graphene can destroy its hexagonal symmetry thereby opening up the band gap at the Fermi level.[16] A combined effect of divacancy and Stone-Wales defect formation in graphene gives rise to a planar semiconductor carbon allotrope, known as octite. It has the structural motif consisting of a central octagon surrounded by hexagons and the neighboring cells are attached through pentagons and octagons. Density functional theory (DFT) calculations demonstrate a band gap of 0.2 eV for octite. The 5-6-8 defect, which is basically the building block of octite, can also be introduced into pristine GNRs. These defective GNRs again preserve the semiconducting behavior of octite.

Mandal *et al.* have explored the possibility of existence of a new 2D carbon nanostructure and its 1D derivatives by means of first-principles calculations. This new carbon allotrope consists of hexagons (H), octagons (O) and pentagons (P) and is designated as HOP graphene (HOPG). The authors have discussed the crystal structure, stability, electronic and transport properties of HOPG and its 1D derivatives. It is notable that, octite is also made of pentagons, hexagons and octagons but HOPG is different from octite in different aspects (see Figure 11.8). First, the unit cell of octite consists of a central octagon, completely surrounded by hexagons, while the rectangular unit cell of HOPG consists of one hexagon, pentagon and parts of a pentagon, hexagon and octagone. Second, HOPG is a covalent metal, while octite is a semiconducting material. Of particular note, HOPG is energetically more stable than graphyne and graphdiyne, the latter one having been synthesized recently. Most of the HOPG ribbons are predicted to be metallic but they possess a gap above the Fermi level, which makes them very useful in nanoelectronics. The HOPG

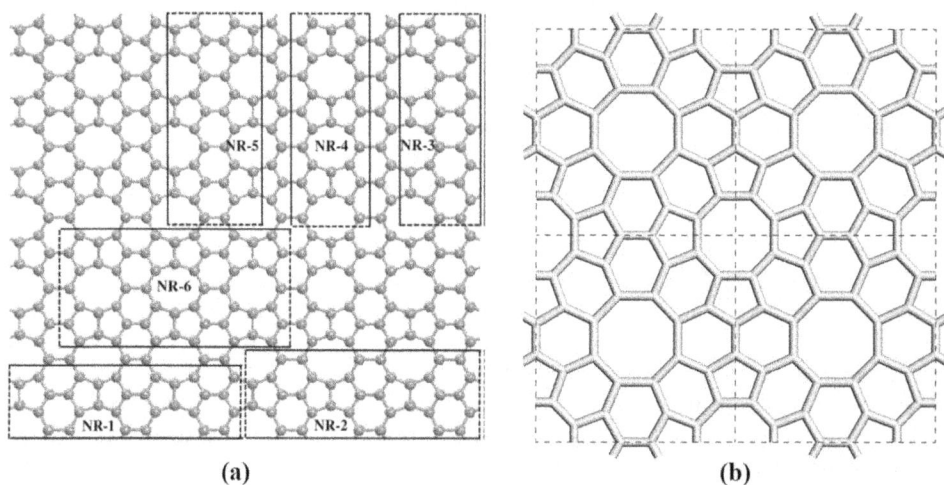

(a)   (b)

FIGURE 11.8 Structural difference between HOPG (a) and octite (b). (Reprinted with permission from B. Mandal *et al.*, Phys. Chem. Chem. Phys., 2013, 15, 21001–21006. @ 2013, The Royal Society of Chemistry; D.J Appelhans *et al.*, Phys. Rev. B., 2010, 82, 073410. @ 2010 The American Physical Society.).

nanoribbons reveal negative differential resistance (NDR) phenomena with a high peak-to-valley ratio and also current rectification properties.

Recently, another 2D metastable carbon allotrope, penta-graphene, comprising only pentagons has been proposed on the basis of high level theoretical calculations. Penta-graphene resembles the Cairo pentagonal tiling. It has been shown that such a 2D carbon sheet can be obtained by chemically exfoliating a single layer from the T12-carbon, phase. First-principles calculations demonstrate that although penta-graphene is energetically metastable compared to graphene, it is dynamically and mechanically stable and can withstand temperatures as high as 1000 K. Due to its unique atomic configuration, penta-graphene has an unusual negative Poisson's ratio and ultrahigh ideal strength that can even outperform graphene. Furthermore, unlike graphene that needs to be functionalized for opening a band gap, penta-graphene possesses an intrinsic quasi-direct band gap as large as 3.25 eV, close to that of ZnO and GaN (see Figure 11.9). When rolled up, it can form pentagon-based nanotubes, which are semiconducting, regardless of their chirality. Penta-graphene nanoribbons (pentaGNRs) are also predicted to be energetically and dynamically stable with wide band gap semiconducting properties. Stacking into different patterns, penta-graphene produces stable 3D twin structures of T12-carbon, which have band gaps even larger than that of T12-carbon. Although penta-graphene has not been synthesized yet, the versatility of the 2D structure and its derivatives is expected to have broad applications in nanoelectronics and nanomechanics.

As penta-graphene is a wide bandgap semiconductor (in between 2.24-4.3 eV), it is necessary to develop gap tuning strategies for tailoring this material for optoelectronic applications. In this sense, the particular topology of penta-graphene, which presents both sp^2 and sp^3-like carbon hybridization in its lattice, raises the question of how its structural and electronic properties might behave by targeting a particular hybridization in the doping processes. It has been investigated that selective doping of penta-graphene structures through engineered line defects can be an effective tool for tuning their electronic behavior. Theoretical calculations have demonstrated that such strategies may bring penta-graphene from large band gap semiconductors to metallic or semimetallic ones.

Rajbanshi *et al.* systematically explored the energetics and electronic structure of pentaGNR to understand the feasibility of its synthesis and its applicability in nanoelectronics, using computational methodologies. They have further explored the possibility of tuning the electronic properties by varying the width of the nanoribbons and also by applying uniaxial strain. The computational results demonstrate that, pentaGNRs are thermodynamically meta-stable with respect to graphene nanoribbons but more stable than α–graphyne nanoribbons. Compared to the edge energies of graphene and many other TMC nanoribbons the N-pentaGNRs have much lower edge energy suggesting that formation of N-pentaGNRs from 2D penta-graphene sheet may be thermodynamically feasible. The pentaGNRs are direct wide band gap semiconductors and the band gap variation with the ribbon width is very small for wider ribbons. This opens up

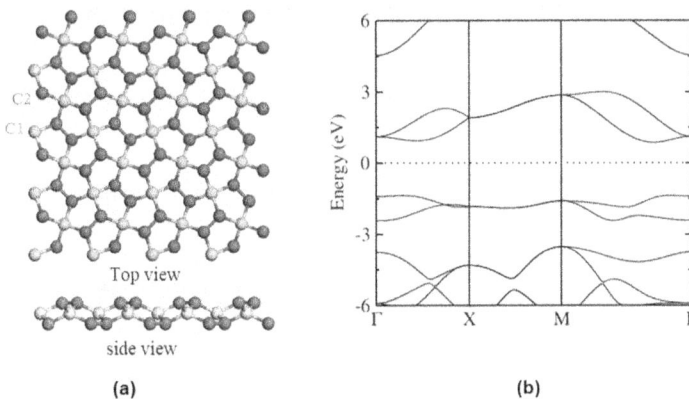

FIGURE 11.9 Geometry (a) and electronic structure (b) of penta-graphene monolayer. (Reprinted with permission from B. Rajbanshi *et al.*, Carbon, 2016, 100, 118–125. @ 2016, Elsevier B.V.).

a scope for the experimentalists to synthesize a stable wide pentaGNR with large band gap. On application of uniaxial strain, the band gap of pentaGNRs decreases continuously, providing a strain-tunable band gap for pentaGNRs.

11.7 White Graphene

Another interesting material in the era of quasi-two-dimensional material is hexagonal boron nitride (h-BN), commonly referred as 'white graphene' due to its highly transparent nature. It is an interesting insulating analog to semimetallic graphite or graphene. As its name implies, at the atomic level this material resides in a hexagonal pattern with alternating atoms of boron and nitrogen, creating the hexagon. The system is isoelectronic with graphene and the crystal structure very closely resembles that of the hexagonal crystal lattice of graphene. On its own, h-BN has some very interesting characteristics; as mentioned, it is almost completely transparent, absorbing very little visible light. The material is also quite strong, is an excellent thermal conductor but an electric insulator and has also been shown to provide many advantages over other materials in similar applications.

11.7.1 Geometry and Crystal Structure

h-BN forms graphytic-like crystals in 3D bulk environment. It is noteworthy, besides the layered h-BN structure, there are varieties of 3D bulk cubic crystals of BN (c-BN); the most common phases are wurtzite BN (wz-BN), and zinc-blende BN (zb-BN). However, the equilibrium cohesive energies of h-BN is the highest and hence it is the most energetic bulk structure. On the other hand, the cubic BN (c-BN) is the most stable in its wurtzite form and is known to be the second hardest material of all. Boron nitride in this form has a variety of interesting properties, including being ultrahard in nature, having a high melting point and high thermal conductivity. Cubic BN has a similar structure to diamond, although, unlike diamond, c-BN is not reactive to ferrous materials even at high temperatures, and it has the ability to form both *n*- and *p*-type semiconductors when doped appropriately. c-BN is superior to diamond for industrial tools used in cutting and shaping iron- or nickel-containing materials, as the c-BN blades do not deteriorate as a result of reactions with materials of these types. Cubic boron nitride thin films are also known.

h-BN monolayers are fascinating, atomically thin materials consisting of sp^2-hybridized boron (B) and nitrogen (N) atoms with a 1:1 stoichiometry arranged in a honeycomb lattice. As already mentioned, these sheets are isostructural and isoelectronic to graphene, featuring eight valence electrons per BN that engage in strong in-plane σ and weaker π bonds. However, owing to the electronegativity difference between B and N atoms, there involves partial electron transfer from B to N and as a consequence, in contrast to purely covalent bond in graphene the bonding between B and N gains an ionic character. The charge transfer from B to N dominates several properties of 2D h-BN including the opening of the band gap. In this respect the h-BN honeycomb structure is complementary to graphene. In analogy with the carbon-based relatives, 2D h-BN sheets also form BN-nanotubes and fullerene-like structures. Furthermore, 1D h-BN nanoribbons and porous 2D networks are also extensively studied on the basis of first-principles calculations.

11.7.2 Electronic Structure of h-BN

Three-dimensional h-BN, wz-BN and zb-BN crystals are indirect band gap semiconductors with calculated band gaps in the range of 4.5 to 5.5 eV. For graphite-like 3D h-BN having 2D BN atomic layers (in *xy* plane), the band structure is composed of the band structures of these individual atomic layers with hexagonal symmetry; however, they are slightly split due to weak interlayer interaction. The valence band top (VBT) has N-p_z character, while the conduction band minimum (CBM) is formed from B-p_z orbitals. The overall features of the total density of states (TDOS) look similar to that of other 3D crystal structures. VBT consists of two parts separated by a wide intraband gap where the lower energy band is projected mainly to N-*s* and partly to N-*p* and B-*s* orbitals and the higher energy band arises mainly due to N-*p* and partly due to B-*p* orbitals. zb-BN and wz-BN crystals also have similar characteristics but the main

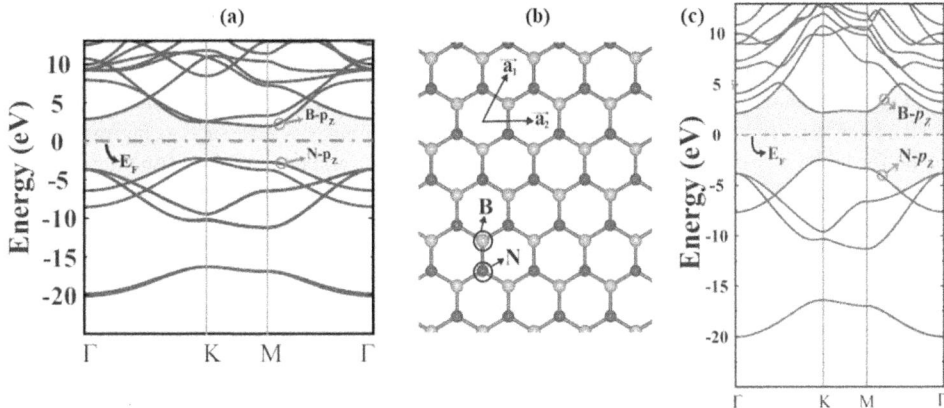

FIGURE 11.10 (a) Electronic band structure of 3D bulk h-BN (b) Geometry of 2D h-BN and (c) electronic band structure of 2D h-BN. (Reprinted with permission from M. Topsakal *et al.*, Phys. Rev. B, 2009, 79, 115442. @ 2009 American Physical Society.).

points of differences of these three 3D crystals are pronounced in the lower part of the conduction band. As previously mentioned, there involves a charge transfer interaction between B and N atoms in the same layer. The amount of charge transfer from B to N, ΔQ for h-BN, wz BN, and zb BN are of the order of 0.3–0.4 electrons. ΔQ of h-BN crystal is significantly larger than that of of zb BN and wz BN crystals, which may be attributed to the shorter B-N bond length in h-BN crystal. In 3D bulk crystal, the h-BN monolayers are weakly bound by long range cohesive forces, maintaining an equilibrium distance of 3.3 Å. This exhibits a clear difference between the electronic structures of 3D and 2D h-BN crystals. The chemically inequivalent sublattices make 3D h-BN an insulator with a band gap of more than 5.0 eV, while the 2D h-BN sheet is semiconducting in nature.

2D h-BN is a semiconductor having hexagonal lattice structure. The π and π^* bands of graphene which cross at the K and K' points of the Brillouin zone open a gap in the 2D h-BN due to bonding and antibonding combinations of N-p_z and B-p_z orbitals. The contribution of N-p_z is pronounced for the filled band at the edge of valence band. The calculated band gap is indirect and reported to be around 4.5 eV. The representative band structures of 3D and 2D h-BN crystals are shown in Figure 11.10.

11.7.3 General Properties and Applications of h-BN

The binary elemental composition of h-BN introduces striking physico-chemical differences in it as compared to the isoelectronic C allotrope, graphene. A considerable charge transfer from B to N, triggered by the difference in electronegativity between B and N, induces ionic character in h-BN sheet. Additionally, h-BN features a low dielectric constant, large breakdown voltage, low defect density, high chemical and thermal stability, and excellent thermal conductivity. Furthermore, absence of dangling bonds at the edges of h-BN makes it behave differently in magnetic character in comparison to that of graphene. Monolayer h-BN prevails in different aspects; as for example, high quality h-BN nanosheets on copper provide a strong corrosion resistance in NaCl or NaOH solution, a single h-BN sheet can sustain temperatures up to 1123 K, even when heated in air thereby out-performing graphene regarding thermal stability. On the basis of such unique characteristics, h-BN is recognized as a relevant 2D material, complementing the library of 2D materials and constituting a highly important building block for different stacked heterostructures.

A wealth of h-BN applications in diverse fields have been reported. Ultrathin h-BN is successfully implemented as insulating material and gate dielectric layer in field effect transistors, sensor material, protective cover in devices, pore material for DNA detection, and buffer layer in film growth. Additionally, different forms of h-BN such as thin films, powders, and nanocrystals are used in deep-UV photonic devices including lasers and neutron detectors. h-BN has been successfully used in high power electronic devices, electrodes in Li-ion batteries, low friction materials, ceramics, heterogeneous catalysis and anti-cancer drug delivery. The good environmental compatibility and non-toxicity of h-BN makes it very useful in green technologies. h-BN is applied to acrylic sheets as fire retardant and anti-microbial agent. It has also been used for developing graphene-based nanoelectronic devices, such as high speed transistors, calls for high electron mobilities. The sensing capability of graphene is quite stunning and is being leveraged in advanced futuristic micro-device designs, particularly in biological applications. However, this atomic level sensing performance may be further enhanced when graphene layer is combined with h-BN.

11.8 Boron Nitride Nanoribbons (BNNRs)

Similar to graphene, two unique orientations in the 2D h-BN yield boron nitride nanoribbons (BNNRs) with uniform edges. These are armchair edged BNNRs (aBNNRs) and zigzag edged BNNRs (zBNNR). The profile of the atomic configuration of B and N at both edges of the nanoribbon determines their electronic and magnetic properties. The properties can, however, be modified by the chemical functionalization as well as by changing the width of the ribbons. Bare and hydrogen-passivated aBNNR are wide band gap semiconductors. Unlike zGNR, hydrogen-passivated zBNNRs are also semiconductors. Band gaps of both types of nanoribbons depend on their widths, w, which is defined by the number of BN pairs, n, in the primitive unit cell. The variation in the band gap (E_G) as a function of n is shown in Figure 11.11. In general, the properties of the nanoribbons approach those of the 2D h-BN as the width $n \to \infty$. However, the band gap of zBNNR approaches a certain value, which is smaller than the band gap of h-BN, may be due to the localized edge states. As shown in Figure 11.11, for narrower aBNNRs ($n < 8$), the band gaps vary with n, but for $n > 8$, the band gap approaches the bulk value.

It should be mentioned here that the atomic and electronic structures of bare and hydrogen-passivated aBNNR are a little bit different. The edge atoms of the bare aBNNRs are reconstructed; the B and N atoms are aligned in 2 planes. As a result of that, there appear two edge states just below the conduction band edge. These two states are normally degenerate when n is large, but the degeneracy is lifted when two edges couple, especially for the narrower ribbons. Further, because of these edge states, the band gap is indirect for bare aBNNRs. Upon hydrogen passivation, these edge states disappear and thus the band gap of H-passivated aBNNR becomes direct. The electronic and magnetic states of zBNNR also depend upon whether their edges are passivated with hydrogen atoms or not. Bare zBNNRs are magnetic metallic metal but H-passivation converts them into nonmagnetic wide band gap semiconductors. However, zBNNRs with H-passivated N-edges only (B-edges remaining bare) are AFM semiconductors, the AFM edge state being localized at the B side. On the other hand, when only the B-edges are passivated with hydrogen atoms, leaving N edges bare, the magnetic edge states are localized at the N sides and the corresponding

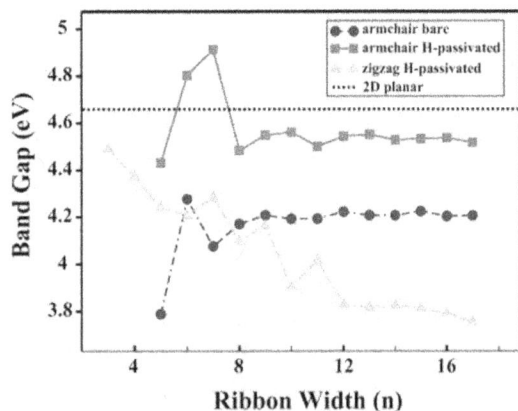

FIGURE 11.11   Variation of band gap for bare and hydrogen-passivated h-BNRs. (Reprinted with permission from M. Topsakal *et al.*, Phys. Rev. B, 2009, 79, 115442. @ 2009 American Physical Society.).

ground state of the zBNNR is ferromagnetic in nature. In a nutshell, the nearest-neighbor N-N interaction is ferromagnetic and the B-B interaction is antiferromagnetic. Thus, as a whole, when both edges are passivated with H atoms, zBNNR becomes non-magnetic.

11.9 Boron Nitride Nanotubes (BNNTs)

As the structural analog of CNT, BN nanotube (BNNT) was first predicted in 1994. Since then, BNNTs have become one of the most intriguing non-carbon nanotubes. Contrary to metallic or semiconducting CNTs, BNNTs are electrical insulators with a high band gap of about 5 eV, irrespective of the tube diameter and chirality. Furthermore, BNNTs are chemically highly stable and possess excellent mechanical properties, and high thermal conductivity. Such unique properties make BNNTs promising in a variety of potential applications such as in optoelectronics nanodevices, functional composites, hydrogen accumulators, electrically insulating substrates, etc.

11.9.1 Morphology and Crystal Structure of BNNTs

BNNTs may crystallize in single- and multi-walled structures; the corresponding 2D sheets may also be mono- or multi-layered thick. Unlike the CNTs, which are most popular in their single-walled form, single-walled BNNTs are rather rarely observed and studied. Single-walled BNNTs are typically unflavored due to peculiar B–N stacking characteristics. Compared to the covalent C-C bonding in CNTs and graphene, the B–N bonding possesses a partially ionic character which leads to the prominent, so-called 'lip–lip' interactions between neighboring BN layers. The B and N atoms are in succession superposed along the chiral axis of the nanotube. It has been reported that the interlayer spacing of multi-walled BNNTs is slightly larger than that of bulk h-BN, which might be attributed to the inner stresses within the bent walls.

Chiralities of BNNTs are similar to that of CNTs. Boundary conditions after one rotation around a flat sheet axis give a limited number of choices for the helicity of a hexagonal layer relative to the tube axis. Thus, the tube helicity may be zigzag or armchair type in which the [10,10] or [11,20] directions of the hexagonal sheet are parallel to the tube axis, respectively. Besides these, there are other types of helicities with varying chiral angles, which have already been discussed about in the case of CNTs. In the case of multi-walled BNNTs, a striking character is that all of the layers within an individual nanotube prefer to have the same layouts. In fact, crystallization of the individual layers follows the same tendency to have the atomically perfect B-N stacked consecutive layers as those in 3D bulk h-BN crystal. In contrast, multi-walled CNTs do not have that tendency, rather, they experience a relative freedom in a rotational disorder between neighboring graphytic sheets, which leads to a wide variety of helicities. Such a characteristic feature of grouped selective helical angles within a multi-walled BNNT and the symmetry of multilayer sheets gives rise to the two unique stacking orders, either hexagonal or rhombohedral stacking patterns.

11.9.2 Electronic Structure of BNNTs

The tight-binding method has been employed to study the electronic structure of BNNTs, which reveals that the tubes are semiconducting in nature with direct or indirect band gaps. Depending upon the tube diameter, the band gap may vary from $\sim$2 eV, in the case of very tiny tubes, to 5 eV for wider tubes. According to band-folding analysis, BNNTs have a direct wide gap for zigzag $(n,0)$ tubes and an indirect gap for armchair (n,n) tubes.

Various methods are attempted to tune the electronic structure of BNNTs, such as application of an electrical field, introducing defects and doping, applying strain, or modifying the tube surface. These can directly reduce the band gap of BNNTs either by introducing localized energy states in between the band gap region or by splitting the degeneracy of the bands (Stark effect). Chemical functionalization also enriches the properties of BNNTs; such as, the polarization field induced by chemical adsorption, has a great influence on the electronic structure of BNNTs.

In the case of double-walled BNNTs, as there involves hybridization between π and σ states of inner and outer tubes, the VBT and CBM are localized on the outer and inner tubes, respectively. Due to confinement effect, the band gap of the outer tubes is slightly smaller than that of the inner tubes. Now, within the interwall region, there exists almost free electronic states, which induce peculiar charge redistribution therein. When a strong electron withdrawing group like fluoride is doped, it significantly modifies the interwall interactions, which leads to both layers coming into effective conduction channels. Furthermore, there are strong interactions between electrons and holes in BNNTs, which alter their optical response too.

11.9.3 General Properties and Applications of BNNTs

For a long time, people were much more interested in the astonishing properties of graphytic 3D h-BN such as low density, electrical insulation but high thermal conductivity, superb oxidation resistance, non-reactive towards acids and melts, and low frictional coefficient. Surprisingly, BNNTs possess all those advantageous properties. Compared to metallic or semiconducting CNTs, BNNTs are wide band gap semiconductors. As mentioned in the last sections, BNNTs have a rigid electronic structure, which is almost independent of the tube diameter and morphology. However, proper doping, application of strain, and/or chemical functionalization on BNNTs can also make them narrow band n- or p-type semiconductors. Thermal conductivity of BNNTs is also comparable to that of CNTs, which are supposed to be excellent thermal conductors. Most importantly, unlike CNTs, BNNTs exhibit distinguishable chemical stability. In fact, they are inert to most acids and bases. Thus, BNNTs are superior to CNTs for practical use in nanotube-based nanodevices or protective shields on various nanomaterials, especially for those performing at high temperatures and in hazardous conditions.

BNNTs exhibit excellent luminescence at around 230 nm, which is attributed to the excitonic effects as discussed earlier. This indicates the potentiality of BNNTs used for optical nanodevices working in the UV regime. The photoluminescence quantum yield of BNNTs is reported to surpass that of CNTs. Apart from these, BNNTs exhibit excellent elastic properties; in fact, the mechanical stiffness of BNNTs rivals that of CNTs. BNNTs are possibly the stiffest insulating fibers ever known. There have been many other interesting physico-chemical properties of BNNTs, which include piezoelectricity, immobilization of ferritin proteins on BNNT surfaces for some prospective medical and nanobiological applications, irradiation stabilities of BNNTs, and so on.

Multi-walled BNNTs are also found to be valuable for the reinforcement and/or increasing thermal conductivity of insulating polymeric films and fibers. It has been reported that the yield stress and/or elastic modulus of insulating polymer films may be increased up to 30–50% by loading with rather modest weight fractions of BNNTs only. In such cases, the BNNTs do not change either their alignment, the axes are aligned along the fiber axes during electrospinning. Furthermore, since the tubes do not absorb a visible light due to a large band gap ($\sim$5.0 eV), the film becomes entirely transparent. However, the thermal conductivity of such film increases more than three-fold, along the aligned BNNTs' axes direction, due to the effect of BNNTs.

11.10 Phosphorene

Phosphorene is one of the several inorganic analog of graphene, which is obtained by mechanical exfoliation of black phosphorus (BP). 3D black phosphorus is a layered material consisting of puckered layers of phosphorus, held together by van der Waals forces. A single layer of black phosphorus was successfully exfoliated and isolated from its bulk structure in 2014. It is well-known that graphene is a gapless semiconductor, while its inorganic analogs like transition metal dichalcogenides (TMDs) are large band gap semiconductors. Phosphorene fills the gap between graphene and TMDs with numerous unique properties. High electron mobility, strong photo-response, high Seebeck coefficient, excellent electrical resistivity and many other properties of phosphorene have valued this material in between those of graphene and TMDs. These make phosphorene suitable for several applications in thermal imaging, thermoelectrics, fiber optics, etc.

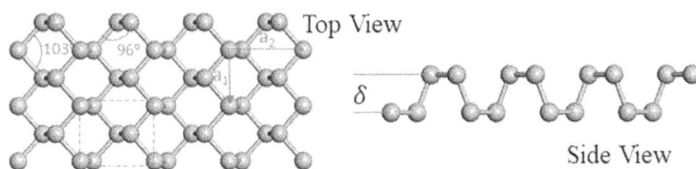

FIGURE 11.12 Geometry of 2D phosphorene monolayer, indicating its puckered structure.

11.10.1 Geometry and Crystal Structure

Unlike the planar structure of graphene, single layered phosphorene possesses a honeycomb-like lattice with puckered structure. It has an orthorhombic crystal structure, where the phosphorus atom is covalently bonded to three other neighboring atoms. The lattice parameters are reported to be $a_1 = 3.35$ Å and $a_2 = 4.62$ Å. As the phosphorus atom has five valence electrons ($3s^2 3p^3$), each phosphorus atom undergoes sp^3 hybridization in order to maintain the puckered structure. Figure 11.12 depicts the puckered structure of phosphorene, where each monolayer consists of two atomic layers of P including two kinds of interatomic bonds. The two atomic planes are separated by 2.5 Å, known as the buckling parameter (δ). Noticeably, the intralayer P–P bonds are shorter than the interlayer P–P bonds. The P-P-P bond angles of phosphorene are 96.34° and 103.09° are much closer to that of a perfect tetrahedra structure (109.5 °), which indicates high structural stability of the quasi-2D crystal lattice. Phosphorene possesses an interesting structural anisotropy, which makes it distinguished from other 2D layered materials. Along the armchair direction of the hexagonal lattice, phosphorene has a puckered structure but along the zigzag direction it appears as a bilayer configuration. Such a structural arrangement in phosphorene is the basis for the anisotropic behavior in its physico-chemical properties.

11.10.2 Mechanical Properties

Phosphorene shows interesting anisotropic mechanical properties. It can bear high tensile strain (~30 and 27% along armchair and zigzag directions, respectively). However, phosphorene has a superior flexibility along with a much smaller Young's modulus (44 GPa in armchair direction and 166 GPa in zigzag direction) than graphene. Therefore, ductility of phosphorene is more along armchair direction than that along zigzag direction. Furthermore, it possesses an unusual negative Poisson's ratio. Such distinct mechanical properties of phosphorene arise mainly from two reasons; first, the P–P bond strengths are relatively weak and second, under the application of tensile strain, the bond angle in the puckered direction can easily change, rather than the bond stretching, which subsequently reduces the required strain energy. The smaller Young's modulus of phosphorene along armchair direction as compared to the zigzag direction is also attributed to the puckered structure. Young's modulus of phosphorene can be modified by chemical functionalization; however, the anisotropy still exists as an inherent property.

11.10.3 Electronic Structure and General Properties of Phosphorene

Phosphorene is assigned to be an intrinsic semiconductor with a direct band gap, which is, however, tunable with applied strain, chemical functionalization, doping and of course on layer thickness. Photolumines-cence study as well as theoretical calculations reveal that the band gap of single layer phosphorene is in between 1.5 and 2.0 eV. Theoretically predicted band structures of bulk black phosphorus and phosphorene are shown in Figure 11.13.

Phosphorene exhibits a strong dependence of band gap on the layer thickness, which is beneficial for its potential applications in photonics and optoelectronics. The band gap of phosphorene gradually decreases with increasing the number of layers and it converges to that of bulk black phosphorus (0.3 eV). As outlined before, there are several other factors that can also control the band gap of phosphorene. As for example, phosphorene can be converted from a direct band gap semiconductor to an indirect band gap semiconductor, semimetal and even metal by applying axial strain. Furthermore, displacement of few

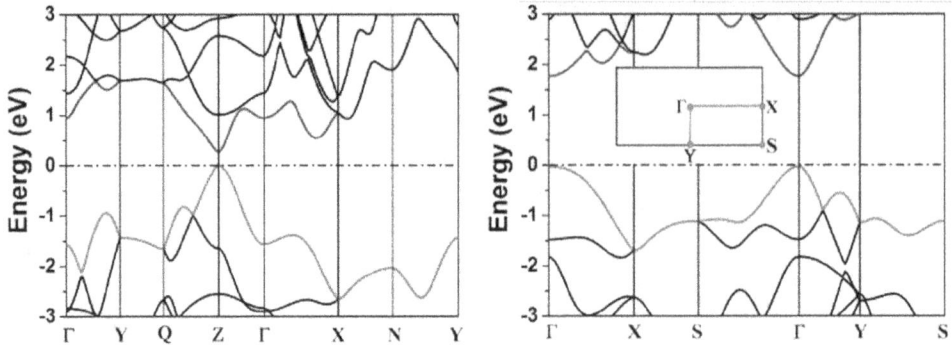

FIGURE 11.13　Electronic band structure of 3D bulk (left panel) and 2D phosphorene monolayer (right panel). (Reprinted with permission from H.Y. Lv *et al.*, arXiv:1404.5171, 2014.).

atoms in monolayer phosphorene results in a honeycomb structure, which is composed of two graphene-like sublattices, appearing as an out-of-plane buckled monolayer in a zigzag manner. Such an arrangement practically produces a new allotrope of phosphorene, known as blue phosphorene. Blue phosphorene is predicted to be an indirect band gap (2 eV) semiconductor. A couple of other thermodynamically stable allotropes of monolayer phosphorene such as γ-phosphorene, δ-phosphorene, etc. have been recently predicted from theoretical studies. The possibility of combining these different polymorphs of phosphorene into a stacked heterostructure paves the way to create novel materials with desired band gaps for potential applications in future generation opto-electronic devices.

As already mentioned, phosphorene has the intermediate electron mobility and on/off ratio in comparison to those of graphene and TMD like MoS_2. Furthermore, it also shows high anisotropy in its electronic mobility, much higher mobility along the armchair direction than that in zigzag direction at room temperature. However, phosphorene has a major disadvantage; it is prone to degradation in air. Nevertheless, such degradation in air can be prevented by hetero-structure encapsulation technique, which in turn can increase the mobility too. Besides high mobility values, both *n*- and *p*-type transistor operations can be realized within a single layer phosphorene, which is quiet uncommon for other 2D semiconductors known to date. Thus, phosphorene can be used in ambipolar field effect transistors.

11.11 Transition Metal Dichalcogenides (TMDs)

Transition metal dichalcogenides (TMDs) are another class of important materials in the quest of emerging 2D materials. They are receiving more and more attention, nowadays, because of their extraordinary properties, arising from the vacant *d*-orbital on the metal centers. TMDs are generally of MX_2-type, where *M* is a transition metal, especially from the groups of 4, 5, 6, 7 and 10 of the periodic table and *X* is a chalcogen (S, Se or Te). Depending on the nature of *M* and *X*, materials properties of TMDs may vary from semiconducting (e.g. MoS_2, WS_2), to semimetallic (e.g. WTe_2), metallic (e.g. NbS_2) or even superconducting (e.g. $NbSe_2$, $TaSe_2$). As already mentioned, due to the presence of unsaturated *d*-orbitals in the transition metal centers, there involves a large number of magnetic and electronic interactions operate providing the materials with a wealth of interesting electronic, optical and magnetic properties. Furthermore, a controlled filling of such unsaturated *d*-orbitals may also be the basis of engineering such properties. Additionally, isovalent doping and alloying, by means of both the transition metals and chalcogenides, provide a way to tailor their electronic structures. In the bulk environment, most of the TMDs form stacked layered structures, strongly bound by van der Waals forces. Although such strong interlayer stacking interactions make them difficult candidates for thermal exfoliation. Unlike graphene or white graphene (h-BN), 2D TMDs can be directly synthesized by chemical vapor deposition (CVD) or two-step thermolysis.

Among all the TMDs, molybdenum dichalcogenides, particularly MoS_2 has been one of the most studied material due to its natural abundance as molybdenite. MoS_2 has shown its potential application in diverse fields like optoelectronics, photovoltaics, energy storage and heterogeneous catalysis, especially for the hydrogen evolution reactions. Additionally, broken inversion symmetry of single layer MoS_2 makes it a novel platform for studying 'spintronics' or 'valleytronics'. Such a property of MoS_2 leads to the coupling of spins and the valley degree of freedom in the momentum space. Other molybdenum dichalcogenides, $MoSe_2$ and $MoTe_2$ are also potential candidates to be used in lubricating, electrocatalysis, battery industry as well as in fabricating supercapacitors, and field effect transistors.

11.11.1 Geometry and Crystal Structure of TMDs

Bulk TMDs have layered structures having individual MX_2 layers bound by strong cohesive forces, where each layer consists of three atomic sublayers of *TM* (one) and *M* (two). The *TM* layer looks sandwiched between two *M* layers, involving covalent interactions between *TM* and *M*. In general, the *TM* has +4 and the *M* has −2 oxidation states. Despite the diverse possible choices of *M* and *X*, one of the unique features of TMDs is their capability to have different crystal phases for a particular 2D TMD. As for example, MoS_2 crystallizes in five different crystal structures, naming $1H$, $1T$, $1T'$, $2H$ and $3R$, depending on the different coordination models between Mo and S atoms and/or stacking orders between the atomic sublayers (see Figure 11.14).

The geometry and electronic structure of 2D TDMs could be understood on the basis of simplified model of crystal field theory. Two well studied crystal structures of single layer MoS_2 are trigonal prismatic $1H$ and octahedral $1T$, which belong to D_{3h} and D_{3d}, symmetry groups, respectively. Obviously, the number of *d*-electrons has a direct influence on the type of coordination sphere around the transition metal center and hence the crystal structure of TDMs. On the other hand, the *p*-orbitals of chalcogens, which lie relatively at lower energy position, hardly have any influence on the crystal structure. For the $1H$ phase, the five *d*-orbital splits into three sets as d_{z^2}, $d_{x^2-y^2,xy}$ and $d_{xz,yz}$; where the energy gap between d_{z^2} and $d_{x^2-y^2,xy}$ sets is almost 1 eV. Similarly, for $1T$ phase the *d*-orbitals split into two sets, t_{2g} ($d_{xy,yz,xz}$) and e_g ($d_{x^2-y^2,z^2}$). The TMDs with group 4 and 6 transition metals, which have *d*-electronic configurations as d^0 and d^2, respectively, have trigonal prismatic phase, while, the TDMs with group 5 transition metal (having d^1 electronic configuration) can have both trigonal prismatic and octahedral phases. Group 7 (d^3) and group 10 (d^6) metal containing TMDs have distorted octahedral ($1T'$) and octahedral crystal phases, respectively. Because trigonal prismatic phase of MoS_2 is energetically more stable than the octahedral phase, it is the most favorable phase and thus naturally abundant. Stacking of $1H$ phase of MoS_2 in *ABA BAB* sequence results in $2H$ crystal structure which has D_{4h} symmetry, while *ABA CAC BCB* type of stacking produces

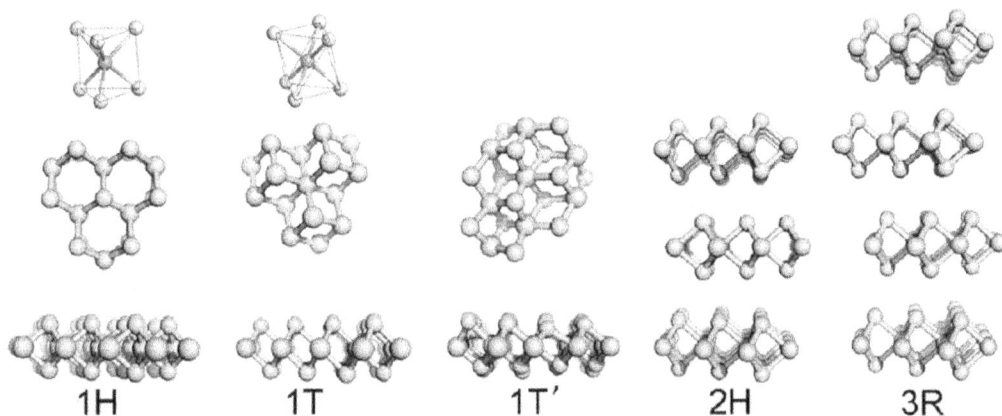

| 1H | 1T | 1T′ | 2H | 3R |

FIGURE 11.14 Structures of different types of TMDs.

3R phase, which has C_{3v} symmetry. For 2D MoS$_2$, the calculated Mo–S bond length is 2.4 Å, while the layer thickness, i.e. the distance between the upper and lower S atomic layers is about 3.0 Å.

11.11.2 Mechanical Properties

Monolayer MoS$_2$ has effective Young's modulus 270±100 GPa, which is higher than that of steel. The corresponding elastic moduli are E_x = 197.9±4.3 GPa and E_y = 200.3 ± 3.7 GPa. Noticeably, monolayer MoS$_2$ is even stronger than the corresponding bulk material. MoS$_2$ can tolerate an axial strain as high as 11% without having any structural deformation. Interestingly, it can be bent to a curvature of radius of 0.75 mm without losing its electronic properties significantly. Such exciting mechanical properties make MoS$_2$ a promising material for future generation of flexible electronics. MoS$_2$ also shows piezoelectric behavior, thus, it can be utilized as sensitive mechanical transducers. Besides these, MoS$_2$-based nanomechanical resonators are used to detect high to very high frequency (HF, 3–300 MHz) bands.

11.11.3 Electronic Structure of TMDs

The periodic multilayered 3D bulk MoS$_2$ provides an indirect band gap of 1.29 eV, locating VBT at the Γ point and CBM at an intermediate position of Γ and K points. Theoretical calculations have shown that with decreasing number of layers, the CBM moves upward without much affecting the VBT. Consequently, the band gap increases for fewer-layered MoS$_2$ systems. Noticeably, the upward shifting of CBM is much higher at the Γ point than that at K point, which is attributed to the fact that at the K point, the CBM is mainly contributed from the d-orbitals of the Mo atoms, which are relatively unaffected by interlayer stacking interactions. However, near the Γ point, the CBM is contributed both from the Mo-d-orbitals and S-p_z-orbitals, the latter ones are highly affected by the interlayer stacking. As a result of that, CBM appears at K point for the bilayered MoS$_2$. However, the system preserves its indirect semiconducting property since VBT lies at Γ point and CBM is shifted to K point. Ultimately, monolayer MoS$_2$ becomes a direct band gap semiconductor with the estimated band gap at K being 1.9 eV. Such a shifting in the property of the material viz. from an indirect to a direct band gap semiconductor, makes MoS$_2$ a material of special interest.

As MoS$_2$ can tolerate a high degree of strain, the variation of electronic properties with strain might be of special attention, especially for optoelectronic and piezoelectric applications. It has been observed that the band gap of monolayer MoS$_2$ decreases with the increasing applied strain. Moreover, it involves a transition from direct to indirect band gap character of MoS$_2$ on an application of small tensile strain (<2%). Surprisingly, it becomes metallic at a higher tensile strain of 9%. Other than applied strain, doping or alloying can also tune the electronic properties of TMDs. As already mentioned, electronic properties of TMDs are sensitive to the nature of TM elements. So, it is quite feasible to design an alloyed material with tunable electronic properties by varying the alloy composition. This can be achieved by isovalent doping or alloying with transition metals and/or chalcogens. In fact, composition-dependent Mo$_{1-x}$W$_x$S$_2$ monolayers have been achieved through layer exfoliating strategy. A mixed chalcogen (MoS$_{2x}$Se$_{2(1-x)}$) nanosheet with composition dependent tunable elctronic and optical properties has also been reported in recent time.[17–18]

11.11.4 Optical Properties of TMDs

Thin films of TDMs show high absorption capabilities throughout the entire zone of visible spectra, rendering them as potential candidates in photovoltaic devices. A 300 nm thin TMD film can absorb almost 95% of the incident visible light. Two prominent absorption peaks are identified in reflection spectra for ultrathin MoS$_2$ layer at 1.85 eV and 1.98 eV, designated as the direct excitonic transitions at the K point of the Brillouin zone. Photoluminescence (PL) studies reveal that the optical band gap of monolayer MoS$_2$ is around 1.9 eV. Notably, the presence of direct band gap in monolayer MoS$_2$ provides a radiative recombination thereby producing much higher PL quantum yield than that of bulk or multilayer MoS$_2$. On increasing the temperature, the PL spectra of monolayer MoS$_2$ undergoes red shift. Furthermore, band gap engineering through applied strain and varying alloy composition can successively tailor the optical properties of monolayer MoS_2, which render its specific application in optoelectronic devices.

11.12 Pristine and TM-doped PtSe₂ Monolayers

Most of the 2D materials such as graphene, white graphene, phosphorene, phosphorus carbides, transition metal mono and dichalcogenides, etc. are intrinsically non-magnetic in nature. However, incorporation of magnetism in 2D materials is crucial and even challenging for nanoscale spintronic applications. Extensive studies have been made to induce magnetism in 2D materials by various factors such as substitutional doping, adatom adsorption, defect formation, edge effect creation and finally applying strain. Defect and edge effect induced magnetism in graphene and boron-nitride monolayer are well reported but the most promising way is the introduction of one or more transition metal (TM) or non-metal atoms into such 2D layers for developing their magnetic character. As already introduced, TMDs have been the focus of extensive research in the field of magnetic materials but the low carrier mobility restrict their applications in designing electronic or spintronic devices. Recently, a new kind of 2D material, platinum diselenide (PtSe₂) has been synthesized by the direct selenization of the Pt (111) substrate, which shows promise to find its applications in valleytronics. The most interesting fact is that PtSe₂ has much better air stability (over one year) as compared to black phosphorus and the carrier mobility of 2D PtSe₂ is also higher than that of 2D phosphorene. However, the magnetism of 2D PtSe₂ remained an open question until recently Sarkar's group has studied it extensively with the help of density functional theory.[19] It has been revealed that the magnetic states of TM-doped PtSe₂ layers are very sensitive to the nature of TM. For instance, low concentration doping of groups 4 and 10 TMs in PtSe₂ monolayer turns it into non-magnetic semiconductor, while some other groups of TMs make it half-metallic. This half-metallicity and magnetic semiconducting characteristics suggest that such doped PtSe₂ layers can be used as a new kind of dilute magnetic semiconductor. Surprisingly, Fe and Ru-doped PtSe₂ monolayers show robust ferromagnetism well above the room temperature, together with half-metallic characteristics which earmark the 2D PtSe₂ layer as a promising material for nanoscale spintronic applications.[19]

11.13 Other Nanomaterials: Special Emphasis on Nanoclusters or Quantum Dots (QDs)

So far we have discussed the properties of various 2D or quasi-2D materials and their 1D or 0D analogs, which of course belong to the family of nanomaterials. As already discussed, the properties of nanomaterials are governed by the quantum mechanical principles. Reducing the dimension of the particle from bulk to nanometer size leads to discrete energy levels of the particles, known as quantum confinement effect. Obviously, this effect is most prominent for the zero-dimensional materials, commonly known as nanoclusters or quantum dots (QDs). Nanoclusters show dramatically distinct electrical, optical and chemical properties from those of the respective bulk materials. In some instances entirely new behaviors are observed for the clusters, which are not seen in the corresponding bulk materials. The ability of a nanocluster to react with any species should depend on cluster size and shape. The size also controls the stability and magnetic behavior of the nanocluster. They play a significant role especially in the field of energy conversion and storage. Over the last few decades a large number of works have been performed, showing the efficacy of different nanoclusters in solar to electrical energy conversion, catalysis and photocatalysis, sensing, biological activities and so on.

11.13.1 Classification of Nanoclusters

Depending upon the morphology and composition, nanoclusters are classified into different groups each of which has different characteristics as mentioned below.

Metal nanoclusters:
They include

(i) *s*-block metals, e.g. alkali and alkaline earth metals. Here, the bonding is metallic, involving mainly the valence *s*-orbitals.
(ii) *sp*-metals, e.g. aluminum nanocluster for which the bonding involves some covalent character.
(iii) Transition metal nanoclusters, where a greater degree of covalency and directionality are observed in the bonding.

Semiconductor nanoclusters:
These are composed of elements like C, Si, Ge, etc., which are also known as bulk semiconductors. However, these elements form semiconductor nanoclusters comprising single element or heteroatomic elements, e.g. Ga_xAs_y. The bonding in between the elements here is mainly polar covalent bond. Besides, semiconductor nanoclusters include different metal or non-metal chalcogenides like II-VI and IV-VI semiconductors. Semiconductor metal oxides are of special interest in this series, the most common examples being TiO_2, Al_2O_3, SnO_2, ZnO, etc.

Ionic nanoclusters:
Heteroatomic nanoclusters composed of elements with large difference in electronegativity form this type of nanocluster. Here, bonding is predominately ionic (electrostatic); e.g. alkali metal halides.

Indeed, in many ways, these nanoclusters represent a unique state-of-matter of their own. The structures of small metal clusters remain, however, largely a mystery and very little is known of the complex relationship between geometrical structure and reactivity. Owing to their reactivity and unusual properties, nanoclusters are of tremendous technological interest in numerous areas of applied science like optoelectronics, energy conversion and catalysis.

11.13.2 Reactivity of Nanoclusters

As particle dimensions reduce towards the nanoscale, the surface-to-volume ratio increases enormously and thus the nanomaterials become extraordinarily reactive. The novel physico-chemical properties of nanomaterials can be applied in the field of catalysis that we have already mentioned. Nanocatalysis plays an important role for different chemical transformations, especially in some reactions where normally high energy barriers are encountered. Now, the major goal for catalysis researches in the current century is to design new catalysts with desirable activity and higher selectivity. For a chemical reaction with multiple possible products, a catalyst may promote the production of a particular product by its unique feature, called selectivity. Nanocatalysts can perform in both heterogeneous and homogeneous conditions. Proper tuning of size, shape and chemical composition of nanomaterials may offer next-generation catalytic systems for energetically challenging reactions, high selectivity to valuable products and with extended life times. One of the striking examples demonstrating the exceptional catalytic activity of nanomaterials is the catalytic gold nanoparticles (in regime of 5 nm) dispersed on a titania support. This catalyst has been shown to provide high activities for hydrocarbon epoxidation and CO oxidation at room temperatures.

This discovery has triggered extensive research in searching for novel nanocatalysts for various catalytic reactions with poor reactivity, such as activation of saturated hydrocarbons in reforming reactions, oxygen reduction reactions in fuel cells, etc. It has been reported that small metal clusters with positive or negative residual charges exhibit facile activation and dehydrogenation of methane molecules. C–H, N–H bond activation, H_2O splitting and different organic transformation reactions by various neutral or charged metal and metal oxide clusters are also reported. Subnanometer platinum clusters (Pt_{8-10}) are shown to provide high catalytic activities for the oxidative dehydrogenation of propane. Although bulk aluminum metal is totally inactive towards carbon–halogen bond, selected Al nanoclusters (size ranging from 3 to 20 atoms) show significantly high activity towards the oxidative addition reactions. Small sized aluminum clusters are also extremely reactive and show strong affinities to adsorb gaseous molecules such as H_2, D_2, O_2, N_2 and H_2O.

Another versatile material in this series is titanium dioxide (TiO_2), which has a wide range of applications in many fields, e.g. photocatalysis, solar cell devices, gas sensors, biomaterials to dimensionally stable electrode. Among different semiconductor metal oxides, TiO_2 appears to be a distinct material not only for its high chemical and optical stability but also for its nontoxicity, low cost, and corrosion resistivity. There are mainly four different naturally occurring polymorphs of TiO_2, which are anatase (tetragonal), rutile (tetragonal), brookite (orthorhombic) and TiO_2 (B) (monoclinic). Additionally, two other high pressure forms have been synthesized from the rutile phase including TiO_2 (II) with a similar structure as that of PbO_2, and TiO_2 (H), which has a hollandite structure. However, rutile phase of TiO_2 is predicted to be the most stable almost at all temperatures. On the other hand, anatase TiO_2, having a tetragonal crystalline structure, exhibits higher activities in many cases, which is attributed to the relatively larger tetragonal distortion in the TiO_6 octahedron in the crystal structure. Moreover, although rutile phase is the most stable one, TiO_2 nanocrystals appear to form metastable anatase form for the particle size up to ~14 nm. This is explained by the lower average surface energy of the anatase polymorph, whose equilibrium crystal structure is largely dominated by the very stable (101) surface. However, it is reported that the majority anatase (101) surface is mostly unreactive; for example, unlike rutile (110) termination, instant water adsorption on anatase (101) surface is purely molecular in nature.

11.14 Nanocomposites or Nanohybrid Materials

Nanocomposites are materials that incorporate nanosized particles into a matrix of standard material or hybrid materials consisting of different nanoscale objects. They can be broadly classified into inorganic-organic, organic-organic or inorganic-inorganic composites, where at least one of the components is in the nanoscale diameter.[20–22] The inorganic components vary widely and may be QDs of different types, nanowires/nanorods or even nanosheets, while the organic component may include small or polymeric organic molecules, fullerenes and their derivatives, CNTs, etc. Some representative structures of composite materials are shown in Figure 11.15. The most important thing is that the resultant materials provide a drastic improvement in properties that includes mechanical strength, toughness and electrical or thermal conductivity. In mechanical terms, nanocomposites differ from conventional composite materials due to the exceptionally high surface to volume ratio of the reinforcing phase and/or its exceptionally high aspect ratio. From the electronic energy term, these materials show extraordinary behavior due to incorporation of a new energy band in between the precise energy gap of the constituent materials. Thus, nanocomposites

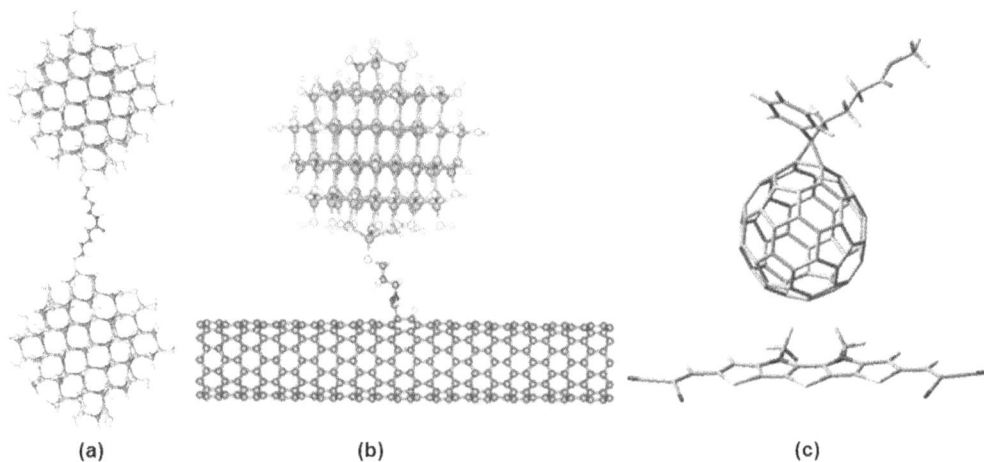

(a)   (b)   (c)

FIGURE 11.15 Different types of composite nanomaterials: (a) inorganic-inorganic; (b) inorganic-organic; (c) organic-organic.

are promising materials for solar energy harvesting processes, further details of which would be given in the forthcoming section.

11.15 Nanomaterials for Energy Conversion Processes

The current energy demand and depletion of fossil fuel have forced mankind to think of economically viable renewable energy sources for their sustainability. So far, some renewable energy sources have been identified and some of them have already been implemented for the generation of energy, such as solar, wind, hydro, biomass, geothermal, etc. However, the most promising one is the solar energy, which can be easily harnessed at domestic and commercial level. The direct conversion of solar energy into electrical energy is done by photovoltaic cell or solar cell. Hence, people have sought to utilize solar energy directly to generate electricity through efficient and cost-effective device fabrication. This goal has a long history starting from the discovery of the photoelectric effect by Einstein. However, the practical implementation started during the last few decades of the last century. The first generation solar cells were made of monocrystalline silicon but their high fabrication cost and low efficiency made them commercially inaccessible. The thin film solar cells or commonly known as the second generation solar cells are mainly made of amorphous silicon materials, although, cadmium telluride, copper indium gallium selenide are also used as thin films for fabricating such devices. The main obstacle of the thin film solar cells is again their low efficiency. Dye sensitized solar cell (DSSC), which was first fabricated by O'Regan and Grätzel in 1991 is referred to as the third generation solar cell whose efficiency is in between thin films and crystalline solar cells. Although the efficiency of DSSCs is not up to the mark, many advantageous features, such as low cost, easy preparation process, etc. make them quite attractive. Furthermore, there remain enormous possibilities for improving the performance of DSSCs through tuning the optoelectronic properties of various dyes. The latest addition in this avenue is the perovskite materials based solar cells whose efficiency is reported to be the highest. For improving the efficiency of solar cell, one requires a concrete idea about the key factors, which basically control the efficiency of a solar cell. In the next paragraph, we will provide such detail, which will further explain the specific role or nanomaterial, especially the composite material for improving the efficiency.

As introduced by Kamat, the nanostructures can help in the solar energy harvesting process in three different ways.[23] First, they can form donor-acceptor (*D–A*) type of composites, which effectively capture photons as that in photosynthesis. Second, semiconducting nanoparticles can act as photocatalysts to convert CO_2, H_2O, etc. to solar fuels. Third and the most appealing one is the semiconductor-based solar cells where the nanostructures act as effective charge carriers. Semiconductor quantum dots (QDs) basically act as the sensitizer in a photovoltaic cell, where it absorbs sunlight and generates photoexcited electron-hole pairs, commonly known as excitons. Obviously, the efficacy of QDs primarily depends on their ability of charge separation and transfer thereafter to another material before the recombination of the charge carriers. It has been revealed that the charge transfer and charge recombination take place in an ultrafast time scale and dynamics of these two competing processes basically control the overall performance of the solar cell device. In a single component device, the recombination process is much faster and there lies the actual role of composite or hybrid material.

In a hybrid solar cell, the initially generated bound electron-hole pair is dissociated at the interface of the hybrid material and consequently they are separated at the two different entities. This practically reduces the chance of electron-hole recombination in a substantial amount and being the major advantage of hybrid solar cell. Now, the separated electron and hole are possibly carried by two different components thereby producing external current. However, in between, there occur several other competing processes, which are also very fast. The overall processes occurring in a typical sensitized (dye) solar cell are shown in Figure 11.16. Note that in dye sensitized solar cell (DSSC), an organic dye serves the role of sensitizer. Besides the dye, a DSSC consists of a semiconductor coated photo anode, a counter electrode (cathode), and an electrolyte as could be found in Figure 11.16. The photoexcited dye produces electron-hole pair (exciton), which subsequently dissociates into charge carriers to produce a hot electron at the lowest unoccupied molecular orbital (LUMO) leaving a hole at the highest occupied molecular orbital (HOMO) of the

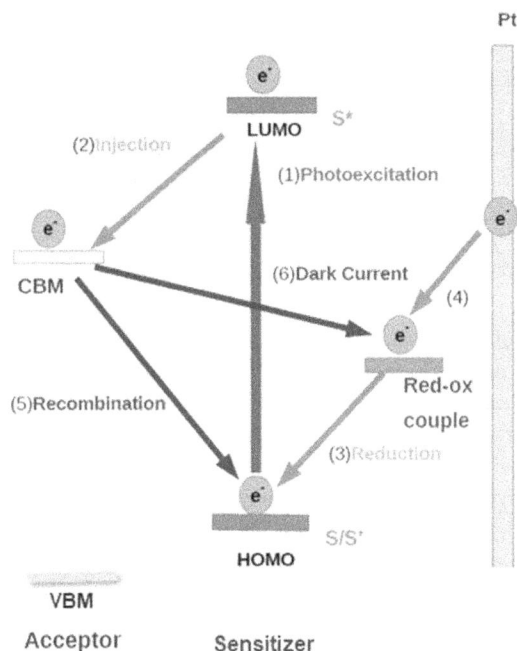

FIGURE 11.16 Schematic representation of a typical DSSC, showing the mechanism of solar to electrical energy conversion.

dye. The minimum energy required to dissociate the exciton is called exciton binding energy. However, the most crucial step is to inject the hot electron to a suitable energy band of the other component of the composite material, possibly the conduction band of a semiconductor. Next, the injected electron passes through the semiconductor material toward the transparent conductive electrode, and reaches an external load in the form of electrical energy. Finally, the cycle is completed via the regeneration of the electron in the dye through the counter electrode and electrolyte.

Note that the semiconductor material should have some special features like high surface area, high porosity and most importantly, its conduction band (CB) edge should lie in between the HOMO and LUMO of the dye, forming a type II band alignment. This is required in order the make the electron injection process thermodynamically feasible. The thermodynamic free energy change of this process is the difference in energy between the LUMO level of the dye and the CB level of the semiconductor surface. TiO_2 is chosen as the semiconductor for fabricating DSSC in many cases; however, ZnO, SnO_2, MgO, Al_2O_3 are also suitable semiconductors for this purpose. Apart from these, different kinds of fullerene and non-fullerene molecules are also used as acceptors in DSSC. The dye sensitizer should have a carboxyl, hydroxyl or any suitable group, which can bind with the semiconductor. At the same time, the sensitizer should show maximum absorbance within the visible to near infra red region of solar spectra and it should not degrade fast. Other material characteristics may be found elsewhere for designing an efficient solar cell.[24]

It is noteworthy that, the mode of nanojunction is also very important for absorbing sunlight and controlling the dynamics of electron transfer, which in turn determines the power conversion efficiency. Researchers have introduced the concept of bilayer solar cell where two conjugated small molecular layers (thin films) are used for absorbing the solar energy.[25] For a complete absorption of the incident radiation, it requires a sufficiently thick layer, which, however, brings its limitation because most of the excitons are destroyed during the diffusion period before reaching the interface. As an alternative, the concept of the bulk heterojunction (BHJ) model has been introduced where donor and acceptor materials are mixed together to form a heterogeneous mixture. BHJ model has now become a standard architecture for fabricating organic donor-acceptor-based solar cells.

The research on nanoscale materials to understand the feasibility of their applications as building blocks in solar cell and further exploiting their activities is an active field of research. In this context, hybrid nanostructures have been the major focus because of their many useful attributed as discussed above. Both experimental and theoretical researches are essential in order to understand the underlying mechanism, especially the dynamics of charge carriers for improving the efficiency of solar cell. The experimental techniques include time-resolved ultra-fast spectroscopy. On the other hand, theoretical researchers try to understand the materials properties on the basis of absorption/emission, energy band alignment, charge transfer and associated dynamical behaviors. As already discussed, the primary mechanism of the solar-to-electric energy conversion is the photo-induced separation of electrons and holes at an interface of hybrid nanostructure and subsequent electron injection to the acceptor material. These processes are intrinsically non-equilibrium in nature and thus classical reaction rate theories based on the quasi-equilibrium assumption are no longer applicable to them. The commonly utilized Börn-Oppenheimer (adiabatic) approximation is also not applicable in such cases. Thus, explicit time-domain molecular dynamics simulations are required to account for the dynamical behavior. Furthermore, for mimicking the real experimental situations, one must include the environmental effects such as solvation, temperature, etc. The efficient approach to study such non-radiative processes is based on mixed quantum-classical dynamics, where the nuclear motions are described by classical trajectories and electronic motions are treated quantum mechanically. For the propagation of the classical nuclear trajectories, the forces as well as the non-adiabatic coupling parameters are required, which can be calculated by several approximations.

The group of Oleg Prezhdo[26] has developed a semi-classical method for simulating different nonequilibrium phenomena, such as charge transfer energy transfer, and thermal relaxation of excited electron in nanoscale regime. It implements a number of basic and more advanced functionalities during photoexcitation as a result of light matter interaction, which includes fewest-switches surface hopping (FSSH), decoherence-induced surface hopping (DISH), multi-electron adiabatic representation of the time-dependent Kohn-Sham (TD-KS) equations. The classical path approximation (CPA) also simplifies the model to a large extent. However, the *ab initio* DFT methods for electronic structure calculations limits its use in larger systems, demanding high computational cost. The method has recently been further extended, incorporating tight-binding formalism, which can handle larger systems consisting of up to thousand atoms. Thus, a contentious effort is still ongoing to understand the material properties in the way of solar energy harvesting process.

Concluding, we have systematically traced the development of low-dimensional materials, starting from graphene, white graphene, phosphorene to transition metal dichalcogenides and the latest hybrid materials, their general properties, the origin of such properties, i.e. the underlying electronic structures and applicability of such materials in catalysis to photocatalysis and energy conversion process. Smart 2D materials are the main focus for future generation light electronics and spintronics industry. They are also the basis for new magnetic materials. Furthermore, for the sustainability of the modern civilization, the development of smarter nanohybrid materials for energy storage and conversion has become a challenge that must be won.

REFERENCES

1. Y.-W. Son, M. L. Cohen and S. G. Louie, 'Half-metallic graphene nanoribbons', *Nature,* 444, 347 (2006).
2. R. Landauer, 'Electrical resistance of disordered one-dimensional lattices', *Philos. Mag.,* 21, 863–867 (1970).
3. S. Datta, *Electronic Transport in Mesoscopic Systems*, Cambridge University Press (1997).
4. A. Pramanik, S. Sarkar and P. Sarkar, 'Charge transport through nanocontacts', *Chem. Model.,* 15, 70–130 (2020).
5. A. Pramanik, S. Sarkar and P. Sarkar, 'Doped GNR p-n junction as high performance NDR and rectifying device', *J. Phys. Chem. C*, 116, 18064–18069 (2012).
6. J. R. Williams, L. DiCarlo and C. M. Marcus, 'Quantum hall effect in a gate-controlled p-n junction of graphene', *Science*, 317, 638–641 (2007).

7. C. L. Kane and E. J. Mele, 'Size, shape, and low energy electronic structure of carbon nanotubes', *Phys. Rev. Lett.*, 78, 1932 (1997).
8. Z. Z. Zhang, Kai Chang and F. M. Peeters, 'Tuning of energy levels and optical properties of graphene quantum dots', *Phys. Rev. B*, 77, 235411 (2008).
9. A. D. Guclu, P. Potasz and P. Hawrylak, 'Excitonic absorption in gate-controlled graphene quantum dots', *Phys. Rev. B*, 82, 155445 (2010).
10. C. Berger *et al.*, 'Electronic confinement and coherence in patterned epitaxial graphene', *Science*, 312, 1191–1196 (2006).
11. Y. Liu, G.Wang, Q. Huang, L. Guo and X. Chen, 'Structural and electronic properties of T graphene: A two-dimensional carbon allotrope with tetrarings', *Phys. Rev. Lett.*, 108, 225505 (2012).
12. X.-L. Sheng, H.-J. Cui, F. Ye, Q.-B. Yan, Q.-R. Zheng and G. Su, 'Octagraphene as a versatile carbon atomic sheet for novel nanotubes, unconventional fullerenes, and hydrogen storage', *J. Appl. Phys.*, 112, 074315 (2012).
13. C. Su, H. Jiang and J. Feng, 'Two-dimensional carbon allotrope with strong electronic anisotropy,' *Phys. Rev. B*, 87, 075453 (2013).
14. C.-P. Tang and S.-J. Xiong, 'A graphene composed of pentagons and octagons', *AIP Adv.*, 2, 042147 (2012).
15. X.-Q. Wang, H.-D. Li and J.-T. Wang, 'Prediction of a new two-dimensional metallic carbon allotrope', *Phys. Chem. Chem. Phys.*, 15, 2024 (2013).
16. D. J. Appelhans, Z. Lin and M. T. Lusk, 'Two-dimensional carbon semiconductor: Density functional theory calculations,' *Phys. Rev. B*, 82, 073410 (2010).
17. H. Li *et al.* 'Growth of alloy $MoS_{2x}Se_{2(1--x)}$ Nnnosheets with fully tunable chemical compositions and optical properties', *J. Am. Chem. Soc.*, 136, 10, 3756–3759 (2014).
18. B. Rajbanshi, S. Sarkar and P. Sarkar, 'The electronic and optical properties of $MoS_{2(1-x)}Se_{2x}$ and $MoS_{2(1-x)}Te_{2x}$ monolayers', *Phys. Chem. Chem. Phys.*, 17, 26166–26174 (2015).
19. M. Kar, R. Sarkar, S. Pal and P. Sarkar, 'Engineering the magnetic properties of $PtSe_2$ monolayer through transition metal doping', *J. Phys.: Condensed Mater.*, 31, 145502 (2019).
20. M. Kar, B. Rajbanshi, R. Sarkar, S. Pal and P. Sarkar, 'Periodically-ordered one and two dimensional CdTe QD superstructures: A path forward in photovoltaics', *Phys. Chem. Chem. Phys.*, 21, 19391–19402 (2019).
21. S. Sarkar, S. Saha, S. Pal and P. Sarkar, 'Electronic structure of Thiol-capped CdTe quantum dots and CdTeQD-carbon nanotube nanocomposites', *J. Phys. Chem. C*, 116, 40, 21601–21608 (2012).
22. S. Biswas, A. Pramanik, S. Pal and P. Sarkar, 'A theoretical perspective on the photovoltaic performance of S,N-heteroacenes: An even-odd effect on the charge separation dynamics', *J. Phys. Chem. C*, 121, 2574–2587 (2017).
23. P. V. Kamat, 'Meeting the clean energy demand: Nanostructure architectures for solar energy conversion', *J. Phys. Chem. C*, 111, 7, 2834–2860 (2007).
24. S. Sarkar, S. Saha, S. Pal and P. Sarkar, 'Exploring the electronic structure of nanohybrid materials for their application in solar cell', *Chem. Model.*, 13, 27–71 (2016).
25. Z. Tan *et al.*, 'Efficient all-polymer solar cells based on blend of tris(thienylenevinylene)-substituted polythiophene and poly[perylene Diimide-alt-bis(dithienothiophene)]', *Appl. Phys. Lett.* 93, 073309 (2008).
26. S. Pal, D. J. Trivedi, A.V. Akimov, B. Aradi, T. Frauenheim and O.V. Prezhdo,'Nonadiabatic molecular dynamics for thousand atom systems: A tight-binding approach toward PYXAID', *J. Chem. Theory Comput.*, 12, 4, 1436–1448 (2016).

12

Energy Materials

12.1 Introduction

Broadly speaking, energy materials belong to a collection of many classes of materials that find applications in production, transformation (conversion), transmission and storage of energy. Energy is a critical consumable component of growth and development. The demand for this key component of development is growing fast all over the world and will continue to do so in the foreseeable future. It has become critically important that improved technologies for production, conversion, storage and transmission of energy are developed for the most efficient use of energy and energy saving. The required technological improvements, however face certain roadblocks in the form of unavailability of appropriate materials for implementing improved technologies. For example, electric automobiles (cars) and portable electronics have now become popular due to vast improvements in both of the technologies. The rapid growth in the demand for electric cars and portable electronics has catalyzed the search for energy storage devices like batteries, super capacitors with high power and energy density, which depends significantly, if not critically, on the availability of new energy materials for storage. Energy storage materials play a crucial role for providing clean, efficient and versatile use of energy and for efficient exploitation of all the renewable energy sources at our disposal.[1–3] The storage systems can be thermal, electromagnetic, mechanical or hydrogen-based, requiring different types of energy materials. However, we will restrict ourselves to only a few of them, namely, hydrogen storage materials primarily for on-board applications, non-linear optical materials and photonic materials for laser and optical communication, photovoltaic and thermoelectric materials for clean energy conversion systems and so on, for conserving space.

12.2 The Looming Energy Crisis

Energy, environment and sustainable development are critical issues. It will not be an exaggeration to say that the very survival of the present civilization delicately depends on our ability to address all the energy related issues adequately, sensitively and promptly.[1]

The human civilization is staring at a looming energy crisis. The root of impending crisis lies in our overdependence on and unmindful exploitation of fossil fuels, the reserve of which is finite and fast depleting without any prospect of replenishments. The large scale consumption of fossil fuels for our commercial and domestic needs has been responsible for releasing huge amounts of carbon-dioxide in the atmosphere everyday. The amount released is so huge that the natural carbon cycle can not handle it and restore equilibrium any longer. The amount of CO_2 in the atmosphere has therefore been steadily rising, bringing in its wake global warming and catastrophic change of climate across the globe. The primary need of the hour is to move away from fossil fuels and start exploiting alternative, renewable, non-polluting or environmentally benign sources of energy. There are diverse such sources of energy available to us. The sun, which appears to be an almost inexhaustible source of energy, is one such source. It is an endless but dilute source of energy. The technology to harness solar energy is still developing. The presently available technology is not yet adequate to meet the huge demand for power for static as well as on-board applications. The imperative is that energy technologies must be refashioned to enable us to make use of solar and other renewable sources like the wind, hydroelectric, geothermal and nuclear energy. Apart from solar, the other renewable sources are suitable mainly for static application. For on-board applications alternative sources are necessary. Is it possible to replace the fossil fuels with some other 'fuel' or an 'energy carrier', which can be generated and used in closed cycles? Presently, the only material that fits the bill is hydrogen – the lightest element in the periodic table. The fossil fuels produce energy when they are oxidized by oxygen in air. For a hydrocarbon having the molecular formula C_pH_q (p, q are integers), the ideal oxidation reaction proceeds as

$$C_pH_q + (p + q/4)O_2 = pCO_2 + (q/2)H_2O + Energy$$

If an alcohol is used, the corresponding ideal oxidation takes place as follows:

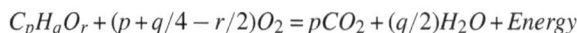

$$C_pH_qO_r + (p + q/4 - r/2)O_2 = pCO_2 + (q/2)H_2O + Energy$$

The oxidation products are carbon-dioxide and water (steam). Water is environmentally benign while carbon-dioxide is not, and must be captured, which requires new materials and technology. If the fossil fuels are replaced by hydrogen, the oxidation product is water only. It is environmentally benign, and there is no scope for pollution. Hydrogen is an excellent carrier of energy. It has the highest energy density per unit mass, 142 MJ/Kg, which is much higher than the commonly used liquid hydrocarbons; but hydrogen does not exist in the free state on the earth. It exists however in a chemically combined state with oxygen in the form of water in abundant quantities. Water can be split by electricity or by photons in the presence of a suitable photocatalyst into hydrogen and oxygen thereby making the generation and consumption of hydrogen in closed cycle possible. There is one difficulty though. Being the lightest element in the periodic table, hydrogen occupies a huge volume per unit mass under ambient pressure and temperature. Moreover, it forms an explosive mixture with oxygen. Thus for mobile application, hydrogen must be scrupulously oxygen-free. The oxygen-free compressed hydrogen gas may be stored in cylinders or as a cryogenic liquid in specially designed vessels – the hydrogen storage tanks for mobile applications. We note, however that the boiling point of liquid hydrogen is 20.3K. A very significant part of the energy ($\approx 30\%$) carried by hydrogen must then be spent in achieving and maintaining the low temperature in storage tanks. For compressing the gas to 75 MPa for storage in gas cylinders requires almost half the energy carried by hydrogen. The energy expenditure not withstanding, both the modes of storage are currently being used in mobile applications. The heavy and bulky storage tanks and cylinders are, however impediments as they add to the mass and consume space of the vehicles running on hydrogen fuel. The problem would be resolved if hydrogen could somehow be stored as a solid at the ambient temperature. It is not that the use of 'solid-H_2' is being proposed. What is being proposed involves storing hydrogen gas (reversibly) in suitable solids, preferably light and porous, from which the gas may be recovered on demand without any serious technical difficulty. We will briefly review various materials that have been investigated and even used for the purpose.

12.3 Materials for Hydrogen Storage

It has been well-known since ages that hydrogen is 'occluded' by palladium (Pd) at ambient pressure and temperature, the hydrogen uptake being 900 times the volume of palladium. The hydrogen so 'occluded' is discharged on heating. The act of occlusion can be regarded as a hydriding reaction

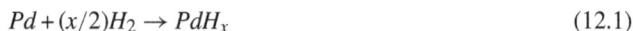

$$Pd + (x/2)H_2 \rightarrow PdH_x \qquad (12.1)$$

while the act of hydrogen recovery can similarly be regarded as the dehydriding reaction

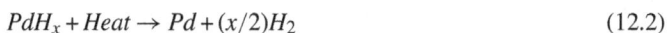

$$PdH_x + Heat \rightarrow Pd + (x/2)H_2 \qquad (12.2)$$

The palladium-hydrogen system is an example of 'solid state hydrogen storage system'. Any other solid that reversibly takes up hydrogen either in the form of hydrogen atoms or hydrogen molecules can be used to compress hydrogen into the solid to the desirably high level of storage density. Apart from palladium, another solid state hydrogen system that has attracted wide attention is $LaNi_5$, which too, stores hydrogen in atomic form (as $LaNi_5H_x$). It turns out that both Pd and $LaNi_5$ can store hydrogen at atomic densities that far exceed the atomic density in liquid hydrogen.[2−4] The heavy mass of the host lattice in either case is a distinct disadvantage, which is further compounded by the cost of the metal hosts (Pd, La are costly). These two adverse factors stand in the way of large scale practical use of these storage systems in actual transportation systems. However, there are many other classes of materials (other than metals/alloys), which have been the subject of serious investigation and evaluation as probable solid state storage systems for hydrogen. They can be broadly classified into three categories:

1. Microporous materials.
2. Interstitial hydrides.
3. Complex metal hydrides.

Of these, microporous materials have high storage capacities, albeit at low temperatures (at liquid nitrogen temperature, for example), the interstitial hydrides can absorb and desorb hydrogen at ambient temperatures while the complex metal; hydrides generally require elevated temperatures in the range of 100-300 °C for desorption to take place efficiently. The categorization outlined above is operational. The mechanism of hydrogen uptake can be another basis for the classification of solid state hydrogen storage materials. Thus the microporous materials take up hydrogen into their pores as hydrogen molecules (H_2), which get bound to the surface of the host by low energy interactions, which are typically in the range 10-100 meV. This mode of hydrogen uptake is called physisorption. Interstitial hydrides absorb hydrogen in its atomic form (H). These H atoms move into the interstitial voids of the solid host and eventually diffuse throughout the solid. This mode of hydrogen uptake has been traditionally known as chemisorption. Apparently, the hydrogen molecules dissociate into H atoms on interacting with the metal surface, which is followed by the absorption of hydrogen atoms. The binding energy typically ranges from 2-4 eV accounting for the requirement of elevated temperatures for the dehydriding step. It is possible that the metal-hydrogen interaction just weakens the H-H bond instead of breaking it and the weakened H_2 molecules are taken up by the host lattice. This mode of hydrogen uptake has become known as molecular chemisorption. Molecular chemisorption provides energetically the most facile pathways for absorption and desorption dynamics of hydrogen on appropriate solid state storage materials and is therefore the most favored mechanism of hydrogen storage for practical applications.

Complex metal hydrides, on the other hand, may have both covalently and ionically bonded hydrogen atoms integrated into the structure, which disintegrates upon desorption of hydrogen. In what follows, we will consider several examples of each kind of hydrogen storage material and examine their viability. Further details can be found elsewhere.[2-4]

12.3.1 Microporous Materials for H_2-Storage

These materials absorb molecular hydrogen in their network of micropores with narrow width. That leads to an increase in the density of hydrogen molecules compared to the gas phase density. The adsorption occurs at relatively low, subambient temperatures. The kinetics of hydrogen absorption by microporous materials is rapid, which is advantageous for storage, but low temperatures are required to be maintained to achieve a significant level of absorption at useful storage pressure.

12.3.2 Carbon-Based Solid State Materials for Hydrogen Storage

Microporous templated carbons had been an attractive choice as a material for hydrogen storages. Activated carbon, which can be nano, meso or microporous, tend to have slit-shaped pores with wide pore size distributions. This is in sharp contrast with crystalline adsorbents like the zeolites or metal organic frameworks. Jorda-Beneyto[5] strongly advocated for the use of activated carbon monoliths as storage materials for hydrogen gas, because microporous activated carbon in these systems has higher bulk density than the same carbon in powder form. The H_2-sorption capacities of carbon nanostructures such as carbon nanotubes and nanofibers have been seriously investigated in recent years. Dillon *et al.*[6] claimed to have achieved room temperature sorption capacities of the order of 5-10 wt%, which raised expectations.

Hirschers *et al.*[7] debunked the claim and showed that the observed high capacities are due to metal nano-particles deposited during purification of the nanostructures (tubes). Carbon nanofibers are made up of graphene layers stacked together in various orientations with respect to the fiber axis. These nano-fiber materials were initially touted as the panacea of all the solid state hydrogen storage systems. However, the expectations were belied in practice and the idea of carbon nanofiber-based hydrogen storage systems has practically been abandoned. Fullerenes and other carbon nanomaterials have been investigated both theoretically and experimentally without any significant outcome or promise. On the other hand templated carbons, especially those for which templating has been done using a carbide precursor are still receiving enthusiastic attention of researchers and are still in contention. The carbide-derived carbons (CDCs) have been shown to reach upto 4.7 wt % of H_2 at a pressure of 60 atmospheres and temperature of 77 K.[7] It appears that there is enough opportunity to tune and optimize the pore sizes and achieve better storage performance.

12.3.3 Zeolites

They are microporous complex aluminosilicates formed by SiO_4 and AlO_4 tetrahedra. These materials appear in a variety of crystallographic structures. Their ordered crystalline nature ensures uniform cavities and channels with dimensions typically in the microporous regime. The framework structures of zeolites are rigid. They have high surface area per unit mass with large pore volumes. Their properties are tunable to an extent by controlling the Si to Al ratio. Despite having many positive attributes, zeolites have rather disappointing hydrogen – uptake ability, the highest uptake reported so far is 2.55 wt% only in Na-X at 77 K and under 4.0 MPa pressure. Theoretical calculations do not brighten the picture at all. Thus, van den Berg *et al.*[8] theoretically found that the H_2-uptake could reach upto a level of (4.8 ± 0.5) wt %. Their modeling was based on the molecular mechanics method and they assumed the sodalite structure for the zeolite.

Vitillo *et al.*[9] however predicted a much lower hydrogen uptake by zeolites based on calculations done by the same molecular mechanics methods. The maximum capacity was predicted to be in the range of 1.9 wt.% or so. Van den Berg *et al.*[10] carried out grand canonical Monte Carlo simulations, which also led to much lower H_2-uptake capacities for the zeolites compared to their previous predictions. Calculations by Song[11] however suggested higher uptake values in the range of 4.45 wt% for Mg-X. Zeolites have a number of significant practical advantages over other microporous adsorbates, the most significant being their thermal stability. However, the H_2-uptake saturation achieved is not yet upto the requirements of industry. Even then, light weight zeolites continue to be in the contention and are being seriously investigated as solid state hydrogen storage materials.

12.3.4 Metal Organic Frameworks (MOFs)

MOFs are another class of solids receiving a great deal of attention as potential hydrogen storage materials of the microporous category. They are crystalline inorganic-organic hybrid solids consisting of metal ions or metal clusters linked by organic bridges. A typical example is provided by Zn_4O (bdc) where bdc stands for an organic bridging component called benzene dicarboxylate, the material being commonly known as MOF-5. It consists of zinc oxide clusters joined by the benzene linker. The metallic clusters are recognized as secondary building units while the primary building units are provided by the organic linkers. On examination it is found that the structure of MOF-5 produces a highly porous cubic network, the pores providing space for hydrogen adsorption. Rosi *et al.*[12] were the first to measure and report the hydrogen uptake capacity of MOF-5. The reported value of 4.5 wt% at 77 K under a hydrogen pressure of 0.07 MPa was an overestimation as found later. The first report, however, generated great enthusiasm and many framework materials were designed, synthesized and put to test. The maximum hydrogen uptake was reported to be 7.5 wt% for what has now become known as MOF-177. It also is based on zinc oxide clusters with another organic linker called 1,3,5-benzene tribenzoate (btb), the structural code being Zn_4O(btb). The maximum uptake was achieved at 77 K under 7.0 MPa hydrogen pressure. In room temperature studies, the maximum hydrogen uptake of 1.4 wt% under a pressure of 9.0 MPa was recorded for the MOF Mn(btt), btt being the organic linker named benzene tris-tetrazolate. The same MOF at 77 K recorded uptake of 6.9 wt% under 9.0 MPa of hydrogen pressure. It is not difficult to see that a very large number of possible MOFs can be generated from different SBUs and different organic linkers leading to microporous solids with varieties of pore geometries. The fact that there are metal sites exposed within the pore has enhanced the interest in the MOFs many-fold as they offer tunability of the operating temperature and of the maximum storage capacity at near ambient temperatures. To increase the operating temperature or increase the maximum storage capacity at or around the ambient temperature, the strength of interaction between the incoming hydrogen molecules and the surface of the adsorber must be increased. In this context we may point out that molecular hydrogen is known to form molecule-metal complexes with virtually all the transition metals. These complexes are known as Kubas complexes. If the interaction is strong enough H_2 molecule breaks down into hydrogen atoms giving rise to interstitial hydrides. Since metal atoms are known to be exposed inside the pores of MOFs, the Kubas mechanism[13a] can be exploited to develop MOFs with enhanced storage properties.[13b]

A rather interesting feature of some of the MOFs is the flexibility of the framework structures. Zhao *et al.*[14] [2004] reported two MOFs, which displayed hysteretic hydrogen adsorption triggered by the flexibility of the framework structure. That means, at a given temperature the material desorbs at a pressure that is higher than the pressure at which adsorption of H_2 takes place at the same temperature. There are however many other MOFs that behave perfectly reversibly at any given temperature in so far as hydrogen absorption and desorption or adsorption-desorption of any other guest molecules are concerned. The framework flexibility displayed during the adsorption/desorption of hydrogen by some of the MOFs may or may not be accompanied by chemical bond breaking events occuring in the framework structure. The flexibility can lead to certain structural transformation involving stretching, rotational, breathing or scissoring motion.

The enthalpy of hydrogen adsorption by MOFs is generally accepted as a measure of the strength of interaction of hydrogen with the surface or the pore structure. The measured isosteric enthalpy of adsorption has been generally found to lie in the range of 3.8 kJ mol^{-1} for MOF-5 to 12.3 kJ mol^{-1} for a mixed zinc/copper metal organic material (M′MOF). MOFs continue to hold out more promise as potential solid hydrogen storage materials compared to zeolites. This is because of their structural flexibility and exposed metal sites (inside pores), which offer scope for further development and tuning. They are, however much less robust (in terms of thermal stability) than zeolites and microporous carbon-based materials. Nevertheless, their commercialization is already underway. We will address some issues concerning the theoretical design of MOFs with enhanced performance features in Chapter 13.

12.3.5 Organic Polymers for Hydrogen Storage

A number of microporous organic polymers have recently emerged as potential candidates for use as a solid state hydrogen adsorber. Three categories of such materials have been generally recognized and experimented with[3]

1. Polymers, which are intrinsically microporous, acronymed PIMs.
2. Hyper cross linked polymer or HCPs.
3. Covalent organic framework or COFs.

Type 1 and type 2 materials turn out to be amorphous under X-ray and neutron diffraction diagnostics. They are solids with disordered structure, and resemble activated carbons more than crystalline microporous materials like zeolites or MOFs or COFs. Type 1 polymers or the PIMs are macromolecules formed from organic fused-ring systems as components. They have rigid structures with contortions and are unable to pack the available free space efficiently, creating big voids in the macromolecular structures. They have large BET surface area. The HCPs too, possess a high degree of microporosity and high BET surface area because of the high density of cross-links and covalent bonds between macromolecules in the polymeric material.

The COFs on the other hand, have typical crystalline networks of elements like carbon, boron, silicon and oxygen, which are connected by strong covalent bonds like B-C, B-O, Si-C, C-C, etc. They have light weight networks and resemble MOFs. It will be interesting now to compare the hydrogen uptake capacities of the three kinds of organic polymers on the basis of data available both from experiments and simulations.

A tryptycene-based polymer of the PIM category (trip PIM) has been found to have a hydrogen uptake capacity of 2.7 wt% at 77K and 1 MPa hydrogen pressure. A higher uptake of 3.68 wt% has been reported for an HCP type of material formed from 4-4'-bis(chloromethyl)1-1' biphenyl at 77K and under a pressure of 1.5 MPa. The maximum hydrogen uptake ability of these materials turns out to be rather modest. Interest in these materials still persists because of their light-weight characteristics and the hope that it would be possible to design much better performing materials belonging to HCP or PIM category. It is necessary to mention here one distinct drawback of the HCP or PIM group of materials. It is their comparatively much lower thermal stability vis-a-vis other microporous solid state storage materials.

The H_2-uptake capacities of the COFs are much higher, be it the experimentally measured quantity or the ones predicted by MD-simulations. Thus, COF-102, a self-condensation polymer formed by tetra (4-di-hydroxy bory phenyl) methane is also known by the acronym TBPM. COF-103 is the silicon analogue of COF-102 and is known by acronym TBPS. Furkawa and Yaghi[15] experimentally obtained the gravimetric H_2-uptake capacity of 72.4 mg g^{-1} for COF-102 at 77 K and 9.0 MPa pressure. For COF-103, the value reported was 70.5 mg g^{-1}. Simulation studies generally echo the experimental findings. Thus, the simulated total hydrogen uptake of COF-105 and COF-108 were around 18 wt% at 77 K and under 10 MPa pressure. Their volumetric capacities are much lower as is expected because the polymers have large free volumes.

It turns out that polymeric materials that possess the largest gravimetric hydrogen uptake capacities also have the lowest enthalpy of hydrogen adsorption. This inverse correlation may not be unexpected. These polymers have larger pore volumes and dimensions, which permit high overall hydrogen uptake; but at the same time the larger pore dimension leads to a reduction of the effective adsorption potential, which in turn is reflected in the lower adsorption enthalpy.

12.3.6 Interstitial Hydrides

A metallic element or an intermetallic compound very often reacts with gaseous hydrogen under appropriate conditions to produce binary, ternary or higher hydrides. The reaction is mediated by the surface leading to the dissociation of H_2 molecules into hydrogen atoms, which then move into the interstitial voids of the solid and diffuse through out the solid in the long run. A majority of the metals absorb hydrogen in the manner described above, at appropriate temperatures and hydrogen pressures, forming binary hydrides (MH_2). Unfortunately, binary hydrides are either too stable or unstable. In either case, their practical use as hydrogen storage materials is not viable. If the hydride is too stable, the pressure at which the corresponding metal can reversibly absorb and desorb hydrogen is too low at practical temperatures maintainable in mobile applications. If the hydride is too unstable, the required hydrogen pressure is too high at the operating temperatures. Thus a metallic compound for which the enthalphy for reversible hydrogen sorption is neither too high, nor too low has to be designed, that means the material must have intermediate value of enthalphy of hydrogen sorption. As a rule of thumb, it is therefore desirable that a metal, which forms a very stable hydride should be combined with a metal that forms a relatively unstable hydride to form an intermetallic compound, which could have an enthalphy of hydrogen absorption in the required range.[2-3]

The history of metal hydrides is rather old. Palladium hydride (PdH_x) was discovered by Thomas Graham more than a century ago. But the real journey of metal hydrides began in 1960 when nickel-metal hydride batteries were very successfully commercialized.[16] The cathodes in these materials were made of an intermetallic compound. To date, this has been the most successful commercialization of intermetallic hydrides. We may note here that many intermetallic hydrides have a bit too low value of gravimetric capacity for hydrogen storage. The interest in these compounds still continues unabated because they are known to have favorable features for reversible hydrogen absorption and desorption energetics and kinetics. The intermetallic compounds are therefore projected as important materials for sustainable hydrogen economy that may replace the fossil fuel economy of the present era.

12.3.7 Intermetallic Compounds

How to choose the intermetallic compounds to make a good hydrogen storage materials? We have already referred to a thumb rule that can possibly be a useful guide. Let us look into it a little more closely.

Let AH_x and BH_y be the two binary hydrides that the metals A and B form respectively, with enthalphy of formation ΔH_A and ΔH_B. Let AH_x be a stable binary hydride and AH_y be an unstable hydride (relatively). Now, if A and B form an intermetallic A_mB_n, which in turn captures hydrogen to form the hydride $A_mB_nH_z$ with formation enthalphy equal to ΔH_{AB}, an inequality as stated below is expected to hold:

$$\Delta H_A < \Delta H_{AB} < \Delta H_B$$

By varying the composition ratio of the intermetallic (m/n), it is possible to push ΔH_{AB} either to the left or the right. Also, if A and B are replaced by atoms of the same size and chemical features, the inequality continues to hold. A large number of combinations of A and B are now known to form reversible hydriding materials although only a few of them absorb and desorb hydrogen reversibly at practically useful pressure and temperature. A fairly large compilation of ΔH_{AB} values of intermetallics has been done by Griessen and Riesterer.[17] We will briefly review some of the listed materials that have still retained their practical relevance.

12.3.7.1 AB₅ Intermetallics

We have already referred to the most well known AB_5 intermetallic viz $LaNi_5$. $LaNi_5$ readily absorbs hydrogen forming the intermetallic hydride $LaNi_5H_x$ at ambient temperature and under a moderate hydrogen pressure. The hydrogen uptake capacity exceeds what $LaNi_5H_6$ represents which comes to a reversible gravimetric capacity of 1.25 wt%. $LaNi_5$ like many other AB_5 intermetallic, has a tendency to suffer significant disproportionation reactions and loose its reversible hydrogen uptake capacity substantially during hydriding-dehydriding cycles. This tendency to undergo disproportionation can be curbed very effectively by introducing a third component in the $LaNi_5$ intermetallic. It turns out that Sn is the most effective metal for such partial substitution. In fact $LaNi_{5-x}H_x$ with $x = 0.2$, is a viable material for hydrogen storage and mobile applications.[18a−c]

12.3.7.2 AB₂ Intermetallics

Typically, the A atoms belong to Gr IV (Ti, Zr, Hf) while B can be a metal of transition series or even a non-transition metal. The preferable choice for B is V, Cr, Mn, Fe and Ni. These intermetallics crystallize in either the hexagonal (C14) or cubic (C15) Laves phase structures. Depending on the elemental composition, these compounds display a fairly wide range of hydriding characteristics just as the AB_5 compounds do. The hydriding property of AB_2 intermetallics can be tuned by introducing partial elemental substitution. In addition to partial elemental substitution, the introduction of sub or superstoichiometric composition can be used to form new multicomponent AB_2 compounds with altered hydriding properties. Thus, derived from the parent $ZnMn_2$, the substoichiometric compound $ZrMn_{2-x}(x > 0)$ has a significantly lower reversible hydrogen storage capacity and lower plateau pressure than $ZnMn_2$. The superstoichiometric $ZrM_{2+x}(x > 0)$, on the other hand, has higher plateau pressure and only slightly lower reversible hydrogen capacity than the parent compound. It may be mentioned here that a non-stoichiometric intermetallic of the AB_2 class viz. $Ti_{0.98}Zr_{0.02}$-$Cr_{0.05}V_{0.43}Fe_{0.09}Mn_{1.5}$ has been successfully used as the hydrogen storage material for automobiles.[2−3] The intermetallic referred to display very fast kinetics and quite good long term cycling stability. However, the gravimetric storage capacity of this material is just 1.8 wt%. The cost of materials is quite high preventing its large scale use at this stage. The alloy, Ti_2CrV has recently been evaluated for solid state hydrogen storage application. The alloy is highly stable with storage capacity going upto 4 wt% under subatmospheric pressure. The peak desorption temperature is 180°C.[18d] It appears that AB_2 non-stoichiometric compounds like the AB_5 materials are potentially and operationally viable hydrogen materials and provide excellent examples of the interstitial metal hydrides that performs well in practical applications. They also satisfy acceptability criteria except cost.

12.3.7.3 AB – Intermetallics

Only a very few of intermetallic compounds of the AB-type have been of interest in the context of the search for potential solid state hydrogen materials. $ZrNiH_3$ is the earliest reported ABH_x-type of interstitial hydride. When it comes to finding an AB-type of intermetallic, which is practically usable as a solid state hydrogen storage material for mobile applications, TiFe is the only AB intermetallic that has evinced keen interest. It has an ordered BCC structure and the solid solution alloy Ti-V-Cr-Fe reaches its maximum reversible capacity for a (BCC) lattice parameter value of 3.036 Å. In addition, the electron to atom ratio (e/a) also appears to play an important role. The critical (e/a) value seems to be 5.25, where 'e' is the number of valence electrons and 'a' is the number of atoms. For values higher than 5.25, the storage capacity registers a sharp decline.[2–3]

12.3.8 Modified Binary Hydrides

MgH_2 has attracted some attention because of its high gravimetric hydrogen storage capacity of 7.66 wt %. Its thermodynamic stability, however is too high ($\Delta H = -75 \text{ kJ mol}^{-1}$ of H_2) and kinetics, too slow to make it useful. Mechanical milling, with or without additives improves the kinetics significantly although the temperature for desorption remains too high. With regard to additive-induced or catalytic enhancement of absorption-desorption properties of MgH_2, the most successful catalyst seems to have been Nb_2O_5. Many other catalysts, e.g. V_2O_5, Mn_2O_3, Al_2O_3, CuO, SiO_2, etc. have been experimental with, but none performs as well as Nb_2O_5 does. The reasons for the enhancement of adsorption desorption rates are not fully understood. It could be that the additives induce further reduction in the grain or particle size. The size reduction appears to bring in some changes in the thermodynamics of the system as well. *Ab initio* HF and DFT calculations lend support by revealing that the stability of MgH_2 is significantly reduced for clusters of very small sizes (< 1.3 nm or so). The theoretical results raise hope that the nanoscale magnesium hydride may perhaps provide required hydrogen storage capacities in a practically useful temperature range.[4]

12.3.9 Quasi-crystalline Materials

Quasi-crystals (QCs) were discovered only only in 1984. Since then, QCs have been found to exist naturally as well, and more than a hundred of QCs have been synthesized. They have long range order but no translational symmetry. Several quasi-crystalline materials have been investigated to assess their hydrogen storage capabilities. The interest in QCs from the point of view of their hydrogen storage properties gained traction because QCs could provide many interstitial sites, which may lead to enhanced hydrogen storage capacities. The investigated QCs are ternary alloys of Ti-Zr-Ni, Ti-Hf-Ni and Mg-Al-Zr, but there is little clarity in the data to make any unambiguous conclusion.[19a] More work, theoretical as well as experimental is needed to draw concrete conclusions.

The important signatures of hydrogenation discernible in the electronic structure of metals may be mentioned here. They can be broadly classified into four types:

1. An expansion of the crystal lattice on hydrogenation, reducing band widths and modifying the symmetry of the electronic states.
2. Appearence of a new band – the M-H bonding band below the metal d-d bands. Electrons are transferred from the s-d band to this new band and some metal states appear below the Fermi level.
3. If the metal hydride has more than 1 electron per unit cell, the H-H interaction brings in some new attributes in the lower portion of the density of states.
4. An upward shift of the Fermi level is generally observed.

These attributes can serve as features in a theoretical design of metal-based hydrogen storage materials.

12.3.10 Complex Hydrides

These are salt-like materials of the general formula $A_xD_yH_z$ in which A is usually group I or group II atoms, the central D sites are occupied by either boron or aluminium (D = B, Al). Hydrogen is covalently linked to the central atom 'D' producing the complex anions like $[BH_4]^-$ or $[AlH_4]^{-3}$, which is bonded to the cations by ionic bonding. The salt-like ionic crystal structure is formed by A^+ and the complex anions of boron or aluminium. The central atom in the complex anion often forms the basis of classification of these materials as metal borohydrides, aluminium hydrides, etc.

Unfortunately, the kinetic barrier to decomposition, releasing hydrogen is high. Thus dehydrogenation requires elevated temperatures, making them unsuitable for reversible hydrogen storage materials. The high kinetic barrier to decomposition can however, be lowered by adding suitable catalysts, which make the complex metal hydrides respond reversibly with respect to hydrogen uptake and release. The catalytic lowering of the barrier to decomposition brings down the temperature at which dehydrogenation takes place smoothly and makes rehydrogenation feasible. This possibility was first realized in the case of sodium aluminium hydride ($NaAlH_4$) with the addition of $TiCl_3$ as catalyst. Of all the complex metal hydrides, only ($NaAlH_4$) meets the requirements for use as solid hydrogen storage material for mobile application with on-board rehydrogenation. The complex sodium-aluminium hydride decomposes in two steps:

1. $3\,NaAlH_4 \rightarrow Na_3AlH_6 + 2\,Al + 3\,H_2$ at temperature 210–220 °C.
2. $Na_3AlH_6 \rightarrow 3\,NaH + Al + 1.5\,H_2$ at temperature 250 °C.

The first step is endothermic with $\Delta H = 37$ kJ mol^{-1} H_2 and releases 3.7 wt % of hydrogen. The second step is also endothermic with $\Delta H = 47$ kJ mol^{-1} H_2 and releases 1.8 wt % of hydrogen. The dehydrogenation of NaH does not take place before the temperature is raised to $450^o C$. Doped sodium alanates (dopant $TiCl_3$) no doubt have kinetics and cyclic stability. Nevertheless, the storage capacity for low temperature fuel-cell applications is far below requirements. The other aluminium hydrides are not reversible under technologically acceptable conditions.

Borohydrides have the highest gravimetric hydrogen storage capacities among the complex hydrides. $LiBH_4$ contains 18.5 wt% of hydrogen, which is released through one of the following two reactions:

1. $LiBH_4 \rightarrow Li + B + H_2$
2. $LiBH_4 \rightarrow LiH + B + 3/2\,H_2$

The decomposition temperature is too high for practical use although the dehydrogenation is fully reversible. The rehydrogenation, however requires elevated temperature and pressure (e.g. 35.0 MPa at 600°C). A number of alkali metal or alkaline earth metal borohydrides have high gravimetric and volumetric capacities, but their practical utility has not been demonstrated. Lithium imide also has attracted some attention, mostly theoretical, in this context.[19b] Combining several complex hydrides into one storage system may improve the overall performance and storage characteristics. The reactions taking place in such system are however, rather complex and must be fully characterized kinetically and thermodynamically before technological applications can be attempted.

The so called first-principles computational approach has been extensively used to assess the H_2 absorption and desorption process in complex hydrides and their decomposition pathways. First-principles DFT calculations based on local density approximation and its improved variants like the generalized gradient approximation are now accepted as reliable theoretical tools that can explain and even predict properties of materials and phenomena taking place in them.

12.4 Optical Properties of Materials and Lasers

When light travels from one medium into another, say from air into a solid material, some of the light may get transmitted through the substance, a part of the light may also get absorbed and another part of it may be reflected at the interface between the two media. The intensity of the beam of light (I_0) incident on the surface of the solid must be equal to the sum of the intensities of the transmitted (I_t), absorbed (I_A) and reflected beams (I_R). That means

$$I_0 = I_t + I_A + I_R \tag{12.3}$$

and

$$T + A + R = 1 \tag{12.4}$$

where $T = \frac{I_t}{I_0}$ is the transmissivity, $A = \frac{I_A}{I_0}$ is the absorptivity and $\frac{I_R}{I_0}$ is the reflectivity (R) of the material. Materials that do not allow the transmission of visible light are called opaque while materials capable of transmitting light with little or no absorption are called transparent. There are also materials that allow light to be transmitted diffusively. In that case, the incident light gets scattered inside the material to such an extent that objects viewed through the material can not be clearly distinguished. Such materials are called translucent. Metals in bulk are opaque to light in the visible part of the spectrum while electrical insulators can be made transparent. Some semiconducting materials, on the other hand are opaque, while some others are transparent.

The electromagnetic radiation interacts with the atoms, ions and electrons in the material causing electronic polarization and transitions of electrons between energy levels that the material supports. One component of light is a fast oscillating electric field, which interacts with the electronic charge density that surrounds each atom lying on the path of the light as it passes through the material. The interaction produces electronic polarization and the polarization density starts oscillating with the frequency of the incident electric field of light. As a consequence some of light may be absorbed by the material while part of it passes through giving rise to the phenomenon of refraction of light.

The absorption (or emission) of electromagnetic radiation may be associated with electronic transitions from one energy level (E_i) to another (E_j) in an isolated atom or a molecule with discrete energy levels, an essential condition for such a transition to occur is the energy matching condition which requires (for $E_j > E_i$)[20–21]

$$\Delta E = (E_j - E_i) = \hbar\omega \tag{12.5}$$

where $\omega = 2\pi\nu$ is the circular frequency of the electromagnetic radiation interacting with the material. The above condition means that an electromagnetic wave while interacting with matter can only transfer energy in bundles called photon with energy $\hbar\omega$. It is necessary to keep in mind that the entire photon energy is absorbed in each excitation event. We have implicitly assumed that the transition from state i to j is an allowed transition under the electric dipole selection rule. The electron that is promoted to the excited state drops back to its initial level after a while with the emission of electromagnetic radiation. The time τ during which the stimulated electron resides in the excited level is called life time of the excited state. In general the decay to the lower state can occur by different paths. Whatever path is followed the law of conservation of energy must be obeyed for all absorption and emission events.

12.4.1 Optical Properties of Metals and Nonmetals

12.4.1.1 Metals

Let us try to understand the optical properties of metals in terms of their band structure and absorption/emission of light. Isolated atoms or molecules have discrete energy levels, but in solids like metals, the discrete energy levels are replaced by energy bands (see Chapters 5 and 8).

In metals, there is no band gap, but the highest energy band is only partially filled with electrons upto the Fermi level. Why do metals appear opaque? They do so because the incident electromagnetic radiation with frequencies lying in the visible range of the spectrum, excites electrons into unoccupied energy states above Fermi level (Figure 12.1) leading to the absorption of radiation, which however, takes place within a very thin outer layer (thickness $\approx 0.1\mu$m) of the metal. So, metallic films having thickness $< 0.1\mu$m are capable of transmitting visible light and appear transparent.[21] Notice the continuously available empty electron states in metals (Figure 12.1), which means electronic transition displayed in the same figure can take place at all frequencies. Most of the absorbed radiation is re-emitted from the surface of the metal as

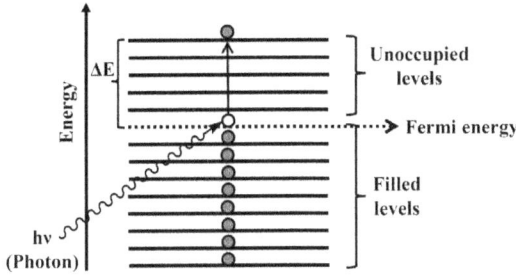

FIGURE 12.1 Pictorial representation of photon absorption by a metal.

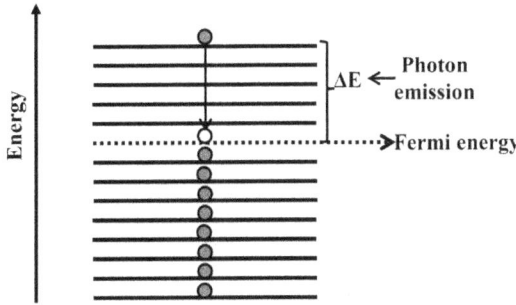

FIGURE 12.2 Pictorial representation of emission of a photon.

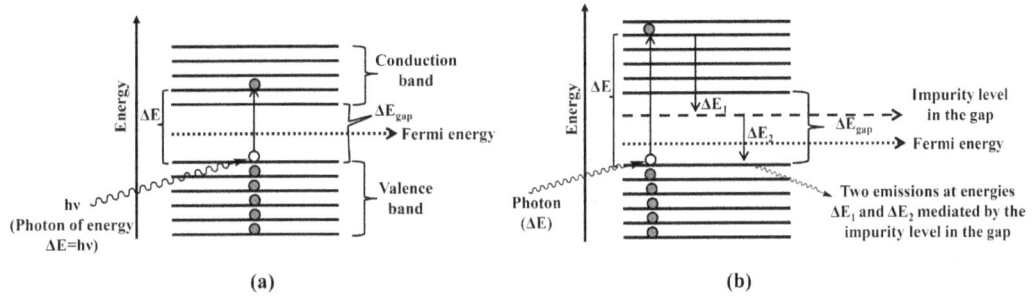

(a)

(b)

FIGURE 12.3 Light absorption/emission in non-metallic material (a) when there is no mid-gap state, (b) when an impurity creates a mid-gap state.

visible light of the same wavelength (Figure 12.1), the reflectivity being almost 90–95%. A small part of the energy from the decay is dissipated as heat. Since metals are opaque and highly reflective, the apparent color of the metal depends on the distribution of the wavelengths in the radiation that is reflected. If the number and frequency of the re-emitted photons in the reflected beam are approximately the same as in the incident beam, the metal takes a bright silvery appearence (e.g. Al, Ag). 'Cu' and 'Au' appear orange-red and yellow, respectively because some of the energy associated with high frequency photons is not re-emitted as visible light.

12.4.1.2 Non-metals

The energy band structures of non-metals are associated with a large band gap (Figure 12.3) preventing absorption of visible light (energy matching is not possible) – so they may appear transparent in visible light. For such materials, the phenomenon of transmission and reflection of light need careful consideration. It is well-known that the refractive index 'n' of the material is equal to the ratio of the velocity of light in vacuum to the velocity 'v' in the given material.

$$n = \frac{c}{v} \qquad (12.6)$$

The value of 'n' clearly depends on the wavelength of light. Now the velocity 'c' of light in vacuum is a fundamental quantity defined by the relation

$$c = \frac{1}{\sqrt{\epsilon_0 \mu_0}} \qquad (12.7)$$

where ϵ_0 and μ_0 are the permittivity and permeability of free space. Likewise v can be defined as

$$v = \frac{1}{\sqrt{\epsilon \mu}} \qquad (12.8)$$

ϵ and μ being the permittivity and permeability of the material. Using Eqs. 12.6, 12.7 and 12.8 we have

$$n = \sqrt{\epsilon_r \mu_r} \qquad (12.9)$$

where ϵ_r is the dielectric constant and μ_r is the relative magnetic permeability of the material. For non-magnetic or only slightly magnetic material $\mu_r \approx 1$ so that

$$n^2 = \epsilon_r \qquad (12.10)$$

For a transparent material, there is thus a relation between its dielectric constant and refractive index. That is not unexpected. Refraction of light has its origin in the electronic polarization of the material at high frequencies of the electromagnetic waves that visible light consists of. So the equality in Eq. 12.10 implies that the electronic part of the dielectric response of the material can be obtained from the measured value of refractive index of the optically transparent material. Since electronic polarization is responsible for the refraction of light, the value of refractive index of a material is markedly influenced by the size of atoms/ions in the material. The larger the size of atoms/ions, the larger the electronic polarization, and the higher the refractive index of the material. Thus, heavily leaded glass (98% by wt of Pb) has its RI value raised to 2.1 vis-a-vis the RI value of 1.5 for soda-lime glass. A similar increase of RI takes place when PbO is replaced by BaO. Crystalline ceramics with cubic crystal structures and glasses are optically isotropic, i.e. the RI is independent of the crystallographic direction that light takes. For non-cubic crystal structures, RI is anisotropic, having the greatest values along the directions having the largest concentration of atoms/ions.[20–21]

Opaque non-metallic materials can absorb light by two basic mechanisms, (i) electronic polarization, (ii) VB $\rightarrow$ CB electron transition. The first mechanism becomes operationally important only when the frequency of light and relaxation frequency of the atoms/ions making up the material nearly match. The operation of the second mechanism depends on the energy band structure of the ceramic or semiconductor materials. Here, absorption of a photon (energy = $\hbar\omega$) can take place by the promotion of an electron from the nearly filled VB into an empty state in the CB right across the band gap, creating a hole in the VB and a free electron in the CB (Figure 12.3a). The frequencies (ω) of light for which VB$\rightarrow$CB transition can take place is related to the gap energy ΔE_{gap} by the following inequality:

$$\hbar\omega \geq \Delta E_{gap} \qquad (12.11)$$

The shortest wavelength of the visible light is approximately 0.4 μm and the longest is $\approx 0.7 \mu$m. It is now easy to see that no visible light will be absorbed by a non-metallic material having band gap energy greater than 3.1 eV and all visible light will be absorbed by VB$\rightarrow$CB transition if the material (non-metallic) has a band gap less than ≈ 1.8 eV. If the band gap of the material lies between 1.8-3.1 eV, only a part of the visible spectrum will be absorbed, thereby imparting a color to the material concerned.

The VB$\rightarrow$CB electron transition is not the only process by which a dielectric may absorb light. If impurities (donors or acceptors) or defects are present, electronic energy levels may appear in the gap region. The appearance of these mid-gap levels provides an additional avenue for electron transition as shown in Figure 12.3b where the VB$\rightarrow$CB transition is followed by emission at two different wavelengths mediated by a single impurity level in the mid-gap region.

The energy absorbed in VB$\rightarrow$CB transition may also dissipate by a direct recombination of e^- and hole producing energy. There are other possibilities as well, which have not been discussed here.[22]

12.5 Photonic Materials

Electrons promoted to excited states in a material falls back to lower states or to the ground state emitting electromagnetic (EM) radiation. The emission event takes place spontaneously and randomly without any coordination whatsoever. They are purely independent events. The emitted radiation is incoherent and the light waves (assuming that the emitted electromagnetic radiation is in the visible part of the spectrum), are completely out of phase with each other. Is it possible at all to get materials that emit coherent light (or EM radiation) when appropriate conditions are created? The answer is in affirmative with the stipulation that the emission process must be 'stimulated' and that only certain specific materials (called gain materials) will support the process. Such materials are called laser materials, laser being the acronym for 'light amplification by stimulated emission of radiation'. Since all the emitted radiations are in phase and involves the same transition, the intensity of the light emitted is very high (light amplification). It is monochromatic as well as coherent. The materials for lasing may be a solid, liquid or a gas. We will however, consider only the solid state lasers and their mode of action briefly. It turns out that the energy level structure or the energy band structure plays a very crucial role in deciding if lasing process is realizable. An essential condition for lasing to happen is that a 'population inversion' must take place. Let us try to explain the phenomenon with reference to the solid state ruby laser.[21–22]

Ruby, a beautiful red natural gem-stone is now made synthetically. Technically, it is a dielectric material called sapphire (transparent Al_2O_3 with a large band gap) into which Cr^{+3} ions have been doped ($\approx 0.05\%$). The Cr^{+3} ions impart red color to ruby and provide the mid-gap states in sapphire crystals, which are instrumental for the generation of laser radiation. For this purpose, the ruby crystal (a highly polished ruby rod which is totally silvered at one end and only partially silvered at the other) is illuminated with a Xe-flash lamp. The electrons in Cr^{3+} ions that were all in their electronic ground state absorb the 0.56 μm light from the flash and are promoted to the higher energy states of Cr^{3+} ion in saphire (Figure 12.4) (more precisely, these are the crystal field states of Cr^{3+} ion in $3d^3$ configuration) in the octahedral environment of saphire crystal). It turns out that these electrons can decay back to their ground state by two different paths (Figure 12.4). One of this is the spontaneous and fast (in about 10^{-9} s) fall back (from excited E to the ground E_0 of Cr^{3+} ion) with incoherent emission of photons, which do not constitute a part of the laser emission. The second path involves an initial fast, non-radiative passage into a metastable intermediate state (E'), which has a much longer life time of approximately 3 ms. The much longer life time of (E') state allows a large number of such states in the ruby crystal to be occupied by electrons leading to a population inversion. A few of these states start decaying into the ground state by spontaneous emission of photons with energy ($E' - E_0$) = ω, it stimulates an avalanche of emissions of photons from the electrons still occupying the metastable E' state. The radiant energy carried by these photons correspond to the wavelength $\lambda = 0.694$ μm, which is in the red part of the visible light. Some of the photons that are emitted parallel to the long axis of the ruby red are transmitted through the partially silvered end; others

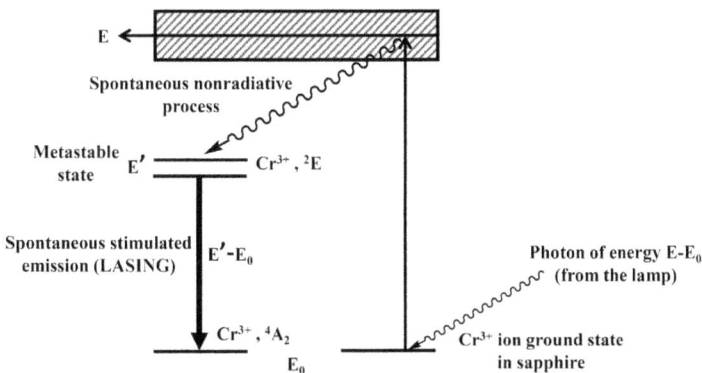

FIGURE 12.4 Lasing in ruby laser.

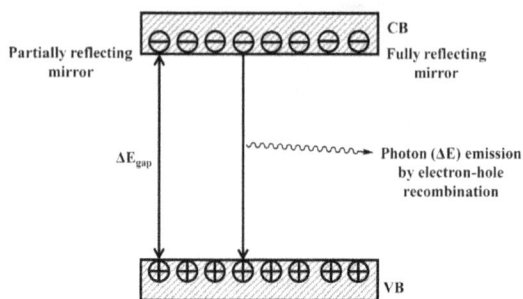

FIGURE 12.5   Semiconductor: electron-hole recombination mediated lasing.

are reflected from the totally silvered end and the beam of emitted photons keeps traveling back and forth along the axis of the ruby rod between the partly silvered and completely silvered end. During this back and forth travel, many more emissions are stimulated, the beam continues to gain in intensity and ultimately an intense and a highly collimated, monochromatic ($\lambda = 0.6943$ μm, red) coherent beam of light comes out through the partly silvered end of the ruby rod. That is how a solid state material like ruby generates laser beam. There are other similar solid state lasers, e.g. Nd-YAG laser ($\lambda = 1.06$ μm), Nd-glass ($\lambda = 1.06$ μm), etc.

Semiconductors with relatively wider band gaps have been exploited for the production of coherent light. The band gap ΔE_{gap} needs to be large enough so that a CB $\rightarrow$ VB transition is associated with wavelengths (λ) that fall in the visible part of the spectrum. That means λ associated with ΔE_{gap} must be such that it falls between 0.4 and 0.7 μm. Gallium arsenide is one of the semiconducting materials with a band gap in the appropriate range.[20] The mechanism of production of coherent light from GaAs is based on electron-hole recombination process that occurs in such materials. A constant voltage applied to the material excites electrons from the VB into the CB (Figure 12.5), simultaneously creating holes in the VB. After a while there will be many excited electron-hole pairs simultaneously existing in the sample of GaAs. However, they can not do so for ever. Some of the excited electrons and holes recombine after a while and each such recombination results in the emission of a photon with wavelength λ. One such photon stimulates the recombination of additional excited electron-hole pairs leading to the production of more photons, each with the same wavelength (or energy) and the same phase. These processes add up to multiply the number of emitted photons and the emission gains in intensity, which is further amplified by allowing the emitted coherent and monochromatic beam to travel back and forth between the totally reflective and partly reflective ends of the semiconductor material, inducing more recombination events. So, what comes out through the partly reflective end is an intense monochromatic and coherent beam of light – the laser beam. Since a constant voltage is maintained, there is always a steady supply of electrons and holes so that the beam that comes out is continuous. GaAs is not the only semiconducting material capable of producing laser beams. There are many others. In fact by choosing materials with appropriate band gaps, it is possible to design structures capable of producing lasers with different wavelengths.[23–24]

12.6 Photovoltaic Materials

The photonic materials as we have seen in Section 12.5 convert electrical energy into light energy (photons). They are one kind of energy conversion material. The reverse process (photovoltaic effect or PVE) in which light energy (solar light) is converted into electrical energy is also mediated by semiconducting materials. They are called photovoltaic materials. PVE has indeed become the cheapest source of energy in arid regions where there is abundant sunshine round the year. In the global context, PVE has already become the third largest producer of renewable energy, the first two being the hydroelectric and wind energy-based power systems. Solar PV materials have evolved through several generations. A brief review of the evolution follows.

12.6.1 Generation I Materials

Single crystalline Si modules have been the most widely studied solar PV material. Multicrystalline PV cells are simpler and easier to fabricate. They are also cheaper compared to the monocrystalline Si modules. However, the multicrystalline PV cells tend to be less efficient at approximately 13.2% compared to the mono-crystalline Si-cells, which have an average efficiency of 14%. A single (*p-n*) junction solar cell has been shown to have a maximum efficiency of 33.7% at a band gap of 1.3 eV. For Si, the band gap is 1.1 eV and the maximum achievable efficiency is 32%. This limit has become known as Shockley-Queisser limit arrived at on the basis of a detailed balance analysis.[25]

12.6.2 Generation II Materials

These are compound semiconductors (see Chapter 8), CdTe-based thin film solar cells being the most rapidly growing solar PV-technology. This new thin film device is also bound by a Shockley-Queisser efficiency limit that is similar to what was theoretically demonstrated for silicon. CdTe-based solar PV technology is however cheaper. Copper indium galium diselenide (CIGS) thin film-based solar cells have evolved from copper indium diselenide (CIS) semiconductor. Soda-lime glass provides the substrate, a layer Mo is used for the back contact. CIS/CIGS layers constitute the energy absorption medium while cadmium sulphide acts as the buffer layers. The front contact is provided by ZnO:Al. These cells have solar power conversion efficiency of 7.8-9.1 %. The design and materials are still evolving.

12.6.3 Generation III Materials

The 3rd generation of solar photovoltaic systems are not constrained by the Shockley-Queisser limit that single junction solar cells were theoretically shown to conform to. Two promising technologies that are coming up are based on mixed sulphides of copper, Zinc and tin (Cu_2ZnSnS_4) and Zinc phosphide (Zn_3P_2) as solar energy converters. Organic-inorganic hybrid solar cells made of methyl ammonium lead sulphide perovskites as solar energy absorber material have also shown promise.[26]

The efficiency already achieved with single crystalline photovoltaic materials like silicon or gallium arsenide has still some scope for improvement. It is not yet possible to conclude whether polycrystalline thin films of CdTe, $CuInGaSe_2$ could match or even exceed the efficiencies of Si or GaAs-based solar cells. A lot more work on the development of new photovoltaic materials and deeper understanding of the mechanism of energy conversion may be anticipated in the coming years.

12.7 Materials that Change Light

Electromagnetic radiation or light is an important form of energy in the context of modern technology of telecommunication, computing, fast switching, data storage, etc. An optical beam has several controllable parameters like amplitude, phase, polarization and frequency. Non-linear optical (NLO) materials have the special ability to alter one or more of the parameters of the optical beam and are often loosely classified as 'materials that can change light'. When light passes through such an NLO medium, the dielectric polarization of the medium responds non-linearly to the intensity of the electric field of light. The NLO response becomes measurable when the intensity of the light, i.e. the intensity of the electric field of the propagating light is very high and competes with the interatomic electric field operating in the material. Such intensities can readily be realized in lasers and the field of NLO response rapidly grew after laser was discovered.

Ever since the discovery of optical second harmonic generation (SHG) in a natural material, i.e. quartz crystals, a number of optical mixing phenomena have been discovered in a number of dielectric media. Such phenomena have become collectively known as NLO processes and the dielectric materials in which such processes occur have been collectively designated as NLO materials or materials that change light. The list of NLO processes is fairly long and diverse, encompassing in its fold a host of light mediated processes beginning with Raman scattering discovered in the pre-laser era to two and multiphoton absorption, optical harmonic generation (second and higher), etc. of the post-laser era. The NLO phenomena can be and has been exploited technologically. Their technological relevance has fuelled the search for new and better NLO materials. Among the technological applications of NLO materials are the optical second harmonic and third harmonic generation processes, which are now routinely used in frequency up-conversion in lasers. More advanced application of NLO processes envisage applications in target recognition and image reconstruction via generation of a time-reversed beam by phase conjugation, in optical switching by exploiting intensity-dependent refractive index of non-linear materials, in optical communication and more commonly in sensors and eye protection devices via optical limiting, etc.[27-28]

12.8 Non-Linear Optical Response of Materials

The dielectric polarization density of a material ($P(t)$) produced by an electric field $E(t)$ can be expressed in a power series in $E(t)$, assuming that the applied field is not too large. Thus we can write

$$P(t) = \epsilon_0 \left[\chi^{(1)} E(t) + \chi^{(2)} E^2(t) + \chi^{(1)} E^3(t) + \cdots \right] \tag{12.12}$$

In Eq. 12.12, we will treat $E(t)$ as a scalar for the sake of simplicity. ϵ_0 is the dielectric permitivity of vacuum, while $\chi^{(n)}$ is the nth order susceptibility of the material and is the key quantity that models the nth order non-linearity of the material or the medium. $\chi^{(n)}$ is a tensor of rank $(n+1)$ and encodes the symmetry of the medium and the parametric nature of the interactions present in the system. We will now try to understand and identify the kinds of non-linear response or processes that can arise from the susceptibility of a particular order. Again, to keep things simple, we will consider only the second order response. The second order non-linear polarization density ($P_{nl}^{(2)}$) is

$$P_{nl}^{(2)} = \epsilon_0 \chi^{(2)} E^2(t) \tag{12.13}$$

where $E(t)$ is assumed to be a two color field given by

$$E(t) = E_1 \cos(\omega_1 t) + E_2 \cos(\omega_2 t) \tag{12.14}$$

$E(t)$ thus causes the medium to oscillate at frequencies ω_1 and ω_2. Expressing the cosine terms in terms of complex exponential, we can write[27]

$$E(t) = \frac{E_1}{2}(e^{-i\omega_1 t} + e^{i\omega_1 t}) + \frac{E_2}{2}(e^{-i\omega_2 t} + e^{i\omega_2 t}) \tag{12.15}$$

Using Eq. 12.15 in Eq. 12.13 we find that

$$P_{nl}^{(2)} = \frac{\epsilon_0 \chi^{(2)}}{4} \left[\left\{ |E_1|^2 e^{-2i\omega_1 t} + |E_2|^2 e^{-2i\omega_2 t} + 2E_1^* E_2 e^{-i(\omega_1 + \omega_2)t} \right. \right.$$
$$\left. \left. + 2E_1 E_2^* e^{-i(\omega_1 - \omega_2)t} \right\} + \left(|E-1|^2 + |E_2|^2 \right) + A \right] \tag{12.16}$$

The first two terms on the right hand side of Eq. 12.16 represent the process of second harmonic generation (SHG) at frequencies $2\omega_1$ and $2\omega_2$, respectively, the next two terms models the sum frequency generation (SFG) and difference frequency generation (DFG) processes while terms in the parenthesis correspond to the process of optical rectification. 'A' represents the complex conjugation of terms inside the square brackets. Thus, we have been able to identify all the non-linear optical processes linked to the second order susceptibility $\chi^{(2)}$ of the material. In a similar fashion we can also identify and characterize the non-linear processes associated with the third order susceptibility $\chi^{(3)}$. Our main concern in this section will be to review the types of materials that are available or being used for the most common second order process, namely second harmonic generation (SHG). A simple argument based on symmetry reveals that the second order susceptibility $\chi^{(2)}$ must be zero, if the material does not crystallize in a non-centro symmetric (NCS) crystal class, i.e. the crystal must be acentric. Absence of the center of symmetry is essential, but the material must possess, in addition, certain other qualities for being practically useful as a source for generating coherent radiation at a frequency 2ω, ω being the fundamental frequency of the source laser. The additional properties required for a functionally useful materials for SHG and also for the 4th and higher harmonics by cascaded frequency conversion are the following:

(i) The material must have a wide transparency range. It must not absorb the higher harmonics.

(ii) The SHG coefficient should be as large as or larger than that of KH_2PO_4 (KDP), currently one of the most widely used material for second harmonic generation.

(iii) It must also have a moderate birefringence with $\Delta n = |n(\omega) - n(2\omega)| > 0.07$-$0.10$ to enable phase matching of an optical direction in the crystal where $N(\omega) = q(2\omega)$.

(iv) It must be easy to grow large and high quality single crystals of the material.

(v) The chemical stability and damage threshold under strong illumination must be high.

Although KH_2PO_4 and $LiNbO_3$ are the most widely used inorganic NLO material for SHG, quite a few new inorganic materials for UV and deep UV applications have been explored.[27–28] Some of these recently discovered materials are listed here:

(i) $MMgCO_3F$ where M is either potassium (K) or rubidium (Rb);

(ii) $MM'CO_2F$ where M may be either K or Rb while M' may be either strontium (Sr) or calcium (Ca)

(iii) MBO_3F where M may either be magnesium (Mg) or zinc (Zn);

(iv) $MZn_2BO_3X_2$ where M is either K or Rb and X may be Al or Br;

(v) $M_3M'_3Li_2Al_4B_6O_{20}F$ with M being an alkali metal atom, M'_3 being an alkaline earth metal, etc. All these compounds form crystals belonging to the NCS class. In addition, some materials have been theoretically designed, e.g. $MAlCO_3F_2$ (M = alkali metal) but are yet to be synthesized. There are only a few organic non-linear materials that display SHG efficiency larger than that of KDP crystals or equal to it. L-arginine maleate dihydrate (LAMD) and L-methionine L-methioninium hydrogen maleate are the examples. The former has SHG efficiency equal to 1.68 times that of KDP and its crystals can be grown from solutions by solvent evaporation. Crystals of the second compound have been grown from aqueous solutions containing L-methionine and maleic acid. The SHG efficiency of this compound is equal to that of KDP. A lot of research has been carried out on organic intramolecular charge transfer (ICT) chromophores which display high second and third-order non-linear optical response. Some of them also have strong two-photon absorption. If these ICT chromophores could be converted into materials with high non-linear optical response, we would have at our disposal a well-defined route to fabricate materials with desired NLO properties. The molecular chromophores that have evoked keen interest are organic π-electron systems endcapped with an electron donor (D) and an electron acceptor (A).[27] These $D - \pi - A$ chromophores are known as organic push-pull systems. The donor-acceptor interactions in such systems are mediated by the π-electron bridge and are responsible for imparting distinct optoelectronic properties to these $D - \pi - A$ molecular chromophores. It turns out that the donor-acceptor interaction, leads to the appearance of a 'new' low energy $\pi-$ molecular orbital (MO) delocalized over the entire $D - \pi - A$ framework. Visible light easily excites electrons into these low energy MOs, imparting color to these chromophores and causing the appearance of a low energy absorption band (in the visible). The ICT process also strongly polarizes the electronic charge distribution in the $D - \pi - A$ system, generating non-zero dipole moment. The 'charge-transferred' configuration ($D^+ - \pi - A^-$) is expected to contribute more to the excited state than to the ground state implying that the change in dipole moment ($\delta\mu$) associated with the 'low energy' electronic transition would be large. These two special attributes have been shown to be responsible for generating a non-zero first hyperpolarizability (β) of the ICT molecules. The picture that emerges is highly simplified, but provides easy designing clues for ICT chromophores that could lead to ICT-based organic NLO materials.[27] It needs to be emphasized at this point that there is a big gap between the synthesis of an ICT chromophore, which displays large NLO-response (first hyperpolarizability) and fashioning it into a material (a solid) that responds similarly to the electric field of light. The solid must have an NCS crystal structure, must be mechanically robust, optically transparent, and have high resistance to photodamage. These requirements have so far impeded the journey from organic NLO chromophores to organic NLO materials. The technologically important NLO materials for second harmonic generation, for high speed all optical switching, in optical storage and computing that are being used are all inorganic in origin. We can however, foresee the coming of organic NLO materials in the near future in view of their flexibility and durability.[28]

12.9 Thermoelectric Materials

Is it possible to convert thermal energy directly into electrical energy. It turns out that such a possibility exists if a thermal gradient is created in a metal or a semiconducting material, which leads to the appearance of a thermoelectric effect (Seebeck effect). It is also possible to induce liberation or absorption of heat when an electric current flows through a contact between two conductors even when no thermal gradient exists along the inhomogeneous material (Peltier effect). In a current carrying conductor, the presence of a temperature gradient causes an additional heating effect over and above the Joule's heating proportional to the square of the current (J^2). The additional heat is proportional to ΔT and is known as Thomson heat. Seebeck, Peltier and Thomson effects together constitute the thermoelectric phenomenon.

Thermoelectric cells, however were never in serious contention as useful power converting devices till in 1929 Ioffe pointed out that the semiconducting materials can be used in a technologically useful way to fabricate thermoelectric cells.[29a] Since then, the search for good quality thermoelectric materials and effort to fabricate thermoelectric devices gained momentum and thermoelectricity has now become a viable source of electrical energy. The thermoelectric materials have indeed become an important energy conversion material and are being pursued in the context of utilizing low grade waste heat into electricity.

Why is it that semiconductors but not metals are viable materials for thermoelectricity? This behavior can be understood on the basis of certain distinctive features of band structures and density of states of metals and semiconductors. The density of states in semiconductors appear to be better disposed for thermoelectricity than metals. The Fermi energy (ϵ_F) in semiconductor lies below the CB. It causes the DOS to be asymmetrically distributed around the Fermi energy. The average energy of electrons in the CB is higher so that the carrier motion into the lower energy states is favored. In metals, the Fermi level lies in the CB and the DOS is more symmetric about ϵ_F. The average energy of electrons in the CB is closer to ϵ_F so that the driving force for charge transport is subdued. The carrier density in semiconductors is much lower compared to metals. The figure of merit (thermoelectric efficiency) ZT of the thermoelectric material is given by[29b]

$$ZT = \sigma S^2 T / \kappa \tag{12.17}$$

where σ is the electrical conductivity, S is the Seebeck coefficient, κ is the thermal conductivity and T is the material temperature. Eq. 12.17 provides some designing clues. It appears from Eq. 12.17 that materials with high electrical conductivity (σ) and low thermal conductivity (κ) will have higher thermoelectric efficiency. Thermal conductivity term (κ) can be split into an electronic and a phonon mediated thermal conductivity terms ($\kappa = \kappa_{electron} + \kappa_{phonon}$). In metals $\kappa_{electron}$, which increases with increase in σ dominates κ. So, $\frac{\sigma}{\kappa}$ ratio is practically fixed in metals. In semiconductors, κ_{phonon} dominates thermal conductivity and must be reduced along with increase in electrical conductivity. So, a better figure of merit for semiconductors can be realized by heavily doping the semiconductor to produce carrier density in the range of $10^{19} - 10^{21}/cm^3$, which enhances σ and by suppressing κ_{phonon}, say by providing a glassy environment in which there is heavy scattering of phonons, reducing contribution from κ_{phonon}.

Bismuth chalcogenides (Bi_2Te_3, Bi_2Se_3 etc.) and nanostructures derived from them have provided some of the best performing thermoelectric materials. When nanostructured to produce superlattices with alternating layers of Bi_2Te_3 and Sb_2Te_3, the electrical conductivity remains good, but the thermal conductivity perpendicular to the layers becomes low, thereby increasing the figure of merit. Other materials of interest and promise include the half-Heusler alloys, oxide compounds (ZnO, MnO_2, NbO_2 etc.), $NaCo_2O_4$, homologous oxide compounds of the general formula $(SrTiO_3)_n(SrO)_n$ and Si-Ge alloys. Amorphous materials formed in Cu-Ge-Te, In-Ga-Zn-O, Zr-Ni-O, Zn-Ni_2Sn and Ti-Pb-V-O systems have been the focus of a number of studies devoted to the search for new classes of thermoelectric materials.[30–31] The driving force for these studies can be traced to a suggestion that systems with phonon mean free path much larger than the mean free path of charge carriers can lead to better thermoelectric efficiency.[32] A lot of development can be expected in this area over next few years.

Organic materials like the conjugated polymers have been used in organic electronic devices like the light weight photovoltaics or flexible display board because of their easy processability. Such materials are now increasingly being targetted for developing wearable thermoelectric heating/cooling devices and for near room temperature power generation. Hybrid organic-inorganic thermoelectric materials[33] are fast approaching the figure of merit achieved in inorganic thermoelectric materials like Bi_2Te_3.

There are several fundamental issues posed by these organic and hybrid organic-inorganic materials with regard to their thermoelectric properties. The issues are related to their processing and doping. The coming years will witness breathtaking development in this area and growth for the use of thermoelectric materials for converting low grade waste heat into electric power.

REFERENCES

1. D. E. Newton, *World Energy Crisis, A Reference Handbook* (ContemporaryWorld Issues), ABC clio (2012).
2. D. P. Brown, *Hydrogen Storage Materials: The Characterization of their Storage Properties*, Springer-Verlag, Berlin (2011).
3. M. Hirscher, Ed., *Handbook of Hydrogen Storage: New Materials for Future Energy Storage*, Wiley-VCH, Berlin (2010).
4. Y.-P. Chen, Sajid Bashir and Jingbo Louise Liu, Eds., *Nanostructured Materials for Next Generation Energy Storage and Conversion: Hydrogen Production, Storage and Utilization*, Springer-Verlag, Berlin (2017).
5. M. Jorda-Beneyto, F. Suarej-Garcia, D. Lozano Castello, D. Cazorla Amoros and A. Linares Solano, 'Hydrogen storage on activated carbons and carbon nanomaterials at high pressures', *Carbon*, 45, 293–303 (2007).
6. A. C. Dillon, K. M. Jones, T. A. Bekkudahl, C. H. Kiang, D. S. Bethune and M. J. Heben, 'Storage of hydrogen in single walled carbon nanotubes', *Nature*, 386, 377–379 (1977).
7. M. Hirscher *et al.*, 'Hydrogen storage in sonicated carbon nanotubes', *Appl. Phys. A*, 72, 129–132 (2001).
8. A. W. C. Van den Berg, S. T. Bromkey, J. C. Wojdel and A. C. Jansen, 'Adsorption of H_2 in microporous materials: A molecular mechanics investigation', *Mesoporous Mater.*, 78, 63–71 (2005).
9. J. G. Vitillo, G. Riachiardi, G. Spoto and A. Zechina, 'Theoretical maximal storage of hydrogen in zeolitic framework', *Phys. Chem. Chem. Phys.*, 7, 3948–3954 (2005).
10. A. W. C. Van den Berg *et al.*, 'Adsorption isotherm of hydrogen in microporous materials with SOD structure- a grand canonical Monte Carlo study', *Microporous Mesoporous Mater.*, 87, 235–242 (2006).
11. M. K. Song and K. T. No, 'Molecularmulation of hydrogen adsorption in organic zeolite', *Catalysis Today*, 120, 374–382 (2007).
12. N. L. Rosi, J. Eckert, M. Eddaoudi, D. T. Dodak, J. Kim, M. O'Keefe, O. M. Yaghi, 'Hydrogen storage in micropurus metal-organic frameworks', *Science*, 300, 1127–1129 (2003).
13. (a) G. J. Kubas, 'Fundamentals of H_2 binding and reactivity on transition metals underlying hydrogenase function and H_2 production and storage', *Chem. Rev.*, 107, 4152–4205 (2007). (b) T. K. A. Hoang and D. M. Antonelli, 'Exploiting the Kubas interaction in the design of the hydrogen storage materials', *Adv. Mater.*, 21, 1787–1800 (2009).
14. X. Zhao, B. Xiao, A. J. Fletcher, K. M. Thomas, D. Bradshaw and M. J. Rosseinsky, 'Hysteretic adsorption and desorption of hydrogen by nanoporous metal organic framework', *Science*, 306, 1012–1015 (2004).
15. H. Furukawa and O. M. Yaghi, 'Storage of hydrogen, methane and carbon-dioxide in highly porous covalent organic frameworks for clean energy applications', *J. Am. Chem. Soc.*, 131, 8875–8883 (2009).
16. F. Feng, M. Geng and D. O. Northwood, 'Electrochemical behaviour of intermetallic based nickel metal hydride batteries: A review', *Int. J. Hydrogen Energy*, 26, 725–734 (2001).
17. R. Griessen and T. Riesterer, 'Heat of formation models', In L. Schlapbach, Ed., *Topics in Applied Physics* (pp. 219–284), Volume 67: *Hydrogen in Intermetallic Compounds I. Surface and Dynamic*

Properties, Applications, Springer-Verlag, Berlin (1988).

18. (a) S. Luo, J. D. Clewley, T. B. Flanagon, R. C. Bowman Jr. and L. A. Wade, 'Further studies of the isotherms of $LaNi_{5-x}Sn_xH$ for x = 0.0-0.05', *J. Alloy Compd.*, 267, 171–181 (1998). (b) R. C. Bowman *et al.*, 'The effect of tin on the degradation of $LaNi_{5-y}Sn_y$ metal hydrides during thermal cycling', *J. Alloy Compd.*, 217, 185–192 (1998). (c) D. Chaudra *et al.*, 'Metal hydrides for vehicular application: The state of the art', *JOM*, 58, 26–32 (2006). (d) S. Banerjee, P. Das, A. Kumar and R. Bhattacharyya, 'Evaluation of Ti_2CrV for solid state hydrogen storage applications', BARC News Letter, January-February, 1–12 (2020).

19. (a) A. Takahasi and K. F. Keltron, 'High pressure hydrogen loading in $Ti_{45}Zr_{38}Ni_{17}$ amorphous and quasi-crystalline powders synthesized by mechanical alloying', *J. Alloy Compd.*, 347, 295–300 (2002). (b) S. Bhattacharyya and G. P. Das, 'First-principles design of complex chemical hydrides as hydrogen storage materials', In *'Concepts and Methods of Modern Theoretical Chemistry* (pp. 415–430), S. K. Ghosh and P. K. Chattaraj, Eds., CRC Press, Boca Raton (2013).

20. A. Javan, 'The optical properties of materials', *Scientific American*, 217, 238–248 (1967).

21. W. D. Callister Jr., *Materials Science and Engineering - An Introduction*, Sixth Edition, Wiley, India (2006).

22. L. V. Azavoff and J. J. Brophy, *Electronic Process in Materials*, McGraw-Hill, New York (1963).

23. T. Suhara, *Semiconductor Laser Fundamentals*, CRC Press, Boca Raton (2004).

24. C. Harper, *Electronic Materials and Processes Handbook*, Third Edition, McGraw-Hill, New York (2003).

25. W. Shockley and H. J. Queisser, 'Detailed balance limit of efficiency of p-n junction solar cells', *J. Appl. Phys.*, 32, 510–519 (1961).

26. J. M. C. da Silva Filho, V. A. Ermakov and F. C. Marques, 'Perovskite thin films synthesized from sputtered lead sulphide', *Energy Environ. Sci.,* 10, 2280–2283 (2017).

27. R. Misra and S. P. Bhattacharyya, *Intramolecular Charge Transfer: Theory and Applications*, Wiley-VCH, Berlin (2018) and References cited therein.

28. R. W. Boyd, *Nonlinear Optics*, Third Edition, Elsevier, Netherlands (2008).

29. (a) A. F. Ioffe, *Physics of Semiconductors*, Academic Press, New York (1960). (b) G. J. Synder, 'Figure of merit ZT of a thermoelectric device defined from materials properties', *Energy Environ. Sci.* 10, 2280–2283 (2017).

30. A. P. Goncalves, E. B. Lopes, O. Rouleau and C. Godart, 'Conducting glasses as new potential thermoelectric materials: The Cu–Ge–Te case', *J. Mater. Chem.*, 20, 1516–1521 (2010).

31. Y. Fujimoto, M. Uenuma, Y. Ishikawa and Y. Uraoka, 'Analysis of thermoelectric properties of amorphous InGaZnO thin film by controlling carrier concentration', *AIP Adv.*, 5, 097209 (2015).

32. G. Nolas and H. Goldsmid, 'The figure of merit in amorphous thermoelectrics', *Physica Status Solidi a*, 194, 271–276 (2002).

33. B. Russ, A. Glaudell, J. Urban, M. L. Chabinyc and R. A. Segalman, 'Organic thermoelectric materials for energy harvesting and temperature control', *Nature Rev. Mater.*, 1, 1–14 (2016).

13

Designer Materials

13.1 Introduction: Design by Thumb Rules

Rapid technological evolution demands equally fast evolution of materials needed for implementing such technologies. The discovery of new materials has traditionally been approached from an Edisonian point of view: we try hundreds of things based on experience and intuition; if we are lucky and serendipity blesses us we may get what we are looking for. This 'hit and miss' method or the method of random trials has certainly led to the discovery of many new materials and processes. Behind the random trials of 'hit and miss' approach, certain 'rules of thumb' usually guide the investigator to choose the objects of trial from a pool of virtually infinitely many objects. Thus for designing an alloy having high strength and yet some ductility and toughness the selection of metals is guided primarily by the requirement of strengthening – the ductility factor being sacrificed in the design goal, which in this case is making a strong alloy. The strengthening techniques on the other hand are guided by a simple but scientific rule of thumb – restricting dislocation motion renders a material harder and stronger. That in turn leads to a rule that a fine-grained material will be stronger and harder than a coarse grained material since the former has a greater total grain boundary area to resist the dislocation motion.[1] Thus a correct choice of the alloying component together with a processing technique that ensures formation of grains of smaller diameter will ensure that the alloy produced has the desired property. Instead of hardening by grain-size reduction, one could also adopt another technique of strengthening and hardening metals by alloying with impurity atoms that go into the base (host) metal forming either interstitial or substitutional solid solution. In fact highly pure metals are practically always softer and weaker than alloys made up of the same base metal. An increase in the concentration of the impurity atoms say, nickel in the host say copper, leads to a concomitant increase in tensile and yield strengths. This happens because the impurity atoms that go into solid solution generally cause lattice strains on the surrounding host atoms. The lattice strain field interactions among the impurity atoms and dislocations impedes dislocation movement preventing macroscopic plastic deformation.[1] The alloy becomes stronger and harder in consequence. Assuming that such thumb rules are available not just for mechanical properties, but also for electrical, optical, thermal and magnetic properties of materials, the question arises if we could design a material with a desired or the optimum level of say, electric or magnetic or thermal response. If it becomes possible, we would have at our disposal a method for fabricating designer materials, be it a super conductor, a super magnet, super capacitor or non-linear optical material with high second or third order polarizabilities, to name a few. However, there are serious limitations and road blocks to the success of the 'thumb rule' guided approach. For example, if a metal with low electrical conductivity is alloyed with a metal of high conductivity, the resulting alloy will not have conductivity higher than that of the host metal with low conductivity. The electrons moving through the alloy, will recognize the other atoms – the atoms of the high conductivity metal as impurity atoms and undergo scattering. These scatterings will raise the resistivity of the alloy to a value higher than the resistivities of both components of the alloy. The resistivity ρ_{AB} of a binary alloy having the composition $A_x B_{1-x}$ is therefore never equal to $x\rho_A + (1-x)\rho_B$ where ρ_A, ρ_B are the resistivities of pure A and pure B, respectively. The designing of an alloy with a targeted conductivity is a more difficult problem. If one is trying to design a ternary alloy or a quaternary alloy, the problem becomes much more involved. It is not that the problem exists only in the prediction of the resistivity of an alloy. Predicting thermal expansivity[2] of a solid alloy $A_x B_{1-x}$ can illustrate the dilemma of the designer very clearly. If we try to model the expansivity (σ) of a one-dimensional alloy ($A_x B_{1-x}$) in terms of expansivities of the

DOI: 10.1201/9781003244882-13

different bonds that exist in the alloy and the composition variable (x) we have

$$\sigma_{alloy} = x^2 \sigma_{AA} + (1-x)^2 \sigma_{BB} + 2x(1-x)\sigma_{AB} \tag{13.1}$$

where σ_{AA}, σ_{BB}, σ_{AB} denote the thermal expansivities of A-A, B-B and A-B bonds, respectively which are expected to exist in the alloy. If the A-B bond is assumed to respond like an average of the A-A and B-B bonds, we can replace σ_{AB} by $\frac{1}{2}(\sigma_{AA} + \sigma_{BB})$ and replace Eq. 13.1 with

$$\sigma_{alloy} = x\sigma_{AA} + (1-x)\sigma_{BB} \tag{13.2}$$

Eq. 13.1 can sometimes be useful in predicting the expansivity of certain alloys, but its severe limitation is exposed when it comes to modeling the expansivity of INVAR – an alloy of iron and nickel with the composition $Fe_{0.64}Ni_{0.36}$. The expansivity of this alloy is known to be practically zero contradicting the prediction from Eq. 13.1, which predicts $\sigma_{alloy} >> 0$ whatever be the value of the composition variable. This counter intuitive response of the alloy arises from the fact that the anharmonicity in the pair-potential describing the Fe-Ni bond is very different from the anharmonicities in the Fe-Fe and Ni-Ni pair potentials, causing a cancellation of diagonal and off diagonal bond expansivities over a range of temperatures, leading to the observed zero-expansion of the alloy over a range of temperatures. Thus, designing an alloy or a material with targeted properties requires a detailed knowledge and understanding of the different interactions operating in the system and how they modulate or influence the property of the material. A few thumb rules expressing broad correlations among specific properties and some transferable features of the atoms or molecules of which the materials are made of, are not generally adequate for the task of designing materials with targeted properties.

13.2 Materials by Design: Beyond Thumb Rules

Chemical scientists synthesize, characterize, produce materials and assemble them and construct with them structures at various length scales, ranging from molecules at the nanometer level, to polymers and electronic devices at the submicron level and to ceramics in large scale structures. Understanding materials primarily requires understanding the substances of which they are made of (atoms, molecules), knowing how they interact and how to organize or assemble them into useful structures. All the interesting assemblies whether non-living or living, have at their core, materials with specially tailored properties. The broad goal of material science is to design and control through synthesis and processing and explore all the relationships among structures, properties, composition as well as processing, that are supposed to determine the technologically useful behavior of all such materials.

The goal thus defined poses a difficult challenge for materials scientists. The first step in this task is to investigate the chemistry of a single molecule (or atom) and its properties (calls for use of atomic and molecular quantum mechanics). The next step calls for comparing the individual behavior with the average molecular behavior in an assembly, may be in solution or in a condensed phase of the molecules (calls for use of statistical mechanics). The collective properties of materials in condensed phases may have many surprising features that can be exploited to develop useful materials. Most of the interesting materials, as we may appreciate today are functional systems, which have their roots firmly entrenched in our gradually evolving knowledge of structure-activity relationships in chemistry. The catalysts for example are surprising realizations of chemically functional materials and the design and development of catalytic materials constitute an important branch of the science of functional materials. It is not that materials can only be chemically functional. There are smart or responsive materials (Chapter 9), which either individually or often in combination with others materials, impart different kinds of functionalities such as sensing or actuation. The challenge to design materials with a targeted set of functionalities - designer materials, so to say, is a harsh one. From a chemist's point of view it is important to understand the properties of materials in the context of the organization of its components. Chemists have long been designing and implementing schemes capable of placing atoms in a structure with subnanometer precision. Material scientists are now aiming to do the same thing at much larger scales. The challenge now is to achieve macromolecular structural control (and hence control of functionalities) at the nanometer length scale. To meet the challenge, experiment and theory must act in unison and synergy. Let us see how and in what ways this can be made possible. The clue lies in the deft use of computer science, computational materials science, artificial intelligence, data bases and machine learning.

13.3 Designing Materials: Beyond Thumb Rules

Let us consider a material M made up of atoms or molecules $A_1, A_2, A_3, \ldots, A_{n-1}, A_n$. The composition variable x_i denotes, let us say, the mole fraction or percentage by weight (wt %) of each component so that M can be represented as follows:

$$M = (A_1)x_1(A_2)x_2 \cdots (A_{n-1})x_{n-1}(A_n)x_n \tag{13.3}$$

Let us suppose, the property of M can be represented by a unique function of the composition variables (x_i) and we know the function $P(M) = f(x_1, x_2, \ldots, x_n)$ explicitly. The task before us is to find that set of $x_1, x_2, \ldots, x_n$, which will produce the targeted property value p_0 of P in the designed material. Note that the composition variables can assume any value between 0 and 1 (i.e. they are positive quantities) under the stipulation that

$$\sum_{i=1}^{n} x_i = 1 \tag{13.4}$$

What would be the target of the search? Clearly it can be cast as a constrained minimization of an objective function $O(x)$ where

$$\begin{aligned} O(x) &= |P(M) - p_0| \\ &= |f(x_1, x_2 \ldots x_n) - p_0| \end{aligned} \tag{13.5a}$$

under the constraint

$$\sum_{i=1}^{n} x_i = 1 \tag{13.5b}$$

Needless to mention, there can be many other forms of the objective function $O(x)$ that could have been chosen. For the time being, we will not consider any alternative form of $O(x)$. We can conveniently replace the problem of constrained minimization of $O(x)$ with an unconstrained version of it by defining

$$O(x) = |f(x_1, x_2 \ldots x_n) - p_0| + \lambda |(1 - \sum_{i=1}^{n} x_i)| \tag{13.6}$$

λ is a penalty weight factor, which controls the severity with which the constraint is imposed on the minimization problem. Now the task at hand is completely defined: find out an optimal set of n composition variables $x_1^*, x_2^*, \ldots x_{n-1}^*, x_n^*$ for which $|f(x_1^*, x_2^*, \ldots, x_{n-1}^*, x_n^*) - p_0|$ is minimum under the constraint that $\sum_{i=1}^{n} x_i^* = 1$, and $0 < x_i^* < 1, i = 1, 2, \ldots n$. In general, the function $f(x_1, x_2, \ldots, x_n)$ is expected to be a complicated non-linear function of the composition variables x_i. Analytical solution of the problem is therefore ruled out. The solution of the problem must therefore be sought numerically. Several options exist and may be tried.

1. A random search for $\{x_i^*\}_{i=1,n}$ can be made taking care to see that $0 < x_i < 1$ is maintained at every step. If $\sum_i x_i \neq 1$, the randomly generated values of the composition variables x_i must be renormalized so that the constraint Eq. 13.5b is satisfied at each step of the search. Since x_is can vary continuously between 0 and 1, the number of possibilities is combinatorially huge and therefore optimization of composition variables by direct enumeration of the objective function for the problem is not computationally feasible. Instead of randomly searching the space spanned by the composition variables, we have to adopt a search technique that has random search elements along with a built-in strategy that helps the search to quickly home-on to the promising region of the search space, i.e. where the optimum lies. A viable set of methods that have

the ability to explore the search space and at the same time exploit the information already available in the search space efficiently are the well-known evolutionary algorithms, which are computational transcription of the Darwinian law of evolution and population genetics.[5–7]

We will briefly introduce the oldest version of the evolutionary algorithm – the so called genetic algorithm (GA). GAs work with a population (n_p) of trial solutions of the problem, each solution being represented as a string of binary or floating point numbers. In our case we can represent the strings $(S_1, S_2, ..., S_{n_p})$ as a collection of the composition variables $x_1, x_2, ..., x_n$. Then, we have the strings $(S_1, S_2, ..., S_{n_p})$ defined as

$$S_1 = (x_1^1, x_2^1, ..., x_i^1, ..., x_{n-1}^1, x_n^1)$$

$$S_2 = (x_1^2, x_2^2, ..., x_i^2, ..., x_{n-1}^2, x_n^2)$$

$$\cdots \qquad (13.7)$$

$$\cdots$$

$$S_{n_p} = (x_1^{n_p}, x_2^{n_p}, ..., x_i^{n_p}, ..., x_{n-1}^{n_p}, x_n^{n_p})$$

For each string, the condition $\sum_i x_i^p = 1$, $p = 1, 2, ..., n_p$ has been imposed. That means, all the trial strings are consistent with the constraint imposed on the search. The targeted property for each string can be calculated easily by feeding the composition variables in that string to the function $f(x_1, ..., x_n)$ (let us call it the property modeling function or the property modeling PROGRAM). The n_p number of property values can be represented as an array of numbers

$$P(1) = P(S_1) = p_1$$

$$P(2) = P(S_2) = p_2$$

$$\cdots \qquad (13.8)$$

$$\cdots$$

$$P(n_p) = P(S_{n_p}) = p_{n_p}$$

These n_p number of property values, each attached to the corresponding string S_k, can be used to define a fitness value F_k for the kth string. The definition is not unique. It is a number that represents the degree or extent of acceptability of the string as a solution of the problem. That means, the problem of finding the material with the targeted property (p_0) now becomes the problem of maximizing a fitness function 'F' that takes the objective function $O_i(x)$ as input. If we assume $\sum_i x_i = 1$ is satisfied, the fitness values can be generated as follows:

$$\text{Fitness function } F = \frac{1}{C + |P(s) - p_0|} \qquad (13.9a)$$

$$\text{Fitness values for the } k^{th} \text{ string } F_k = \frac{1}{C + |P(s_k) - p_0|} \qquad (13.9b)$$

where C is a small positive number. Clearly, the closer the trial string is to the desired solution string that generates the property value p_0, the higher the value of fitness F_k is. Since $|P(s_k)) - p_0| \to 0$ as $s_k \to s_*$ the function F could blow up, which is prevented by the small number C in the denominator of the equation 13.9. Let the fitness values of the strings in the initial population thus calculated be $F_1, F_2, ..., F_k, ..., F_{np}$, respectively. In the parlance of GA, the initial population is now subjected to a fitness proportional 'selection' method (mimicking the Darwinian dictum of the survival of the fittest) that generates a 'breeding pool' for the generation of new solution strings by the application of certain genetic operators called crossover and mutation operators (they mimic the chromosomal crossover and mutations occuring in the living systems). The operation of selection, crossover and mutation can be carried out in a large number of ways. We will describe one simple realization of each of the operations in the context of the specific example case.

Selection:

The most simple and much used selection operation is based on the slotted Roulette wheel method. The wheel has slots with widths proportional to fitness values of strings in the population and is provided with an indicator arrow. The wheel is spun exactly n_p number of times. After each spin the position of the arrowhead of the pointer is noted. If the arrowhead points to the slot with fitness F_i, then the string S_i is selected and copied into the breeding pool. It is quite reasonable to expect on the basis of selection process that the number of copies of strings with higher fitness values will be larger. Selection thus tends to overemphasize the strings (potential or feasible solution) with higher fitness values and underemphasize the ones with lower fitness. Thus the breeding pool is expected to have a higher average fitness than the pre-selection pool of strings. Implementing the roulette wheel method of selection on the computer is simple. The interested reader may consult references[5-7]. Notice that the breeding pool does not have any new strings – only the frequency of distribution of the old strings has changed. Now, GAs operate by applying the crossover and mutation operators on the 'post-selection' pool of population (i.e. the breeding pool). Once again, many different realizations of these two operators have been tried and experimented with. The construction of these operators depends on whether binary, floating point, integer or character, strings are used to represent the population. In what follows, we will be using floating point strings. So, we will briefly introduce what are known as arithmetic crossover and mutation operators for floating point strings. The crossover process leads to an exchange of genetic information among the members of the breeding pool in a pair-wise fashion. Suppose that the strings S_k and S_l have been chosen for undergoing crossover. The implementation requires us to select a crossover site on the chosen strings randomly with a probability p_c (say $p_c = 0.8$). Let the site chosen for crossover be the ith site. Then the arithmetic crossover operator acts on

$$S_k = (x_1^k, x_2^k, \ldots, x_i^k, x_{i+1}^k \ldots, x_{n-1}^k, x_n^k)$$

and

$$S_l = (x_1^l, x_2^l, \ldots, x_i^l, x_{i+1}^l \ldots, x_{n-1}^l, x_n^l)$$

to generate a pair of new strings S_k' and S_l' where the new strings are

$$S_k' = (x_1^k, x_2^k, \ldots, x_i^k, x_{i+1}^{k\prime}, x_{i+2}^{k\prime} \ldots, x_n^{k\prime})$$

and

$$S_l' = (x_1^l, x_2^l, \ldots, x_i^l, x_{i+1}^{l\prime}, x_{i+2}^{l\prime} \ldots, x_n^{l\prime})$$

Note that in each of the two new strings generated by crossover, all the entries upto the ith state (the chosen crossover site) are unchanged. The entries at all the sites from $i + 1$ to n has changed under arithmetic crossover and the altered values are given by[7]

$$(x_p^k)' = a \cdot x_p^k + (1 - a)x_p^l$$
$$(x_p^l)' = (1 - a)x_p^k + a \cdot x_p^l; \quad p = i + 1, i + 2, \ldots, n$$

where $0 \leq a < 1$ is the amplitude of mixing of information present in the two strings of composition variables undergoing crossover. There are many other versions of crossover operation on floating point strings. We have described the simplest one, which, according to our experience, works well. Next, another pair is selected from the original breeding population and subjected to the same crossover operation and then another one and so on, till exactly n_p-number of new strings have been generated. These new strings are sequestered into a temporary population of n_p number of post-crossover strings S_i. Every string in the population is now subjected to another genetic operation called 'mutation'. The arithmetic mutation operator works in the following fashion, in one of the most simple realization.

Mutation Operation:

Let the mutation probability be p_m ($<< p_c$) and let the kth string of the new population of n_p strings be selected for mutation. One of the n-sites in the kth string is then chosen randomly with a probability p_m for undergoing mutation. Let it be the lth site. The mutation is then represented by the transformation[3]

$$x_l'' = x_l' + (-1)^L \Delta m \times r \tag{13.10}$$

where L is a randomly generated integer, Δm is the intensity of mutation and r is a random floating point number ($0 \leq r \leq 1$). The mutation intensity is a small quantity ($\approx 0.01 - 0.1$), which may be kept fixed or dynamically adjusted. Each one of the n_p number of strings $\{S'_k\}_{k=1,n_p}$ in the post-crossover population is allowed to undergo mutation. The set of n_p number of post-crossover and post-mutation strings $\{S''_k\}_{k=1,n_p}$ are evaluated for their fitness values. These fitness values $\{f''_i\}$ values are compared with the fitness values f_i of the initial strings. If $f''_i \geq f_i$, then S''_i replaces S_i in the original population; if not S_i is retained in the original population. Once the screening is over, the set of n_p strings, some of them new and some of them old are accepted in the new population. This completes one step of evolution. The whole process is repeated over and over again till there is no further improvement in the fitness values of the best string in the population or in the average fitness value of the population. The string with the highest fitness in the population is accepted as the solution of the problem. Although attaining the global minimum (or optimum) is not guaranteed, GA navigates through the search space well, exploring and exploiting the information already available and generally hits, at least a good robust solution. Suppose that this string is

$$S_* = \{x_{1*}, x_{2*}, \ldots, x_{k*}, \ldots, x_{n*}\}$$

It can be immediately seen that the string S_* translates into a new predicted alloy (or an intermetallic compound) with the formula

$$(A_1)_{x_{1*}}(A_2)_{x_{2*}} \cdots (A_n)_{x_{n*}}$$

which is expected to have the targeted value of the property p_0 as closely as possible. Genetic algorithms can therefore guide the search to discover a designer material avoiding costly trials and tribulations of the hit and miss method. The compositions of different trial strings or materials are not conjured up randomly. One makes an educated guess and the GA learns adaptively and predicts possible compositions as the evolution comes to an end.

13.4 The Advent Computational Material Science

In our elucidation of a GA guided approach to the job of designing an alloy or an intermetallic with a targeted property, we have tacitly assumed that given all the composition variables, we have at our disposal a certain function 'f' of the composition variables, which can and does, calculate the property p_0 value easily. There are certain possibilities for constructing such a function. One of course is by regression analysis of data already available. In this case $f(x_1, \cdots, x_n)$ is just a non-linear least-squares fitting function (generally) or an interpolating function based on the available data. It is possible that the function $f(x)$ may contain, in addition to the composition variables, certain additional inputs, representing different features and attributes of the individual elements making up the material. Such attributes may include nuclear charge of each atom (Z_{A_i}), their atomic or ionic radii (r_{A_i}), electronegativity (χ_{A_i}), hardness (η_{A_i}) and possibly many others. Thus, in general, we may write

$$p = f(x_1, x_2, \ldots, x_n; r_1, r_2, \ldots, r_n; \chi_1, \chi_2, \ldots, \chi_n; \eta_1, \eta_2, \ldots, \eta_n)$$

where it is assumed that the functional form of 'f' has been carefully chosen on the basis of mathematical reasoning, experience and intuition.

There could be an alternative way to represent the function 'f' as a function of all the variables and attributes. This mode of representation does not use a predetermined functional from explicitly. The available data is used to train an artificial neural network (ANN), which non-linearly models the data in terms of many neurons, their connection weights and thresholds. During the training process, the network senses various hidden connections among the input and output data and accordingly assigns the connection weights and threshold values. The ANN therefore represents a formless many dimensional interpolation formula capable of predicting the property values of a material in a matter of seconds, once the composition variables and atomic attributes or features are specified. The third alternative is to use the methods of computational quantum chemistry, specially variants of density functional theory, which have been already implemented in a number of sophisticated quantum chemistry or computational material science packages. Essentially, these packages take the coordinates and nuclear charges of various atoms in the molecule or material of interest as inputs and uses appropriate methods to solve the quantum many body problem. That amounts to solving the relevant many electron Schrödinger equation either for a single free molecule or the same equation for a crystalline solid. In either case, appropriate basis functions are required and these packages have many built-in options for choosing the basis-sets according to the level of accuracy required in a specific calculation. For solids, DFT has become almost the panacea for all the computational structure and property calculation problems. Solids having a few thousand atoms are now routinely studied by the DFT-based computing technologies. The most widely used method for electronic structure calculation of materials with fixed geometry is based on the use of delocalized plane-wave basis or localized basis sets. The calculation of ground state total energy leads to the estimate of forces acting on individual atoms in the material, which is essential to carry out *ab initio* molecular dynamics simulation via the Car-Parrinello[8−9] recipe realized within the framework of the density functional theories. The dynamical simulation leads to the construction of energy landscape and enables the user to follow different dynamical events, and evolution of various properties of interest. Such a computing tool of electronic structure calculation are available in software packages like VASP, SIESTA, MATERIALS STUDIO, etc. The framework outlined for designing new materials by automated, intelligent or semi-intelligent search through the space of composition variables by invoking evolutionary algorithms like a simple GA, can be represented in a step-wise manner as follows:

1. A database of the already known materials of the type being searched for, along with the values of the particular property for each is created. Regression is done to find a property-composition fitting function.

2. A set of n_p number of materials each represented by a string of composition variables $S_i = (x_1^i, ..., x_k^i, ..., x_n^i)$ are selected from the database along with their respective property values (p_i) are fed as inputs to the GA-driven search.

3. Genetic operators like selection, crossover and mutations are used as 'moves' in the space of composition variable to generate n_p number of new strings, i.e. new materials, which are evaluated for their respective 'fitness' values (property desired), by using the property-composition fitting function.

4. A suitable screening scheme is used to choose from the pool of n_p old (parent) and n_p new (offspring) strings a set of n_p new materials or strings. That completes one generation.

   (a) If any one of the new string have a remarkably better property value, it is copied onto the database and the database thus extended is used to recalibrate the property-composition mappings. The new population is fed into step (3).

   (b) If no remarkably new material has been generated, the algorithm checks if the number of evolutionary steps has already crossed a pre-set limit (n_{max}). If it has exceeded (n_{max}), the material strings are sent to an output routine, which prints them. Some of the predicted new materials are then sought to be synthesized and technologically evaluated. Otherwise, the population is fed back to step 3 for further evolution. Several points may be noted.

      (i) Instead of a regression-based fitness function or a property-composition relationship we could have used a back propagation neural network to represent the property-composition relationship in an abstract, but a very powerful way.

      (ii) We have used 'moves', i.e. genetic operators in the space spanned by the composition variables ($x_1, ..., x_n$). We could have generalized the scheme to generate 'moves' in the 'chemical space' as well. Thus a generalized mutation operator could be designed to change the atom A_k in the *k*th location in a given string to another atom say B_k with the restriction that the B atom must be chemically similar to the A atom (belong to the same group in the periodic table, have similar atomic radius, electronegativity, etc.). This way the search conducted could be given an extra dimension to look for new materials not just in terms of composition variables, but also in terms of atom types of which the material is made of.

      (iii) We have so far assumed that 'materials' or strings produced by genetic operators or moves are energetically stable. This may not be so. Accordingly along with the property-composition relation, a fair, accurate and rapid method of calculating the energy of the new materials in terms of bonds should be available. If not, quantum chemical calculations using DFT need to be undertaken to estimate thermal stability. The high energy strings may preferably be discarded during the search. The thermal stability estimate becomes essential if 'mutation' operator changes the atom type.

(iv) When generalized mutation operators are used the fitness function needs to be constructed just not in terms of composition variables but also in terms of 'atomic' descriptors like atomic radius, electronegativity, polarizability of the atom etc.

(v) If we are trying to design a material with more than one type of targeted properties the problem becomes a multi-objective optimization problem, which then is handled by Pareto-optimization technique.[10]

Having laid out the basic strategy of developing designer materials by going beyond thumb rules, let us now review some of their realizations already made.

13.4.1 Designing Hard and Superhard Materials

Hard and superhard materials are in great demand in industry. Mining, defence, aviation and space industries are constantly looking for harder and still harder materials. Nature has given us the hardest substance in the form of diamond. It is, however a very costly material and a rarity. Synthetic diamonds are used in many applications, but that is not enough to meet the constantly rising demand of hard and superhard materials. A superhard material has Vickers hardness (H_V) greater than 40 GPa, the hardest known material (diamond) has Vickers hardness in the range 60–120 GPa. Cubic boron nitride has Vickers hardness equal to 40 GPa. Many transition metal carbides, borides and nitrides are known to be hard. Tungsten carbide (WC) for example, is already in use in drilling machines. If the crystal structure of a material is known and its elastic constants have been calculated, its Vickers hardness (H_V) index can be calculated quite accurately by the correlation[11]

$$H_V = 2(k^2 G)^{0.585} - 3$$

where G is the shear modulus of the material, k is the push ratio ($= G/B$) and B is the bulk modulus. The problem is that for many borides and carbides of transition metals, it is difficult to determine the correct crystal structure and even the composition, by X-ray diffraction. With so much uncertainty in structure, the calculated mechanical properties are vitiated. Lyakhov and Oganov proposed[12] a theoretical model that takes the crystal structure as input and estimates the maximum hardness in a given class of system by electronegativity based hardness model augmented with bond valence model and graph theory. The crystal structure itself is calculated by invoking universal structure prediction by evolutionary crystallography (USPEX) method. This non-empirical approach to the task of designing materials with targeted properties combines a coevolutionary search adapted to a carefully restructured Mendelevian chemical space, energy filtering and Pareto optimization[13,14] of the targeted properties and energetic stability. The search ensures that the discovered materials have the required properties and are stable enough to be synthetically realizable. Now to search for the targeted materials the algorithm must be able to move from one region of the search space to another. To this end in view, they have carefully designed what they call variation operators (akin to the genetic operators introduced in the previous section), which act in the space of two critically important elemental or atomic properties, namely electronegativity and atomic radii. The operators are designed so as to ensure that the newly sampled compounds are based on elements that have properties similar to the elements present in compounds with desirable properties already sampled or discarded. The evolutionary search devised by these authors has been called Mendelevian search (MendelevS) as it has built in ability to move across the periodic table. The 'variation operators' by their constructions ensure that the Mndelevian search learns in a step by step fashion, what the promising regions of the 'chemical space' are for the materials to be hard or superhard and emphasizes on them at the cost of unpromising regions.[12] The authors used their MendS to search through the entire chemical space of all the elements in the periodic table excluding the rare earths, the noble gases and elements heavier than Pu. The hardness was computed by the Lyakhov-Oganov method and stability calculated by the Maxwell's Convex Hull method. The multiobjective Pareto optimization and energy filtering, i.e. rejecting all structures generated by the search with energy above the Convex Hull by 0.50 eV/atom ensured that the algorithm generated hard phases that have low energies, too. The success of the method was impressive. All the known hard materials already reported in the literature were discovered along with materials that are claimed to be potentially hard with correct relative ordering in hardness. In the tungsten-boron system, WB_5 was predicted to have high hardness, fracture toughness and thermodynamic stability. TiB_2, ZrB_2, VB, V_2B_6 and many others were predicted to be low pressure phases that have high hardness and roughness. Interestingly, the authors also concluded that TiO, in none of its polymorphic form can be one of the hardest oxide as its Vickers hardness was predicted to be less than 17 GPa. A material is characterized as hard if its Vickers hardness is ≥ 20. The work by Lyakhov and Oganov paves the way to the development of many other computational strategies for designing materials with targeted properties. These new strategies may not involve evolutionary computing but exploit some other technique of constrained global optimization empowered with adaptive learning protocols in some form with a view to accelerating targeted materials discovery avoiding the pitfalls of Edisonian approach to the problem. In what follows, we will discuss in some detail a recent work by D. Xue *et al.* (2016).[15] These authors demonstrate the power of an adaptive designing approach to the discovery of new materials with targeted properties.

13.4.2 Adaptive Design in Materials Discovery

There is a class of materials called shape memory alloys (see Chapter 9). Xue *et al.*[15] addressed the problem of designing new Ni- and Ti-based shape memory alloys with very low thermal hysteresis. Shape memory alloys are functional materials, their basic functionalities have roots in their super-elasticity and shape memory effect. It is now well-known that these properties arise from the reversible martensitic transformation between a high temperature austenite and a low temperature martensitic phase. It turns out that heating and cooling across the martensitic transformation temperature leads to hysteresis (measured by ΔT) because the transformation temperatures fail to coincide. After several cycles of heating and cooling, the transformation temperature undergoes a substantial shift, which indicates that the material is prone to suffer from fatigue. In fact in the $Ni_{50}Ti_{50}$ alloys, 60 cycles of heating and cooling was found to shift the transformation temperature by almost 25 K. The probability of the material developing fatigue after use is therefore quite high. How does one reduce the fatigue generation probability? One simple route is to design an alloy or an intermediate based on Ni and Ti with substitution added such that the designed material has much lower hysteresis, i.e. has a low ΔT value. Conventional wisdom (or thumb rule) suggests that we can substitute the nickel site with other transition metals. That opens up a huge number of possibilities – a combinatorially explosive situation that can not be negotiated with the Edisonian philosophy. To narrow down the search space, the authors of Ref. 15 restricted their search to the family of alloys represented by the formula $Ni_{50-(x+y+z)}Ti_{50}Cu_xFe_yPd_z$. x, y and z represent the composition variables with necessary restrictions placed on them. The minimizable quantity is ΔT, which is supposed to be a function of the composition variables, i.e. x, y, z. The problem is that $(x,y,z) \rightarrow \Delta T$ mapping is unknown. The family of alloys explored by Xue *et al.*[15] can undergo a cubic to rhombohedral ($B_2 \rightarrow R$), cubic to orthorhombic ($B_2 \rightarrow B_{19}$) and cubic to monoclinic ($B_2 \rightarrow B'_{19}$) crystallographic transitions. Which of the transformations actually takes place depends on x, y and z values. Their adaptive design approach has been to find the values of x_*, y_* and z_* that globally minimize ΔT. The task is a challenging one as the algorithm must be able to find out alloys undergoing the ($B_2 \rightarrow R$), ($B_2 \rightarrow B_{19}$) and ($B_2 \rightarrow B'_{19}$) transition[15] and while doing so, also locate the global minimum in ΔT. It was estimated that the constrained pseudo quaternary composition space could accommodate 797504 candidate alloys. Screening them directly by high throughput experimental techniques or by invoking *ab initio* quantum chemical tools was impossible. The adaptive design strategy advanced by Xue *et al.*[15] with feedback from experiments and/or theory was able to discover the alloys in the sixth, out of a total of nine iterative loops, with $\Delta T = 3.14K$, being the lowest value of ΔT in the initial database. The best they succeeded in finding was the alloy $Ti_{50}Ni_{46.7}Cu_{0.8}Fe_{2.3}Pd_{0.2}$ ($\Delta T = 1.K$). It is undoubtedly an accelerated discovery of a new material with very low ΔT value and therefore highly resistant to the development of fatigue. The details of the adaptive designing strategy can be found in Ref. 15. However, for the readers we will outline some of the salient techniques adopted by Xue *et al.*[15] for the global search.

1. **Features:** Each alloy is represented by a collection of several features $\{f_i\}$, each representing some aspect of structure, chemistry, bonding of the atoms in the alloy. The authors used Pauling electronegativity, metallic radii, valence electron number, Weber-Chromes pseudopotential radii, atomic radii of Clementi, and Pettifor chemical scale, as features.

2. **Inference drawing:** The task here is to establish from data available initially how ΔT varies with the input features and predict ΔT for newly discovered alloys. Several well-known regression techniques including a Gaussian process model and a support vector regression, which aims at constructing (learning) a non-linear function by mapping onto a high-dimensional feature's space were used.[15] These regressions generally tended to produce a non-convex fitness function (features to property mapping), which could have multiple local optima. The fitness landscape of the problem is thus expected to be a rather complex one. It can not be completely explored by a regressor alone, which is likely to get stuck at a local optimum.

3. **Selectors:** The task here is to provide guidance to generate the next set of alloys. It requires the search to combine both exploration and exploitation of the search space, which was made possible with the adoption of different design functions or selectors based on the efficient global optimization (EGO) method of Jones *et al.*[16] or the knowledge gradient algorithm (KGA) of Powell and Ryzhov. Since there is no universal optimizer for all optimization problems as emphasized in the 'no free lunch theorem', it was necessary to experiment with several regressor-selector combinations. The authors[15] concluded that SVR-KGA performed the best. Xue *et al.*[15] studied one kind of material only but their method opens up the possibility to adopt the strategy for designing different kinds of new materials with targeted properties, and may be augmented with machine learning.

13.4.3 Accelerated Discovery of New Magnetic Materials

It would not be an exaggeration to state that modern technologies of data storage, energy conversion, contactless sensing, etc. are critically dependent on the availability of appropriate magnetic materials. The availability is, however severely limited. There are about two dozen magnetic materials that are in use in mainstream technological applications. Adding a new magnetic material to the list of already available materials is a time-consuming job. Sanvito *et al.*[18] developed a theoretical systematism to design novel magnetic materials. The method displayed its ability to accelerate the speed of discovery of new magnetic materials with high throughput. The application concentrated on the alloys of the Heusler family. Heusler alloys, named after Fritz Heusler, the discoveree of the first alloy of its kind, are intermetallic compounds made up of non-magnetic component elements. The chemical formula of these compounds are XYZ or X_2YZ where X, Y are transition metals and Z may also be an element of the p-block. The surprise lies in the fact that these alloys or intermetallics are magnetic while the component elements are not. The first alloy of the family discovered was Cu_2MnSn.[19]

Before proceeding further, it is necessary to consider the types of constraints that the search for new magnetic material must keep in view. Whatever may be the field of technological application be, the new magnet should be able to operate in a temperature range of $-50^\circ C$ to $120^\circ C$ with magnetic ordering temperature $<T_c>$ being $\geq 300^o C$. Magnetic electrodes in a high performance tunnel junction must additionally have a band structure suitable for spin-filtering. If the same junction is used in a magnetic memory system (based on spin transfer torque), the magnet must have a low Gilbert damping coefficient and high spin polarization. The task of designing a new magnetic material is therefore complex and involves many dimensional search. It is rather unlikely that we proceed by Edisonian trial and error method and succeed. Sanvito *et al.*[18] designed a strategy that involved large scale application of the advanced electronic structure theory, large scale creation of databases and search adopting the 'high throughput' highway to computational materials design[19] developed earlier. The approach can be broken down into the following steps:

1. Creation of an extensive database containing the computed electronic structures of potential novel magnetic materials including those of the Heusler family and extend it to the AFLOW.org set of repositories. A rough stability analysis, which involves evaluation of the formation enthalpy against reference single phase compounds is made, providing the first screening of the database.

2. A more precise analysis of the thermodynamic stability of only the Heusler family alloys that survive the first screening is made. The analysis involves calculation of enthalpy of formation of all the decomposition products of a given Heusler alloy (HA).

3. In the final step, the magnetic orders of the thermodynamically stable magnetic intermetallic HAs predicted in step 2 are determined and their T_cs estimated by a regression trained with available magnetic data.

4. Finally the theoretical screening is validated by actually synthesizing some of the predicted alloys and measuring their T_cs.

The HA library was created by considering all possible three-element combinations of atoms from the $3d$, $4d$ and $5d$ periods and some elements from group III-VI, deliberately excluding the rare earths. A total of 236,115 prototypical three-element HAs including the regular, inverse and half-HAs, were considered. The electronic structure of each was computed by DFT in the generalized gradient approximation of the exchange correlation functional (DFT-GGA) parametrized by Perdew-Burke-Ernzerhof. The VASP platform was used for DFT calculation. To cut down the computational labor, only the HAs involving $3d$, $4d$ and $5d$ metals were actually analyzed computationally by the authors,[18] which reduced the number of compounds for actual computation to 36,540. The Convex-Hull analysis showed only 248 of them to be thermodynamically stable out of which 22 had a magnetic ground state consistent with the type of unit cells used. A total of 20 novel magnetic HAs belonging to the Co_2YZ, Mn_2YZ and X_2MnZ classes were established to be stable against decomposition. Magnetic T_cs of these compounds were predicted by a simple machine learning (ML) method that correlated certain features of the compounds provided by microscopic electronic structure calculation, with experimentally measured macroscopic properties of the material. This way they finally succeeded in discovering a high magnetic T_c ferromagnet (Co_2MnTi, $T_c = 938K$) and a tetragonally distorted antiferromagnet (Mn_2PtPd), which were already synthesized and experimented with. The work opens up possibility for designing materials for energy, data storage and even futuristic spintronics applications. ML methods are fast becoming an integral part of materials science especially in the context of designing new functional materials with targeted properties. Machine learning methods can be exploited to guide experiments or computations to concentrate on the promising regions of the design-space where materials with targeted properties are more likely to be found. That avoids a lot of costly experimentation and computation. Progressively materials informatics is fast becoming an essential tool in the search for novel materials with desired functionalities for specific or more general applications. Materials informatics, which is a confluence of many scientific disciplines including large data and information science, machine learning and optimization coupled with large scale electronic structure and property calculations and experimentation for validation, can be a potent tool in the discovery of new materials. In what follows, we briefly review an important work[19−20] that underscores the power of the approach.

13.4.4 Materials Informatics in the Search for Novel Materials

Non-centrosymmetric (NCS) materials constitute a technologically important class of functional materials. They can be chiral, piezoelectric, polar or ferroelectric. Many oxide materials have important technological relevance; but when it comes to NCS materials, not too many NCS oxide materials are known to date. Barium titanates, lead zirconate, lead titanate as well as barium ferrate are already known polar oxide materials that are important in critical technologies (see Chapter 8). In view of their importance search for new NCS-oxide materials assumes great significance. It may appear at the first sight that the task is simple. Going by conventional wisdom it is sufficient to choose polar or chiral building blocks (BBs) and assemble them into the desired NCS structures. However, it is usually impossible to predict *a priori*, if the BBs will organize into a non-centrosymmetric crystal structure. For inorganic oxides the design of

NCS materials relies heavily on BBs that have metal centres with d^0 electronic configuration expecting that the second order Jahn-Teller or pseudo Jahn-Teller effects will destroy the center of symmetry. That expectation is belied in most of the cases and the oxides frequently organize themselves into close packed arrangement of atoms without loss of inversion symmetry. The inversion symmetry can also be broken by another mechanism involving the coupling of non-polar lattice modes with a polar lattice mode. Even if the second mechanism is operative, there is no thumb rule to provide guidance to the task of fabricating new, hybrid improper ferroelectric materials, for example without undertaking exhaustive electronic structure calculations to map out the entire chemical (elemental) and energy landscapes. The potential energy surfaces of complex oxides are rather difficult to search through as phonon instabilities arise at the high symmetry points in the irreducible Brillouin zones (away from the Γ point).[19] That causes the unit cell to grow many-fold leading to a large increase in the system size and possible atomic arrangements. Performing accurate and exhaustive electronic structure calculations on such systems is a formidable task even though high throughput first-principles calculations have succeeded in the design of NCS $\frac{1}{2}$ - Heusler alloys. The difficulty multiplies in the case of oxide materials as in the chemical search space atoms with incompletely filled d or f orbitals are present and in many instances, competing ground states also exist. Thus, the power of modern electronic structure calculation tools (packages) alone proves inadequate to handle such problems unless the power of ML, data search, informatics, global optimization techniques are integrated into the strategy.

The recent works[19–20] developed a predictive paradigm that integrated a data-driven computational strategy with group theory, informatics and *ab initio* electronic structure calculation for designing novel NCS materials from among the Ruddlesden-Popper (RP) oxide structures. RP oxides have the general formula $A_{n+1}B_nX_{3n+1}$ or $A_nA'B_nX_{3n+1}$ where A, A' are alkali or alkaline earth or rare earth metals, B represents a transition metal and X is oxygen. The A cations are located in the perovskite layer with a coordination number of 12 while the A' cations are located at the boundary between the perovskite layer and the intermediate block layer with a coordination number of 9. The B cations are located outside the anionic octahedral pyramids and squares. The authors applied their computational strategy to the $n = 1$ family of RP oxide structures for which few are known to grow into NCS class of crystals. Even with the $n = 1$ restriction imposed, the chemical search space still remains huge with 30 A site elements and 19 B site elements in the reckoning. The authors deftly used informatics to screen the chemical space and select 242 RP oxide compositions that would be more likely to have NCS ground state structure. Several stannates and ruthenates turned out to be the materials holding promise; stannates: $NaASnO_4$ and ruthenates: $NaARuO_4$ with A = La, Pr, Nd, Gd and Y. The informatics guided RP oxides were then validated by DFT calculations. It is worth summarizing their results for the stannates and ruthenates.

13.4.4.1 Stannates

The scientists[22] have found two competing NCS ground state phases – one piezoactive ($P_4^-2_1m$ space group) and the other chiral plus piezoactive. The band gaps were calculated (in $P_4^-2_1m$ symmetry) revealing optical transparency in the visible region with their piezoelectric response depending on the size of the A-cations. These materials would thus be good candidates for fabricating sensors and transparent conducting oxide films.

13.4.4.2 Ruthenates

The NCS ruthenates revealed rather contrasting features. They turned out to be magnetic with metallic, half-metallic or insulator type of electronic structures. The NCS ground states were found to be either piezo-active (P_42_1m) when A = La, Pr, Nd or polar (PC_a2_1) when A = Gd, Y. They also revealed, depending on the A-cation size transitions from ferromagnetic to metallic (A = La), or half-metallic (A = Pr, Nd) to antiferromagnetic (A = Gd, Y). That conjures up a rather intriguing property landscape for the NCS ruthenates. They can be metallic or half-metallic with piezoactive symmetry or even antiferromagnetic insulators with polar symmetry, which means they may be viable candidates for multiferroic materials. The authors also reported several other RP oxide structures predicted by their screening based on informatics. These structures have different cations occupying *B*-sublattice. Seven among them were validated by DFT calculations, which confirmed that they possess NCS ground state. These NCS RP oxides are $NaLaZrO_4$, $NaLaHfO_4$, $KBaNbO_4$, $NaLaIrO_4$, $SrYGaO_4$ and $SrLaInO_4$. The work reviewed here strongly suggests that the materials informatics in conjunction with group theoretical tools and DFT-based high end computational electronic structure calculation can be a viable route to designing new materials with targeted properties even if the target is as abstract as NCS ground state.

13.4.5 Computational Design of New MOF-Based Material for Hydrogen Storage

In Chapter 12, we briefly reviewed the status of metal organic frameworks (MOFs) in the hydrogen-based energy applications. Thousands of MOFs have been synthesized and structurally characterized. The information is available in a number of databases. The question that comes to mind concerns the possibility of exploiting the huge information about the MOFs and their H_2 uptake capacities to design theoretically new MOFs with better hydrogen storage capabilities. Ahmed *et al.*[21] in a recent publication addressed the issue and was able to come up with a number of MOF-based materials with exceptionally high usable hydrogen storage capacities, by screening nearly half a million MOF-based materials. The work is reviewed in some detail in view of its importance in the possible realization of technologically feasible and efficient solid state hydrogen storage systems both for static and mobile applications.

As we had seen in Chapter 12, hydrogen adsorbents usually fail to strike a balance between high usable volumetric and gravimetric storage capacities (VSC and GSC). The MOFs can offer viable solutions that tend to bridge the VSC and GSC gap. The volumetric ceiling in 2019 appeared to be at 40 g H_2 L^{-1} and the task ahead called for MOFs that breached the ceiling. Which features could be tuned for design? Ahmed *et al.*[21] found that usable capacities in the highest capacity H_2 storage materials correlates negatively with density and volumetric surface area. In contrast, the capacity was found to be maximizable by increasing the gravimetric surface area and porosity of the material.

The authors[21] examined a database of 493,458 MOF crystal structures obtained by collating information from 11 published databases of which 15,235 were real MOFs, 478, 205 were hypothetical MOFs to which 18 new MOFs designed by them were added. Initial screening was done by calculating capacities by the semiempirical Chahine rule.[22] It turned out that 43,777 MOFs have storage capacities that exceeded the capacity of the benchmark MOF (MOF-5) at 77 K and 35 bar pressure. All these molecules were evaluated for their usable hydrogen storage capacities by grand canonical Monte Carlo calculations using the pseudo-Feynman-Hibbs potential, assuming an isothermal pressure swing between 5 and 100 bar at 77 K. MOFs predicted to have usable capacities exceeding those of MOF-5 (4.5 wt % + 31.1 g H_2L^{-1}) and IRMOF-20 (5.7 wt% + 33.4 g H_2L^{-1}) were identified and assessed experimentally. Three MOFs with usable capacities surpassing that of IRMOF-20 were identified and their viability demonstrated. These materials viz. SNV-70, VMCM-9 and PCN-610/NV-100 appear to define a new high point for usable hydrogen capacities for MOFs. It is quite possible, still better MOFs may exist in other databases, suggesting the need to explore further.

13.4.6 Miscellaneous Materials Designing Approach

Piezoelectric materials introduced in Chapter 9 are essential components for electromechanical transducers or pressure sensors and are in high demand. The search for better piezoelectric materials is therefore a continuous process. Perovskite oxides are a common class of piezoelectric materials. Theoretical analysis seems to indicate that high piezoelectric activity in perovskite oxides may be traced to the existence of a flat thermodynamic energy landscape that connects two or more ferroelectric phases. Fei *et al.*[23] suggested a new approach to design ceramics materials with ultrahigh ferroelectric response. The authors made use of phenomenological theories and phase-field simulations and proposed an alternative to the commonly adapted strategy involving morphotropic phase boundaries to further flatten the energy landscape. This new strategy involved judicious introduction of local structural heterogeneity with a view to manipulating the interfacial energies. They validated their approach with the synthesis of rare earth-doped $Pb(Mg_{1/3}Nb_{2/3})O_3$-$PbTiO_3$ code named PMN-PT. The rare earth dopants tend to change the local structures of lead-based perovskite ferroelectrics and enhance the ferroelectric response. Indeed they were able to achieve high piezoelectric coefficient (d_{33}) upto pCN^{-1} and dielectric permitivity $\epsilon_{33} > 13,000$ in a samarium doped PMN-PT ceramic oxide, which was shown to have a Curie temperature of 89°C. It remains to be seen whether the approach marks the beginning of a new paradigm in the designing of new functional materials.

Another novel paradigm[24] in the context of realizing new materials by design refers to the emerging field of anion-engineering in oxide-based materials, that seeks to tune the physical properties of the parent oxide-based material by incorporating additional anions of different sizes, electronegativities and charge. The oxychalcogenides, oxynitrides, oxyhalides and oxypnictides, for example, may exhibit entirely new or enhanced response not easily predicted from the parent simpler homoanionic (oxide) materials. The mixed anionic compounds have anions with contrasting polarizabilities. It brings them closer to the ionic-covalent bonding boundary and produces novel materials with enhanced response. The presence of multiple anions in these heteroionic materials enable them to span a much larger and more complex atomic structural design and interaction space than the parent homoanionic (oxide) materials. Harada *et al.*[24] in a recent review reflected on the well established atomic and electronic principles in the context of rationally designing heteroanionic materials with targeted properties and described the synergy between the quantum theoretical methods and laboratory experiments to provide guidance to realizing designer materials with superior properties.

Z. Fang *et al.*[25] (2014) explored the possibility of designing non-linear optical (NLO) crystals for second harmonic generation (SHG), by a specially designed genetic algorithm (GA), which searched for stable non-centrosymmetric structures of inorganic crystals with known compositions and calculating their SHG coefficients. DFT was used to calculate the energetic as well as SHG efficiency. The novel feature of the GA-method of Fang et al. lies in the use of the coordination pattern of the building blocks to construct the individuals (structures) of the evolving population. The GA search led to two crystals based on the two tertiary compounds $AgGaS_2$ and $LiAsSe_2$.

We will conclude this chapter with a brief reference to the 'quantum design of semiconductor active materials for laser and amplifier applications'. In a recent publication, Moloney *et al.*[26] has given an overview of a novel first-principles-based quantum theoretical approach to designing as well as optimizing semiconductor quantum-well-based systems for generating coherent light with targeted wavelengths. Using only microscopic parameters as inputs and building blocks they were able to predict the light – current characteristics for a low power *InGaPAs* ridge laser without having to use any adjustable fitting parameters. The authors demonstrated how the microscopic parameters could be used for designing and optimizing high power laser structures. As an explicit example, they designed a vertical external cavity semiconductor laser acronymed VECSEL.

The first step in their design and optimization involves calculation of the relevant part of the band structures of the materials providing a detailed description of the valence and conduction bands of the gain materials and the correct wavefunctions. The wavefunctions are used to calculate the transition dipole matrix elements, which determine the strengths of optically allowed interband transitions. In the second step, semiconductor Bloch equations are used to calculate the optical response of the material. In the third step, the density, temperature- and frequency-dependent decoherence rates are calculated. Finally the theoretically predicted gain spectrum is compared with the experimentally measured spectrum. Semiconductor lasers based on InGaAs quantum wells sandwitched between AlGaAs barriers are commercially and widely used for coherent emission at 980 nm. The theoretically predicted gain spectrum for InGaAs/AlGaAs system matches well with the measured gain spectrum validating the calculational scheme suggested by the authors.

We have given glimpses of the research and development taking place in the realm of realizing designer materials or materials with preset properties. The dream of designer materials is fast becoming a reality, thanks to the rapid development of techniques of quantum mechanical calculations of properties of materials, sophisticated optimization strategies, artificial intelligence, data mining, machine learning, etc. More spectacular development will take place over the next decade and a half.

REFERENCES

1. W. D. Callister Jr., *Materials Science and Enginnering – An Introduction*, Volume 5, Sixth Edition, Wiley, India (2011).
2. M. de Podesta, *Understanding the Properties of Matter,* Second Edition, Taylor and Francis, Washington (2002).
3. J. D. Robert and M. C. Caserio, *Basic Principles of Organic Chemistry*, Second Edition, W. A. Benjamin, Menlo Park, CA (1977).
4. C. D. Johnson, *The Hammett Equation*, Cambridge University Press (1975).
5. J. H. Holland, *Application in Natural and Artificial Systems: An Introductory Analysis with Applications to Biology, Control and Artificial Intelligence*, MIT Press, Ann Arbor, Michigan (1975).
6. D. E. Goldberg, *Genetic Algorithm*, Pearson Education, London (2006).
7. K. Sarkar and S. P. Bhattacharya, *Soft-Computing in Physical and Chemical Sciences : A Shift in Computing Paradigm*, CRC Press, Taylor and Francis, Boca Raton (2018).
8. R. Car and M. Parrinello, 'Unified approach for molecular dynamics and density functional theory', *Phys. Rev. Lett.*, 55, 2471–2474 (1985).
9. T. D. Kuhne, 'Second generation Car-Parrinello molecular dynamics', *WIRE, Comput. Mol. Sci.*, 4, 391–406 (2014).
10. K. Deb., A. Pratap, S. Agarwal and T. Meyarivan, 'A fast and elitist multiobjective genetic algorithms: NSGA-I', *IEEE Trans. Evol. Comput.'*, 6, 182–197 (2002).
11. X. Q. Chen *et al.*, 'Modeling ithardness of polycrystalline materials and bulk metallic glasses', *Intermetallics*, 19, 1275 (2011).
12. A. O. Lyakhov and A. R. Oganov, 'Evolutionary search for superhard materials: Methodology and applications to forms of carbon and TiO_2', *Phys. Rev. B*, 84, 092103 (2011).
13. Z. Allahyari and A. R. Oganov, 'Coevolutionary search for optimal materials in the space of all possible compounds', NPJ *Comput. Mater.*, 6, 1–10 (2020).
14. Z. Allahyari and A. R. Oganov, 'Computational Discovery of Hard and Superhard Materials', *J. Appl. Phys.*, 126, 040901 (2019).
15. D. Xue *et al.*, 'Accelerated search for materials with targeted properties by adaptive design', *Nat. Commun.*, 7, 11241 (2016).
16. D. R. Jones *et al.*, 'Efficient global optimization of expensive black box function', *J. Global. Optim.*, 13, 455–492 (1998).
17. W. B. Powell and I. O. Ryzhov, 'Optimal learning and approximate dynamic programming', In *Reinforcement Learning and Approximate Dynamic Programming for Feedback Control* (pp. 410–431), F. L. Lewis and D. Liu, Eds., Wiley Series in Probability and Statistics, Wiley, New Jersey (2013).
18. Stefano Sanvito *et al.*, 'Accelerating discovery of new magnets in the Heusler alloy family', *Sci. Adv.*, 3, e1602241 (2017).
19. P. V. Balachandram *et al.*, 'Learning from data to design functional materials without inversion symmetry', *Nature Commun.*, 8, 14282 (2017).
20. P. V. Balachandram, 'Machine Learning (ML) guided design of functional materials with targetted properties', *Comput. Mater. Sci.*, 164, 82–90 (2019).
21. A. Ahmed *et al.*, 'Exceptional hydrogen storage achieved by screening nearly half-a-million metal organic frameworks', *Nature Commun.*, 10, 1568 (2019).
22. J. Goldsmith *et al.*, 'Theoretical limit of hydrogen storage in MOFs: Opportunities and trade-offs', *Chem. Mater.*, 25, 3377–3382 (2013).
23. F. Li *et al.*, 'Ultrahigh piezoelectricity in ferroelectric ceramics by design nature materials', *Nature Mater.*, 17, 349–354 (2018).
24. J. K. Harada *et al.*, 'Heteroanionic materials by design: Progress towards targeted properties', *Adv. Mater.*, 31, 1805295 (2019).
25. Z. Fang *et al.*, 'Computational design of inorganic NLO crystals based on a genetic algorithm', *CrysEngCom*, 16, 10569–10580 (2014).
26. J. V. Moloney, J. Hader and S. W. Koch, 'Quantum design of semiconductor active materials: Laser and amplifier applications', *Laser and Photon Rev.*, 1, 24–43 (2007).

14

Current Status and Outlook for Future

14.1 Introduction

We have described in some of the preceeding chapters (Chapters 8–13) a host of materials, from metals and alloys to ceramics, from insulators to superconductors, from ferromagnet to ferroelectrics, semiconductors to half-metals and so on. We have tried to understand how and why a material behaves the way it does, on the basis of elementary quantum mechanics and statistical mechanics where appropriates (Chapters 2–7). The electronic band structure of solids turns out to be an important tool for analyzing and understanding many properties of solids. Crystal symmetry and electron interaction as we have seen, can make or break a specific property like the spontaneous electrical polarization in a solid or the emergence of long range magnetic order. We have examined the critical role that impurity atoms, crystal imperfections and scatterings there from plays to shape up important properties like electrical conductivity, superconductivity and semiconducting response. The techniques of handling many electron systems, especially the continuously evolving technology based on the celebrated density functional theory has advanced to a stage that enables one to predict the properties of atoms, molecules and solids quantitatively (though approximately) and to explain and understand what causes the emergence of a specific type of response and even provide designing clues to the synthesis of new materials with targetted properties. It is in this context that we reviewed specific applications involving computational electronic structure theories, artificial intelligence, informatics, that have successfully demonstrated the feasibility of accelerating the search for the discovery[1,2] of new tailor made materials (Chapter 13) we are in a new era of materials discovery where the Edisonian approach is being fast replaced by intelligent designing paradigms. It is no longer a long wait for a new material to appear on the technology horizon by a chance discovery or after a long series of hit and miss experiments. The traditional labor intensive and time-consuming materials discovery method has therefore been slowly driven toward materials exploration through an informatics and computational materials chemistry approach over the last decade or so. The emphasis has shifted and is now dominantly centered around the concept of doing materials exploration through high throughout computation/experimentation. Due to the constraints on resources, the targetted materials discovery by the new methods referred to face certain bottlenecks.[1,2] Even after elimination some of these bottlenecks, the requirement of human intervention at some stage of the exploration remains. A decision about what to do with the information gleaned from the largely automated informatics approach must be taken by the user. Talapatra *et al.*[3] in a recent publication claimed that an alternative paradigm is therefore gaining traction. It poses materials discovery as an optimal experiment designing problem. The new paradigm is evolving and may turn out to be a game changer in the materials discovery process with minimal or perhaps even without any human intervention at all.

14.2 Where Do We Stand?

Quantum mechanics has enabled us to make sense of the microscopic world of atoms and molecules. We have now a fairly complete understanding of the way atoms are expected to behave in a given situation, how they would interact to form bewildering varities of molecules and how the individual molecules would respond to different stimuli. The basic building blocks of the material world are therefore largely understood on the basis of quantum mechanics or the mechanics of the microworld. Our understanding

DOI: 10.1201/9781003244882-14

405

of 'materials' at this critical juncture is however, far from complete. Understanding materials requires not only understanding the building blocks (substances) they are made of, but also understanding how to assemble them into useful structures that are endowed with 'special' or 'targetted' properties so that they are 'functional' in one particular way or the other. Materials with special properties are at the core of all useful assemblies or structures. Chemists synthesize and characterize molecules and carry out 'construction' with them, working at length scales that span a wide range, from the ordinary molecules at nm level to polymers and electronic devices at submicron level.[4] When it comes to ceramic materials, chemists have to deal with materials for constructing large scale structures (macrostructures). That way, materials science can perhaps be viewed as a sub branch of chemical science where the broad mandate is to explore through synthesis as well as processing, the rather complex relationship among structures and properties and come up with new design. Controlling all the available degrees of freedom that determine the useful behavior of materials is a challenging task. The task is by no means easy and our understanding of materials in this context is incomplete or at least less complete than our understanding of atoms and molecules as individual entities. Indeed, one of the challenging tasks is to investigate and understand how and to what extent the properties of individual molecules are modified in an assembly of them in the condensed phases (solids, liquids or solutions) where the average molecular behavior determines the response of the material in question. The collective properties of materials in condensed phases have many complex but controllable features, which is required to be understood more completely. It is well recognized now that the most interesting materials are functional systems that have grown out of our limited (but evolving) understanding of the complex structure activity relationships. Catalysts provide fascinating examples of such functional materials that are so useful. With our evolving understanding of the structure activity relationships, soon a stage will be reached when responsive materials consisting perhaps of softer substances will be designed to produce a host of different functionalities like sensing. It is therefore necessary to seek to understand the properties of materials in which different kinds of organization of components exist. Self-organization or self-assembly will progressively play an important role as nanomaterials and nanotechnology begin to move out from laboratories to manufacturing platforms.[4] It is extremely necessary at this stage to develop better (or more complete) understanding of the phenomenon of self-organization so as to be able to create newer avenues of exercising macromolecular structural control at the nanometer scale and produce novel materials with novel properties almost at will. We are at the moment standing at the crossroad of traditional and modern science of materials. Our understanding of the materials has evolved to a stage where it is increasingly being used in conjunction of computational techniques of materials discovery and design to give shape to designer materials and supermaterials (superhard, supercapacitor, supermagnets, etc.) required by industry and new technologies. The traditional 'Edisonian' approach still continues along with the use of the age-old thumb rules. The advent of the era of 2D materials and nanomaterials has increased the demand for new and more precise approaches to materials design and synthesis in which self-organization plays a key role. The synthetic chemists' ability to place atoms in precise positions at the nm scale to realize functional structures will be replicated more and more in materials science and engineering in the near future, providing new avenues to novel materials.

14.3 Future Outlook

Our understanding of the properties of atoms, molecules and their assemblies in condensed phases along with our ability to exploit the understanding to design new materials will shape the future of materials science and new materials. We expect that attention of researchers will now be increasingly turned on to come up with superior structural materials that are stable at high temperatures, for example in contact with superheated steam. Simultaneously the machinability of such materials will be sought to be improved. For many applications, the materials for future would be lighter, stronger and more and more environment friendly and easily recycled. Energy economy and sustainability will be the key factors for the new materials of future. We can therefore expect invention of materials with improved electrical properties (low resistivity, for example). High temperature superconductors with improved machinability and ease of production can be expected to be discovered in the next five years or so with great impact on energy economy.

High speed telecommunication network is at the heart of modern society and industry. Such networks require ultrafast switches. All optical switching is an ideal solution to the problem of high speed switching. We can expect that the future will see the emergence of new non-linear optical materials with improved optical properties and durability to translate the proposition of all optical switching in ultrafast network into a viable one. Better semiconducting materials (compounds) can be expected to be developed for more efficient integrated circuits and chips.

We can imagine that attention will be progressively directed more and more at understanding nanoscale materials and exploit their properties for practical use (Chapter 11). Non-homogeneous materials could become extremely important in the coming decade. In this context we can foresee a lot of developments in designing and processing materials in which complex structural assemblies occur fast enough without any or with little guidance. We expect such self-assembled materials to lead to the production of durable systems of diverse utility. A lot of developments can be anticipated in the invention and use of materials with high grade actuating response found in living systems, such as muscles. Great advancement in the design and production of biometric materials is expected to take place in the next decade.

2D materials led by graphene will perhaps emerge as a game changer in the world of advanced technological materials (Chapter 11). Graphene is a 2D allotrope of carbon and is the basic building unit of 1D carbon nanotubes and 0D fullerene. It has a single layer of planar hexagonal arrangement of carbon atoms in a honeycomb lattice (Chapter 11). This simple structure of graphene is at the root of many of its exotic properties.[5] Thus, graphene is an exceptionally strong material, almost 200 times as strong as steel; it has low frictional coefficient. Its toughness is phenomenal, yet its stretchability is a high as 20%. All these mechanical properties make it an ideal structural material. It is an electrically conducting with ballistic mode of conduction. Graphene is a zero band gap material and this makes it useful in solar cells while its electrical properties like superconductivity at room temperature makes it an exciting candidate for lossless electronics. Graphene sheets are completely impermeable to particles even as small as helium atoms making it a promising material for desalination. On the basis of many surprising properties of graphene, we anticipate its extensive use in technological applications, which include rechargeable batteries, solar power generators, electric vehicles, industrial lubrication, trapping radioactive wastes, desalination and a host of others. There are a number of graphene-like substances, which too can be visualized to have many useful applications in critical areas. Black phosphorus, for example is an allotrope of phosphorus with puckered honeycomb layers arranged in stacks, held together by weak van der Waals forces. It has a band gap of 0.3 eV for the bulk and 2 eV for a single layer. Thus black phosphorus (BP) bridges the gap between zero band gap of graphene and much larger band gaps of transition metal di-chalcogenides (TMD).[6]

The range of band gap of BP enables it to absorb and emit electromagnetic radiation (light) from infrared to visible range of the spectrum. BP can therefore be imagined to be a potential material for many optoelectronic and photon mediated applications. BP is non-toxic and degrades in our bodies into wall-tolerated phosphates, increasing its future potential for biomedical applications. No wonder that BP and TMDs like MoS_2, $MoSe_2$, WS_2, WSe_2 are being touted as the materials for future technologies. WSe_2, for example, is a promising candidate for spintronics applications in the near future.[7−9]

Materials can often be seen as harbours of quantum many body systems in the sense that the solids accomodate many strongly correlated electrons give rise to many novel functionalities due to the emergence of remarkable collective behavior or phenomena like the Mott transition, high temperature superconductivity, colossal magnetoresistance, giant magnetic resistance, topological insulators, etc. We can anticipate that these phenomena will play an important and perhaps even critical role in shaping new generations of quantum technologies so essential for ensuring sustainable and safe development across the globe.[9−11] Looking ahead, we are tempted to imagine that quantum materials and quantum technologies will move ahead in long strides and quantum computers will become routinely available for large scale, safe information processing and communication. Full scale quantum simulation of strongly correlated many body systems will become a reality, which in turn will make the task of designing new exotic materials and technologies less arduous than it is today, eventually paving the way to a better understanding of the material world. Electronics without energy dissipation, lossless transmission, topological currents, energy harvesting through photovoltaics, thermo and piezoelectrics[9−11] will be the essential ingredients of the revolutionary changes in technology and progress that we envisage at this point of time. Let us hope that it will be a dream come true, not a dream that lapses into nightmare.

REFERENCES

1. A. G. Kvashnin, C. Tantardini, H. A. Zakaryan, Y. A. Kvashnina and A. R. Oganov, 'Computational Search for New W-Mo-B Compounds'. Chem. Mater., 32, 7028–7035 (2020).
2. D. Xue, P. V. Balachandran, J. Hogden, J. Theiler, D. Xue, and T. Lookman. 'Accelerated Search for Materials with Targeted Properties by Adaptive Design'. Nature Commun., 7, 1–9 (2016).
3. A. Talapatra *et al.* 'Experiment Design Framework for Accelerated Discovery of Targeted Materials Across Scales'. Frontiers in Materials, 6, 82 (2019).
4. N. R. Council 'Beyond the Molecular Frontier: Challenges for Chemistry and Chemical Engineering'; The National Academies Press: Washington, DC, 2003.
5. A. C. Neto, F. Guinea, N. M. Peres, K. S. Novoselov and A. K. Geim. 'The Electronic Properties of Graphene'. Rev. Mod. Phys., 81, 109 (2009).
6. K. Novoselov, O. A. Mishchenko, O. A. Carvalho, and A. C. Neto, '2D Materials and van der Waals Heterostructures'. Science, 353, aac9439 (2016).
7. H. S. Nalwa, 'A Review of Molybdenum Disulfide (MoS_2) Based Photodetectors: From Ultra-Broadband, Self-powered to Flexible Devices'. RSC Adv., 10, 30529–30602 (2020).
8. E. C. Ahn, '2D Materials for Spintronic Devices', npj 2D Materials and Applications, 4, 1–14 (2020).
9. Y. Tokura, M. Kawasaki and N. Nagaosa, 'Emergent Functions of Quantum Materials'. Nat. Phys., 13, 1056–1068 (2017).
10. B. Keimer and J. Moore 'The Physics of Quantum Materials'. Nat. Phys., 13, 1045–1055 (2017).
11. D. Basov, R. Averitt and D. Hsieh 'Towards Properties on Demand in Quantum Materials'. Nat. Mater., 16, 1077–1088 (2017).

Index

For Product Safety Concerns and Information please contact our EU
representative GPSR@taylorandfrancis.com
Taylor & Francis Verlag GmbH, Kaufingerstraße 24, 80331 München, Germany

www.ingramcontent.com/pod-product-compliance
Lightning Source LLC
Chambersburg PA
CBHW080139220326
41598CB00032B/5112